T0402349

Yuri P. Raizer
Gas Discharge Physics

Springer
Berlin
Heidelberg
New York
Barcelona
Budapest
Hong Kong
London
Milan
Paris
Santa Clara
Singapore
Tokyo

Yuri P. Raizer

Gas Discharge Physics

With 209 Figures

Springer

Professor Dr. Yuri P. Raizer
The Institute for Problems in Mechanics, Russian Academy of Sciences,
Vernadsky Street 101, 117526 Moscow, Russia

Editor:

Dr. John E. Allen
Department of Engineering Science, University of Oxford, Parks Road,
Oxford OX1 3PJ, United Kingdom

Translator:

Dr. Vitaly I. Kisin
24 Varga Street, Apt. 9, 117133 Moscow, Russia

This edition is based on the original second Russian edition: *Fizika gazovogo razryada*

1st Edition 1991
Corrected 2nd Printing 1997

ISBN-13: 978-3-642-64760-4 e-ISBN-13: 978-3-642-61247-3
DOI: 10.1007/978-3-642-61247-3

Library of Congress Cataloging-in-Publication Data.

Raizer, ÎU. P. (ÎUriĭ Petrovich) [Fizika gazovogo razrîada. English] Gas discharge physics / Yuri P. Raizer. p. cm. "Corr. printing 1997" – t.p. verso. Includes bibliographical references and index. ISBN-13: 978-3-642-64760-4 1. Electric discharges through gases. I. Title.
QC711.R22713 1997 537.5'3–dc21 96-53988

Typesetting: Springer TEX inhouse system
Cover design: *design & production* GmbH, Heidelberg

SPIN 10565824 54/3144 – 5 4 3 2 1 0

Preface

Gas discharges are of interest to physicists and engineers in a number of fields. Several decades ago excellent textbooks were written by von Engel and Steenbeck, Loeb, Brown, Kaptsov and several other authors. These books faithfully served many generations of students, and specialists still refer to them. Nevertheless, their usefulness does suffer from the time elapsed since publication: It is not that the material they present has become obsolete and irrelevant – this has happened to a very minor extent, if at all. Rather, the subject has greatly advanced both in scope and in depth, and its emphases have somewhat shifted. Of course, new books have been written, mostly monographs devoted to narrow branches of gas discharge physics. But these books are typically intended for the specialist and not so much for the novice in the field.

The need for a new textbook that is understandable to a beginner in gas discharge physics, and that conveys the right *amount* of information (even more important: information of the right *kind*) making it also useful to the specialist is apparent. With this in mind, our intention has been to produce a book that serves both as a textbook and a handbook.

From an immense amount of material we have selected, as best we could, the parts that are required for an understanding of the physics and those points that are most frequently needed in research. As a convenient and comprehensive volume, the book contains a maximum of useful data: experimental results, results of calculations, and reference data; formulas required for estimates have been reduced to a form suitable for computations.

This work was published in Russian in 1987 as a substantially larger volume. The English edition has been abridged at the expense of ancillary material concerning collisions, elementary processes, plasma radiation, plasma diagnostics, and other topics, though the chapters dealing with the central themes of discharge physics are retained in full, and even expanded by the addition of new data.

We have decided not to cover actual circuits, techniques, or methods (we will cover the ideas, though) of experiments and measurements; instead we concentrate on the physics of the processes of interest. Purely technical applications of gas discharges are not discussed for the same reason.

It would be impossible to give a comprehensive bibliography when covering such an immensely wide scope of topics; hence, original papers are cited only when recent results are discussed. In all other cases we refer to a book or review paper where more complete references are given.

The author is deeply grateful to Professors A. V. Eletsky and L. D. Tsendin, who read the Russian version of the manuscript, and Professor J. E. Allen, who read the English, for a number of useful comments. In addition, the author would like to thank the translator, Dr. V. I. Kisin, for a fruitful collaboration.

Moscow, April 1991 *Yu. P. Raizer*

Contents

1. Introduction

1.1 What Is the Subject of Gas Discharge Physics

The term "gas discharge" originates with the process of discharge of a capacitor into a circuit incorporating a gap between electrodes. If the voltage is sufficiently high, electric break down occurs in the gas and an ionized state is formed. The circuit is closed and the capacitor discharges. Later the term "discharge" was applied to any flow of electric current through ionized gas, and to any process of ionization of the gas by the applied electric field. As gases ionized to a sufficient degree emit light, it has become customary to say that a discharge "lights up," or is "burning."

As a rule, the flow of electric current is associated with the notion of a circuit composed of conductors. Actually, a closed circuit or electrodes are not needed for a directed motion of charges (electric current) in rapidly oscillating electric fields, and even less so in the field of electromagentic radiation. However, quite a few effects observed in gases subjected to oscillating electric fields and electromagnetic waves (breakdown, maintaining the state of ionization, dissipation of energy of the field) are not different, in principle, from dc phenomena. Nowadays all such processes are referred to as discharges and included within gas discharge physics. The fact that electric current flow in open circuits in the field of electromagnetic waves is of no general significance. In such cases, the dissipation of the energy of the field is described not as the release of the Joule heat by electric current, but as the absorption of radiation.

The modern field of gas discharge physics is thus occupied with processes connected with electric currents in gases and with generating and maintaining the ability of a gas to conduct electricity and absorb electromagnetic radiation.

Gas discharge physics covers a great variety of complex, multi-faceted phenomena; it is full of an enormous amount of experimental facts and theoretical models. Before we begin their analysis, it is expedient to single out the main types of discharge processes and clarify them.

1.2 Typical Discharges in a Constant Electric Field

A relatively simple experiment introduces us to several fundamental types of discharge. Two metal electrodes connected to a dc power supply are inserted into a glass tube (Fig. 1.1). The tube can be evacuated and filled with various

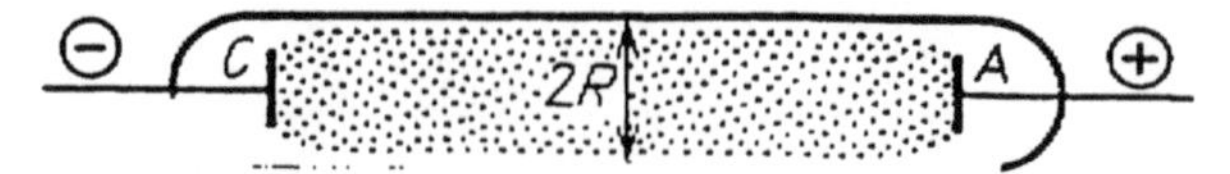

Fig. 1.1. Typical gas discharge tube

gases at different pressures. The quantities measured in the experiment are the voltage between the electrodes and the current in the circuit. This classical device served the study of discharge processes for nearly 150 years, and still remains useful.

If a low voltage is applied to the electrodes, say several tens of volts, no visible effects are produced, although a supersensitive instrument would record an extremely low current, on the order of 10^{-15} A. Charges are generated in the gas by cosmic rays and natural radioactivity. The field pulls them to the opposite-sign electrodes, producing a current. If the gas is intentionally irradiated by a radioactive or X-ray source, a current of up to 10^{-6} A can be produced. The resultant ionization is nevertheless too small to make the gas emit light. A discharge and an electric current that survive only while an external ionizing agent or the emission of electrons or ions from electrodes is deliberately maintained (e.g. by heating the cathode) are said to be *non-self-sustaining*. As the voltage is raised, the non-self-sustaining current first increases because most of the charges produced by ionization are pulled away to electrodes before recombination occurs. However, if the field manages to remove all new charges, the current ceases to grow and reaches saturation, being limited by the rate of ionization.

As the voltage is raised further, the current sharply increases at a certain value of V and light emission is observed. These are the manifestations of *breakdown*, one of the most important discharge processes. At pressure $p \sim 1$ Torr and interelectrode gap $L \sim 1$ cm, the breakdown voltage is several hundred volts. Breakdown starts with a small number of spurious electrons or electrons injected intentionally to stimulate the process: The discharge immediately becomes self-sustaining. The energy of electrons increases while they move in the field. Having reached the atomic ionization potential, the electron spends this energy on knocking out another electron. Two slow electrons are thus produced, which go on to repeat the cycle described above. The result is an electron *avalanche*, and electrons proliferate. The gas is appreciably ionized in 10^{-7} to 10^{-3} s, which is sufficient for the current to grow by several orders of magnitude.

Several conditions determine how the process develops at higher voltage. At low pressure, say 1 to 10 Torr, and high resistance of the external circuit (it prevents the current from reaching a large value), a *glow discharge* develops. This is one of the most frequently used and important types of discharge. It is characterized by low current, $i \sim 10^{-6} - 10^{-1}$ A in tubes of radius $R \sim 1$ cm, and fairly high voltage: hundreds to thousands of volts. A beautiful radiant column, uniform along its length, is formed in sufficiently long tubes of, say, $L \sim 30$ cm at $p \sim 1$ Torr. (This is how glowing tubes for street advertisements are made.) The ionized gas in the column is electrically neutral practically everywhere except in the regions close to the electrodes; hence, this is a *plasma*. The glow discharge plasma is very weakly ionized, to $x = 10^{-8} - 10^{-6}$ (where x denotes the fraction of ionized atoms), and is nonequilibrium in two respects. Electrons that get energy directly from the field have a mean energy $\bar{\varepsilon} \approx 1\,\mathrm{eV}$ and a temperature $T_e \approx 10^4\,\mathrm{K}$. The temperature T of the gas, including the ions, is not much higher than the ambient temperature of 300 K. This state, with widely separated electron and gas temperatures, is sustained by a low rate of Joule heat release under conditions of relatively high specific heat of the gas and high rate of its natural cooling. Also as a result of the high rate of charge neutralization in a cold gas, its degree of ionization is many orders of magnitude lower than the thermodynamic equilibrium value corresponding to the electron temperature.

If the pressure in the gas is high (about the atmospheric level) and the resistance of the external circuit is low (the circuit allows the passage of a high current), an *arc discharge* usually develops soon after breakdown. Arcs typically burn at a high current ($i > 1\,\mathrm{A}$) at a low voltage of several tens of volts; they form a bright column. The arc releases large thermal power that can destroy the glass tube: Arcs are often started in open air! Atmospheric-pressure arcs usually form thermodynamic equilibrium plasmas (the so-called low-temperature plasma), with $T_e \approx T \approx 10^4\,\mathrm{K}$ and the ionization of $x = 10^{-3} \div 10^{-1}$ corresponding to such temperatures. The arc discharge differs essentially from the glow discharge in the mechanism of electron emission from the cathode, which is vital for the flow of dc current of the arc. In the glow discharge, electrons are knocked from the surface of the cold metal by impacts of positive ions. In the arc discharge, the high current heats up the cathode,and thermionic emission develops.

If $p \sim 1\,\mathrm{atm}$, the interelectrode gap $L > 10\,\mathrm{cm}$, and the voltage is sufficiently high, *sparking* occurs. The breakdown in the gap develops by rapid growth of the plasma channel from one electrode to another. Then the electrodes are as if short-circuited by the strongly ionized spark channel. Lightning, whose "electrodes" are a charged cloud and the ground, is a giant variety of the spark discharge. Finally, a *corona discharge* may develop in strongly nonuniform fields that are insufficient for the breakdown of the entire gap: A radiant corona appears at sharp ends of wires at sufficiently high voltage and also around power transmission line conductors.

1.3 Classification of Discharges

Discharges in a dc electric field can be classified into (a) non-self-sustaining and (b) self-sustaining types. The latter are more widespread, more diversified, and richer in physical effects; and they are the subject of this book. Steady and quasi-steady self-sustaining discharges contain (1) glow and (2) arc discharges. We have already mentioned in Sect. 1.2 that the cathode processes of two types differ in principle. A close relation of the glow discharge is (3) Townsend's dark discharge. It proceeds with a cold cathode and at very weak current. The (4) corona discharge, also self-sustaining and also at a low current, is a special case.

Corona has common features with glow and dark discharges. Among transient discharges, the (5) spark discharge stands out sharply, among others.

Many features of purely plasma processes, characterizing breakdown in a dc electric field, as well as the glow and arc discharges, are typical for discharges in rapidly oscillating fields, where electrodes are not necessary at all. It is therefore expedient to construct a classification avoiding the attributes related to electrode effects, and the following two properties will be basic for the classification: the state of the ionized gas and the frequency range of the field. The former serves to distinguish between (1) breakdown in the gas, (2) sustaining nonequilibrium plasma by the field, and (3) sustaining equilibrium plasma. Frequency serves to classify fields into (1) dc, low-frequency, and pulsed fields (excluding very short pulses), (2) radio-frequency fields ($f \sim 10^5 - 10^8$ Hz), (3) microwave fields ($f \sim 10^9 - 10^{11}$ Hz, $\lambda \sim 10^2 - 10^{-1}$ cm), and (4) optical fields (far from infrared to ultraviolet light). The field of any subrange can interact with each type of discharge plasma. In total, we have 12 combinations. All of them are experimentally realizable, and quite a few are widely employed in physics and technology. Typical conditions under which each of the combinations can be observed are summarized in Table 1.1.

Table 1.1. Classification of discharge processes

	Breakdown	Nonequilibrium plasma	Equilibrium plasma
Constant electric field	Initiation of glow discharge in tubes	Positive column of glow discharge	Positive column of high-pressure arc
Radio frequencies	Initiation of rf discharge in vessels filled with rarefied gases	Capacitively coupled rf discharges in rarefied gases	Inductively coupled plasma torch
Microwave range	Breakdown in waveguides and resonators	Microwave discharges in rarefied gases	Microwave plasmatron
Optical range	Gas breakdown by laser radiation	Final stages of optical breakdown	Continuous optical discharge

1.4 Brief History of Electric Discharge Research

Leaving lightning aside, man's first acquaintance with electric discharges was the observation, dating back to 1600, that friction-charged insulated conductors lose their charge. Coulomb proved experimentally in 1785 that charge leaks through air, not through imperfect insulation. We understand now that the cause of leakage is the non-self-sustaining discharge.

Occasional experiments were conducted in the 18th century with sparks produced by charging a body by an electrostatic generator, and with atmospheric electricity, experiments with lightning sometimes having tragic consequences.

Sufficiently powerful electric batteries were developed at the beginning of the 19th century to allow the discovery of the arc discharge. V.V. Petrov, who worked in the Saint Petersburg Medical Surgery Academy in Russia, reported the discovery in 1803. The arc was obtained by bringing two carbon electrodes connected to battery terminals into contact and then separating them. Several years later Humphrey Davy in Britain produced and studied the arc in air. This type of discharge became known as "arc" because its bright horizontal column between two electrodes bends up and arches the middle owing to the Archimedes' force. In 1831–1835, Faraday discovered and studied the glow discharge. Faraday worked with tubes evacuated to a pressure $p \sim 1$ Torr and applied voltages up to 1000 V.

The history of physics of gas discharges in the late 19th and early 20th centuries is inseparable from that of atomic physics. After William Crookes's cathode ray experiments and J.J. Thomson's measurements of the e/m ratio, it became clear that the current in gases is mostly carried by electrons. A great deal of information on elementary processes involving electrons, ions, atoms, and light fields was obtained by studying phenomena in discharge tubes.

Beginning in 1900, J.S.E. Townsend, a student of J.J. Thomson and the creator of a school in the physics of gas discharges discovered the laws governing ionization and the gaseous discharge (known as the Townsend discharge) in a uniform electric field. Numerous experimental results were gradually accumulated on cross sections of various electron-atom collisions, drift velocities of electrons and ions, their recombination coefficients, etc. This work built the foundations of the current reference sources, without which no research in discharge physics would be possible. The concept of a plasma was introduced by I. Langmuir and L. Tonks in 1928. Langmuir made many important contributions to the physics of gas discharge, including probe techniques of plasma diagnostics.

As regards different frequency ranges, the development of field generators and the research into the discharges they produce followed the order of increasing frequencies. Radio frequency (rf) discharges were observed by N. Tesla in 1891. This kind of discharge is easily produced if an evacuated vessel is placed inside a solenoid coil to which high-frequency voltage is applied. The electric field induced by the oscillating magnetic field produces breakdown in the residual gas, and discharge is initiated. The understanding of the mechanism of discharge initiation came much latter, in fact, after the work of J.J. Thomson in 1926–1927. Inductively coupled rf discharges up to tens of kW in power were obtained by G.I. Babat in Leningrad around 1940.

The progress in radar technology drew attention to phenomena in microwave fields. S.S. Brown in the USA began systematic studies of microwave discharges in the late 1940s. Discharges in the optical frequency range were realized after the advent of the laser: A spark flashed in air when the beam of a ruby laser producing so-called giant pulses (of more than 10 MW in power) was focused by a lens, this success being achieved in 1963.

Continuously burning optical discharges, in which dense steady-state plasma is sustained by the energy of light radiation, were first initiated in 1970 by a cw CO_2 laser. Optical discharges (this term reflects a large degree of similarity with conventional discharges) immediately attracted considerable attention. Both microwave and optical discharges have by now been studied with at least the same thoroughness that the discharges in constant electric fields has been during nearly 100 years of research.

The physics of the glow discharge, one of the oldest and, presumably, best-studied fields, has lived through an unparallel revival in the past 15–20 years, and numerous new aspects of this phenomenon have been revealed. This surge of attention was stimulated by the use of glow discharges in electric-discharge CO_2 lasers developed for the needs of laser technologies. Likewise, the application of plasmatrons (generators of dense low-temperature plasma) to metallurgy, plasma chemistry, plasma welding and cutting, etc. provided a stimulus for new extensive, detailed studies of arc plasma at $p \sim 1$ atm, $T \sim 10^4$ K, and of similar discharges in all frequency ranges. These, and many other practical applications of gas discharge physics place it within the range of sciences that lie at the foundation of modern engineering.

1.5 Organization of the Book. Bibliography

A long-standing tradition demands that a general-type book on gas discharges begin with a discussion of elementary processes: possible types of collisions of electrons and ions with atoms and molecules, the fundamentals of kinetic theory of gases, statistical physics, theory of radiation, and so forth. In this book, we mostly ignore these topics, wishing to use to maximum effect the severely limited space; besides, these topics are well represented in the literature, including some general textbooks. The reader is expected to have mastered a university general physics course, although some required information is cited in direct relation to processes to be studied.

The book starts by describing the behaviour of charged particles of an ionized gas in constant and oscillating electric fields. Chapters 2 and 3 treat the behavior of electrons in a field in terms of elementary theory. Its essential feature is that the attention is focused on one "mean" electron. Averaged behaviour of one electron is considered, and when a quantity characterizing the electron gas as a whole is to be calculated, all electrons are assumed to have the same mean free time between collisions. Chapter 4 briefly discusses the processes of creation and removal of charges in a gas placed in the field. This is necessary for avoiding later on an infinite number of digressions while presenting discharge phenomena. As we remarked previously, here we spend little time on physical details of collision processes and reactions. In fact, we have tried to compile a large amount of data useful in discharge research. Chapter 5 elaborates a rigorous approach based on the kinetic equation, to the velocity and energy distribution functions. Chapter 6 is devoted to the fundamental probe method of studying gas discharge plasma. These chapters prepare the ground for the following eight chapters, which discuss systematically and in detail the discharges of various types in fields belonging to different frequency ranges. The order in which this is done is clear from the elaboratory detailed table of contents, a list that hardly requires comment.

For reasons of restricted space, it was necessary to omit mentioning a large number of facts from discharge practice and discharge theory. Quite a few of them are discussed in available books, including the older ones. We will list several popular textbooks and general-type handbooks that treat a number of subjects of discharge physics and some elementary processes [1.1–1.10]. The Russian version of the present book [1.11] contains a modern treatise, dropped from the English edition, of collisions and radiation phenomena in plasmas, useful for discharge research. The kinetics and radiation of low-temperature plasmas ($T \sim 10^4$ K) are treated in [1.12, 13]. A number of fields in the physics of discharges are represented in recent volumes of collected papers [1.14, 15], in which each chapter was written by an appropriate specialist. The data book on electron collisions [1.16] is very useful in discharge work. The list of references to each chapter of the book cites monographs on elementary processes and on various types of discharge.

The book leaves out all aspects of discharges and plasma behavior in magnetic fields. This is also caused by shortage of space and also by the fact that magnetic fields are not much employed in traditional types of discharges; except in magnetohydrodynamic generators which are not discussed here (see [1.17]). Plasmas in magnetic fields as well as high-temperature plasmas for thermonuclear fusion ($T \sim 10^6$ K), have also become objects of a special science, viz. Plasma Physics represented by a copious literature (see, e.g., [1.18, 19]). We do not consider these aspects here, even though it is not very easy to draw a very clear-cut separation of the "spheres of influence" of gas discharge physics and plasma physics. Neither do we treat here the technical applications of gas discharges, except for CO_2 lasers and plasmatrons.

A survey of technical applications of gas discharges can be found in [1.9]; detailed discussion of applications to gas lasers is given in [1.10].

At present, there is a growing interest in radio-frequency discharge. Two major applications have stimulated its study: the use of moderate pressure capacitive discharges ($p \approx 10-100$ Torr) for high-efficiency, reliable and small size CO_2 lasers and the use of low-pressure discharges ($p \approx 10^{-3}-1$ Torr) for etching, deposition and other technologies. A detailed book about the physics and applications of radio-frequency discharges was recently published (see Further Reading [1]).

The units used in the book are traditional for gas discharge physics. Energy of particles is measured in eV, macroscopic energy and power – in J and W, respectively. Electrical quantities are measured in V, A, Ohm, etc.; pressure is measured in Torr (mm Hg) and atm, and temperature in K and eV.

2. Drift, Energy and Diffusion of Charged Particles in Constant Fields

2.1 Drift of Electrons in a Weakly Ionized Gas

In the interval between two collisions, an electron is accelerated along the line of force of the electric field $\boldsymbol{E}$. A collision changes the direction of motion sharply and in a random way, after which the electron is again accelerated, etc. Encounters of charged particles are rare in a weakly ionized gas; electrons mostly collide with neutral molecules. The systematic motion along the direction of the external force amid the random motion background is known as *drift*.

2.1.1 Equation of Averaged Motion

The duration of the act of scattering being very short in comparison with the average time τ_c between collisions, we can write the equation for the true velocity $\boldsymbol{v}_e$ of an electron in the form

$$m\dot{\boldsymbol{v}}_e = -e\boldsymbol{E} + \sum_i m\Delta\boldsymbol{v}_i\delta(t - t_i) , \quad \Delta\boldsymbol{v}_i = \boldsymbol{v}'_e - \boldsymbol{v}_e , \tag{2.1}$$

where $\Delta\boldsymbol{v}_i$ is the change in the velocity vector in the ith collision at a moment t_i, δ is the Dirac delta function, and $\boldsymbol{v}'_e$ is the velocity after collision. The equation has to be averaged because monitoring the trajectory of an individual particle would be a hopeless task. The true velocity $\boldsymbol{v}_e$ is then turned into the average velocity $\boldsymbol{v}$. The sum is also averaged over collision moments t_i and scattering angles θ between the vectors $\boldsymbol{v}'_e$ and $\boldsymbol{v}_e$. It is now interpretable as the mean change of momentum per unit time, $m\langle\Delta\boldsymbol{v}\rangle/\tau_c$. This is the *resistive force* ("*friction*") applied to the electron by the medium.

Let us decompose $\Delta\boldsymbol{v}$ into components that are perpendicular and parallel to the mean velocity $\boldsymbol{v}$ before the collision. In view of the collision symmetry, $\langle\Delta\boldsymbol{v}_\perp\rangle = \langle\boldsymbol{v}'_\perp\rangle = 0$. The electron and molecule masses, m and M, are so vastly different that the electron velocity v is almost unchanged in elastic collisiions. Hence,

$$\langle\boldsymbol{v}_\parallel\rangle = \langle\boldsymbol{v}'_\parallel\rangle - \boldsymbol{v} = \boldsymbol{v}\langle\cos\theta\rangle - \boldsymbol{v} \equiv -\boldsymbol{v}(1 - \overline{\cos\theta}) ,$$

where $\overline{\cos\theta}$ is the mean cosine of the scattering angle. Inelastic scattering events (those that change v) are much less frequent than elastic collisions, and will be neglected here.

As a result, (2.1) yields an equation for the mean velocity

$$m\dot{\boldsymbol{v}} = -e\boldsymbol{E} - m\boldsymbol{v}\nu_{\mathrm{m}} \,, \quad \nu_{\mathrm{m}} = \nu_{\mathrm{c}}(1 - \overline{\cos\theta}) \,, \tag{2.2}$$

where $\nu_c = \tau_c^{-1} = Nv\sigma_c$ is the frequency of collisions of the electron, N is the number of molecules in $1\,\mathrm{cm}^3$, σ_c is the cross section of elastic collisions, and v is the velocity of random motion. It will be shown in Sect. 2.3.6 that v is much greater than the drift velocity. The frequency ν_m is called the *effective collision frequency for momentum transfer,* and $\sigma_{tr} = \sigma_c(1 - \overline{\cos\theta})$ is the momentum transfer cross section. If the scattering is isotropic then, $\overline{\cos\theta} = 0$, $\sigma_{tr} = \sigma_c$, $\nu_m = \nu_c$. If electrons are scattered mostly forward, then $\overline{\cos\theta} \approx 1$, $\nu_m \approx 0$, the momentum remains almost unchanged by the collision, and the resistive force is small. If the scattering is mostly backward, then $\overline{\cos\theta} \approx -1$ and $\nu_m \approx 2\nu_c$: the momentum change is doubled. In most gases at electron energies $\varepsilon \sim 1$–$10\,\mathrm{eV}$, typical for discharges, σ_{tr} is slightly less than σ_c (by up to 10%); at more high energies, it is less by a factor of about 1.5.

2.1.2 Drift Velocity

Integrating (2.2) results in

$$\boldsymbol{v}(t) = -(e\boldsymbol{E}/m\nu_{\mathrm{m}})[1 - \exp(-\nu_{\mathrm{m}}t)] + \boldsymbol{v}(0)\exp(-\nu_{\mathrm{m}}t) \,. \tag{2.3}$$

We find that the initial oriented velocity $\boldsymbol{v}(0)$ of the electron vanishes (is randomized) after several collisions. The mean velocity becomes

$$\boldsymbol{v}_{\mathrm{d}} = -e\boldsymbol{E}/m\nu_{\mathrm{m}} \,; \tag{2.4}$$

this is the *drift velocity*. The electric force applied to drifting electrons compensates for the resistive force. The arguments above are valid for electrons having a definite random velocity v. Usually the cross section and frequency of collisions depend on the electron energy $\varepsilon = mv^2/2$ in a complicated manner [2.1], (Figs. 2.1, 2.2), so that (2.4) must be averaged over the spectrum.

A consistent approach to calculating the drift velocity is based on analyzing the *kinetic equation* for the electron velocity distribution function (Chap. 5). This approach shows how to average a formula of type (2.4) correctly. It is then found that the assumption of the independence of the effective collision frequency on velocity, which is quite acceptable in a number of cases, reduces the rigorous expression for $\boldsymbol{v}_{\mathrm{d}}$ exactly to (2.4); there is no need, then, to average (2.4). This fact is an obvious justification of the broadest use of the simplest formula (2.4) in theoretical models and estimates. The easiest way of numerical evaluation is to make use of the experimental data on $\sigma_{tr}(v)$ [2.2, 4] and refer ν_m to the mean electron energy. Although this energy is field-dependent, the relevant reference data are available (see Sect. 2.3).

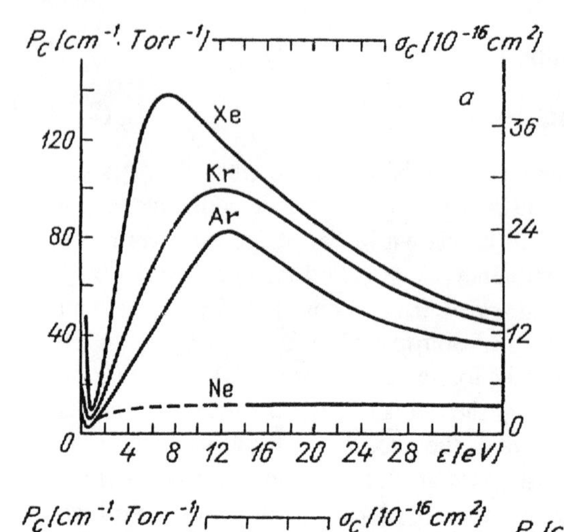

Fig. 2.1a–c. Experimental cross sections and probabilities of elastic collisions of electrons for different gases. The probability P_c is the number of collisions per cm or inverse free path length, at $p = 1$ Torr and $t = 0°$ C. From [2.3]

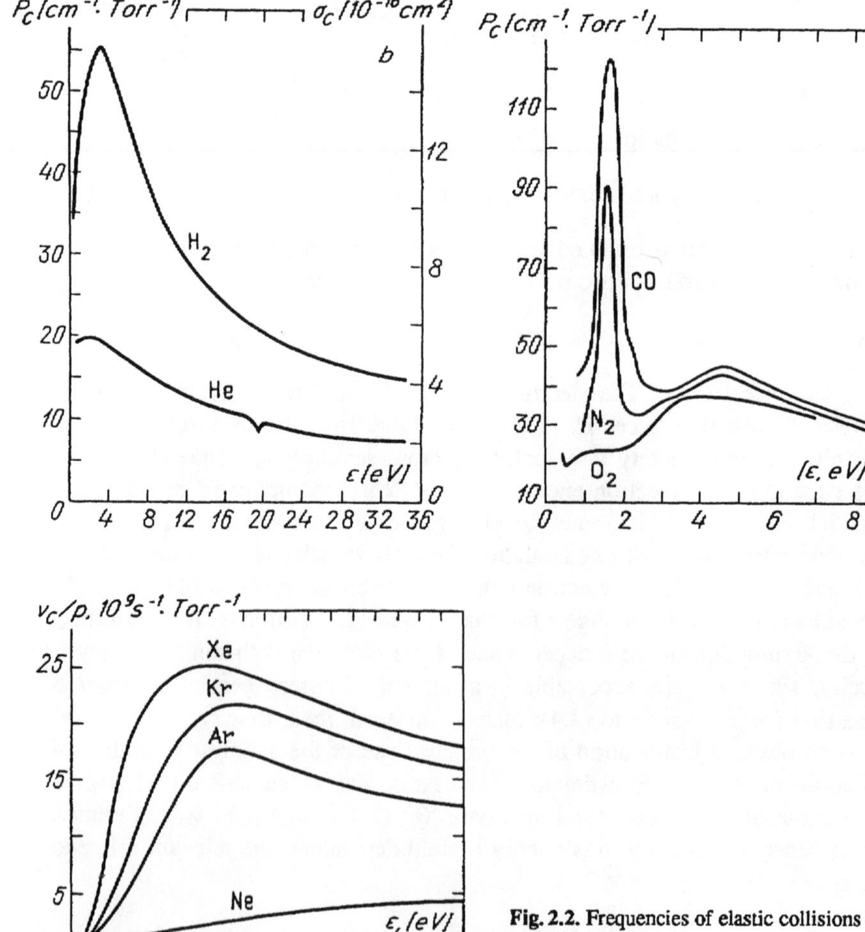

Fig. 2.2. Frequencies of elastic collisions of electrons for different gases

2.1.3 Mobility

Mobility is defined as the proportionality coefficient between the drift velocity of a charged particle and the field. The *mobility* of electrons is

$$\mu_e = \frac{e}{m\nu_m} = \frac{1.76 \cdot 10^{15}}{\nu_m[\mathrm{s}^{-1}]} \frac{\mathrm{cm}^2}{\mathrm{V \cdot s}}, \quad v_d = \mu_e E . \tag{2.5}$$

The mean energy of electrons is field-dependent; hence, v_d is not a strictly linear function of E, and the mobility depends on field strength. However, the convenient linear relation (2.5) with μ_e = const is used for theoretical analysis of various discharge processes. A reasonable effective value of μ_e is chosen for numerical estimates (see Table 2.1). As a rule, this simplification does not interfere with the qualitative validity of the theory; nevertheless, in some cases the nonlinearity of the function $v_d(E)$ causes well-pronounced effects (Sect. 2.4.4).

Table 2.1. Estimated values of electron mobility, effective collision frequency for momentum transfer, conductivity, and mean free path length

Gas	$\mu_e p$, $10^6 \frac{\mathrm{cm^2\,Torr}}{\mathrm{V \cdot s}}$	ν_m/p, $10^9\,\mathrm{s^{-1}\,Torr^{-1}}$	$\sigma p/n_e$, $10^{-13} \frac{\mathrm{Torr \cdot cm^2}}{\mathrm{Ohm}}$	range of E/p, $\frac{\mathrm{V}}{\mathrm{cm \cdot Torr}}$	lp, $10^{-2}\,\mathrm{cm \cdot Torr}$
He	0.86	2.0	1.4	0.6–10	6
Ne	1.5	1.2	2.4	0.4–2	12
Ar	0.33	5.3	0.53	1–13	3
H_2	0.37	4.8	0.58	4–30	2
N_2	0.42	4.2	0.67	2–50	3
air	0.45	3.9	0.72	4–50	3
CO_2	1.1	1.8	1.8	3–30	3
CO	0.31	5.7	0.5	5–50	2

[Mobilities were found by approximating the experimental curves $v_d(E/p)$ with the function $v_d = \mu_e E$; ν_m and σ were calculated using the value of μ_e. Mean free path lengths, $l = (N\sigma_{tr})^{-1}$, refer to electron energies of 1 to 10 eV, typical for the positive column of glow discharges.]

2.1.4 Similarity, Results of Measurements, Drift in Mixtures

The collision frequency ν_m is proportional to the density N of the gas or to its pressure p.[1] If the frequency is constant, $\mu_e \propto p^{-1}$ and $v_d \propto E/p$. The energy spectrum and the mean electron energy also depend on E and N (or p) not independently, but in the combination E/p (Sect. 2.3; Chap. 5). Hence, the drift velocity is invariably a function of the ratio E/p. We will see that *similarity laws* manifest themselves in quite a few characteristics of gas discharge. Their importance is considerable. Similarity serves to reduce the amount of measurements and the results can be plotted not as functions of two variables (say, E and p),

[1] Traditionally, gas discharge physics operates with p instead of N; p is measured in Torr = 1 mm Hg – this is very convenient. At room temperature 20°C, $N = 3.295 \cdot 10^{16}\, p\,[\mathrm{Torr}]\,\mathrm{cm}^{-3}$. If the gas is heated, however, the N–p correspondence is not one-to-one: at low temperatures, $N = 3.3 \cdot 10^{16}\, p\,[\mathrm{Torr}]\,(293/T\,[\mathrm{K}])\,\mathrm{cm}^{-3}$. Hereafter, p in numerical formulae is always expressed in Torr.

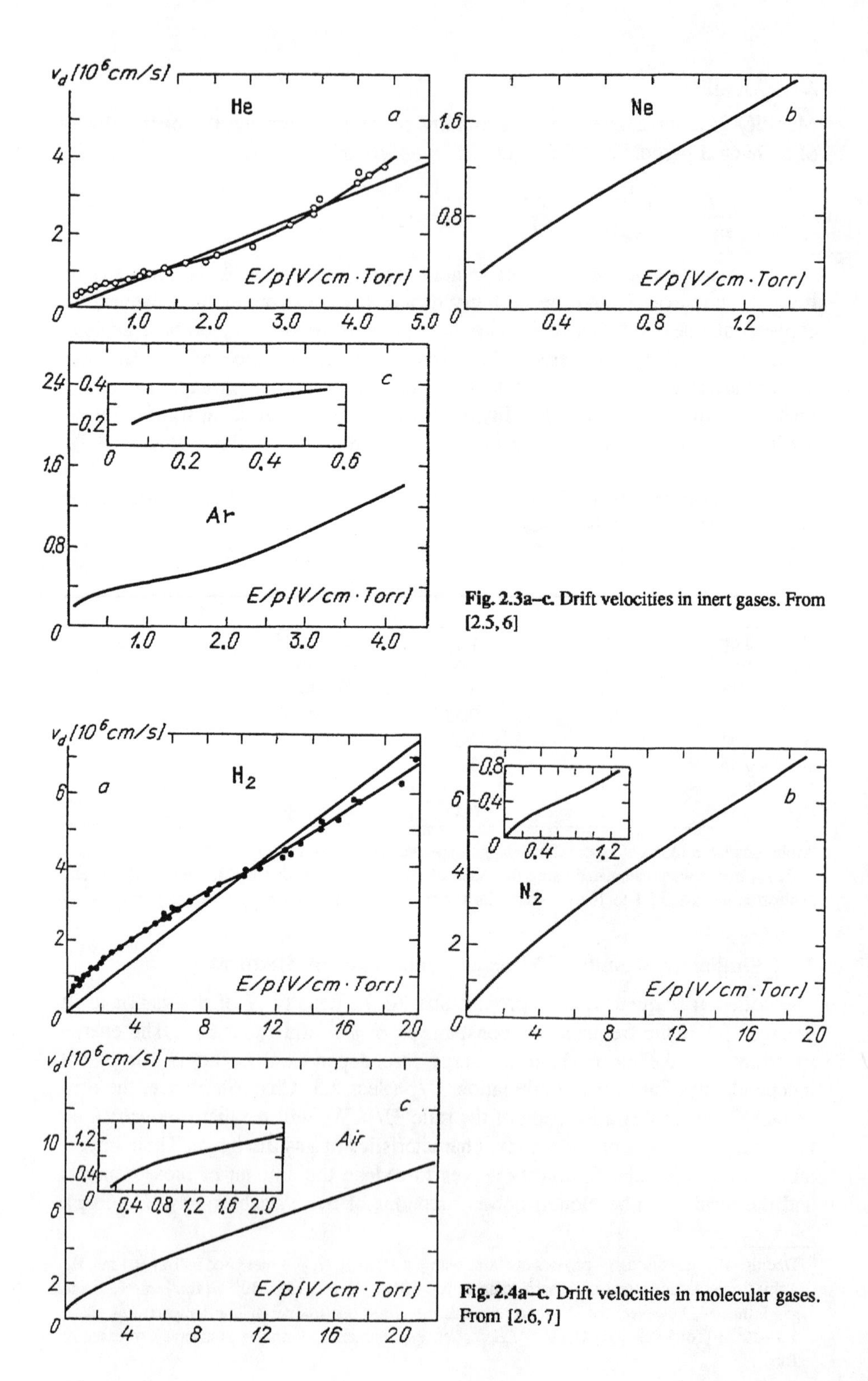

Fig. 2.3a–c. Drift velocities in inert gases. From [2.5, 6]

Fig. 2.4a–c. Drift velocities in molecular gases. From [2.6, 7]

but as functions of, say E/p, like $v_d = v_d(E/p)$. The drift velocity always increases with E/p (Figs. 2.3, 2.4) but this growth is not necessarily close to direct proportionality, since ν_m and v_d depend on the electron energy distribution. For example, anomalously fast drift is observed in argon at those values of E/p at which the characteristic electron energies fall in the range of the Ramsauer minimum of the collision cross section, $E/p \sim 10^{-3} - 10^{-1}$ V/(cm·Torr). When drift velocities are evaluated for gas mixtures, the averaging over percentage contents of the components must be carried out not for velocities or mobilities, but for their inverse values ("resistances") because the quantities that are summed in a mixture are collision frequencies. A small error is inevitable, since the electron spectrum of a mixture is different from those of the component gases.

2.2 Conduction of Ionized Gas

2.2.1 Weakly Ionized Plasma

The mobilities of massive ions are hundreds of times less than for light electrons. The contribution of ions to electric current is thus small, except in those rare cases when the ion densities n_+, n_- exceed by an appropriate number of times the electron density n_e. The current density j and conductivity σ in a plasma with $n_e \approx n_+$ are

$$j = -en_e v_d = en_e \mu_e E = \sigma E \ , \tag{2.6}$$

$$\sigma = e\mu_e n_e = \frac{e^2 n_e}{m\nu_m} = 2.82 \cdot 10^{-4} \frac{n_e[\mathrm{cm}^{-3}]}{\nu_m[\mathrm{s}^{-1}]} \mathrm{Ohm}^{-1}\,\mathrm{cm}^{-1} \ . \tag{2.7}$$

The conductivity of a weakly ionized gas is mostly determined by the degree of ionization n_e/N.

2.2.2 Strongly Ionized Plasma

The scattering of electrons by ions impedes their drift along the field as much as that by molecules. If ionization is not too weak and $n_+ = n_e$, then

$$\nu_m = Nv\sigma_{tr} + n_e v \sigma_{Coul} \ ,$$

where σ_{Coul} is the cross section of electron-ion collisions dominated by Coulomb forces.

The scale of the *Coulomb cross section* is πr_{Coul}^2, where the *Coulomb radius* $r_{Coul} = (2/3)e^2/kT_e$ is found by equating the mean thermal energy of an electron to the energy of its interaction with the ion. An electron passing by the ion at an impact distance $r \lesssim r_{Coul}$ is deflected strongly, and if $r \gg r_{Coul}$, the deflection is small. Nevertheless, trajectories passing at a large distance give appreciable contributions to the momentum transfer cross section, owing to the long-range nature of Coulomb forces. For this reason, πr_{Coul}^2 is multiplied by the so-called *Coulomb logarithm*:

$$\sigma_{\text{Coul}} = \frac{4\pi}{9}\frac{e^4 \ln \Lambda}{(kT_e)^2} = \frac{2.87 \cdot 10^{-14} \ln \Lambda}{(T_e[\text{eV}])^2}\,\text{cm}^2 , \tag{2.8}$$

$$\ln \Lambda = \ln \left[\frac{3}{2\sqrt{\pi}}\frac{(kT_e)^{3/2}}{e^3 n_e^{1/3}}\right] = 13.57 + 1.5 \log\{T_e[\text{eV}]\} - 0.5 \log n_e .$$

For example, if $T_e = 1\,\text{eV}$ (and $n_e = 10^{13}\,\text{cm}^{-3}$, $\ln \Lambda = 7.1$), then $\sigma_{\text{Coul}} \approx 2 \cdot 10^{-13}\,\text{cm}^2$, while $\sigma_{\text{tr}} \approx 10^{-16} - 10^{-15}\,\text{cm}^2$. As a result of a high Coulomb cross section, electron-ion collisions are already appreciable at ionizations greater than 10^{-3}.

These collisions become dominant at still higher degrees of ionization. In this case $\nu_m \propto n_e$, so that conductivity is independent of electron density (or rather, depends only weakly, via $\ln \Lambda$). The conductivity is

$$\sigma = \frac{e^2}{mv\sigma_{\text{Coul}}} = \frac{9(kT_e)^2}{4\pi e^2 mv \ln \Lambda} ; \quad \sigma = 1.9 \cdot 10^2 \frac{\{T_e[\text{eV}]\}^{3/2}}{\ln \Lambda}\,\text{Ohm}^{-1}\,\text{cm}^{-1} . \tag{2.9}$$

The numerical formula was obtained with the mean thermal velocity of electrons and the coeffcient was refined by a factor of about 2 [2.8].

2.2.3 Why Electron-Electron Collisions Do not Contribute to Electric Resistance

The point is that the resistive force ("friction") in the drift of electrons is caused by the loss of momentum along the field in scattering events. As to the total momentum of any pair of interacting electrons, $m\boldsymbol{v}_1 + m\boldsymbol{v}_2$, it is conserved under scattering although the velocity of each one changes both in magnitude and in direction. This means that scattering preserves the total electric current of the pair, which is proportional to $-e\boldsymbol{v}_1 - e\boldsymbol{v}_2$. Hence, the motion of a group of electrons colliding only with electrons would be accelerated on average; this means that the electric resistance would be zero. Note that electron-electron collisions can affect the conductivity indirectly, namely, by changing the energy spectrum of the electron gas (by "Maxwellizing" it).

2.3 Electron Energy

2.3.1 Joule Heat

The work done by an electric field on an electron moving at a velocity $\boldsymbol{v}_e$ is $-e\boldsymbol{E} \cdot \boldsymbol{v}_e$ per unit time. Let us represent the electron's velocity as a sum of random $\boldsymbol{v}$ and drift $\boldsymbol{v}_d$ components: $\boldsymbol{v}_e = \boldsymbol{v} + \boldsymbol{v}_d$. By definition, averaging over a large number of electrons gives $\langle \boldsymbol{v} \rangle = 0$ and $\langle \boldsymbol{v}_e \rangle = \boldsymbol{v}_d$. The mean work of the field equals $-\langle e\boldsymbol{E} \cdot \boldsymbol{v}_e \rangle = eEv_d$. The energy released by the current in $1\,\text{cm}^3$ of the gas in 1 s is $eEv_d n_e = jE$. This is the *Joule heat* of the electric current. The field does work to overcome the resistive force. The Joule heat equals the energy of the field dissipated in response to resistance.

2.3.2 Mean Energy Gained by an Electron in One Effective Collision

This quantity equals

$$\Delta\varepsilon_E = \frac{eEv_\mathrm{d}}{\nu_\mathrm{m}} = \frac{e^2E^2}{m\nu_\mathrm{m}^2} = mv_\mathrm{d}^2 , \tag{2.10}$$

and coincides, in order of magnitude, with the "kinetic energy of drift motion", $mv_\mathrm{d}^2/2$. On average, the total kinetic energy of an electron is composed of the random $\bar{\varepsilon}$ and drift components:

$$\left\langle \frac{m\boldsymbol{v}_\mathrm{e}^2}{2} \right\rangle = \left\langle \frac{m\boldsymbol{v}^2}{2} \right\rangle + \frac{m\boldsymbol{v}_\mathrm{d}^2}{2} = \bar{\varepsilon} + \frac{mv_\mathrm{d}^2}{2} , \quad \langle \boldsymbol{v} \cdot \boldsymbol{v}_\mathrm{d} \rangle = 0 .$$

Qualitatively, the result (2.10) can be given the following interpretation. On average, the velocity of the electron immediately after a collision ("effective" collision) is completely randomized: on average, its vector is zero. By the next collision, the electron has built up the drift velocity along the field, with the corresponding kinetic energy. The collision transfers this new portion of energy into the random part (the electron "heat") and the process starts anew.

2.3.3 True Change of Electron Energy in a Collision

Formula (2.10) reflects only the final result of various, sharply opposing situations that arise in different collisions. In fact, an electron in the interval between two collisions can be accelerated by the field or decelerated, it can store up energy or lose it. It depends on whether the motion after the collision started along or against the direction of the force, at a large or small velocity. A very simple example will be useful.

Let us fix our attention on two electrons that have velocities of identical magnitude v and start moving parallel to the field in opposite directions. Their initial kinetic energies are equal, $mv^2/2$. The electron moving along the force (against the field) reaches the velocity $v + eE/m\nu_\mathrm{m} = v + v_\mathrm{d}$ by the time of the next collision, picking up the additional energy

$$\Delta\varepsilon_+ = \frac{m(v+v_\mathrm{d})^2}{2} - \frac{mv^2}{2} = mvv_\mathrm{d} + \frac{mv_\mathrm{d}^2}{2} .$$

The electron that started moving along the field is decelerated some of the time (or all of it), and reaches the velocity $-v + v_d$ by the moment of collision. The additional energy is

$$\Delta\varepsilon_- = \frac{m(-v+v_\mathrm{d})^2}{2} - \frac{mv^2}{2} = -mvv_\mathrm{d} + \frac{mv_\mathrm{d}^2}{2} ;$$

if $v > v_\mathrm{d}/2$, this increment is negative, that is, the electron loses energy. This is a typical situation: we will see in Sect. 2.3.6 that random velocities v are usually much greater than the drift velocity.

On average, the additional energy gained in these two scenarios is

$$\frac{\Delta\varepsilon_+ + \Delta\varepsilon_-}{2} = mv_\mathrm{d}^2 ,$$

which is independent of v, is always positive, and coincides with (2.10). Clearly, electrons with arbitrary initial vectors $\boldsymbol{v}$ can be divided into similar pairs with oppositely directed velocities, producing the same result, at least to the order of magnitude.

We have already mentioned, referring to Sect. 2.3.6, that the mean random velocity $\bar{v} \gg v_\mathrm{d}$; hence, the resulting average energy gained per one collision is a small net difference between large actual gains and losses; these are of order $|\Delta\varepsilon_\pm| \approx m\bar{v}v_\mathrm{d} \gg \Delta\varepsilon_E = mv_\mathrm{d}^2$. In their turn, the actual changes of energy $|\Delta\varepsilon_\pm|$ are small in comparison with the mean electron energy $\bar{\varepsilon} \approx m\bar{v}^2/2$. The ratio of these quantities is

$$\frac{\Delta\varepsilon_E}{|\Delta\varepsilon_\pm|} \sim \frac{|\Delta\varepsilon_\pm|}{\bar{\varepsilon}} \sim \frac{v_\mathrm{d}}{\bar{v}} , \quad \frac{\Delta\varepsilon_E}{\bar{\varepsilon}} \sim \left(\frac{v_\mathrm{d}}{\bar{v}}\right)^2 . \tag{2.11}$$

2.3.4 Equation of Electron Energy Balance

Electrons gain energy from the field and pass it on to atoms and molecules. The current density and Joule heat released are small when the ionization is low. The gas heats up slightly. But the mean energy and temperature of electrons in the discharge cannot be too low, since electrons would be unable to ionize atoms and sustain the conductive state in the gas. In such cases $T_\mathrm{e} \gg T$ at low pressures, so that the energy exchange in elastic collisions is a one-way process: from electrons to the gas (if $T_\mathrm{e} \sim T$, the exchange is mutual, on average). Let us find the mean energy lost by an electron in an elastic collision with a molecule.

When an electron loses momentum $\Delta\boldsymbol{p}$, the molecule gains the same momentum. If it was "at rest" before the collision, the energy gained is $(\Delta\boldsymbol{p})^2/2M$. The electron loses the same amount of energy; on the average, this amount is $(m^2/2M)\langle(\Delta\boldsymbol{v})^2\rangle$. Since

$$\langle(\Delta\boldsymbol{v})^2\rangle = \langle(\boldsymbol{v}' - \boldsymbol{v})^2\rangle = v^2 - 2v'v\overline{\cos\theta} + v^2 = 2v^2(1 - \overline{\cos\theta}) ,$$

the mean fraction of energy $\varepsilon = mv^2/2$ that the electron loses in an elastic collision is $(2m/M)(1 - \overline{\cos\theta})$. We denote $\delta = 2m/M$ and can write the equation of energy balance for the "mean" electron undergoing only elastic collisions:

$$\frac{d\varepsilon}{dt} = (\Delta\varepsilon_E - \delta\varepsilon)\nu_\mathrm{m} = \left(\frac{e^2E^2}{m\nu_\mathrm{m}^2} - \delta\varepsilon\right)\nu_\mathrm{m} . \tag{2.12}$$

The mean electron energies in discharge plasmas are usually quite low compared to the fairly high potentials of excitation and ionization of atoms, $I \sim 10\,\mathrm{eV}$, and the corresponding energy losses are small. The gas is ionized by "super-energetic" electrons, and these are rare. For this reason, the main mechanism of energy transfer from electrons to the gas is the *elastic loss*. Electrons in molecular gases mostly dissipate energy by exciting the *vibrational* (and *rotational*) energy levels of molecules. This case too can be described by (2.12), but the coefficient δ is not calculated as simply as for elastic losses. Inelastic losses are usually greater than elastic ones by one to two orders of magnitude; nevertheless, the corresponding coefficient δ is small: $\delta \sim 10^{-3} - 10^{-2} (2m/M \sim 10^{-4} - 10^{-5})$.

2.3.5 Mean Energy

The equilibrium value of energy of the "mean" electron corresponds to energy gains compensated for by losses; in the approximation above it can be treated as the mean energy of electrons, $\bar{\varepsilon}$, in the field (the rigorous calculation of this quantity requires the solution of the kinetic equation; see Chap. 5). Assume that the coefficient δ and scattering cross section σ_{tr} are independent of energy, that is, the free path of electrons, $l = 1/N\sigma_{tr}$, is energy-independent. Note that $\nu_m = \bar{v}/l \sim \sqrt{\bar{\varepsilon}}$. We also specify that $m\bar{v}^2 = (16/3\pi)\bar{\varepsilon}$, as in the case of the Maxwellian distribution. Equating the right-hand side of (2.12) to zero, we obtain

$$\bar{\varepsilon} = \frac{\sqrt{3\pi}}{4}\frac{eEl}{\sqrt{\delta}} \approx 0.8\frac{eEl}{\sqrt{\delta}} \approx \frac{e}{\sigma_{tr}\sqrt{\delta}}\frac{E}{N} . \tag{2.13}$$

The mean energy is proportional to E/N and exceeds by a factor $1/\sqrt{\delta}$ the energy eEl that the electron builds up while moving along the direction of the electric force. The assumption of constant free path length corresponds to the square root dependence of drift velocity on the electric field. Indeed, substituting $\bar{v} = (16\bar{\varepsilon}/3\pi m)^{1/2}$ into $\nu_m = N\bar{v}\sigma_{tr}$ and defining $\bar{\varepsilon}$ by (2.13), we find from (2.5) that

$$v_d = \left(\frac{3\pi}{16}\delta\right)^{1/4}\left(\frac{eE}{m\sigma_{tr}N}\right)^{1/2} \approx 0.9\delta^{1/4}\left(\frac{eE}{m\sigma_{tr}N}\right)^{1/2} . \tag{2.14}$$

On the other hand, the assumption of constant collision frequency, with $\mu_e =$ const and $v_d \sim E/N$, corresponds to the quadratic dependence of energy on field strength (provided $\delta =$ const). In this case (2.12) implies

$$\bar{\varepsilon} = \frac{e^2E^2}{\delta m\nu_m^2} = \frac{e^2}{\delta m\tilde{\nu}_m^2}\left(\frac{E}{N}\right)^2 , \quad \tilde{\nu}_m \equiv \frac{\nu_m}{N} . \tag{2.15}$$

The choice of a particular model thus produces a dilemma: assume either

$$\nu_m, \mu_e = \text{const} , \quad v_d \sim E/N, \bar{\varepsilon} \sim \delta^{-1}(E/N)^2 ,$$

or

$$\sigma_{tr}, l = \text{const}, v_d \sim \delta^{1/4}(E/N)^{1/2}, \bar{\varepsilon} \sim \delta^{-1/2}(E/N) .$$

The actual dependence of v_d and $\bar{\varepsilon}$ on E/N found in experiment or obtained by solving the kinetic equation is usually quite complicated and, at best, is well approximated by a specific model within certain intervals of E/N for specific gases. Consequently, for a theoretical analysis of various effects, for revealing characteristics of qualitative behavior and achieving more profound understanding of the physics of processes, one is forced to choose a version that is better suited for analysis. If an effect is determined by drift or electric current, it is expedient to choose constant mobility. If attention is focused on the energy aspects of electron behavior in the field as is the case in this section, it is better to use the approximation $l = \text{const}$ and the clearly descriptive formula (2.13); this is what we propose to do below.

Formula (2.13) gives reasonable numerical results. For instance, in helium ($\delta = 2.7 \cdot 10^{-4}$), the momentum transfer cross section changes little in the characteristic energy range and is approximately $\sigma_{tr} \approx 5.5 \cdot 10^{-16}\,\text{cm}^2$ ($l \approx 0.055/p[\text{Torr}]\text{cm}$); hence we find for $E/N = 3.3 \cdot 10^{-17}\,\text{V} \cdot \text{cm}^2$ [$E/p = 1\,\text{V}/(\text{cm} \cdot \text{Torr})$] that $\bar{\varepsilon} \approx 2.5\,\text{eV}$. Experiment gives $\bar{\varepsilon} \approx 2\,\text{eV}$. It did not prove possible to calculate $\bar{\varepsilon}$ in molecular gases in a simple manner because serious difficulties are encountered in finding the loss coefficient δ. However, this does not make (2.13) useless. Quite the opposite, it makes the estimation of δ possible because an independent way of determining $\bar{\varepsilon}$ is available (see Sect. 2.4). Thus $\bar{\varepsilon} \approx 1.5\,\text{eV}$ in nitrogen at $E/p = 3\,\text{V}/(\text{cm} \cdot \text{Torr})$; the cross section in the characteristic energy range is of order $\sigma_{tr} \approx 10^{-15}\,\text{cm}^2$ ($l \approx 0.03/p[\text{Torr}]\,\text{cm}$), and (2.13) yields $\delta \approx 2.1 \times 10^{-3}$, in agreement with the result obtained by an independent method. The elastic loss coefficient of nitrogen is much smaller, $2m/M = 3.9 \times 10^{-5}$.

2.3.6 Relation of Random to Drift Velocity

It is immediately implied by (2.5) and (2.13) that

$$\frac{v_d}{\bar{v}} = \frac{elE}{m\bar{v}^2} = \frac{\sqrt{3\pi}}{4}\sqrt{\delta} \approx 0.8\sqrt{\delta}\,, \quad \frac{\bar{v}}{v_d} \approx \frac{1.2}{\sqrt{\delta}}\,. \tag{2.16}$$

The random velocity that an electron develops in the field is greater than the drift velocity by a factor of $1/\sqrt{\delta}$, that is, of tens or hundreds. Relation (2.16), which clarifies the physical meaning of the smallness of $v_d/\bar{v}$, is closely tied to the parameters of smallness of the energy characteristics of electrons. In a collision, an electron gains or loses an energy of order $|\Delta\varepsilon_\pm| \approx \sqrt{\delta}\,\bar{\varepsilon} \approx eEl$, which corresponds to the potential difference traversed in an arbitrary direction across one free path length. The energy $\Delta\varepsilon_+$ is slightly greater than $|\Delta\varepsilon_-|$, by the amount $\Delta\varepsilon_E \approx \sqrt{\delta}|\Delta\varepsilon_+|$ that an electron picks up, on average, per one collision.

Now, several words on the limits of applicability of the relations above. On the side of very weak fields, it is restricted by the assumption of one-way energy exchange between electrons and the gas, $\bar{\varepsilon} \gg kT$. Indeed, in the absence of the field, electrons are thermalized and acquire the gas temperature T, unless they perish first! The condition $\bar{\varepsilon} \gg kT$ is usually met in weakly ionized discharge plasmas "many times over." Considerable inelastic losses become important in high-strength fields, at electron energies $\varepsilon \gtrsim 10\,\text{eV}$. Even the formal transition to the limit of maximal possible losses $\delta \sim 1$ in (2.13, 16) shows that the oriented and random components of velocity become comparable. Hence, strong asymmetry in motion and in true energy exchange between electrons and the field must develop. An electron builds up considerable energy between subsequent collisions, excites or ionizes an atom, losing energy and rushes forward again. This is not quite the picture that was taking shape before. Such phenomena may occur in the cathode layer of a glow discharge (Sect. 5.9, 8.4). Sometimes relatively high drift velocities are also observed outside the cathode layer. In low pressure mercury vapour the drift velocity is of one quarter of the random velocity [2.9].

2.3.7 Energy Relaxation; Criteria of Constant and Homogeneous Fields

We refer to the quantity $\nu_u = \delta\nu_m$ as the *frequency for energy losses*. If the field is instantaneously switched off, an electron dissipates its energy in a time of order

$$\tau_u = \nu_u^{-1} = \tau_m/\delta\ , \tag{2.17}$$

that is, after about $1/\delta$ effective collisions. A very slow electron picks up the appropriate energy in about the same time, because in one collision it acquires a fraction δ of it. As follows from (2.12), τ_u characterizes the rate at which equilibrium energy builds up in a given field. This is the *energy relaxation time*. If the field changes little over the time τ_u, the mean energy (and hence the electron energy distribution) track the changing field and is thus *quasi-steady*. If changes are fast, tracking becomes impossible. The field "constancy" criterion can thus be written as $(dE/dt)\tau_u \ll 1$.

On the average, the electrons are systematically displaced in the direction of the electric field, so that energy equilibration proceeds not only in time but in space as well. Over one relaxation time τ_u, electrons drift over a distance

$$\Lambda_u = v_d\tau_u \approx 0.8l/\sqrt{\delta}\ . \tag{2.18}$$

This quantity can be called the *energy relaxation length*. It is greater than the electron mean free path by a factor $1/\sqrt{\delta}$, not by the $1/\delta$ that characterizes the ratio of relaxation time to time between collisions. According to (2.13, 18),

$$\bar{\varepsilon} = eE\Lambda_u\ , \tag{2.19}$$

that is, an electron builds up the mean energy in the potential difference across one Λ_u.

A dc field can be treated as homogeneous if it varies only slightly over distances of order Λ_u, that is, if $(dE/dx)\Lambda_u \ll 1$. The opposite case is that of strongly inhomogeneous fields in which the energy distribution and the mean energy of electrons cease to be functions of only the local ratio E/N. For instance, the energy may be determined by the potential difference crossed by electrons after being emitted from the atoms.

The loss term in parentheses in the energy balance equation (2.12) increases with increasing ε, while the energy gain generally decreases; under the assumption l = const, it varies as $1/\varepsilon$. Therefore, electron energy always tends to the stationary value $\bar{\varepsilon}$. If $\varepsilon < \bar{\varepsilon}$, then $d\varepsilon/dt > 0$; if $\varepsilon > \bar{\varepsilon}$, then $d\varepsilon/dt < 0$. This shows that the stationary state is stable because energy invariably returns to $\bar{\varepsilon}$ (at the relaxation rate) after any random deviation of ε from $\bar{\varepsilon}$.

2.4 Diffusion of Electrons

2.4.1 Equation of Continuity

If the density of particles moving in a gas is spatially noniuniform, a *diffusion flux* appears that tends to level it off. The total flux consists of the drift and diffusion components. The flux densities of positively and negatively charged particles are

$$\boldsymbol{\Gamma}_{\pm} = \pm n\mu \boldsymbol{E} - D\nabla n \,. \tag{2.20}$$

(If the gas flows at a velocity $\boldsymbol{u}$, *convective* components $n\boldsymbol{u}$ are added to $\boldsymbol{\Gamma}$.) Note that subscripts $\pm$ with n, μ, and D are dropped. The diffusion coefficients are

$$D = \langle v^2/3\nu_{\mathrm{m}} \rangle \approx l\bar{v}/3 \,, \quad D \propto p^{-1} \,. \tag{2.21}$$

Particle number densities satisfy the *continuity equations*

$$\frac{\partial n}{\partial t} + \operatorname{div} \boldsymbol{\Gamma} = q \,, \tag{2.22}$$

which generalize the standard diffusion equation; q are the bulk sources of creation or annihilation of particles in $1\,\mathrm{cm}^3\,\mathrm{s}^{-1}$.

2.4.2 Relation Between Diffusion Coefficient, Mobility, and Mean Energy

Assuming that the collision frequency is constant, we find from (2.5) and (2.21) that

$$D_{\mathrm{e}}/\mu_{\mathrm{e}} = m\overline{v^2}/3e = (2/3)\bar{\varepsilon}/e \,, \tag{2.23}$$

where $\bar{\varepsilon}$ is the true mean energy of electrons, regardless of their energy spectrum. If the spectrum is Maxwellian, (2.23) is valid regardless of the dependence of ν_{m} on v. One only needs to substitute for μ_{e} a rigorous expression implied by the kinetic equation (Sect. 5.6.1). This is natural since here $\bar{\varepsilon} = (3/2)kT_{\mathrm{e}}$ and (2.23) reduces to the *Einstein relation*

$$D/\mu = kT/e \,, \tag{2.24}$$

which is thermodynamic in nature. Indeed, there are no fluxes in the state of thermodynamic equilibrium, and charge densities in the field $\boldsymbol{E} = -\nabla\varphi$ (φ is the potential) satisfy Boltzmann's law $n_{\pm} \propto \exp(\mp e\varphi/kT)$. This gives (2.24).

If the electron spectrum is non-Maxwellian and $\nu_m(v) \neq \text{const}$ (this is typical of weakly ionized gases in electric fields), the quantity $(3/2)D_e/\mu_e$ also characterizes the mean electron energy, but does not exactly coincide with it. The ratio D_e/μ_e, corresponding to the electron "temperature", is known as the *characteristic energy*. Like the spectrum, it is a function of E/p.

Experimentally, the ratio D_e/μ_e is measured by determining the spreading, due to diffusion, of electrons drifting in the field E. At a distance $x = v_d t$ from the point where electrons start, the beam has spread in the transverse direction to a radius $r \sim \sqrt{D_e t} = \sqrt{D_e x/v_d} \approx \sqrt{(D_e/\mu_e)(x/E)}$. Experimental curves of

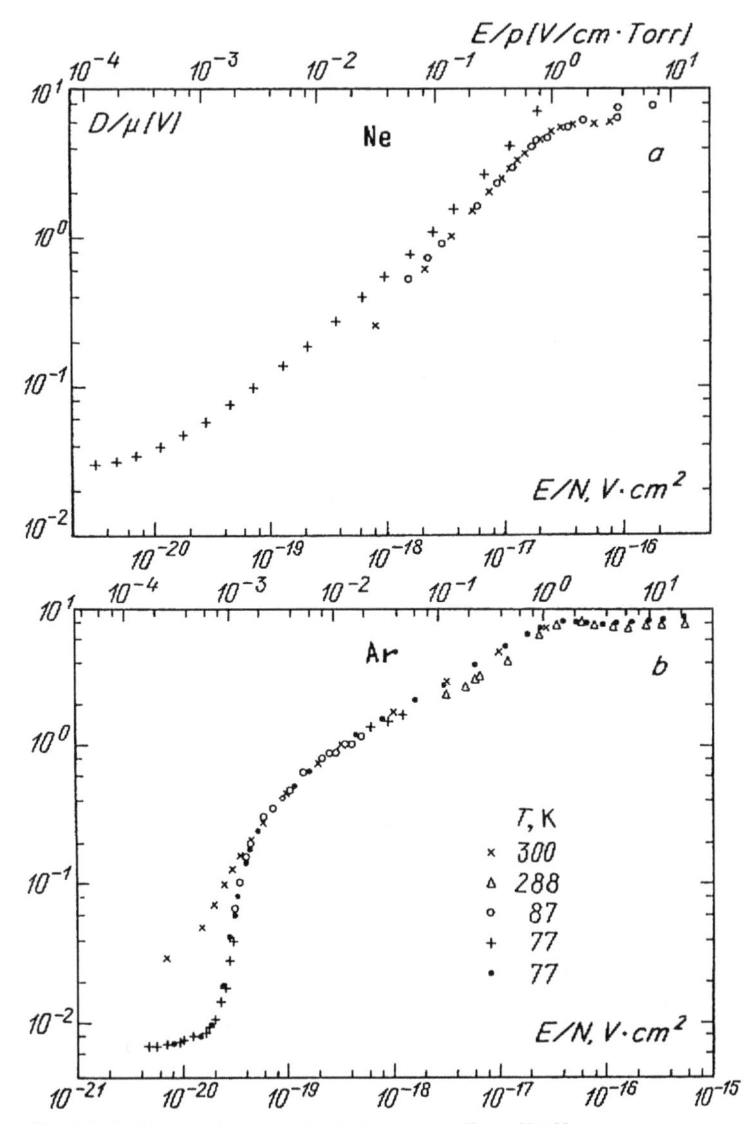

Fig. 2.5a,b. Characteristic energies in inert gases. From [2.10]

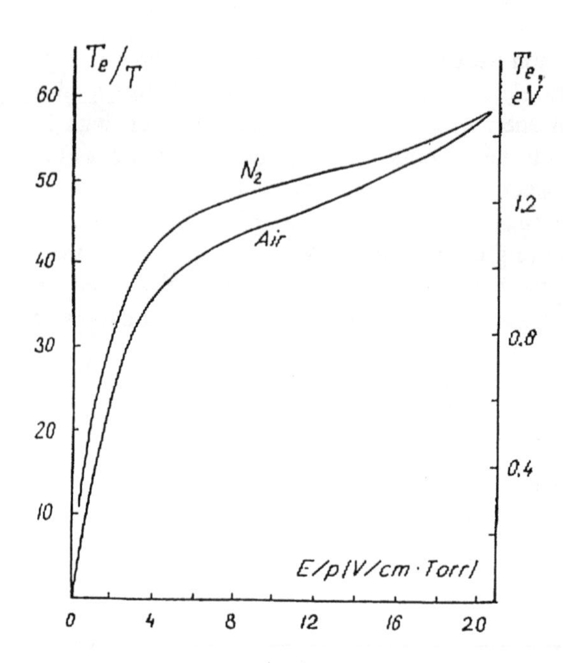

Fig. 2.6. The ratio of electron and gas temperatures in air and N_2. From [2.11]

D_e/μ_e as a function of E/p (Figs. 2.5, 2.6) can be successfully fitted for Ne and H_2, with reasonable accuracy, by a straight line through the origin in the ranges of E/p typical for glow discharges. Formula (2.19) then gives some idea about electron energy relaxation lengths Λ_u. Namely,

$$\text{Ne:}\quad \bar{\varepsilon} \approx 9.7\left(\frac{E}{p}\right)\ \text{eV}\,, \quad \frac{E}{p} \sim 0.1 - 1.2\,\text{V}/(\text{cm}\cdot\text{Torr})\,;$$

$$\Lambda_u \approx \frac{9.7}{p[\text{Torr}]}\text{cm}\,,$$

$$\text{H}_2\text{:}\quad \bar{\varepsilon} \approx 0.17\left(\frac{E}{p}\right)\ \text{eV} \quad \frac{E}{p} \sim 0.5 - 13\,\text{V}/(\text{cm}\cdot\text{Torr})\,;$$

$$\Lambda_u \approx \frac{0.17}{p[\text{Torr}]}\text{cm}\,.$$

2.4.3 Calculation of Diffusion Coefficients

The measurement of drift velocities and characteristic electron energies is perfectly feasible. On the other hand, it is exceptionally difficult to measure diffusion coefficients directly in the presence of an electric field. In fact, only diffusion coefficients of *thermalized* electrons that have come into thermal equilibrium with the gas in zero field have been measured in direct experiments. At room temperature, $D_{e,\text{therm}} = K \cdot 10^5/p\,[\text{Torr}]\,\text{cm}^2/\text{s}$, where

	He	Ne	Ar	H_2	N_2	O_2
$K =$	2	20	6.3	1.3	2.9	12 .

The value of D_e can be estimated using the data on v_d and D_e/μ_e. Thus in Ne at $E/p = 1$ V/(cm·Torr), we have $v_d = 1.46\times10^6$ cm/s, $T_e \approx 5.5$ eV [2.10], whence $D_e \approx v_d T_e/E \approx 8\times10^6/p\,\text{cm}^2/\text{s}$. In air at $E/p = 20$, we find $v_d = 8.5\times10^6$ cm/s, $T_e \approx 1.5$ eV[2.11], from this $D_e \approx 6.3\times10^5/p\,\text{cm}^2/\text{s}$.

2.4.4 Longitudinal and Transverse Diffusion

The ratio D_e/μ_e found from the measured diffusional spreading of electron packets along the direction of drift differs systematically from the results obtained by recording transverse spreading. The physical reason for the difference between the coefficients of *longitudinal* D_L and *transverse* D_T diffusion is the dependence of the collision frequency ν_m on electron energy [2.12]. According to (2.20), the mean electron flux consists of drift and diffusion components:

$$\boldsymbol{v} = \boldsymbol{\Gamma}_e/n_e = \boldsymbol{v}_d - D_e \nabla n_e/n_e \equiv \boldsymbol{v}_d + \boldsymbol{v}_{dif}\,;$$

in the zeroth approximation, D_e is the "ordinary" coefficient characterizing the diffusion perpendicular to the drift direction. If there is a gradient of n_e along the field, the field does additional work per second on the electrons, in comparison with the case of pure drift motion $-e\boldsymbol{E}\boldsymbol{v}_{dif} = eD_e(\boldsymbol{E}\nabla n_e)/n_e$. The mean electron energy $\bar{\varepsilon}$ gets an increment $\Delta\varepsilon$ proportional to the projection of the gradient on the direction of the vector $\boldsymbol{E}, \nabla_{\|} n_e$. If ν_m increases with increasing ε (this is the typical situation), the mobility reduces by $\Delta\mu_e$, which is proportional to $\nabla_{\|} n_e$. The effect of this response is equivalent to the drift velocity remaining unchanged but the velocity due to longitudinal diffusion, $\boldsymbol{v}_{dif,\|} = -D_e \nabla_{\|} n_e/n_e$, getting an increment proportional to $\Delta\mu_e \sim \nabla_{\|} n_e$. In its turn, this effect means a drop in the longitudinal diffusion coefficient. In the first approximation, in a small gradient, we find [2.12] that

$$D_T = D_e,\; D_L = \left(1 - \frac{\hat{\nu}_m}{1+2\hat{\nu}_m}\right) D_e\,, \quad \hat{\nu}_m \equiv \frac{\partial \ln \nu_m}{\partial \ln \varepsilon}\,. \tag{2.25}$$

The logarithmic derivative $\hat{\nu}_m$ characterizes the steepness of the function. Thus for $\nu_m \sim \varepsilon^k$, we have $\hat{\nu}_m = k$. Experimentally, D_L is seen to diminish by a factor of up to 2 in comparison to D_T.

2.5 Ions

2.5.1 Collisions with Molecules

An ion of mass M_i comparable with that of a gas molecule M exchanges larges portions of energy with molecules. If the field is not strong, an ion gets from the field an energy less than kT; ions then reach the gas temperature T. These conditions are typical for the positive column of a glow discharge. The cross section of elastic scattering of slow ions is determined by *polarization forces*. An ion at a distance r from a molecule induces in it a dipole moment $d = \alpha e/r^2$, where α is the polarizability of the molecule, and is attracted to the molecule with a force $2ed/r^3$. The ion is scattered strongly if it passes at a distance $r \lesssim \varrho_0$, where $\varrho_0 \approx (\alpha e^2/2\varepsilon')^{1/4}$ corresponds to the equality of the potential energy of interaction $|U| = \alpha e^2/2r^4$ and the kinetic energy of relative motion of the particles ε'. To an order of magnitude, therefore, the scattering cross section is $\sigma_{\text{tr}} \approx \pi\sigma_0^2$. Including the corrective factor of $2\sqrt{2}$ [2.13], we have

$$\sigma_{\text{tr}} \approx 2\pi\sqrt{\alpha e^2/\varepsilon'} = 2\sqrt{2}\pi a_0^2\sqrt{(\alpha/a_0^3)\,(I_{\text{H}}/\varepsilon')}\,. \tag{2.26}$$

To indicate the scale, we have substituted $e^2 = 2a_0 I_{\text{H}}$ into σ_{tr} (a_0 is the Bohr radius and I_{H} is the ionization potential of the hydrogen atom; see Appendix). In the polarization interaction, $\sigma_{\text{tr}} \propto 1/v'$ and the collision frequency is $\nu_{\text{m}} \propto v'\sigma_{\text{tr}} = \text{const}$, where v' is the relative velocity of the particles.

When the "radius" of polarization foces ϱ_0 becomes less than the molecular size, scattering occurs only when particles come into "contact." The polarization cross section is replaced with the gas-kinetic one, which weakly depends on v'; now $\nu_{\text{m}} \propto v'$. In Ar, N_2, and O_2 (where $\alpha/a_0^3 \approx 10-12$), as well as in He, this occurs when $\varepsilon' > \varepsilon_k \approx 0.5-0.6\,\text{eV}$, in Ne when it is 0.15 eV, and in H_2 when it is 0.9 eV.

Ions moving in their own gases, (e.g., He^+ in He, N_2^+ in N_2) lose momentum intensively via charge transfer. An ion accelerated by the field appropriates an electron from a neutral molecule. This happens so quickly that the new ion (the former neutral) fails to move at all. The charge transfer cross section, $\sigma_{\text{c.t.}}$, is usually even greater than the elastic scattering cross section [2.4] (Fig. 2.7). In the center-of-mass reference frame, the molecule and the ion move towards each other at equal velocities, while after charge transfer the charge moves at the same velocity but in the opposite direction. This is equivalent to scattering by 180°, so that in charge transfer, $\sigma_{\text{tr}} = 2\sigma_{\text{c.t.}}$. Charge transfer considerably reduces the mobility of ions in their own gas.

2.5.2 Drift in Weak and Moderate Fields

In the general case, the mean rate of momentum loss by a particle in a collison (resistive force) is determined by the reduced mass $M' = M_{\text{i}}M/(M_{\text{i}}+M)$ and the relative velocity $\boldsymbol{v}'$. The quantity $M'\langle \boldsymbol{v}'\nu_{\text{m}}(v')\rangle$ is averaged over the velocities of molecules for a fixed ion velocity $\boldsymbol{v}$. The situation looks simpler when electrons are considered (Sect. 2.1.1) because $m \ll M$ and $M' \approx m$, $\boldsymbol{v}' \approx \boldsymbol{v}_{\text{e}}$. There is also no need for averaging if the ion cross section is determined by polarization forces and charge transfer, since $\nu_{\text{m}}(v') \approx \text{const}$ and $\langle \boldsymbol{v}'\nu_{\text{m}}\rangle = \boldsymbol{v}\nu_{\text{m}}$, where $\boldsymbol{v}$ is the mean velocity of ions. Equating the resistive forces and fields, we obtain the drift velocity and ion mobilities similar to (2.4,5):

$$\boldsymbol{v}_{\text{id}} = e\boldsymbol{E}/M'\nu_{\text{m}}\,, \quad \mu_{\text{i}} = e/M'\nu_{\text{m}}\,. \tag{2.27}$$

If no charge transfer occurs and the cross section is purely polarization-induced, then

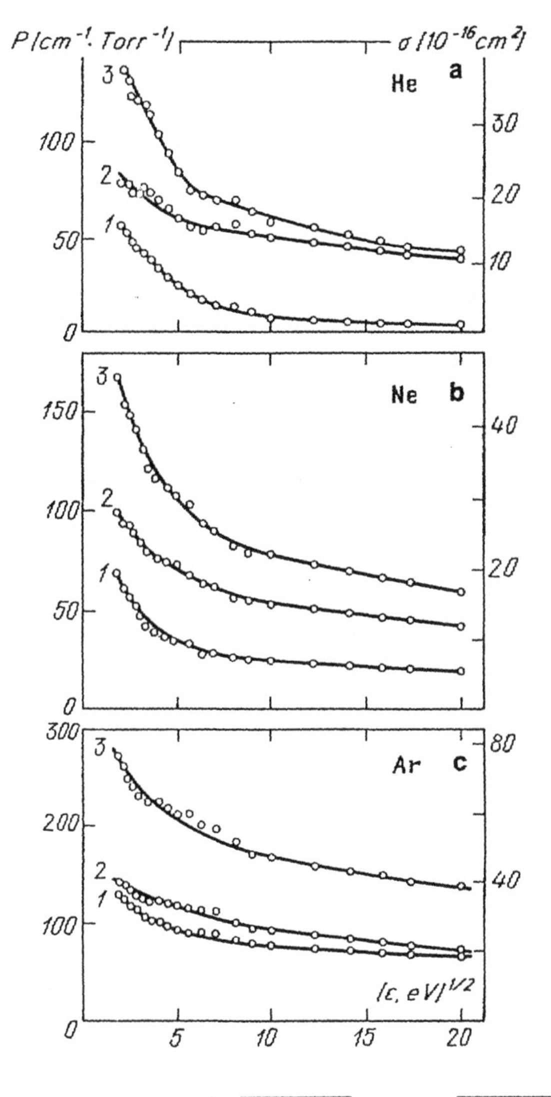

Fig. 2.7a–c. Cross sections and probabilities of collisions of atomic ions in inert gases. (1) – elastic scattering, (2) – charge transfer, (3) — sum of (1) and (2). From [2.14, 15]

$$\mu_{\rm i} = \frac{2.7 \times 10^4 \sqrt{1 + M/M_{\rm i}}}{p\,[\text{Torr}]\,\sqrt{(\alpha/a_0^3)A}} = \frac{36\sqrt{1 + M/M_{\rm i}}}{p\,[\text{atm}]\,\sqrt{(\alpha/a_0^3)A}}\ \frac{\text{cm}^2}{\text{V}\cdot\text{s}}\,, \tag{2.28}$$

where A is the molecular weight of the gas. Formula (2.28), with a slightly different coefficient, was derived by Langevin in 1905. It is in good agreement with experimental data [2.1]. Ions often tend to join with molecules and atoms into complexes of the type N_4^+, O_4^+, Ne_2^+ and He_2^+ (in contrast to Ne_2 and He_2, these complexes are sufficiently stable). This process affects mobility since it eliminates charge transfer (Fig. 2.8). As an example, consider the drift of Ne_2^+ in Ne. For Ne, $\alpha/a_0^3 = 2.76$. From (2.28), we have $\mu_{\rm i} = 4.5 \times 10^3/p\,[\text{Torr}]\,\text{cm}^2/(\text{V}\cdot\text{s})$. The experimental value is $\mu_{\rm i} = 5 \times 10^3/p$. In a field typical of glow discharges,

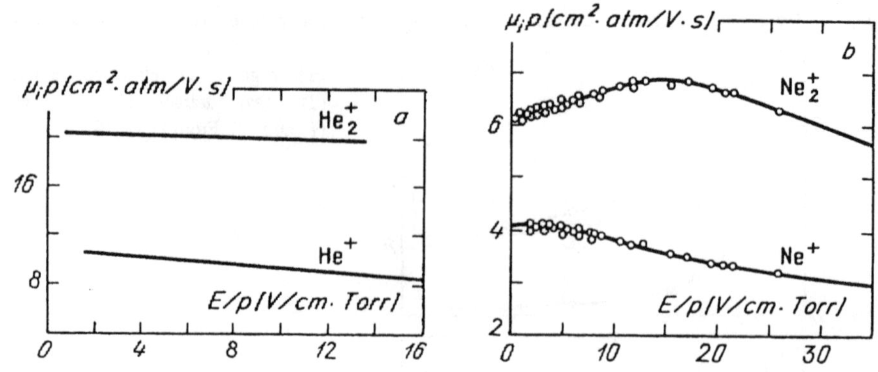

Fig. 2.8. Mobility of (a) Ne^+ and Ne_2^+ in Ne; (b) He^+ and He_2^+ in He; $p = 1$ atm, $T = 300$ K. From [2.16]

$E/p = 1\,\mathrm{V/(cm \cdot Torr)}$, we have $v_{id} = 50\,\mathrm{m/s}$. The corresponding thermal velocity at $T = 300\,\mathrm{K}$ is $v_{iT} \approx 400\,\mathrm{m/s}$.

2.5.3 Mean Energy

The equation for the mean ion energy $\bar{\varepsilon}_i$ in the approximation ν_m = const, in which the problem of averaging over the particle velocities is greatly simplified, is

$$\frac{d\bar{\varepsilon}_i}{dt} = \left[\frac{e^2E^2}{M'\nu_m^2} - \frac{2M'}{M + M_i}(\bar{\varepsilon}_i - \bar{\varepsilon}_M)\right]\nu_m\,, \quad M' = \frac{M_iM}{M + M_i}\,. \tag{2.29}$$

If the mean energy of molecules $\bar{\varepsilon}_M = (3/2)kT \ll \varepsilon_i$ and $M_i \ll M$, then (2.29) is identical to (2.12) for electrons undergoing only elastic energy loss ($\delta = 2m/M$). The equilibrium energy of ions is

$$\bar{\varepsilon}_i = \frac{3}{2}kT + \frac{(1 + M_i/M)^3}{2(M_i/M)}\frac{e^2E^2}{M_i\nu_m^2}\,. \tag{2.30}$$

If the field is not too strong, this energy is only slightly greater than the thermal value $3kT/2$. Electrons exchange energy very poorly with a gas and are thermalized only in extremely weak fields of $E/p \lesssim 10^{-3} - 10^{-2}\,\mathrm{V/(cm \cdot Torr)}$. The masses M_i and M being comparable, ions reach the temperature of the gas in fields that are not necessarily weak, say $E/p \sim 1 - 10\,\mathrm{V/(cm \cdot Torr)}$.

If the field is strong, an ion acquires in a free path length l an energy eEl much greater than $\bar{\varepsilon}_M$, so that its energy $\bar{\varepsilon}_i$ runs much ahead of the thermal energy. The collision cross section becomes close to the gas-kinetic value and $l \approx$ const. For example, if $\sigma = 3 \times 10^{-15}\,\mathrm{cm}^2$, then $l = 10^{-2}/p\,[\mathrm{Torr}]\,\mathrm{cm}$. For room temperature and $E/p > 40\,\mathrm{V/(cm \cdot Torr)}$, we find $eEl > 10\bar{\varepsilon}_M \approx 0.4\,\mathrm{eV}$. Such conditions are characteristic of the cathode layer of a glow discharge. We now approximate (2.30) for the case of l = const and $\nu_m = v/l \sim v$. Replacing v with the approximation $\sqrt{2\bar{\varepsilon}_i/M_i}$, we find

$$\bar{\varepsilon}_i = \frac{(1 + M_i/M)^{3/2}}{2(M_i/M)^{1/2}}eEl\,. \tag{2.31}$$

In the limit $M_i \ll M$, (2.31) transforms into (2.13). In contrast to electrons, an ion with a mass $M_i \sim M$ does not store energy pumped by the field: it sheds it in each collision, so that $\bar{\varepsilon}_i \sim eEl$.

2.5.4 Drift in Strong Field

Assuming $\nu_m = v/l$ and $l = \text{const}$, expressing velocity v through $\bar{\varepsilon}_i$ [as in deriving (2.31)] and substituting it into (2.27), we find v_{id}:

$$v_{id} \approx \left(\frac{M_i}{M}\right)^{1/4} \left(1 + \frac{M_i}{M}\right)^{1/4} \sqrt{\frac{eEl}{M_i}} \,. \tag{2.32}$$

This is proportional not to E/p, as in moderate fields, but to $\sqrt{E/p}$. If $M_i \approx M$, v_{id} is approximately equal to the ion velocity $\bar{v} \approx \sqrt{2\bar{\varepsilon}_i/M_i}$ that corresponds to its mean energy, because the motion of the ion is sharply oriented. Ions in a heavy gas, however, drift more slowly than they move randomly: $v_{id}/\bar{v} = [M_i/(M_i + M)]^{1/2} \sim \sqrt{M_i/M}$ [cf. (2.16)]. Like electrons, they store energy $\bar{\varepsilon}_i \approx (M/M_i)^{1/2} eEl$.

The transition from the mobility law $v_{id} \propto E/p$ to the law $v_{id} \propto \sqrt{E/p}$ is gradual (Fig. 2.9). It begins usually in fields in which ion energies reach about 1 eV and the polarization forces are replaced with short-range forces and the cross section becomes gas-kinetic. If ions move in their own gas and charge transfer dominates, this occurs when the ion energy is appreciably greater than the thermal energy.

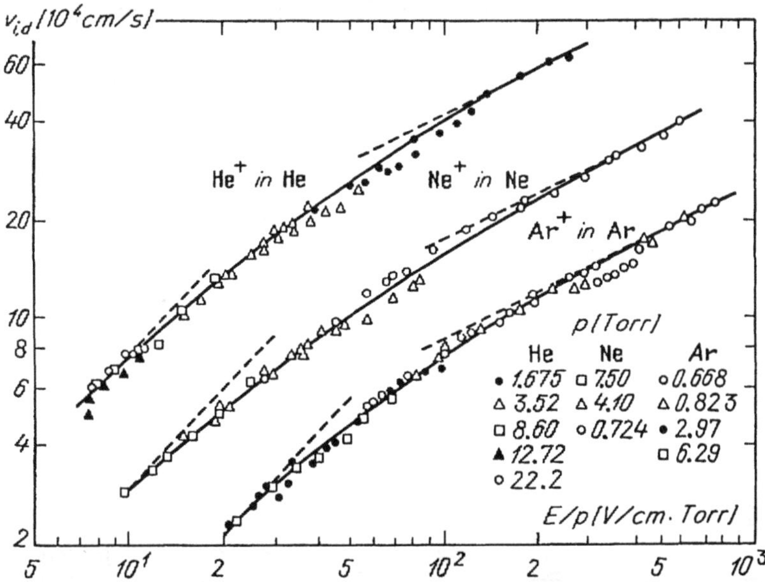

Fig. 2.9. Drift velocities of ions in inert gases. Change in mobility laws: from $v_{id} \sim E$ (*dashed line* on the left) to $v_{id} \sim \sqrt{E}$ (*dashed line* on the right); $T = 300$ K. From [2.17]

2.5.5 Diffusion

The Einstein relation (2.24) is valid in not too strong fields when ions are in thermal equilibrium with the gas and their energy is $\bar{\varepsilon}_{\rm i} \approx 3kT/2$. Direct studies of ion diffusion are much less numerous than mobility measurements, so that the diffusion coefficients $D_{\rm i}$ are typically found in this way. In the case of moderate E/p, D varies with E/p as little as $\mu_{\rm i}$ does. For instance, in the case of nitrogen ions in nitrogen, $\mu_{\rm i} \approx 1.5 \times 10^3/p\,[\text{Torr}]\,\text{cm}^2/(\text{V}\cdot\text{s})$, $D_{\rm i} \approx 40/p\,\text{cm}^2/\text{s}$.

2.6 Ambipolar Diffusion

When the density of charged particles, $n_{\rm e}$ and n_+, is very low, the charges of opposite signs diffuse independently of each other. This is known as *free diffusion*. Electrons are more mobile and thus diffuse faster; if there is a charge density gradient in the plasma, electrons may leave their less mobile partners far behind. If, however, the densities $n_{\rm e}$ and n_+ are not low, a considerable space charge is formed as a result of charge separation, and the generated polarization field impedes further violation of charge neutrality (Fig. 2.10). Charge separation and the field so adjust to each other that the field restrains the run-away electrons and pulls forward the heavy ions, making them diffuse only as a team. This diffusion is known as *ambipolar*; the concept was introduced by Schottky in 1924.

2.6.1 Ambipolar Diffusion Coefficient

Let us turn to the general expressions (2.20) for charge particle fluxes. We will be interested in the cases of no external field or of diffusion in the direction normal to it. Then the field $\boldsymbol{E}$ entering these formulas is connected exclusively with the polarization of the plasma. It satisfies the electrostatics equation

$$\operatorname{div} \boldsymbol{E} = 4\pi e (n_+ - n_{\rm e}) \,. \tag{2.33}$$

Let the separation of charges be small: $|n_+ - n_{\rm e}| \ll n_{\rm e} \approx n_+ \approx n$. For the separation not to grow appreciably, the electron and ion fluxes must be almost

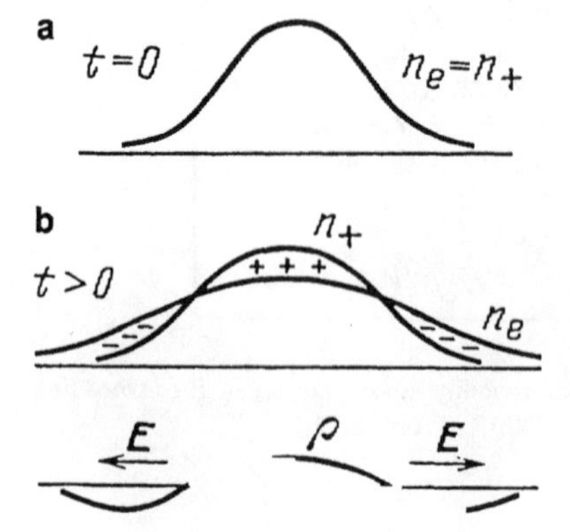

Fig. 2.10a,b. Plasma polarization in the presence of electron and ion density gradients. (**a**) Initial distributions $n_{\rm e} = n_+$; (**b**) distributions $n_{\rm e}$, n_+, and space-charge density $\varrho = e(n_+ - n_{\rm e})$ some time later. *Arrows* indicate the direction of the polarization field

equal: $\Gamma_{ex} \approx \Gamma_{+x} \approx \Gamma_x$, where the x axis is chosen to be perpendicular to the external field, provided it exists. In order to eliminate the polarization field from the expressions

$$\Gamma_x \approx -\mu_e E_x n - D_e \frac{\partial n}{\partial x}, \quad \Gamma_x \approx +\mu_+ E_x n - D_+ \frac{\partial n}{\partial x}, \tag{2.34}$$

we divide the first by μ_e, the second by μ_+, and add up the results. We find that the flux of charged particles of both signs is written in the form standard for diffusion:

$$\Gamma_x = -D_a \frac{\partial n}{\partial x}, \quad D_a = \frac{D_+ \mu_e + D_e \mu_+}{\mu_e + \mu_+}, \tag{2.35}$$

with an effective coefficient D_a, the *ambipolar diffusion coefficient*. Since $\mu_e \gg \mu_+$ and $D_e \gg D_+$, the quantity $D_a \approx D_+ + D_e(\mu_+/\mu_e)$ is greater than D_+ but less than D_e, in accord with the remarks on "speeding-up of the ions" and "restraining" electrons. For an equilibrium plasma, where the electron, T_e, and ion, T, temperatures are equal, the Einstein relations (2.24) yield $D_a = 2D_+$. For a nonequilibrium plasma, where the electron temperature is much higher than that of the ions (the latter is equal to the gas temperature), we have

$$D_a \approx D_e \frac{\mu_+}{\mu_e} = D_+ \frac{T_e}{T} = \mu_+ \frac{kT_e}{e} = \frac{2}{3} \mu_+ \bar{\varepsilon}_e \, [\mathrm{eV}]. \tag{2.36}$$

2.6.2 What Are the Conditions Under Which the Diffusion Is Ambipolar?

This question is very important because the free and ambipolar diffusion coefficients differ by a factor of ten and more. For the flux Γ_{ex} of (2.34) not to exceed Γ_{+x} despite the strong inequalities $D_e \gg D_+$ and $\mu_e \gg \mu_+$, the diffusion and drift terms of Γ_{ex} with opposite signs must compensate each other to within the relatively small flux Γ_{+x}. Hence, the polarization field that appears automatically in ambipolar diffusion equals

$$E_x \approx -\frac{D_e}{\mu_e} \frac{1}{n} \frac{\partial n}{\partial x} = -\frac{kT_e}{e} \frac{1}{n} \frac{\partial n}{\partial x} \sim \frac{kT_e}{eR}, \tag{2.37}$$

where R is the length characterizing the scale of the charge density gradient. This is the distance over which the electron density varies considerably. For example, R is the radius of plasma in a tube because the density at the axis is much greater than at the tube walls, where the charges are neutralized.

The *polarization field* is a result of charge separation; in its turn, the separation is caused by the random (thermal) motion of the faster (electron) component and stems from the electron thermal energy. Indeed, where the polarization field exists, its potential difference over the entire length R is such that the electric energy to which the charge is accelerated over it is of the order of the thermal energy of electrons: $e\delta\varphi \sim eE_x R \sim kT_e$. The polarization field is generated by the space charge $e\delta n = e(n_+ - n_e)$; according to (2.33), this charge is determined, to within an order of magnitude, from the relation $E_x/R \sim 4\pi e\delta n$. Using (2.37), we find

$$\frac{\delta n}{n} \approx \frac{kT_e}{4\pi e^2 n} \frac{1}{R^2} = \left(\frac{d}{R}\right)^2 , \quad d = \left(\frac{kT_e}{4\pi e^2 n}\right)^{1/2} . \tag{2.38}$$

The quantity d is the *Debye radius* of a plasma.[2] It gives the distance characterizing strong charge separation and plasma polarization. If $R \gg d$, that is, if large density differences appear over distances greater than the Debye radius, then $\delta n/n \ll 1$, deviations from charge neutrality are small, and the diffusion is ambipolar. If $R \lesssim d$, electrons and ions diffuse independently. For example, for $T_e = 1\,\text{eV}$, $n_e = 10^8\,\text{cm}^{-3}$, and $R = 1\,\text{cm}$, we have $d = 0.052\,\text{cm}$ and $\delta n/n = 2.5 \cdot 10^{-3}$, that is, the diffusion is clearly ambipolar. If T_e and R are the same but $n_e < 10^6\,\text{cm}^{-3}$, the diffusion of charges is free.

2.6.3 Definition of "Plasma"

The condition $(d/R)^2 \ll 1$, where R is the characteristic size of the region of large density difference, gives the quantitative criterion that distinguishes between *plasma* as an electrically neutral ionized medium and other cases of the presence of charges in a gas.

2.7 Electric Current in Plasma in the Presence of Longitudinal Gradients of Charge Density

2.7.1 Continuity Equation for Charges and Currents

If plasma is placed in an external electric field and if the current is nonzero, then in contrast to Sect. 2.6.1 the densities of electron and ion fluxes are not equal, and the current density is $\boldsymbol{j} = e(\boldsymbol{\Gamma}_+ - \boldsymbol{\Gamma}_e)$. The continuity equation (2.22) for particles of a given species and the fact that positive and negative charges are created and annihilated *only in pairs* (if negative ions are absent, $q_e = q_+$) imply the following continuity equation:

$$\frac{\partial \varrho}{\partial t} + \operatorname{div} \boldsymbol{j} = 0 , \quad \varrho = e\left(n_+ - n_e\right) . \tag{2.39}$$

In electrically neutral media, and under steady-state conditions in all media, the current has no source:

$$\operatorname{div} \boldsymbol{j} = 0 . \tag{2.40}$$

[2] The Debye radius of an equilibrium plasma ($T_e = T$) is less than (2.38) by a factor of $\sqrt{2}$. When considering charge screening in plasmas with $T_+ \ll T_e$, one should not subject the density of low-mobility ions around a charge to Boltzmann's law $n_+ = n_{+\infty} \exp(-e\varphi/kT_+)$, but assume it to be constant, $n_+ = n_{+\infty}$. Then d is given by (2.38).

In the one-dimensional plane case $dj/dz = 0$ along the current direction z, $j(z)$ = const, and the current density does not change. It is usually determined not by local characteristics, but by conditions of the entire system, including the circuit external to the discharge.

2.7.2 Diffusion Current and the Distortion of the Field by Gradients

Under the conditions of quasineutrality ($n_e \approx n_+ \approx n$), formulas (2.20) yield

$$j/e = e(\boldsymbol{\Gamma}_+ - \boldsymbol{\Gamma}_e) = (D_e - D_+)\nabla n + (\mu_e + \mu_+)\boldsymbol{E}n \,, \tag{2.41}$$

$$\boldsymbol{E} = \frac{\boldsymbol{j}}{e(\mu_e + \mu_+)n} - \frac{D_e - D_+}{\mu_e + \mu_+}\frac{\nabla n}{n} \,. \tag{2.42}$$

The electric field is made up of the external one, which drives the current, and the polarization field due to the presence of gradients (in fact, temperature gradients generate additional, thermodiffusion currents but their role is usually minor). Of course, these components are indistinguishable: measuring a field, say, by a probe (Chap. 6), we determine the total field (2.42).

If the charge density falls steeply in the direction opposite to that of the external field the total field may be completely suppressed or may even reverse in direction with respect to the current (such effects are observed in glow discharges, low-voltage arcs, etc.). The electric current cannot be affected by this [see (2.40)] and is carried through by electron diffusion: if $E \approx 0$, then (2.41), taking into account that $D_e \gg D_+$, yields $j \approx D_e \nabla n$. The diffusion is then free (ambipolar diffusion cannot transfer charge). Indeed, electrons in zero field do not pull ions behind. In gases, this situation cannot, however, exist on too long a path, as we see from (2.42). Note that the diffusion coefficient D_e depends on the electron energy distribution, which reaches a state corresponding to the field after an energy relaxation length $\Lambda_u = l/\sqrt{\delta} \gg l$ (Sect. 2.3.7). Strictly speaking, it strengthens the usual condition of applicability of diffusion concepts, namely, a small drop in n over a length l; at any rate, it calls for very careful analysis in treating diffusion in highly nonuniform fields.

2.7.3 Plasma Density Equation

Using (2.42) in the formulas (2.20) for $\boldsymbol{\Gamma}_e$ and $\boldsymbol{\Gamma}_+$ and recalling definition (2.35), we find that

$$\boldsymbol{\Gamma}_e = -D_a\nabla n - \frac{\mu_e}{\mu_e + \mu_+}\frac{\boldsymbol{j}}{e} \,, \quad \boldsymbol{\Gamma}_+ = -D_a\nabla n + \frac{\mu_+}{\mu_e + \mu_+}\frac{\boldsymbol{j}}{e} \,. \tag{2.43}$$

Electron and ion fluxes are made up of identical ambipolar diffusion fluxes (they can also have a component along the current) and fluxes due to the electric current. In contrast to ambipolar fluxes, these latter fluxes differ greatly, by a factor of $\mu_e/\mu_+ \gg 1$. Substituting any of expressions (2.43) into the appropriate continuity equation (2.22) and recalling (2.40), we arrive at *the general balancing equation for the number of charged particles*:

$$\frac{\partial n}{\partial t} - D_a \Delta n = q \,, \quad n \approx n_e \approx n_+ \,. \tag{2.44}$$

This looks like an ordinary diffusion equation (with ambipolar coefficient) for a problem with bulk sources of charge; it manifests no signs of the possibility of electric current in the medium. This current affects charge fluxes (2.43), not the balance. The balance is current-independent because, metaphorically speaking, the amount of electricity flowing into somewhere is exactly equal, in view of (2.40), to the amount flowing out of it.

2.7.4 Charge Neutrality Criterion

This requires different descriptions for two different situations. If the drift current is zero or small in comparison with the electron diffusion current [the polarization field is greater than the external one in (2.42)], we return to the situation treated in Sect. 2.6.2 and the criterion $(d/R)^2 \ll 1$ [see (2.38)]. If the drift current is greater than the diffusion current, then (2.42) implies that $\boldsymbol{j} \approx e\mu_e \boldsymbol{E} n$ and we have, in accord with (2.40, 33),

$$-\boldsymbol{E}\cdot\nabla n = n\,\mathrm{div}\,\boldsymbol{E} = 4\pi e n(n_+ - n_e) = 4\pi e n \delta n\,,$$
$$\frac{\delta n}{n} \sim \frac{E}{4\pi e n L} \sim \frac{kT_e}{4\pi e^2 n}\frac{eE}{kT_e L} \sim \frac{d^2}{L\Lambda_u} \sim \left(\frac{d}{L}\right)^2\frac{L}{\Lambda_u}\,, \tag{2.45}$$

where Λ_u is the electron temperature relaxation length (or that of the mean electron energy) of (2.18) and (2.19), and L is the characteristic length of strong variation of n_e and conductivity. The Debye radius must be compared with the geometric mean of L and Λ_u. The actual conditions decide whether the criterion based on (2.45) is stronger or weaker than that based on (2.38) and whether the gradient along the current or transverse to it violates charge neutrality more strongly.

2.7.5 Ambipolar Flow of Charges Along a Nonuniform Field

In contrast to the diffusion flux, this flow is of a drift nature, and is caused by a space charge [2.18]. Let us use in the continuity equations (2.22) for n_e and n_+ the expressions (2.20) for the fluxes $\boldsymbol{\Gamma}_e$ and $\boldsymbol{\Gamma}_+$. Multiplying the equation for n_e by μ_+/μ_e and assuming, for simplification, that μ_e and μ_+ are constant, we then add the result to the equation for n_+. When summing up the drift fluxes, we retain the "small" difference δn because we are interested in the effect of space charge; this charge may be considerable even in quasineutral plasma (in the sense of $\delta n/n \ll 1$). Substituting $\delta n = (4\pi e)^{-1}\,\mathrm{div}\,\boldsymbol{E}$ and neglecting the terms proportional to μ_+/μ_e (this ratio is less than 10^{-2}), we arrive at the equation

$$\frac{\partial n}{\partial t} + \mathrm{div}\left(-D_a\nabla n + \frac{\mu_+}{4\pi e}\boldsymbol{E}\,\mathrm{div}\,\boldsymbol{E}\right) = q\,, \quad n \approx n_+ \approx n_e\,, \tag{2.46}$$

which is a refinement of (2.44). The ambipolar diffusion flux is now supplemented with the "ambipolar drift" flux.

In the one-dimensional case, the latter term equals $\mu_+/8\pi e \times \partial E^2/\partial x$. It is not related directly to charge density. However, the current is constant along its direction x, $j \approx e n_e \mu_e E = \mathrm{const}$, so that a field gradient is invariably accompanied with a plasma density gradient. The ambipolar drift flux is thus equivalent to a diffusion flux,

$$\frac{\mu_+}{8\pi e}\frac{\partial E^2}{\partial x} \approx -\frac{\mu_+ E^2}{4\pi en}\frac{\partial n}{\partial x} = -D_{\mathrm{E}}\frac{\partial n}{\partial x}, \quad D_{\mathrm{E}} = \frac{\mu_+ E^2}{4\pi en}, \tag{2.47}$$

with effective diffusion coefficient D_{E} to be added to D_{a} [2.19]. The relative roles of the two fluxes are characterized by the ratio of the energy density of electric field to the density of the thermal energy of electrons,

$$D_{\mathrm{E}}/D_{\mathrm{a}} = E^2/4\pi nkT_{\mathrm{e}} = \{E[\mathrm{V/cm}]\}^2/\{1.8 \times 10^{-6} nT_{\mathrm{e}}[\mathrm{eV}]\}\ . \tag{2.48}$$

For example, for $n = 6 \times 10^9\,\mathrm{cm}^{-3}$ and $T = 1\,\mathrm{eV}$, the *ambipolar drift* dominates ambipolar diffusion if $E > 10^2\,\mathrm{V/cm}$ or, in view of the E/p values typical for plasma, 1–10 V/(cm·Torr) if $p > 10^2$–10 Torr. As a result of the two effects, plasma is pumped from regions of weaker field and higher density to those of stronger field and lower density. Details concerning the application of (2.46) are given in Sect. 8.6.6.

2.8 Hydrodynamic Description of Electrons

A partially ionized gas is a three-component mixture of electrons, ions, and neutral particles. Its behavior in a field can be described in terms of the ordinary equations of gas dynamics. We will consider here a simplified and more frequently used (for discharge conditions) version of equations, assumming the gas to be at rest as a whole, weakly ionized, and quasineutral. Actually, it is then sufficient to write the equations only for the electron gas. Its state is characterized by the electron density n_{e}, the vector of macroscopic velocity $\boldsymbol{v}_{\mathrm{e}}$, and temperature T_{e} (or pressure $p_{\mathrm{e}} = n_{\mathrm{e}}kT_{\mathrm{e}}$).

2.8.1 Equations of Continuity and Motion

The former has already been written out, see (2.22). The latter is

$$mn_{\mathrm{e}}\frac{d\boldsymbol{v}_{\mathrm{e}}}{dt} = -n_{\mathrm{e}}e\boldsymbol{E} - \nabla p_{\mathrm{e}} - n_{\mathrm{e}}m\boldsymbol{v}_{\mathrm{e}}\nu_{\mathrm{m}}, \quad \frac{d\boldsymbol{v}_{\mathrm{e}}}{dt} = \frac{\partial \boldsymbol{v}_{\mathrm{e}}}{\partial t} + (\boldsymbol{v}_{\mathrm{e}} \cdot \nabla)\,\boldsymbol{v}_{\mathrm{e}}\ . \tag{2.49}$$

In fact, this result has already been used above. The inertia term can be dropped from (2.49) because of the smallness of the electron mass. Then (2.49) reduces to (2.20) for the flux density $\boldsymbol{\Gamma}_{\mathrm{e}} = n_{\mathrm{e}}\boldsymbol{v}_{\mathrm{e}}$, considered together with (2.21) and (2.24), and differs from (2.20) only in the additional *thermodiffusion* flux proportional to $-\nabla T_{\mathrm{e}}$. This last term is usually smaller than the diffusion term, and has been accordingly neglected.

2.8.2 Energy Equation

This equation is of the form [2.20]

$$\frac{\partial}{\partial t}\left(\frac{3}{2}n_e kT_e\right) + \operatorname{div} \boldsymbol{F} = -n_e e\boldsymbol{E}\cdot\boldsymbol{v}_e - \frac{3}{2}n_e kT_e \nu_u - qI \,, \tag{2.50}$$

$$\boldsymbol{F} = \tfrac{5}{2}n_e kT_e \boldsymbol{v}_e - \lambda_e \nabla T_e \,, \quad \lambda_e = \tfrac{5}{2}kn_e D_e \,. \tag{2.51}$$

The flux density of electron energy $\boldsymbol{F}$ is composed of the hydrodynamic flux of enthalpy and the *heat conduction* flux; λ_e is its coefficient.[3] The terms containing the electron kinetic energy of drift motion, $mv_e^2/2 \ll kT_e$, are neglected; according to (2.16), $v_e \sim v_d \ll \bar{v}$. Electron energy losses in collisions with molecules are written in (2.50) in accordance with (2.12, 17). The last term in (2.50) describes energy spent to create new electrons, q is the resultant creation rate that enters the continuity equation (2.22), and I is the ionization potential.

The combination of (2.22, 49–51) implies that the equation for the rate of heating of a moving particle of the electron gas is

$$\begin{aligned}\frac{3}{2}k\frac{dT_e}{dt} &= -\, e\boldsymbol{E}\cdot\boldsymbol{v}_e - \frac{3}{2}kT_e\nu_u - \frac{1}{n_e}\operatorname{div}(p_e\boldsymbol{v}_e) \\ &+ \frac{1}{n_e}\operatorname{div}(\lambda_e\nabla T_e) - \left(I + \frac{3}{2}kT_e\right)\frac{q_e}{n_e} \,,\end{aligned} \tag{2.52}$$

where dT_e/dt is the total derivative with respect to time, as in (2.49).

Equation (2.52) generalizes (2.12), covering the effects caused by *spatial inhomogeneity*: work of pressure forces and the contribution of heat conduction. The last term describes energy spent on ionization and passing thermal energy to the created electron; it is usually small in comparison with the term proportional to $\nu_u = \delta\nu_m$ in (2.12).

2.8.3 Current-Carrying Plasma

It is convenient to present the velocity $\boldsymbol{v}_e$ or $n_e\boldsymbol{v}_e$ in (2.22, 50–52) in the form (2.43), where the current $\boldsymbol{j}$ satisfying (2.40) is singled out of $\boldsymbol{\Gamma}_e$. Ignoring the ambipolar thermodiffusion flux and taking into account that $\mu_+ \ll \mu_e$, we obtain

$$n_e\boldsymbol{v}_e = -\boldsymbol{j}/e - D_a\nabla n_e \,, \quad \operatorname{div}\boldsymbol{j} = 0 \,. \tag{2.53}$$

According to (2.24, 42), the field $\boldsymbol{E}$ in (2.50, 52) is, to the same accuracy,

$$\boldsymbol{E} = \boldsymbol{j}/e\mu_e n_e - (kT_e/e)\,\nabla n_e/n_e \,. \tag{2.54}$$

The outlined system of equations will be employed in Sect. 9.7.

[3] The expression for $n_e\boldsymbol{v}_e$, derived from (2.49), and expression (2.51) for $\boldsymbol{F}$ and λ_e stem from the kinetic equation for electrons (Chap. 5), assuming Maxwellian distribution functions and $\nu_m(\varepsilon) =$ const. If $l(\varepsilon) =$ const, $\nu_m \sim \sqrt{\varepsilon}$, and the coefficients in $\boldsymbol{F}$ and λ_e, $\frac{5}{2}$, are set equal to 2; the thermal diffusion coefficient is found to be twice as small.

3. Interaction of Electrons in an Ionized Gas with Oscillating Electric Field and Electromagnetic Waves

3.1 The Motion of Electrons in Oscillating Fields

Both the equations of electrodynamics and the equations of motion of electrons are linear with respect to the fields $\boldsymbol{E}$, $\boldsymbol{H}$ and the velocity $\boldsymbol{v}$ of the electron. For this reason, the *superposition principle* holds. Any periodical field can be resolved into harmonic components, so that it is sufficient to consider only the *sinusoidal* field, all the more so because one normally deals with *monochromatic* fields and waves. In the case of nonrelativistic motion, the magnetic force of the wave, $e(v/c)H$, is much less than the electric force eE. Furthermore, the amplitude of electron oscillations in discharge processes is usually small in comparison with wavelength λ. We assume, therefore, that the electron is in a spatially uniform electric field $\boldsymbol{E} = \boldsymbol{E}_0 \sin\omega t$, $\boldsymbol{E}_0 = \text{const}$.

3.1.1 Free Oscillations

Assume that an electron moves without collisions, an assumption that is meaningful if the electron performs a large number of oscillations in the interval between collisions, $\omega \gg \nu_m$. We integrate the equation of collisionless motion,

$$m\dot{\boldsymbol{v}} = -e\boldsymbol{E}_0 \sin\omega t\ , \quad \dot{\boldsymbol{r}} = \boldsymbol{v}\ ,$$

to give

$$\boldsymbol{v} = \frac{e\boldsymbol{E}_0}{m\omega}\cos\omega t + \boldsymbol{v}_0\ , \quad \boldsymbol{r} = \frac{e\boldsymbol{E}_0}{m\omega^2}\sin\omega t + \boldsymbol{v}_0 t + \boldsymbol{r}_0\ . \tag{3.1}$$

An electron oscillates at the frequency of the field; these oscillations are superimposed onto an arbitrary translation velocity $\boldsymbol{v}_0$. The displacement and oscillation velocities are

$$a = \frac{eE_0}{m\omega^2}\ , \quad u = \frac{eE_0}{m\omega}\ . \tag{3.2}$$

The displacement is in phase with the field, while the velocity is out of phase by $\pi/2$. The limiting case of "collisionless" oscillations is approximately realized at optical frequencies, and also at microwave frequencies at low pressures, $p \lesssim 10$ Torr.

3.1.2 Effect of Collisions

Collisions "throw off" the phase, thereby disturbing the purely harmonic course of the electron's oscillations. A sharp change in the direction of motion after scattering stops the electron from achieving the full range of displacement (3.2) that the applied force can produce; the electron starts oscillating anew after each collision, with a new phase and new angle relative to the instantaneous direction of velocity. In order to take this factor into account, we add the rate of loss of momentum due to collisions to the equation of motion of the "mean" electron. As in the case of constant fields (Sect. 2.1.1), we have the equation for the mean velocity:

$$m\dot{\boldsymbol{v}} = -e\boldsymbol{E}_0 \sin\omega t - m\boldsymbol{v}\nu_{\mathrm{m}} , \quad \dot{\boldsymbol{r}} = \boldsymbol{v} . \tag{3.3}$$

The solution of (3.3), valid after several collisions, is

$$\begin{aligned} \boldsymbol{v} &= \frac{e\boldsymbol{E}_0}{m\sqrt{\omega^2+\nu_{\mathrm{m}}^2}} \cos(\omega t + \varphi) , \quad \varphi = \arctan\frac{\nu_{\mathrm{m}}}{\omega} , \\ \boldsymbol{r} &= \frac{e\boldsymbol{E}_0}{m\omega\sqrt{\omega^2+\nu_{\mathrm{m}}^2}} \sin(\omega t + \varphi) . \end{aligned} \tag{3.4}$$

The amplitudes of displacement and velocity of the electron are less by a factor of $\sqrt{1+\nu_{\mathrm{m}}^2/\omega^2}$ than those for free oscillations. The higher the effective collision frequency ν_{m}, the smaller they are (ν_{m} is determined by the velocity of random motion, which is much greater in discharges than the oscillation velocity; see Sect. 3.2). The displacement is shifted in phase relative to the field, the phase shift increasing from 0 to $\pi/2$ as the relative role of collisons ν_{m}/ω increases from 0 to ∞.

The oscillation displacement and velocity (3.4) can always be resolved into two components, one proportional to the magnitude of the field $\boldsymbol{E} = \boldsymbol{E}_0 \sin\omega t$, and the other to its rate of change, $\dot{\boldsymbol{E}} = \omega\boldsymbol{E}_0 \cos\omega t$:

$$\begin{aligned} \boldsymbol{r} &= \frac{e\boldsymbol{E}_0}{m(\omega^2+\nu_{\mathrm{m}}^2)} \sin\omega t + \frac{\nu_{\mathrm{m}}}{\omega} \frac{e\boldsymbol{E}_0}{m(\omega^2+\nu_{\mathrm{m}}^2)} \cos\omega t , \\ \boldsymbol{v} &= \frac{\omega e\boldsymbol{E}_0}{m(\omega^2+\nu_{\mathrm{m}}^2)} \cos\omega t - \frac{\nu_{\mathrm{m}} e\boldsymbol{E}_0}{m(\omega^2+\nu_{\mathrm{m}}^2)} \sin\omega t . \end{aligned} \tag{3.5}$$

The ratio of the components is determined by the relative role of collisions and is unambiguously related to the phase shift φ. This form of presenting the solution adds visual clarity to the results of the subsequent sections.

Expressions (3.4, 5) show that the role of collisions is characterized by the ratio of the effective frequency ν_{m} and the circular frequency of the field $\omega = 2\pi f$, which is greater than the frequency f by nearly an order of magnitude.[1] In the limit $\nu_{\mathrm{m}}^2 \ll \omega^2$, formulas (3.4, 5) are close to (3.1) for free oscillations. To illustrate numerical values, consider an example of microwave radiation at frequency $f = 3\,\mathrm{GHz}$; $\lambda = 10\,\mathrm{cm}$, $\omega = 1.9 \times 10^{10}\,\mathrm{s}^{-1}$. Let $p \approx 1\,\mathrm{Torr}$, then $\nu_{\mathrm{m}} \approx 3 \times 10^9\,\mathrm{s}^{-1} \ll \omega$; $E_0 = 500\,\mathrm{V/cm}$, roughly corresponding to the threshold

[1] When the degree of spatial uniformity of the field is evaluated, the displacement amplitude must be compared not with wavelength $\lambda = c/f$, but with $\lambda\!\!\!^- = \lambda/2\pi$: $a/\lambda\!\!\!^- = eE_0/m\omega^2 \lambda\!\!\!^- = u/c$.

of microwave breakdown at such pressures. Formulas (3.2) show that $a = 2.5 \times 10^{-3}$ cm, $u = 4.7 \times 10^7$ cm/s. We find that $a \ll \lambda\!\!\!^{-} = 1.6$ cm, that is, the field in the electromagnetic wave is "uniform".

3.1.3 Drift Oscillations

In the limit of very frequent collisions or relatively low frequencies, $\nu_m^2 \gg \omega^2$, the oscillation velocity drops to

$$\boldsymbol{v} \approx -\frac{e\boldsymbol{E}_0}{m\nu_m} \sin\omega t = -\frac{e\boldsymbol{E}(t)}{m\nu_m} = -\mu_e \boldsymbol{E}(t) = \boldsymbol{v}_d(t) . \tag{3.6}$$

At each moment of time, the oscillation velocity coincides with the drift velocity that corresponds to the field vector at this moment. For brevity, we refer to such oscillations in the *mobility regime* as *drift oscillations*.

An electron behaves as it would in a constant field, responding to relatively slow changes of the field. Its displacement,

$$\boldsymbol{r} \approx \boldsymbol{A}\cos\omega t , \quad \boldsymbol{A} = \frac{e\boldsymbol{E}_0}{m\nu_m\omega} = \frac{\mu_e\boldsymbol{E}_0}{\omega} , \tag{3.7}$$

has an amplitude A less than that of free oscillations in the same field by a factor of $\nu_m/\omega \gg 1$.

The oscillations of electrons in rf fields (and of course, at lower frequencies) are of drift type. For example, the collision frequency at $f \approx 10$ MHz, $\nu_m \approx 3 \times 10^9 p\,\mathrm{s}^{-1}$, exceeds $\omega \sim 10^8\,\mathrm{s}^{-1}$ even at fairly low pressures of $p \sim 0.03$ Torr. In order to maintain a low-pressure, weakly ionized plasma by an rf field, one usually needs the values of E_0/p of the same order as E/p in a constant field. Therefore, at $f \approx 10$ MHz and $E_0/p \approx 10$ V/(cm · Torr), we have $A \sim 0.1$ cm regardless of pressure.

3.2 Electron Energy

3.2.1 Collisionless Motion

If collisions do not occur, the field does no work, on the average, on an electron; indeed, (3.1) implies that

$$\langle -e\boldsymbol{E}\cdot\boldsymbol{v}\rangle = -\frac{eE_0^2}{m\omega}\langle\sin\omega t\cos\omega t\rangle - e\boldsymbol{E}_0\cdot\boldsymbol{v}_0\langle\sin\omega t\rangle = 0 ,$$

where angle brackets denote time averaging.

The electric field pumps up the motion of the electron only once, when it is switched on; then the electron's energy $mv^2/2$ pulsates but remains unchanged on the average. The time averaged energy $\langle mv^2/2\rangle$ is made up of the energy of translational motion $mv_0^2/2$, corresponding to the mean velocity $\boldsymbol{v}_0 = \langle\boldsymbol{v}(t)\rangle$, and that of oscillations. In the case of free oscillations, the latter energy is

$$\varepsilon_{\text{fr.osc.}} = \frac{e^2 E_0^2}{4m\omega^2} = mu^2/4 \,. \tag{3.8}$$

In the example given at the end of Sect. 3.1.2, $f = 3\,\text{GHz}$, $E_0 = 500\,\text{V/s}$, and the oscillation energy $\varepsilon_{\text{fr.osc.}} = 0.31\,\text{eV}$, which is much less than the mean energy of random motion (1–10 eV) necessary to sustain a discharge. The absence of collisions thus means no dissipation of the energy of the field and no deposition of energy in the matter.

3.2.2 Gaining of Energy From the Field

The collisions lead to a net transfer of energy to the electrons, via the electric field. According to (3.5), the mean work per unit time that the field performs on an electron,

$$\langle -e\boldsymbol{E} \cdot \boldsymbol{v} \rangle = \frac{e^2 E_0^2}{2m(\omega^2 + \nu_{\text{m}}^2)} \nu_{\text{m}} = \Delta\varepsilon_{\text{E}} \nu_{\text{m}} \,, \tag{3.9}$$

is determined by that component of velocity that oscillates in phase with the field and is proportional to ν_{m}. The term shifted in phase by $\pi/2$ does no work, on the average, over one period. The random motion is not associated with any transfer of energy. In one effective collision an electron gains the mean energy $\Delta\varepsilon_E$, equal to twice the mean kinetic energy of oscillations:

$$\Delta\varepsilon_E = \frac{e^2 E_0^2}{2m(\omega^2 + \nu_{\text{m}}^2)} = 2\left\langle \frac{mv^2}{2} \right\rangle = 2\varepsilon_{\text{osc}} \,. \tag{3.10}$$

This result can be given the following interpretation. In the interval between two collisions, the electric field imparts to an electron an average kinetic energy ε_{osc}. If the electron goes through a large number of oscillations in this period, then ε_{osc} is of the order of the free oscillation energy (3.8). An act of elastic scattering of an electron by an atom sharply changes the direction of motion but leaves the absolute value of velocity unaltered. Then the field starts swinging the electron in a new direction with respect to its velocity, that is, imparts to it an energy of order ε_{osc} as if anew. The mean amount of energy gained from the field in each scattering event following the preceding collision is thus transformed into the energy of translational random motion. Microscopically, the field does work on overcoming the friction due to collisions of the electron. Everything proceeds as in a constant field (see Sect. 2.3.2), but the role of the drift energy is played by that of oscillations.

3.2.3 Balancing of the Electron Energy

The balance is made up of gaining energy from the field and transferring it to heavy particles as a result of elastic and inelastic losses. If an electron loses in each collision a fraction δ of its energy ε, then

$$\frac{d\varepsilon}{dt} = (\Delta\varepsilon_E - \delta\varepsilon)\nu_{\text{m}} = \left[\frac{e^2 E^2}{m(\omega^2 + \nu_{\text{m}}^2)} - \delta\varepsilon\right] \nu_{\text{m}} \,, \tag{3.11}$$

where the amplitude E_0 is replaced with the mean-square field E defined by the equality $E^2 = \langle E^2(t) \rangle = E_0^2/2$. If $\omega^2 \ll \nu_m^2$, then (3.11) transforms into (2.12). In the $\omega \to 0$ limit, the R.M.S. of E plays the role of a constant field.

3.2.4 Mean Equilibrium Energy

The mean energy reached by electrons under stationary conditions, when they transfer the entire energy gained from the field, is

$$\bar{\varepsilon} = \Delta\varepsilon_E/\delta = 2\varepsilon_{osc}/\delta = e^2E^2/m\delta(\omega^2 + \nu_m^2) \,. \qquad (3.12)$$

Low-frequency fields ($\omega^2 \ll \nu_m^2$) behave indistinguishably from constant fields; the similarity $\bar{\varepsilon} = f(E/p)$ holds. At high frequencies, $\omega^2 \gg \nu_m^2$, the mean electron energy is independent of ν_m, p, and the similarity in field frequency holds: $\bar{\varepsilon} = f_1(E/\omega)$. If $\delta = \text{const}$, $\bar{\varepsilon} \propto (E/\omega)^2$. In equivalent situations, $E \propto \omega$. This is the reason why gas breakdown at optical frequencies ($\omega \sim 10^{15}\,\text{s}^{-1}$) requires enormous fields ($E \sim 10^7\,\text{V/cm}$) in the light wave, realizable only when giant laser pulses are focused (Sect. 7.6). Indeed, electron avalanches can develop only if the electron energies are of the order of 10 eV.

3.2.5 Actual Change in Electron Energy in a Collision

The situation in ac fields is also similar in this respect to that found in dc fields (Sect. 2.3.3). An electron may either gain energy from the field or lose energy to it, in amounts that much exceed the mean change $\Delta\varepsilon_E$ averaged over a large number of collisions. The relative directions of the motion and the field and the phase of field oscillations at the moment of collision decide whether the electron is to gain or lose energy. This is a fact of fundamental importance, which contains a classical analogue of such purely quantum phenomena as the *true absorption* and *stimulated emission* of photons.

To illustrate this, we calculate directly the change in the energy of an electron in a collision. Let an electron undergo its most recent collision at a moment t_1 and have, immediately after scattering, a velocity $\boldsymbol{v}_1$ and energy $\varepsilon_1 = mv_1^2/2$. Effective collisions occurring at a frequency ν_m each time give the velocity a completely random direction; hence, ε_1 is the energy of random motion at the moment of collision t_1. In the time $t \geq t_1$ and until the next collision, the electron is driven by the force $-e\boldsymbol{E}_0 \sin\omega t$ at a velocity $\boldsymbol{v}(t) = \boldsymbol{u}(\cos\omega t - \cos\omega t_1) + \boldsymbol{v}_1$, $\boldsymbol{u} = e\boldsymbol{E}_0/m\omega$. Its energy $\varepsilon = mv^2/2$ at each moment of this period is

$$\varepsilon(t) = \frac{m}{2}\Big[v_1^2 + 2\boldsymbol{v}_1\boldsymbol{u}\big(\cos\omega t - \cos\omega t_1\big) + u^2\Big(\cos^2\omega t - 2\cos\omega t\cos\omega t_1 + \cos^2\omega t_1\Big)\Big] \,.$$

At the moment t_2 of the next collision, the velocity is again directed in an arbitrary direction but remains virtually unchanged in magnitude. The electron resumes motion at an energy $\varepsilon(t_2)$, which is also the energy of random motion. Therefore, between two collisions the random-motion energy changes by

$$\Delta\varepsilon(t_1, t_2) = \varepsilon(t_2) - \varepsilon_1 \approx m\boldsymbol{v}_1 \cdot \boldsymbol{u}\big(\cos\omega t_2 - \cos\omega t_1\big) \,.$$

This last (approximate) transition takes into account that the random-motion velocity v_1 is much greater than the oscillation velocity u; hence, we can ignore the specific form of the small term of order mu^2.

For the sake of simplicity, assume that collisions are rare: $\nu_m \ll \omega$. Then many oscillations occur in the time $t_2 - t_1$ between two collisions, and the correlation between the field phases at the moments of collisions ωt_1 and ωt_2 vanishes owing to the random nature of collisions, i.e., the phases may be arbitrary. The value of $\Delta\varepsilon$ then varies in this interval from the maximum gain $\Delta\varepsilon_+ = 2mv_1 u$ and the maximum loss $\Delta\varepsilon_- = -2mv_1 u$, the extremal values corresponding to parallel velocities $\boldsymbol{v}_1$ and $\boldsymbol{u}$ and certain phases, ωt_1, ωt_2. When averaged over many collisions, however, that is, over moments t_1 and t_2, an electron gains the energy

$$\Delta\varepsilon_E = \langle \Delta\varepsilon(t_1, t_2) \rangle_{t_1, t_2} = mu^2/2 = 2\varepsilon_{\text{fr.osc.}}$$

which we found above by calculating the mean work done by the field.[2] Since $u/v \ll 1$, the actual changes that the electron energy experiences in collisions, be they positive or negative, are of first order of smallness in u/v, $|\Delta\varepsilon_\pm|/\varepsilon \sim u/v$, while the resulting positive $\Delta\varepsilon_E/\varepsilon \sim (u/v)^2$ is of second order. This latter quantity is a small difference of two relatively large ones; in symbolic form, $\Delta\varepsilon_E \sim (\Delta\varepsilon_+ - |\Delta\varepsilon_-|)$.

3.2.6 Why Electron-Electron Collisions Do not Dissipate the Energy of the Field

Electrons of a weakly ionized gas collide with atoms and molecules. In a strongly ionized gas they collide with ions and other electrons with nearly equal frequency. However, only electron-ion collisions need to be taken into account in considering the effects of the interaction with the field.

To reveal the reason, consider an electron gas (an even more general case can be taken: a gas of particles with an identical e/m ratio) and assume that electrons collide only with electrons. Sum up over all electrons the equation of motion $m\dot{\boldsymbol{v}} = -e\boldsymbol{E}_0 \sin\omega t + \dot{\boldsymbol{p}}_{\text{col}}$, where $\dot{\boldsymbol{p}}_{\text{col}}$ is the rate of change of an electron's momentum due to collisions. As the total momentum of interacting particles is conserved, the total momentum $\sum m\boldsymbol{v}$ of the gas oscillates as $\cos\omega t$, with a $\pi/2$ phase shift. Recalling that the total energy of particles is also conserved under elastic collisions, we can find the rate of change of the gas energy:

$$\frac{d}{dt}\sum \frac{mv^2}{2} = \sum m\boldsymbol{v}\cdot\dot{\boldsymbol{v}}^2 = -\left(\frac{e}{m}\boldsymbol{E}_0 \sin\omega t\right)\left(\sum m\boldsymbol{v}\right)$$
$$\propto \sin\omega t \cos\omega t \propto \sin 2\omega t\ .$$

[2] In the general case of $\nu_m \sim \omega$, only one of the phases is arbitrary because the moments of consecutive collisions are correlated. The probability of the interval $t_2 - t_1$ is $\exp[-\nu_m(t_2 - t_1)]\nu_m dt_2$. Correlations add an additional factor to $\Delta\varepsilon_E$ of (3.10): $\omega^2/(\omega^2 + \nu_m^2)$ [3.1].

The total energy of particles oscillates at double frequency, as in collisionless motion, and remains unchanged on average. No dissipation of the field energy occurs. Recall that electron-electron collisions in a dc field do not contribute to resistance and Joule heat release (Sect. 2.2.3), though as in dc fields, electron-electron collisions may affect dissipation indirectly, by changing the electron energy distribution and ν_m.

3.3 Basic Equations of Electrodynamics of Continuous Media

Sections 3.1 and 3.2 outlined what happens with electrons of an ionized gas placed in an ac electric field. Let us turn to a different aspect of the electron-field interaction: the effect of the ionized state on the behavior of ac fields and the propagation of electromagnetic waves.

3.3.1 Maxwell's Equations

The electromagnetic field and the state of the medium are described in terms of field strengths $\boldsymbol{E}$, $\boldsymbol{H}$ and inductions $\boldsymbol{D}$, $\boldsymbol{B}$. By definition, $\boldsymbol{D} = \boldsymbol{E} + 4\pi \boldsymbol{P}$ and $\boldsymbol{B} = \boldsymbol{H} + 4\pi \boldsymbol{M}$, where $\boldsymbol{P}$ and $\boldsymbol{M}$ are the electric and magnetic moments, respectively, per unit volume. The vectors $\boldsymbol{E}$, $\boldsymbol{H}$, $\boldsymbol{D}$, $\boldsymbol{B}$ satisfy the system of Maxwell's equations:

$$\operatorname{curl} \boldsymbol{H} = \frac{4\pi}{c}\boldsymbol{j} + \frac{1}{c}\frac{\partial \boldsymbol{D}}{\partial t}, \tag{3.13}$$

$$\operatorname{curl} \boldsymbol{E} = -\frac{1}{c}\frac{\partial \boldsymbol{B}}{\partial t}, \tag{3.14}$$

$$\operatorname{div} \boldsymbol{B} = 0, \tag{3.15}$$

$$\operatorname{div} \boldsymbol{D} = 4\pi \varrho. \tag{3.16}$$

This system is not completely closed because the electric current $\boldsymbol{j}$, polarization $\boldsymbol{P}$, and magnetization $\boldsymbol{M}$ generated by the field, depend on material properties. Both experience and theory indicate that direct proportionality reigns in constant fields and fields that vary not too rapidly: $\boldsymbol{j} = \sigma \boldsymbol{E}$, $\boldsymbol{P} = \chi_e \boldsymbol{E}$, $\boldsymbol{M} = \varkappa \boldsymbol{H}$. Instead of the electric χ_e and magnetic susceptibility $\varkappa$, one introduces the permittivity $\varepsilon = 1 + 4\pi\chi_e$ and magnetic permeability $\mu = 1 + 4\pi\varkappa$. Together with the equations

$$\boldsymbol{j} = \sigma \boldsymbol{E}, \quad \boldsymbol{D} = \varepsilon \boldsymbol{E}, \quad \boldsymbol{B} = \mu \boldsymbol{H}, \tag{3.17}$$

where the material constants ε, μ and conductivity σ are assumed to be known, system (3.13–17) is closed. In gases and plasmas, the approximation $\mu = 1$ can be used with extremely high accuracy.

3.3.2 Displacement, Polarization, Conduction, and Charge Currents

The right-hand side of (3.13) can be treated as a current density

$$\boldsymbol{j} + \frac{1}{4\pi}\frac{\partial \boldsymbol{D}}{\partial t} = \boldsymbol{j} + \frac{\partial \boldsymbol{P}}{\partial t} + \frac{1}{4\pi}\frac{\partial \boldsymbol{E}}{\partial t}, \tag{3.18}$$

times $4\pi/c$. Maxwell called the term $(1/4\pi)\partial \boldsymbol{D}/\partial t$, which he postulated should be added to the conduction current, the *displacement current*. Without the displacement current, (3.13, 16) contradict the unassailable law of charge conservation, (2.39).

A variable field changes the polarization of matter with time, namely, displaces negative charges relative to positive ones by applying the electric force. In fact, any displacement of charges in space is a current, so that the term $\partial \boldsymbol{P}/\partial t$ in the displacement current is indeed a current density: that of the *polarization current*. Together with the conduction current $\boldsymbol{j}$, it forms the total charge current $\boldsymbol{j}_t$. The term $(1/4\pi)\partial \boldsymbol{E}/\partial t$ is in no way connected with the motion of charge and therefore, is not literally a current. (Its meaning will be discussed in Sect. 13.5.)

The total charge current was divided into the conduction and polarization components only to facilitate the application of the equations to ideal dielectrics, where $\sigma, \boldsymbol{j} = 0$. This partition is by no means mandatory. The total polarization vector $\boldsymbol{P}_t$ can be defined for any electrically neutral medium; $\boldsymbol{P}_t$ is related to the total charge current $\boldsymbol{j}_t$, so that it is sufficient to operate with just one of these quantities. Indeed, by definition

$$\boldsymbol{P}_t = \sum e_i \boldsymbol{r}_i\,, \quad \sum e_i = 0\,, \quad \boldsymbol{j}_t = \sum e_i \boldsymbol{v}_i = \frac{\partial \boldsymbol{P}_t}{\partial t}\,, \tag{3.19}$$

where $\boldsymbol{r}_i$ is the radius vector of the charge e_i, $\boldsymbol{v}_i = \dot{\boldsymbol{r}}_i$ is its velocity, and summation is extended to absolutely all charges (free, bound, electrons, nuclei) within a unit volume.

3.3.3 Expansion into Harmonics

Equations (3.17) fail in rapidly varying fields. Owing to the inertia of the processes that produce polarization and current, they cease to track the variations of the field. The polarization and current at t_1 are now determined less by the value of $\boldsymbol{E}(t_1)$ than by the evolution of $\boldsymbol{E}(t)$ in the preceding period $t < t_1$. For instance, if the field $\boldsymbol{E}$ pointed for a long time in one direction and then was suddenly reversed, the current would flow for some time in the former direction, against the new field, until the charges are brought to rest.

Obtaining the material equations (like $\boldsymbol{D} = \varepsilon \boldsymbol{E}$) that take into account retardation effects is greatly facilitated because the motion of charges in matter is described by equations linear in $\boldsymbol{E}$, $\boldsymbol{r}$, $\boldsymbol{v}$. Since Maxwell's equations are also linear, all time-dependent quantities can be expanded into Fourier series or integrals; in view of the superposition principle, one can operate only with harmonic components, as we have already done in Sects. 3.1, 2. Three parameters completely determine the evolution of harmonic quantities: amplitude, frequency, and phase, so that the entire retardation effect is contained in the relation between these parameters for material characteristics and the field. If the field for some harmonic is $\boldsymbol{E}_\omega = \boldsymbol{E}_{\omega 0} \sin \omega t$, then the total current is $\boldsymbol{j}_{t\omega} = \boldsymbol{j}_{t\omega 0} \sin(\omega t + \varphi_\omega)$, with $\boldsymbol{j}_{t\omega 0}$ and φ_ω being the functions of $\boldsymbol{E}_{\omega 0}$ and ω.

3.3.4 Material Equations for Harmonic Components

These can be given a convenient and lucid form if we retain the concepts of conductivity and dielectric permittivity, which are familiar from working with constant fields. The total current $\boldsymbol{j}_{t\omega}$ is given by a linear combination of $\sin\omega t$ and $\cos\omega t$, which corresponds to a linear combination of $\boldsymbol{E}$ and $\partial\boldsymbol{E}/\partial t$. Returning to the original concepts of conduction current $\sigma\boldsymbol{E}$ and polarization current $\partial\boldsymbol{P}/\partial t$, which is equal to $[(\varepsilon-1)/4\pi]\partial\boldsymbol{E}/\partial t$ if the field varies slowly, and equipping the new coefficients σ and ε with the subscript ω (because now they are frequency-dependent), we rewrite the material equation in the form

$$\boldsymbol{j}_{t\omega}=\sigma_\omega\boldsymbol{E}_\omega+\frac{\varepsilon_\omega-1}{4\pi}\frac{\partial\boldsymbol{E}_\omega}{\partial t}\,,\qquad \boldsymbol{E}_\omega=\boldsymbol{E}_{\omega_0}\sin\omega t\,. \tag{3.20}$$

The quantities σ_ω and ε_ω are called the *high-frequency conductivity* and *dielectric permittivity* of the medium. It is these characteristics of the medium that affect the behavior of variable fields in it.

3.3.5 Energy Equation

We now form the scalar products of (3.13) with $\boldsymbol{E}$ and of (3.14) with $\boldsymbol{H}$, and subtract the resulting equations from each other. In view of the fact that $\boldsymbol{H}\,\mathrm{curl}\,\boldsymbol{E}-\boldsymbol{E}\,\mathrm{curl}\,\boldsymbol{H}=\mathrm{div}[\boldsymbol{E}\times\boldsymbol{H}]$, assuming the relations between $\boldsymbol{D}$ and $\boldsymbol{E}$, $\boldsymbol{B}$ and $\boldsymbol{H}$ are linear, we obtain the relation

$$\frac{\partial}{\partial t}\frac{\boldsymbol{E}\cdot\boldsymbol{D}+\boldsymbol{H}\cdot\boldsymbol{B}}{8\pi}+\mathrm{div}\frac{c}{4\pi}[\boldsymbol{E}\times\boldsymbol{H}]=-\boldsymbol{j}\cdot\boldsymbol{E}\,. \tag{3.21}$$

This formula expresses the law of conservation of energy of the electromagnetic field. The quantity $\boldsymbol{E}\cdot\boldsymbol{D}/8\pi$ is the electric energy density, and $\boldsymbol{H}\cdot\boldsymbol{B}/8\pi$ is that of magnetic energy;

$$\boldsymbol{S}=\frac{c}{4\pi}[\boldsymbol{E}\times\boldsymbol{H}] \tag{3.22}$$

is the electromagnetic energy flux density (Poynting vector); and $\boldsymbol{j}\cdot\boldsymbol{E}$ is the energy released per second in $1\,\mathrm{cm}^3$ of the medium, equal to the decrease in electromagnetic energy. As a time average, the dissipation of energy in harmonic fields in a plasma is caused only by the conduction current. The polarization current does not result in dissipation because it is shifted by $\pi/2$ with respect to the field, and $\langle\sin\omega t\cos\omega t\rangle=0$ (cf. the results of Sects. 3.2.1, 2).

3.4 High-Frequency Conductivity and Dielectric Permittivity of Plasma

The results of Sect. 3.1 permit the immediate calculation of these quantities, but two qualifications will be made first. We assume that ions do not move and make a negligible contribution to conduction and polarization currents. Next, we single out the term $\varepsilon_M - 1$ in $\varepsilon_\omega - 1$ due to the electrons bound in molecules and ions. This term is of the same order of magnitude as in nonionized gases (unless the polarizability of the excited molecules in the plasma is greater than that of nonexcited ones). Under normal conditions, $\varepsilon_M - 1 = 5.28 \times 10^{-4}$ in air, 2.65×10^{-4} in H_2, and 0.67×10^{-4} in He; it is even smaller at low pressure because $\varepsilon_M - 1$ is proportional to density. These figures refer to the visible part of the spectrum and to lower frequencies. The contribution of the molecular part to the bulk polarizability of plasma is very small for any appreciable ionization.

3.4.1 Calculation of σ_ω and ε_ω

First we substitute the general expression (3.5) of the mean electron velocity $\boldsymbol{v}$ into (3.19) for the charge current $\boldsymbol{j}_t$. Replacing the summation over all electrons with the multiplication by n_e (the term with random velocity vanishes upon averaging) and comparing the obtained expression with (3.20), we find:

$$\sigma_\omega = e^2 n_e \nu_m / m \left(\omega^2 + \nu_m^2\right) , \tag{3.23}$$

$$\varepsilon_\omega = 1 - 4\pi e^2 n_e / m \left(\omega^2 + \nu_m^2\right) . \tag{3.24}$$

These formulas are of fundamental importance for the physics of interaction between plasma and electromagnetic fields. The ratio of the amplitudes of conduction and polarization currents,

$$\frac{j_{\text{cond},0}}{j_{\text{polar},0}} = \frac{4\pi\sigma_\omega}{\omega|\varepsilon_\omega - 1|} = \nu_m/\omega , \tag{3.25}$$

is determined by the ratio of collision and field frequencies.

3.4.2 The High-Frequency Limit (Collisionless Plasma)

This regime is reached when $\omega^2 \gg \nu_m^2$, that is, at not particularly high frequencies, i.e., the microwave or very far IR range, even at atmospheric pressure. In fact, the molecular polarizability of most dielectrics and nonionized gases retains the value typical for dc fields up to optical frequencies. In the high-frequency limit,

$$\sigma_\omega = \frac{e^2 n_e}{m\omega^2}\nu_m , \quad \varepsilon_\omega = 1 - \frac{4\pi e^2 n_e}{m\omega^2} , \tag{3.26}$$

that is, conductivity is proportional to collision frequency, while dielectric permittivity is independent of ω. According to (3.25), the conduction current is small compared with the polarization current. This limit corresponds to the collisionless plasma model (Sects. 3.1, 3.2.1).

3.4.3 Static Limit

If $\omega^2 \ll \nu_m^2$, then

$$\sigma_\omega = \frac{e^2 n_e}{m\nu_m} , \quad \varepsilon_\omega = 1 - \frac{4\pi e^2 n_e}{m\nu_m^2} . \tag{3.27}$$

The conductivity is indistinguishable from the ordinary dc conductivity of the ionized gas. The dielectric permittivity also reaches a value independent of frequency. The polarization current is small in comparison with the conduction current, and vanishes completely in the limit $\omega \to 0$.

3.4.4 Why Dielectrics Usually Have $\varepsilon > 1$, and Plasma $\varepsilon < 1$

Electrons in atoms and molecules are bound, while in plasmas (and metals, where also $\varepsilon < 1$) some of them are free. An absolutely free electron, moving without collisions, oscillates in phase with the field [see (3.1)]. It shifts away from the center of equilibrium along $\boldsymbol{E}$, against the direction of the electric force; having a negative charge, it induces the negative polarizability of the medium, so that $\varepsilon < 1$.

Electrons in molecules are like particles that feel an elastic restoring force in response to displacement. If ω_0 is the frequency of natural vibrations of an elastically bound electron, then

$$\ddot{\boldsymbol{r}} + \omega_0^2 \boldsymbol{r} = -\frac{e\boldsymbol{E}_0}{m} \sin \omega t , \quad \boldsymbol{r} = -\frac{e\boldsymbol{E}_0}{m(\omega_0^2 - \omega^2)} \sin \omega t .$$

The displacement in a static field and at frequencies less than the natural frequency (the latter usually lie in the optical range) is directed against $\boldsymbol{E}$, so that polarizability is positive (the situation at $\omega > \omega_0$ is reversed; this results in the anomalous dispersion of light). In solid and liquid dielectrics ε has usually a value between 1 and 10.

3.5 Propagation of Electromagnetic Waves in Plasmas

In ideal dielectrics, where the conduction current is zero and free charges are absent, the system of Maxwell's equations yields equations for $\boldsymbol{E}$ and $\boldsymbol{H}$ that describe the propagation of electromagnetic waves. Similar equations are obtained for a monochromatic field in an electrically neutral conducting medium such as a plasma. The simplest way to do this is to represent harmonic quantities in complex form. In general, this is expedient in the presence of phase shifts, because it removes the need to operate constantly with combinations of sines and cosines.

3.5.1 Complex Dielectric Permittivity

In the case of a monochromatic field [$\boldsymbol{E}$, $\boldsymbol{H} \propto \exp(-\mathrm{i}\omega t)$], the first Maxwell equaiton (3.13), taken together with the material equation (3.20), transforms to

$$\operatorname{curl} \boldsymbol{H} = (4\pi/c)\sigma \boldsymbol{E} - \mathrm{i}(\omega\varepsilon/c)\boldsymbol{E} , \quad \sigma \equiv \sigma_\omega , \quad \varepsilon \equiv \varepsilon_\omega \tag{3.28}$$

(hereafter we drop the subscript ω with σ_ω, ε_ω). It is convenient to eliminate the conduction current in (3.28), introducing the complex dielectric permittivity

$$\varepsilon' = \varepsilon + \mathrm{i}4\pi\sigma/\omega . \tag{3.29}$$

The resulting equation,

$$\operatorname{curl} \boldsymbol{H} = -\mathrm{i}(\omega\varepsilon'/c)\boldsymbol{E} , \tag{3.30}$$

is apparently identical to the corresponding equation for dielectrics, where $\sigma = 0$, $\varepsilon' = \varepsilon$.

3.5.2 Plane Electromagnetic Wave

First we substitute $\partial \boldsymbol{B}/\partial t = -\mathrm{i}\omega \boldsymbol{H}$ into (3.14) and take the curl of the resulting equation, using the formula curl (curl $\boldsymbol{E}$) = $-\Delta \boldsymbol{E}$ + grad(div $\boldsymbol{E}$), remarking that for $\varrho = 0$ (3.16) implies that div $\boldsymbol{E} = 0$, and replacing curl $\boldsymbol{H}$ with (3.30), we arrive at the equation

$$\Delta \boldsymbol{E} + \left(\varepsilon'\omega^2/c^2\right) \boldsymbol{E} = 0 . \tag{3.31}$$

In fact, this is a result of substituting $\exp(-\mathrm{i}\omega t)$ for $\boldsymbol{E}$ into the wave equation. Eliminating $\boldsymbol{E}$, instead of $\boldsymbol{H}$, from (3.13, 14) and using (3.15), we obtain for $\boldsymbol{H}$ an equation similar to (3.31). Equation (3.31) admits a travelling-wave type solution, $\boldsymbol{E}, \boldsymbol{H} \propto \exp(-\mathrm{i}\omega t + \mathrm{i}\boldsymbol{k}\cdot\boldsymbol{r})$, where $\boldsymbol{k}$ is the wave vector. The pre-exponential coefficients are certain complex numbers that characterize the amplitudes of the fields and the phase shifts between them. Substitution of these expressions into the original equations (3.13) or (3.30) and (3.14) yields

$$[\boldsymbol{k} \times \boldsymbol{H}] = -(\omega\varepsilon'/c)\boldsymbol{E} , \quad [\boldsymbol{k} \times \boldsymbol{E}] = (\omega/c)\boldsymbol{H} . \tag{3.32}$$

These equalities imply that if $\varepsilon' \neq 0$, all three vectors $\boldsymbol{E}$, $\boldsymbol{H}$, and $\boldsymbol{k}$ are mutually perpendicular, that is, the wave is transverse. If $\varepsilon' = 0$ (meaning that $\varepsilon = 0$ in weakly conducting media, where $\sigma \approx 0$), the equations admit the existence of longitudinal purely electric waves with $\boldsymbol{H} = 0$, $\boldsymbol{E} \| \boldsymbol{k}$: these are the plasma waves (see Sect. 3.6.3).

3.5.3 Refractive Index and Attenuation of Waves

Equations (3.32) imply that the wave vector is a function of frequency (the dispersion relation) and give a relation between the complex amplitudes of the fields,

$$k = (\omega/c)\sqrt{\varepsilon'} , \quad H = \sqrt{\varepsilon'} E . \tag{3.33}$$

The wave vector is a complex quantity because ε' is complex. In order to find $k = k_1 + \mathrm{i}k_2$, we introduce the dimensionless numbers n and $\varkappa$ via the formula $ck/\omega = n + \mathrm{i}\varkappa = \sqrt{\varepsilon'}$. Squaring this quantity, substituting ε' from (3.29), and equating the real and imaginary parts, we find that

$$n^2 - \varkappa^2 = \varepsilon , \quad 2n\varkappa = 4\pi\sigma/\omega ,$$

$$n = \sqrt{\frac{\varepsilon + \sqrt{\varepsilon^2 + (4\pi\sigma/\omega)^2}}{2}} , \quad \varkappa = \sqrt{\frac{-\varepsilon + \sqrt{\varepsilon^2 + (4\pi\sigma/\omega)^2}}{2}} . \tag{3.34}$$

The physical meaning of the quantities n and $\varkappa$ follows from the expression for the travelling wave:

$$E,\, H \propto e^{-i\omega t+ikx} = \exp\left[-i\omega\left(t - n\frac{x}{c}\right) - \varkappa\frac{\omega}{c}x\right] .$$

The number n determines the *phase velocity* c/n and wavelength $\lambda = \lambda_0/n$ in the medium ($\lambda_0 = 2\pi c/\omega$ is the wavelength in vacuum) and corresponds to the *refractive index*. The number $\varkappa$ characterizes the *attenuation* of the wave: its amplitude is reduced by a factor e over the path length $\Delta x = \lambda_0/2\pi\varkappa$ or by a factor $e^{\varkappa}$ over a path length $\bar{\lambda}_0$. The numbers n and $\varkappa$ determine the relation between the amplitudes of the field and the phase shift between them:

$$H = (n + i\varkappa)E = \sqrt{n^2 + \varkappa^2}\, e^{i\psi} E\,, \quad \psi = \arctan(\varkappa/n)\,.$$

3.5.4 The Law of Attenuation of the Energy Flux

Only the value of the energy flux density averaged over one period is of practical importance. In order to calculate the mean value of two harmonic variables given in complex form, we have to multiply one variable by the complex conjugate of the other and divide by two.[3] In a homogeneous medium, the energy flux of a wave decays exponentially,

$$S = \frac{c}{4\pi}\frac{1}{2}\mathrm{Re}\{EH^*\} = \frac{cn}{8\pi}EE^* = \frac{cn}{8\pi}|E(0)|^2 e^{-\mu_\omega x}\,, \tag{3.35}$$

where the averaging subscript with S is dropped, $E(0)$ is the amplitude at the point $x = 0$, and

$$\mu_\omega = 2\varkappa\omega/c = 4\pi\sigma/nc \tag{3.36}$$

is the absorption coefficient. The energy flux decreases by a factor e over the length μ_ω^{-1}.

It follows from (3.35) and also from the general equation for energy (3.21), that if (3.35) is averaged over one period of time, the Bouguer law is valid:

$$\frac{dS}{dx} = -\mu_\omega S\,. \tag{3.37}$$

The electromagnetic energy dissipated per second in 1 cm^3 (the energy deposited in the medium) is $\langle \boldsymbol{j} \cdot \boldsymbol{E}\rangle = \sigma\langle E^2\rangle = \mu_\omega S$. The proportionality between the absorption coefficient and conductivity is in perfect agreement with the proportionality of the rate of Joule heat release to the conductivity of the medium.

---

[3] Indeed, if $A = A_0\cos(\omega t + \varphi_A)$ and $B = B_0\cos(\omega t + \varphi_B)$, then $\langle AB\rangle = (1/2)A_0B_0 \cdot \cos(\varphi_B - \varphi_A)$. In the complex representation, $A = A_0\exp(-i\omega t + i\varphi_A)$ and $B = B_0\exp(-i\omega t + i\varphi_B)$. Hence,

$$\tfrac{1}{2}\mathrm{Re}\{AB^*\} = \tfrac{1}{2}A_0B_0\,\mathrm{Re}\{\exp[-i(\varphi_B - \varphi_A)]\} = \tfrac{1}{2}A_0B_0\cos(\varphi_B - \varphi_A) = \langle AB\rangle\,.$$

In the limit of a nonconducting medium ($\sigma = 0$), the dielectric permittivity $\varepsilon' = \varepsilon$ is a real quantity, $n = \sqrt{\varepsilon}$, $\varkappa = 0$, and $\mu_\omega = 0$, i.e., the medium is absolutely transparent. Waves in it are not damped, because on the average the polarization current does not dissipate any Joule heat within one period.

3.5.5 Wave Absorption Coefficient in Plasma

If the imaginary part of ε' is much less than the real part, $4\pi\sigma/\omega\varepsilon \ll 1$, and ε is positive, then equations (3.34) yield $n \approx \sqrt{\varepsilon}$ and $\varkappa \approx 2\pi\sigma/\omega n \ll 1$. The refractive index of the medium is typical for a dielectric, but the absorption is weak, in the sense that the wave is only slightly attenuated over a path length of order λ. The absorption coefficient is given by (3.36), where we use $n = \sqrt{\varepsilon}$. This situation is normally realized in the propagation of light, and partly in the propagation of microwave radiation in laboratory plasmas.

According to (3.23, 36), the absorption coefficient of an electromagnetic wave in an ionized gas is

$$\mu_\omega = \frac{4\pi e^2 n_e \nu_m}{mc(\omega^2 + \nu_m^2)} = 0.106 n_e \frac{\nu_m}{\omega^2 + \nu_m^2} \mathrm{cm}^{-1} . \tag{3.38}$$

Here we assume $n \approx \sqrt{\varepsilon} \approx 1$, since this is the most typical case under the conditions (weak ionization, weak absorption) in which (3.38) is definitely valid. The absorption coefficient is proportional to the electron density. In the high-frequency limit ($\omega^2 \gg \nu_m^2$), absorption is characterized by inverse square frequency dependence: $\mu_\omega \propto \omega^{-2} \propto \lambda^2$; hence short waves are better transmitted through plasma than long waves.

3.5.6 Quasistationary Field and the Skin Layer

Assume that the imaginary part of ε' is much greater than the real part (or more accurately, than the absolute value of the real part, because typically for such cases $\varepsilon < 0$). This is the case, for example, in good conductors if the field frequency is not excessively high. The conduction current then dominates the displacement current:

$$\frac{|j_{\mathrm{cond}}|}{|j_{\mathrm{disp}}|} = \frac{\varepsilon'_{\mathrm{im}}}{|\varepsilon'_{\mathrm{re}}|} = \frac{4\pi\sigma}{\omega|\varepsilon|} \gg 1 .$$

In this limiting case, (3.34) gives $n \approx \varkappa \approx \sqrt{2\pi\sigma/\omega}$, and the field is strongly damped over a distance of order λ. It is then meaningless to speak of a travelling wave or wave propagation, although these, formally, still exist. Electromagnetic waves are possible because of the displacement current. In the absence of this term, Maxwell's equation (3.13) is identical to the equation for the magnetic field of a dc current. This gives the limit of the quasistationary field.

The effective depth of penetration of the quasisteady field into a conductor can also be found from the formulas for the electromagnetic wave, as the distance over which the amplitude of a wave with purely imaginary ε' is reduced by a factor e

$$\delta = \frac{c}{\omega \varkappa} = \frac{c}{\sqrt{2\pi\sigma\omega}} = \frac{5.03}{\{\sigma[\mathrm{Ohm}^{-1}\,\mathrm{cm}^{-1}]f[\mathrm{MHz}]\}^{1/2}}\mathrm{cm}\,. \tag{3.39}$$

This quantity is called the *skin depth*, and ac current in good conductors flows only in this surface (skin) layer. For example, in copper ($\sigma = 6 \times 10^5\,\mathrm{Ohm}^{-1}\,\mathrm{cm}^{-1}$) at $f = 10\,\mathrm{MHz}$, we have $\delta = 2 \times 10^{-3}\,\mathrm{cm}$. However, the limiting case we discuss here is realized not only in metals but also in plasmas in the rf range (Sect. 11.3).

3.6 Total Reflection of Electromagnetic Waves from Plasma and Plasma Oscillations

3.6.1 Nonabsorbing Medium with Negative Dielectric Permittivity

Let a medium have $\varepsilon < 0$ and the conductivity σ be, if not zero, then so small that $4\pi\sigma/\omega|\varepsilon| \ll 1$. As follows from (3.34), this medium has $n \approx 0$, $\varkappa \approx \sqrt{|\varepsilon|}$. Electromagnetic waves cannot penetrate into such a medium, as in the case of purely imaginary ε', albeit for a different reason. The phase velocity and wavelength tend to infinity as $\sigma \to 0$, the field oscillates only in time, and its amplitude decreases exponentially into the medium. However, the energy of the field is not dissipated, in contrast to the case of a good conductor, where the amplitude decreases away from the boundary owing to the strong absorption of energy. The depth of penetration into a medium with $\varepsilon < 0$ and $\sigma \approx 0$ is independent of σ and equals $\lambdabar_0/\sqrt{|\varepsilon|}$. This situation corresponds to the total reflection of the electromagnetic wave and is frequently realized in the collisionless plasmas when $\nu_\mathrm{m}^2 \ll \omega^2$. For example, let the wave frequency be $f = 3\,\mathrm{GHz}$, $\lambda_0 = 10\,\mathrm{cm}$, $p = 0.1\,\mathrm{Torr}$ ($N = 3.3 \times 10^{15}\,\mathrm{cm}^{-3}$). In the case of weak ionization, $\nu_\mathrm{m} \approx 3 \times 10^8\,\mathrm{s}^{-1}$ and $\nu_\mathrm{m}^2/\omega^2 \sim 10^{-4}$.

3.6.2 Critical Electron Density

Let us rewrite the dielectric permittivity (3.24) for a collisionless plasma in the form

$$\varepsilon = 1 - \omega_\mathrm{p}^2/\omega^2\,, \quad \omega_\mathrm{p} = \sqrt{4\pi e^2 n_\mathrm{e}/m} = 5.65 \times 10^4 n_\mathrm{e}^{1/2}\,\mathrm{s}^{-1}\,. \tag{3.40}$$

It is negative if $\omega < \omega_\mathrm{p}$ or if the electron density is greater than the critical value

$$n_{\mathrm{e,cr}} = m\omega^2/4\pi e^2 = 1.24 \times 10^4\{f[\mathrm{MHz}]\}^2 = 1.11 \times 10^{13}\{\lambda_0[\mathrm{cm}]\}^{-2}\,\mathrm{cm}^{-3}\,. \tag{3.41}$$

In the example above, $n_{\mathrm{e,cr}} = 1.1 \times 10^{11}\,\mathrm{cm}^{-3}$. The wave with $\lambda_0 = 10\,\mathrm{cm}$ cannot penetrate the region with high electron density: it undergoes total reflection.

If n_e in a plasma increases in the direction of the x axis, and an electromagnetic wave propagates through the plasma in the same direction, it will reach roughly the point of $n_\mathrm{e} = n_{\mathrm{e,cr}}$, be reflected, and travel back. If plane symmetry holds, then the geometrical optics law is obeyed: the angle of incidence is equal

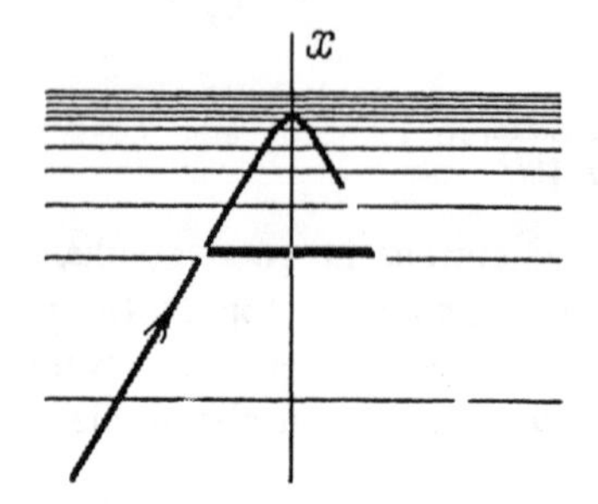

Fig. 3.1. Reflection of electromagnetic waves from plasma. Increasing electron density is shown by crowding of *horizontal lines*. The beam is turned back approximately at the point where the electron density reaches the critical value for the given wave frequency

to that of reflection (Fig. 3.1). This effect is of enormous practical importance and is widely used in experimental work and in technology. It lies at the foundation of one of the most efficient methods of both laboratory and ionospheric diagnostics of plasmas. The plasma is irradiated with signals of various frequencies, and one records which frequencies are transmitted ($\omega > \omega_p$) and which are stopped ($\omega < \omega_p$). The value of n_e is found from (3.40) once the cutoff frequency $\omega = \omega_p$ is known. A low-pressure (collisionless) laboratory plasma with $n_e \sim 10^{11}$–$10^{15}\,\mathrm{cm}^{-3}$ may be investigated in this way with microwave radiation of $\lambda_0 \sim 10 - 0.1\,\mathrm{cm}$.

3.6.3 Plasma Frequency

Longitudinal electric waves with $\boldsymbol{E} \| \boldsymbol{k}$ and $\boldsymbol{H} = 0$ in collisionless plasma are realized if $\varepsilon' = \varepsilon = 0$; from (3.40), this corresponds to $\omega = \omega_p$. The bulk charge density ϱ also undergoes oscillations. Indeed, if the polarization due to free electrons (which is described by oscillations of ϱ) is eliminated from $\boldsymbol{D}$, we find from (3.16) that $(\boldsymbol{k} \cdot \boldsymbol{E}) = 4\pi\varrho/\varepsilon_{\text{bound}}$, where $\varepsilon_{\text{bound}}$ is the dielectric permittivity due to bound electrons. The frequency ω_p defined by (3.40) is called the *plasma*, or *Langmuir, frequency*. This is the natural oscillation frequency of electrons in plasma established by *Tonks* and *Langmuir* in 1929. Strictly speaking, ω_p corresponds to the oscillations of the gas as a whole, that is, to waves of "infinite" wavelength.

Let all the electrons be initially shifted, for some reason, to the right with respect to the ion, which we assume to be at rest (Fig. 3.2). The separation of charges produces the attractive force that tends to return the electrons to where they "belong". Being accelerated by this force, the electrons overshoot the equilibrium position and move to the left of the ions, and so forth. If Δx

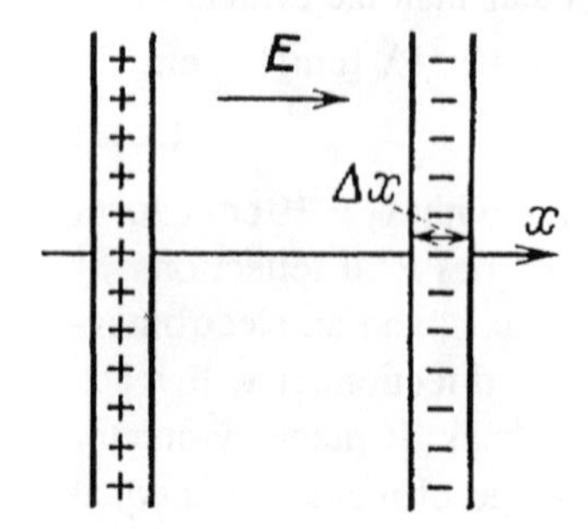

Fig. 3.2. Plasma oscillations. Electrons are displaced rightward with respect to ions by a distance Δx

is the displacement of electrons from the equilibrium, then the surface charge density at the boundaries of the plasma layers is $en_e \Delta x$, the strength of the field of polarization being $E = 4\pi e n_e \Delta x$. Therefore, the equation of electron motion,

$$m(\Delta \ddot{x}) = -eE = -4\pi e^2 n_e \Delta x \ , \quad (\Delta \ddot{x}) + \omega_p^2 \Delta x = 0 \ ,$$

describes harmonic oscillations at a frequency ω_p as implied by (3.40). The plasma frequency, Debye radius, and mean electron velocity are related by the formula

$$\omega_p d = \left(\frac{4\pi e^2 n_e}{m} \cdot \frac{kT_e}{4\pi e^2 n_e} \right)^{1/2} = \left(\frac{kT_e}{m} \right)^{1/2} = 0.62\bar{v} \ .$$

The velocity of the oscillating electrons is randomized at a rate given by the collision frequency ν_m, which determines the rate of damping of free oscillations. The notion of oscillations remains meaningful as long as $\nu_m \ll \omega_p$ or $n_e \gg 2 \times 10^9 \ (p \ [\text{Torr}])^2 \ \text{cm}^{-3}$. *Plasma oscillations* were discovered experimentally by *Penning* in 1926.

4. Production and Decay of Charged Particles

4.1 Electron Impact Ionization in a Constant Field

4.1.1 Ionization Frequency

Ionization of atoms and molecules by electron impact is the most important mechanism of charge generation in the bulk of a gas discharge. The rate of this process, $(dn_e/dt)_i = \nu_i n_e = k_i N n_e$, is characterized by the *ionization frequency* ν_i, that is, the number of ionization events performed by an electron per second, or by the *reaction rate constant* k_i. If $n(\varepsilon)$ is the electron energy distribution function (density-normalized), and $\sigma_i(\varepsilon)$ is the ionization cross section of atoms in their ground state (Fig. 4.1), then

$$\nu_i = N \int n(\varepsilon) v \sigma_i(\varepsilon) d\varepsilon \Big/ \int n(\varepsilon) d\varepsilon = N\langle v\sigma_i\rangle \equiv N k_i \,. \tag{4.1}$$

The electron energy distribution of a weakly ionized plasma in an electric field depends on a number of elastic and inelastic collision processes. Under these conditions, the ionization frequency is found either by solving the kinetic equation for $n(\varepsilon)$ (Chap. 5) or experimentally. If the ionization by electrons proceeds under unvarying conditions, so that ν_i = const., and the removal of electrons can be

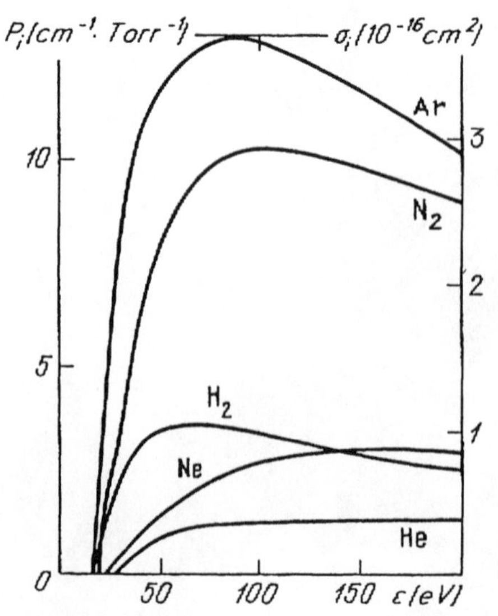

Fig. 4.1. Cross sections and probabilities of electron impact ionization. From [4.1]

neglected, electrons proliferate exponentially: $n_e = n_e(0)\exp(\nu_i t)$: an *electron avalanche* develops.

4.1.2 Maxwellian Distribution

This distribution arises when the degree of ionization is not too small and electron-electron collisions are important. The dependence of ν_i on the field and collisions with molecules in the case of a Maxwellian distribution (see Appendix) is contained implicitly in the electron temperature that determines the ionization frequency. As a rule, the temperature of a gas discharge plasma is substantially lower than the ionization potential I, because strong ionization occurs when kT_e is less than I by a factor in the range of 5 to 10. Atoms are ionized by high-energy electrons in the tail of the Maxwellian distribution. In this energy range, $n(\varepsilon) \propto \exp(-\varepsilon/kT_e)$ falls off steeply, so that a linear function can be used for the cross section, $\sigma_i(\varepsilon) = C_i(\varepsilon - I)$, in integral (4.1); the linearity is valid if ε is slightly greater than the ionization threshold I, which gives

$$\nu_i = N\bar{v}C_i\left(I + 2kT_e\right)\exp\left(-I/kT_e\right), \quad \bar{v} = \left(8kT_e/\pi m\right)^{1/2}. \tag{4.2}$$

For example, in argon, $C_i = 2 \cdot 10^{-17}\,\mathrm{cm^2/eV}$. If $T_e = 1\,\mathrm{eV}$, then $\bar{v} = 6.7 \cdot 10^7\,\mathrm{cm/s}$ and $k_i = \langle v\sigma_i\rangle = 3 \cdot 10^{-16}\,\mathrm{cm^3/s}$. If $p = 50\,\mathrm{Torr}$ and $T = 300\,\mathrm{K}$, then $N = 1.7 \cdot 10^{18}\,\mathrm{cm^{-3}}$. This gives $\nu_i = 510\,\mathrm{s^{-1}}$. At these T_e and N, the equilibrium degree of ionization is $(n_e)_{eq}/N = 0.021$. The values of C_i for several other gases (in $10^{-17}\,\mathrm{cm^2/eV}$) are

$$\mathrm{He} - 0.13\,,\ \mathrm{Ne} - 0.16\,,\ \mathrm{Hg} - 7.9\,,\ \mathrm{N_2} - 0.85\,,\ \mathrm{O_2} - 0.68\,,\ \mathrm{H_2} - 0.59\,.$$

4.1.3 Townsend's Ionization Coefficient

An electron avalanche generated by an electron in a dc field evolves not only in time but also in space, along the direction of drift of the knocked-out electrons. It is more convenient, therefore, to characterize the rate of ionization not by frequency $\nu_i\,\mathrm{s^{-1}}$, but by the *ionization coefficient* $\alpha\,\mathrm{cm^{-1}}$, that is, the number of ionization events performed by an electron in a 1 cm path along the field. Obviously,

$$\alpha = \nu_i/v_d\,, \quad \nu_i = \alpha v_d\,. \tag{4.3}$$

Note that the primary and complete characteristic of the rate of ionization is the frequency ν_i, not α. The distribution function gives us this frequency, as well as the drift velocity. The ionization coefficient α is a derived quantity, found from (4.3). Actually, α is not very meaningful in fast-oscillating fields. However, dc measurements give us α, not ν_i.

4.1.4 Measurement of α and Similarity Laws

If we place plane electrodes at a separation d, apply a voltage V, and irradiate the cathode with UV light, knocking out $\mathcal{N}_0$ electrons in 1 s, the number of electrons in the avalanche grows towards the anode:

$$d\mathcal{N}/dx = \alpha\mathcal{N}\,, \quad \mathcal{N}(x) = \mathcal{N}_0\exp(\alpha x)\,. \tag{4.4}$$

The electron current at the anode is $i = e\mathcal{N}_0 \exp(\alpha d)$. In the steady state, the positive ions produced in the discharge gap arrive at the cathode in the same numbers as the electrons at the anode. The current in the closed circuit is everywhere identical and equal to i. The low-mobility ions accumulate in the gap between the electrodes in much larger numbers than the electrons, which are removed more quickly by the field. As a result, the gap contains positive space charge. However, this space charge causes little distortion of the field at small currents, and the field is known: $E = V/d$. If i is measured with varying d, and $\ln i$ is plotted as a function of d for constant E, the coefficient is found from the slope of the straight line $\ln i = \text{const} + \alpha d$.

The energy distribution, mean electron energy, and drift velocity are functions of the ratio E/p. Hence, a similarity law of the type $\alpha = pf(E/p)$ holds for both ν_i and α. Experimentally, therefore, one can vary p instead of d, keeping E/p constant: $\ln i = \text{const} + (\alpha/p)(pd)$. Experimental data show that points fall on straight lines quite well up to a certain limit of pd (see Sect. 4.7.1); the values of α for a number of gases were measured in this way (Figs. 4.2–5).

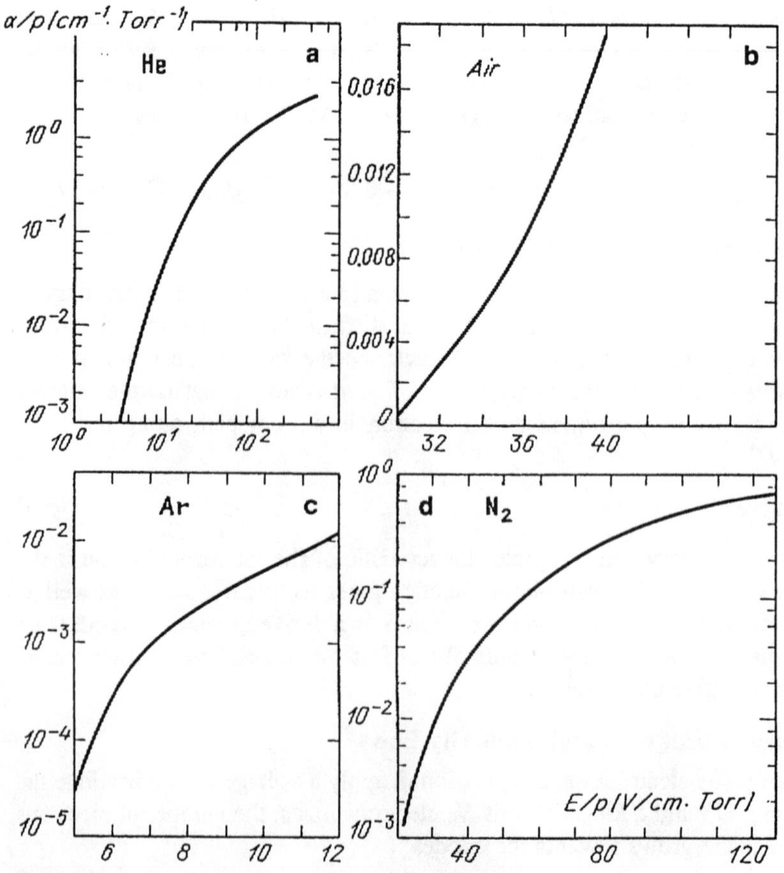

Fig. 4.2. Ionization coefficient in a) He, b) air, c) Ar, d) N_2. From [4.2]

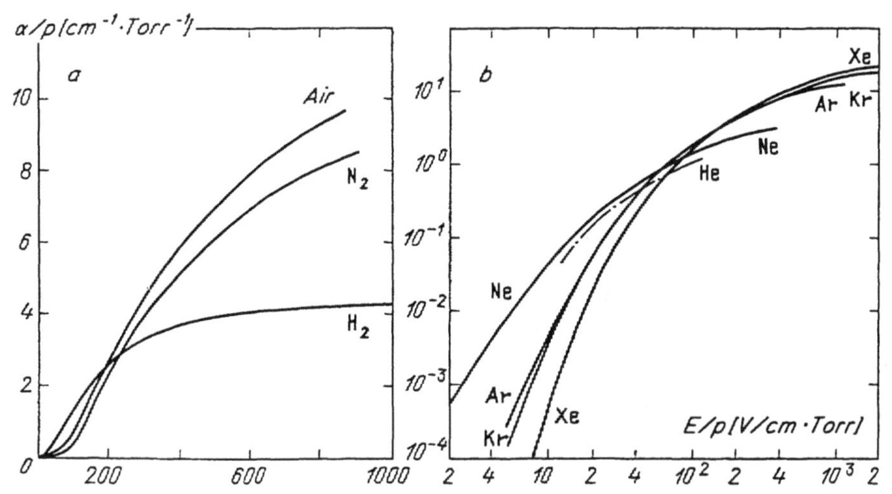

Fig. 4.3. Ionization coefficients for a wide range of E/p values (**a**) in molecular gases, (**b**) in inert gases. From [4.3]

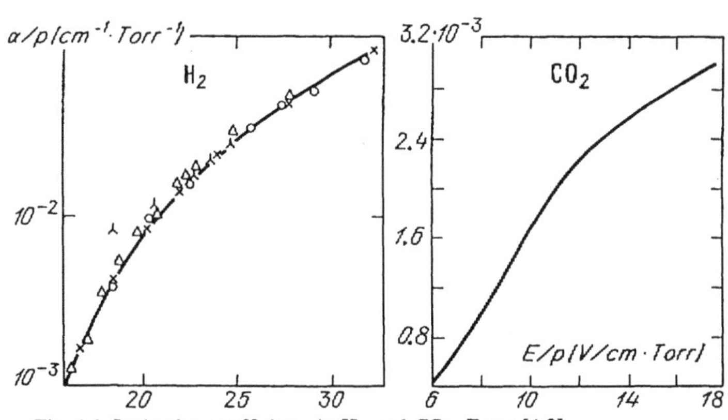

Fig. 4.4. Ionization coefficients in H_2 and CO_2. From [4.2]

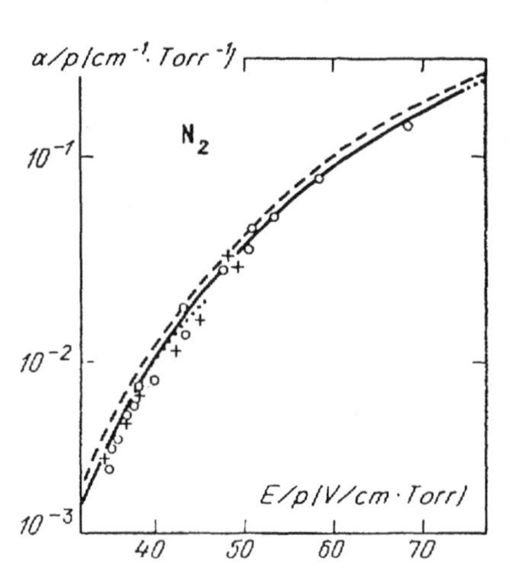

Fig. 4.5. Ionization coefficient in N_2 from the data of a number of authors. From [4.2]

4.1.5 Interpolation Formula for α

The theoretical and numerical analysis of discharges widely uses a conventional empirical formula suggested by Townsend:

$$\alpha = Ap\exp(-Bp/E) . \qquad (4.5)$$

The constants A and B are determined by approximating the experimental curves (Table 4.1). In a number of cases the relation (4.5) can be attributed a certain physical meaning. Assume, for example, that an electron undergoes only ionizing collisions. (This assumption may be realistic at high E/p and moderate energies.) The energy picked up by an electron along a free path length x is slightly greater than the ionization potential I. The probability that it will move the distance $x = I/eE$ without collisions and then be involved in an ionizing collision in a distance dx is $\alpha\, dx = dx\, l^{-1}\exp(-I/eEl)$, where $l = l_1/p$ is the mean-free-path length. This gives us (4.5) with $A = l_1^{-1}$, $B = I/el_1$. If $\sigma = 5 \cdot 10^{-16}\,\mathrm{cm}^2$, then $l_1 = 0.06\,\mathrm{cm}\cdot\mathrm{Torr}$; if $I = 15\,\mathrm{eV}$, then $A = 17$, $B = 250$, which is quite close to tabulated values.

The fraction of electrons in a Maxwellian spectrum that are capable of ionizing an atom is proportional to $\exp(-I/kT_e)$. If $T_e \propto E/p$ (see Sect. 2.3.5), we again arrive at a dependence of ν_i and α on E of type (4.5), but now the constant B has a different meaning. It will be shown in Sect. 7.4.7 that an approximate solution of the kinetic equation that takes into account the large role of inelastic losses of electron energy on excitation also leads to a relation of type (4.5), but again with a changed meaning of B. For inert gases, the formula

$$\alpha = Cp\exp\left[-D(p/E)^{1/2}\right] \qquad (4.6)$$

Table 4.1. Constants in the formulas for the ionization coefficient, and regions of applicability [4.4, 5]

Gas	A	B	E/p	C	D	$E/p <$
	$\mathrm{cm^{-1}Torr^{-1}}$	$\mathrm{V/(cm\cdot Torr)}$	$\mathrm{V/(cm\cdot Torr)}$	$\mathrm{cm^{-1}Torr^{-1}}$	$\mathrm{V/(cm\cdot Torr)^{1/2}}$	$\mathrm{V/(cm\cdot Torr)}$
He	3	34	20–150	4.4	14	100
Ne	4	100	100–400	8.2	17	250
Ar	12	180	100–600	29.2	26.6	700
Kr	17	240	100–1000	35.7	28.2	900
Xe	26	350	200–800	65.3	36.1	1200
Hg	20	370	150-600			
H_2	5	130	150-600			
N_2	12	342	100–600			
N_2	8.8	275	27–200			
Air	15	365	100–800			
CO_2	20	466	500–1000			
H_2O	13	290	150–1000			

[4.4] is sometimes used since it gives a better fit to experimental data then (4.5), even though it is less convenient for the theory (Table 4.1). Note also a useful empirical formula for air at relatively high E/p (see also Table 12.1):

$$\begin{aligned} \alpha/p &= 1.17 \cdot 10^{-4}(E/p - 32.2)^2\,\mathrm{cm}^{-1}\,\mathrm{Torr}^{-1}\,, \\ E/p &\approx 44 - 176\,\mathrm{V/(cm \cdot Torr)}\,. \end{aligned} \qquad (4.7)$$

4.1.6 Optimal Conditions for Ionization

An electron passing through a potential difference of 1 V generates α/E electrons (pairs of ions). In order to create one pair, it must be accelerated by the field to an energy $W = eE/\alpha$. The function $W(E/p)$ has a minimum, which, when using approximation (4.5), is given by $W_{\min} = \bar{e}eB/A$ at $(E/p)_{\mathrm{m}} = B$, where $\bar{e} = 2.718\ldots$. Even under these most favorable conditions for proliferation, the creation of one pair of ions consumes the energy $W_{\min}$ (*Stoletov's constant*), which is several times the ionization potential. Electrons have to devote much energy to the excitation of atoms: In air, $W_{\min} = 66\,\mathrm{eV}$/pair of ions for $(E/p)_{\mathrm{m}} = 365\,\mathrm{V/(cm \cdot Torr)}$.[1]

Note that (4.5) implies that as $E/p \to \infty$, $\alpha \to \mathrm{const} = Ap$. Indeed, at $E/p > 2000 - 3000$, α decreases as E/p increases, because the ionization cross section at high energies falls off with increasing ε (Fig. 2.1). However, such high values of E/p are not typical for discharges.

4.1.7 Stepwise Ionization

The atoms of a weakly ionized gas are mostly ionized from the ground state. Many excited atoms and molecules may be formed if the gas is highly ionized, and stepwise ionization may be predominant. Atoms are first excited by electron impact and then ionized by subsequent collisions. Long-lived metastable excited particles play an important role in this process (Table 4.2); their ionization cross sections are rather high (Fig. 4.6).

4.2 Other Ionization Mechanisms

4.2.1 Photoionization

This mechanism cannot compete with electron impact ionization under discharge conditions. Sometimes, however, it supplies seed electrons that start electron avalanches, as they do in streamer propagation (Chap. 12). Photoionization cross sections close to the threshold are rather high (see Table 4.3) but as a rule, a gas has few quanta of $\hbar\omega > I$ capable of photoionization.

[1] This figure must not be confused with another frequently encountered quantity: 33 eV/ion pair. This the energy dissipated by a fast electron with energy $\varepsilon > 4\,\mathrm{keV}$ when it is being stopped in air. Fast electrons have smaller energy losses.

Table 4.2. Energies of the lower resonance and metastable levels, lifetimes of metastable states, and excitation cross sections

Atom, molecule	Excitation energy, E^* eV (metastable levels: *)	Lifetime, s	Interpolation of the total excitation cross section at the threshold, $\sigma^* = C^*(\varepsilon - E^*)$	
			C^*, 10^{-18} cm^2/eV	E^*_{eff}, eV
H(2s)	10.20*	0.142	25	10
H(2p)	10.20			
He(2^3S_1)	19.82*	$6 \cdot 10^5$		
He(2^1S_0)	20.6*	$2 \cdot 10^{-2}$	4.6	20
He	21.21			
Ne	16.62* 16.7* 16.85		1.5	16
Ar$(4^3P_2^0)$	11.55* 11.61 11.72*	> 1.3 > 1.3	7	11.5
H_2	8.7* 11.5		7.6	8.7
N_2 $(A^3\Sigma_u^+)$	6.2*	1.3–2.6		
N_2 $(a^1\Sigma_u^-)$	8.4*	0.5		
O_2 $(^1\Delta_g)$	0.98*	$2.7 \cdot 10^3$		
O_2 $(b^1\Sigma_g^+)$	1.64*	12		
Hg(6^3P_0) (6^3P_l)	4.65* 4.87 5.4* 6.7		$\sigma^*_{\text{max}} = 1.7 \cdot 10^{-16}$ cm^2 for $\varepsilon = 6.5$ eV	

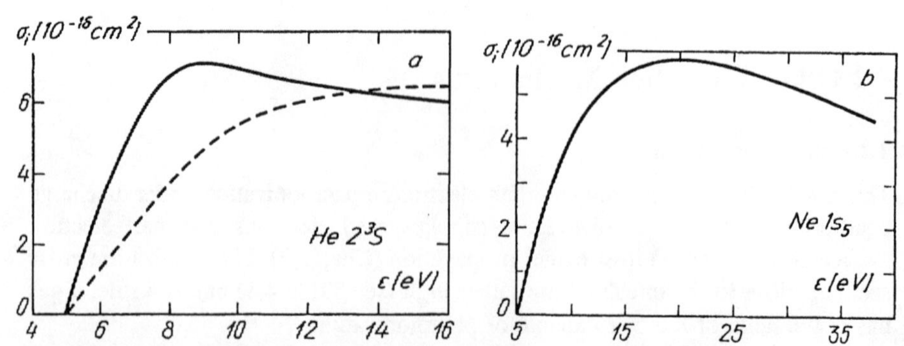

Fig. 4.6. Cross sections of ionization of excited metastables by electron impact: (**a**) He 2^3S; experimental data [4.6] – *solid curve*; theory [4.7] – *dashed curve*; (**b**) Ne $1s_5$ – theory. From [4.7]

Table 4.3. Cross sections of photoionization of atoms and molecules from the ground state close to the threshold

Gas	$\hbar\omega = I$, eV	λ, Å	σ_ν, 10^{-18} cm^2
H	13.6	912	6.3
He	24.6	504	7.4
Ne	21.6	575	4.0
Ar	15.8	787	35
Na	5.14	2412	0.12
K	4.34	2860	0.012
Cs	3.89	3185	0.22
N	14.6	852	9
O	13.6	910	2.6
O_2	12.2	1020	~ 1
N_2	15.58	798	26
H_2	15.4	805	7

4.2.2 Ionization by Excited Atoms

Even the high kinetic energy of slow heavy particles is not effective in ionization processes. Ionization requires the velocities of atoms and molecules to be comparable to the electron velocity in atoms, 10^8 cm/s, which corresponds to energies of 10 to 100 keV, not realizable in discharge conditions. On the other hand, the atomic excitation energy E^* is easily spent on liberating an electron from another atom, provided, of course, that it exceeds the ionization potential I. Resonance-excited atoms are especially effective in this respect. Thus the ionization cross sections of Ar, Kr, Xe, N_2, and O_2 in impacts by He(2^1P) atoms with $E^* = 21.2$ eV is $\sigma \approx 2 \cdot 10^{-14}$ cm^2, which is much greater than the gas-kinetic value [4.8]. Cross sections for ionization by metastable atoms, also with $E^* > I$ (*Penning effect*), are smaller but metastable atoms are much more numerous than short-lived resonance-excited atoms. Cross sections for ionization of Ar, Xe, N_2, CO_2 by metastable He(2^3S) atoms with $E^* = 19.8$ eV reach 10^{-15} cm^2, and that of Hg is exceptionally large: $1.4 \cdot 10^{-14}$ cm^2 [4.8].

4.2.3 Associative Ionization

This process of type $A + A^* \rightarrow A_2^+ + \mathrm{e}$, discovered by Hornbeck and Molnar in 1951, is sometimes important in inert gases. The separation of an electron is facilitated by the release of a small binding energy of order 1 eV in the association of an ion and an atom into a molecular ion. A reaction in helium involves atoms excited to states with the principal quantum number $n = 3$; their electron binding energies are from 1.52 to 1.62 eV. The binding energy of He_2^+ is somewhat higher, 2.23 eV, so that the electron can be ejected. At $T = 400$ K, the reaction cross sections are $2 \cdot 10^{-16} - 2 \cdot 10^{-15}$ cm^2. The *associative ionization* in mercury vapor involves two excited atoms,

$$\mathrm{Hg}\left(6^3P_1,\ E^* = 4.9\,\mathrm{eV}\right) + \mathrm{Hg}\left(6^3P_0,\ E^* = 4.7\,\mathrm{eV}\right) \rightarrow \mathrm{Hg}_2^+ + \mathrm{e}\ ,$$

the first atom being in a resonance and the second, in a metastable state. The total energy is 9.6 eV, less than that required to ionize an Hg atom (I_{Hg} = 10.4 eV); together with the binding energy of an Hg_2^+ molecular ion, however (0.15 eV), it is sufficient to ionize the molecule (I_{Hg_2} = 9.7 eV).

4.3 Bulk Recombination

4.3.1 Decay of Plasma

In the absence of an electric field, the charge densities $n_e = n_+$ in a plasma without electronegative components decay with time according to the law

$$\left(\frac{dn_e}{dt}\right)_r = -\beta n_e n_+ \,, \quad n_e = \frac{n_e^0}{1+\beta n_e^0 t} \underset{t\to\infty}{\longrightarrow} \frac{1}{\beta t} \,. \tag{4.8}$$

For example, if the *electron-ion recombination coefficient* $\beta = 10^{-7}\,\mathrm{cm^3/s}$ and the initial plasma density $n_e^0 = 10^{10}\,\mathrm{cm^{-3}}$, then the characteristic decay time $\tau_r = (\beta n_e^0)^{-1} = 10^{-3}$ s. The recombination coefficient can be determined experimentally, by measuring $n_e(t)$ and plotting n_e^{-1} as a function of t. The slope of the straight line gives β.

4.3.2 Dissociative Recombination

This mechanism follows the scheme $A_2^+ + e \to A + A^*$. This is the fastest mechanism of bulk recombination in weakly ionized plasma, for example, in a glow discharge. In this case the gas is cold and the plasma usually includes molecular ions. The released energy is mostly transformed into the excitation energy of the atom. The dissociative recombination coefficients $\beta_{dis} \sim 10^{-7}\,\mathrm{cm^3/s}$; at temperatures from room to several kK, β_{dis} decreases with increasing T_e as $T_e^{-1/2}$, and as $T_e^{-3/2}$ at still higher temperatures (Fig. 4.7). This is the way recombination proceeds, even in weakly ionized inert gases. Molecular ions are formed from atomic ones, generated in the course of the *conversion reaction* $A^+ + A + A \to A_2^+ + A$. The rate of conversion, $(dN_{A_2^+}/dt)_{conv} = k_{conv} N_{A^+} \cdot N_A^2$, is far from small (Table 4.4). Thus the atomic ion lifetime at p = 10 Torr with respect to the conversion is $\tau_{conv} = (k_{conv} N_A^2)^{-1} \sim 10^{-4}$ s, and at 100 Torr it is 10^{-6} s. If $n_e = 10^{10}\,\mathrm{cm^{-3}}$ and $\beta_{dis} = 10^{-7}\,\mathrm{cm^3/s}$, then $\tau_{conv} \ll \tau_{dis,r} = 10^{-3}$ s. Conversion replenishes the amount of molecular ions almost instantaneously, without impeding the dissociative recombination. In helium, β_{dis} is less by a factor of 10 to 100 than in other gases. Conversion can also produce *complicated ion complexes* O_4^+, N_4^+, and some others that have large coefficients up to $10^{-6}\,\mathrm{cm^3/s}$.

4.3.3 Radiative Recombination

Cross sections of the process $A^+ + e \to A + h\nu$ are very small: $\sigma_c \sim 10^{-21}\,\mathrm{cm^2}$. The recombination coefficient is correspondingly small [4.9]

$$\beta_{rr} = \langle v\sigma_c \rangle \approx 2.7 \cdot 10^{-13} \{T_e[\mathrm{eV}]\}^{-3/4}\,\mathrm{cm^3/s} \sim 10^{-12}\,\mathrm{cm^3/s} \,. \tag{4.9}$$

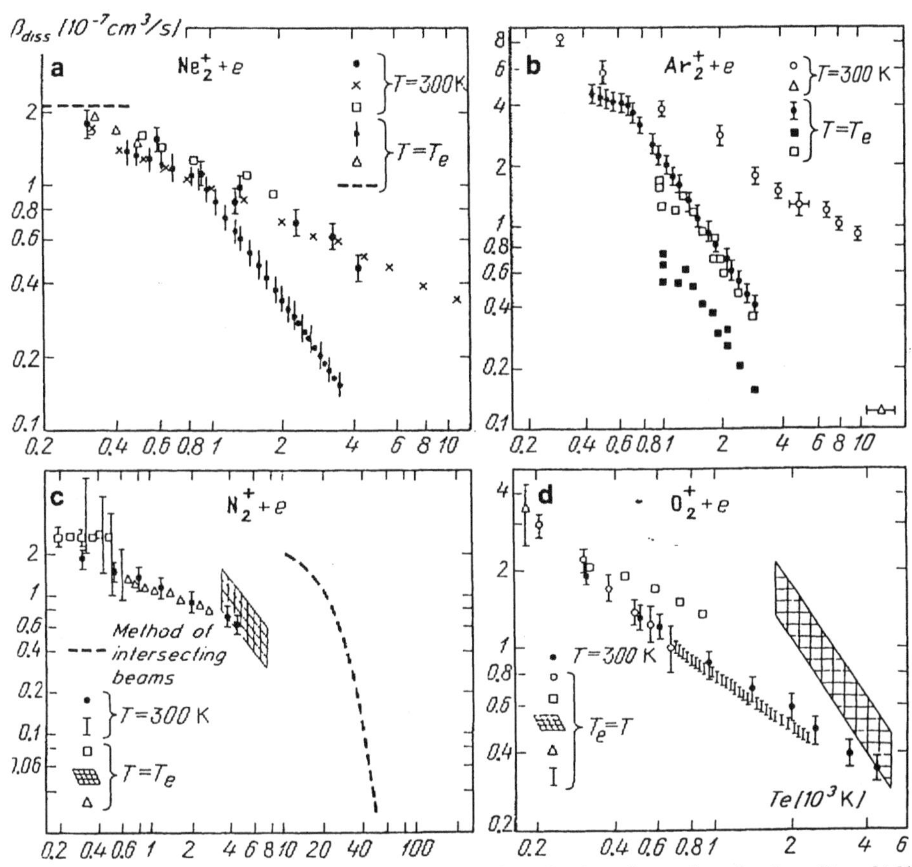

Fig. 4.7a–d. Dissociative recombination coefficients from the data of a number of authors. From [4.8]

Table 4.4. Measured reaction rate constants for the conversion of A^+ into A_2^+ in triple collisions with atoms A at $T = 300$ K [4.8]

Gas A	k_{conv}, 10^{-31} cm^6/s
He	0.63–1.15
Ne	0.42–0.79
Ar	1.46–3.9
Kr	1.9 –2.7
Xe	3.6
Hg	1 ($T = 700$ K)
Cs	150

Electrons are more often captured by the ground state of an atom, emitting a quantum $\hbar\omega \approx 10$ eV in the VUV range, $\lambda \lesssim 1000$ Å. However, capture by excited states, with subsequent photon emission in the visible at $\lambda \approx 4000 - 7000$ Å is also possible. Radiative recombination in gas discharge plasma may happen to be important, not as a channel for electron removal but as a mechanism for light emission.

4.3.4 Radiative Recombination in Three-Body Collisions

This process follows the scheme $A^+ + e + e \to A + e$; it is the main process in high-density low-temperature equilibrium plasma where $T \approx T_e \sim 10^4$ and the concentration of molecular ions is too low for dissociative recombination to be significant. In three-body collisions, electrons are captured by ions to form very high by excited atoms with a binding energy of order kT. An excited atom is then gradually deactivated by subsequent electron impacts, it "cascades" down the level staircase, and finally falls to the ground state from the lower excited state by radiative transition. This completes the process of recombination; its coefficient is [4.9]

$$\begin{aligned}\beta_{\mathrm{crr}} &= 8.75 \cdot 10^{-27} \{T[\mathrm{eV}]\}^{-9/2} n_e \\ &= 5.2 \cdot 10^{-23} \{T[\mathrm{kK}]\}^{-9/2} n_e \,\mathrm{cm}^3/\mathrm{s}\,. \end{aligned} \tag{4.10}$$

According to (4.9, 10), β_{crr} exceeds the radiative recombination coefficient if

$$n_e > 3.1 \cdot 10^{13} \{T[\mathrm{eV}]\}^{3.75} = 3.2 \cdot 10^{9} \{T[\mathrm{kK}]\}^{3.75} \,\mathrm{cm}^{-3}\,. \tag{4.11}$$

The recombination rate constant of triple collisions involving an atom as a third particle, β/N, is less than β_{crr}/n_e of (4.10) by a factor of $10^7 - 10^8$. This process is not typical for discharge conditions and can manifest itself only at very weak ionization and high pressures.

4.3.5 Ion-Ion Recombination

This is the main mechanism of charge neutralization in gases where electron attachment is important. In this process, $(dn_-/dt)_r = (dn_+/dt)_r = -\beta_i n_- n_+$. If $n_e \ll n_-$, then $n_- \approx n_+ \approx n_+^0/(1 + \beta_i n_+^0 t)$. Recombination at low pressure takes place through binary collisions of the type $A^- + B^+ \to A + B^*$ (Table 4.5). The process is similar to charge transfer. The energy thus released goes to excite the former ion B, the excitation later being released in collisions.

Table 4.5. Coefficients of binary ion-ion recombination at room temperature [4.8]

Ions	β_i, $10^{-7}\,\mathrm{cm}^3/\mathrm{s}$
$H^+ + H^-$	3.9
$O^+ + O^-$	2.7
$N^+ + O^-$	2.6
$O_2^+ + O_2^-$	4.2
$N_2^+ + O_2^-$	1.6
$O^+ + O_2^-$	2.0
$O_2^+ + O^-$	1.0
$NO^+ + O^-$	4.9
$NO^+ + NO_2^-$	5.1–1.8
$SF_5^+ + SF_6^-$	0.39

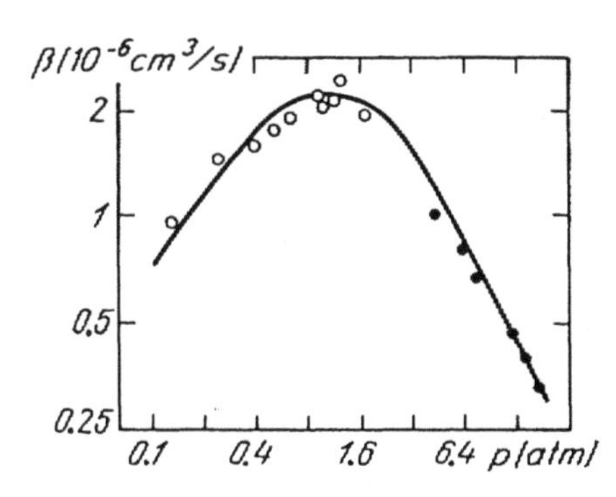

Fig. 4.8. Coefficient of ion-ion recombination in air. From [4.8]

At moderate pressures, recombination proceeds through triple collisions of the type $A^- + B^+ + C \to A + B + C$ (Thomson's theory, developed in 1924). The recombination coefficient is $\beta_i = k_{ii} N_c \propto p$. The recombination rate constant for the ions O_2^- and O_4^+ in oxygen is $k_{ii} \approx 1.55 \cdot 10^{-25}\,\mathrm{cm^6/s}$ at $p \sim 100\,\mathrm{Torr}$. For the ions NO^+ and NO_2^-, $k_{ii} \approx 3.4 \cdot 10^{-26}$ in oxygen and $1.0 \cdot 10^{-25}$ in N_2 (both at $T = 300\,\mathrm{K}$). Frequent collisions of ions with molecules at high pressures impede an ion from approaching another ion with opposite charge required for mutual neutralization. The $\beta_i \propto p$ law is replaced with the $\beta_i \propto p^{-1}$ law (Langevin's theory, developed in 1903). The maximum $\beta_{i,\max} \sim 10^{-6}\,\mathrm{cm^3/s}$ is reached at $p \sim 1\,\mathrm{atm}$ (Fig. 4.8).

4.4 Formation and Decay of Negative Ions

4.4.1 Attachment

Atoms and molecules, such as O, H, O_2, H_2O, Hg, Cs, halogens, Cl, Cl_2, or halogen-containing compounds CCl_4, SF_6, have an electrons affinity of 0.5–3 eV. *Attachment* is an important, sometimes the main, mechanism of removing electrons in *electronegative* gases and gases with electronegative additives. Attachment impedes break-down and makes it difficult to sustain the ionized state and the current. Sometimes this may be useful: to improve the insulating properties of the gas, or to speed up the removal of electrons in counters of nuclear particles.

The process in cold air in the absence of an electric field is $e + O_2 + M \to O_2^- + M$ ($M = O_2, N_2, H_2O$), with reaction rate constants $k_M = k_{O_2} = 2.5 \cdot 10^{-30}\,\mathrm{cm^6/s}$, $k_{N_2} = 0.16 \cdot 10^{-30}$, $k_{H_2O} = 14 \cdot 10^{-30}$ at $T = T_e = 300\,\mathrm{K}$. The electron density decreases according to the law $(dn_e/dt)_a = -\nu_a n_e$, $n_e = n_e^0 \exp(-\nu_a t)$. The attachment frequency of electrons in dry air at $p = 1\,\mathrm{atm}$ is $\nu_a = k_{O_2} N_{O_2}^2 + k_{N_2} N_{N_2} N_{O_2} = 0.9 \cdot 10^8\,\mathrm{s^{-1}}$. The electron lifetime with respect to attachment is $\tau_a = \nu_a^{-1} = 1.1 \cdot 10^{-8}\,\mathrm{s}$.

When an electron attaches itself to a complex molecule, the binding energy is immediately distributed over its vibrational degrees of freedom. As a result, each electron capture in a binary collision forms a stable negative ion.

For an electron energy $\varepsilon = 0.05\,\mathrm{eV}$, most active molecules of CCl_4 and SF_6 have $\sigma_{capt} \approx 1.2{\cdot}10^{-14}\,\mathrm{cm}^2$, and $\sigma_{capt} \propto 1/\varepsilon$. The attachment rate constant corresponding to this is $k_a = \nu_a/N \approx 1.6{\cdot}10^{-7}\,\mathrm{cm}^3/\mathrm{s}$. The dissociation potentials of halogen molecules are very low (1.5–2.5 eV). The electron affinity energy is sufficient for the dissociative attachment $\mathrm{e} + A_2 \rightarrow A + A^-$. In iodine at 300 K, the attachment cross section is $\sigma_a \approx 3.2 \cdot 10^{-15}\,\mathrm{cm}^2$ and $k_a \approx \bar{v}_e\sigma_a \approx 3.4 \cdot 10^{-8}\,\mathrm{cm}^3/\mathrm{s}$. In triple collisions, in which $\nu_a \sim p^2$, attachment can exceed the dissociative recombination with $\nu_a \sim p$ only at $p \gtrsim 100\,\mathrm{atm}$.

In contrast, the molecules O_2, CO_2, H_2O are strongly bound; a fairly high energy is then required for the dissociative attachment of an electron:

$$\mathrm{e} + O_2 + 3.6\,\mathrm{eV} \rightarrow O + O^- ,$$

$$\mathrm{e} + CO_2 + 3.85\,\mathrm{eV} \rightarrow CO + O^- ,$$

$$\mathrm{e} + H_2O + 4.25\,\mathrm{eV} \rightarrow OH + H^- ,$$

$$\mathrm{e} + H_2O + 3.6\,\mathrm{eV} \rightarrow H_2 + O^- ,$$

$$\mathrm{e} + H_2O + 3.2\,\mathrm{eV} \rightarrow H + OH^- .$$

However, an electric field produces enough energetic electrons in discharges, so that such processes are usually faster than attachment in triple collisions (Figs. 4.9–11). Attachment in triple collisions involving a second electron and radiative attachment ($\sigma_{ra} \sim 10^{-21} - 10^{-23}\,\mathrm{cm}^2$) both play insignificant roles in laboratory plasmas.

4.4.2 Attachment Coefficient

Like ionization, the attachment of electrons in dc fields occurs in the course of drift. The attachment coefficient $a = \nu_a/v_d$, similar to α, gives the number of attachment events per 1 cm of path along the field. Dissociative attachment mostly occurs in not too weak fields and obeys the same similarity law, $a = pf(E/p)$. In the case of triple collisions, dominating in very weak fields, $a =$

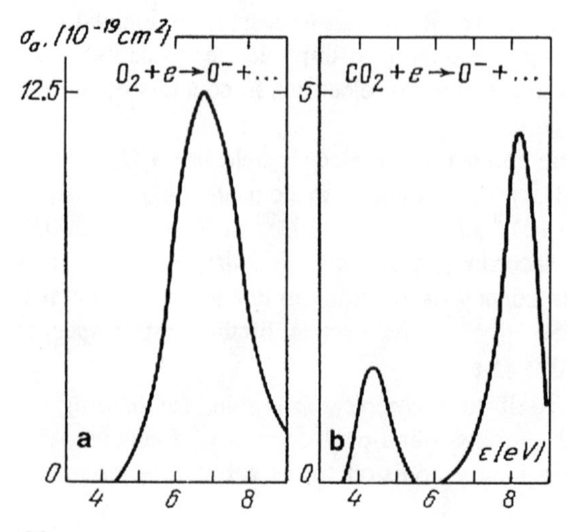

Fig. 4.9. Dissociative attachment cross section of electrons in **a)** O_2 and **b)** CO_2. From [4.10]

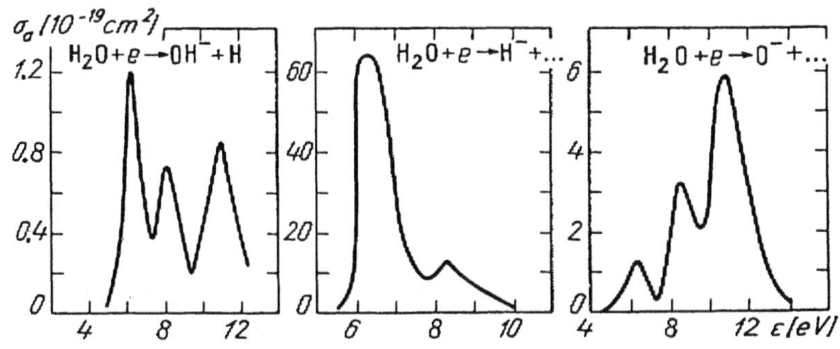

Fig. 4.10. Dissociative attachment cross sections of electrons in H_2O for three possible channels. From [4.11]

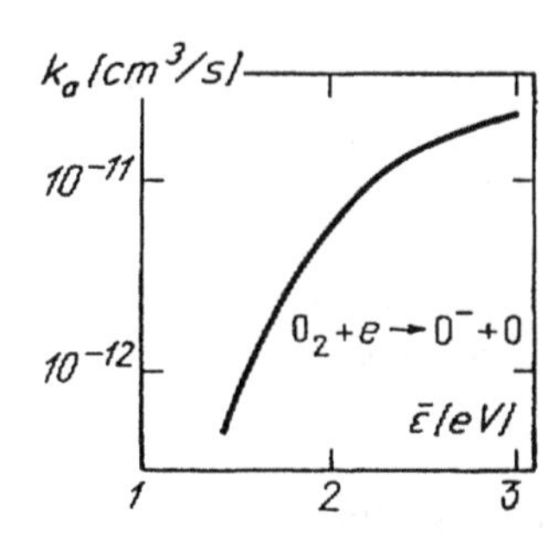

Fig. 4.11. Dissociative attachment rate constant of O_2 as a function of mean electron energy. From [4.12]

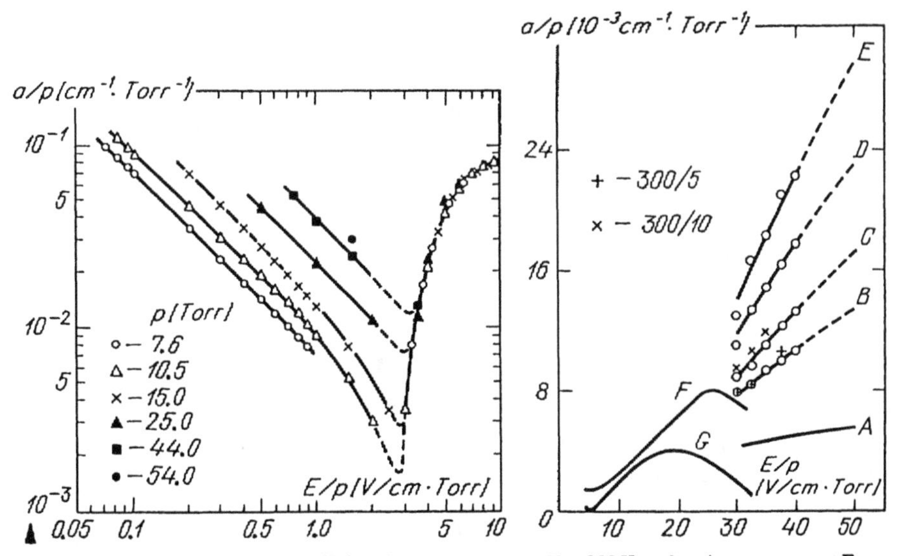

Fig. 4.12. Electron attachment coefficient in pure oxygen at T = 300 K and various pressures. From [4.12]

Fig. 4.13. Electron attachment coefficient in moist air, for various air humidity values: A: dry air; B: total pressure 150 Torr, water vapor pressure 2.5 Torr (150/2.5); C (150/5); D (150/9); E (150/15); F and G – air with negligible amount of water vapor. From [4.13, 14]

$p^2 f_1(E/p)$ (Figs. 4.12, 13). A change of mechanism is apparent in Fig. 4.12. The multiplication of electrons in an avalanche is determined by the effective coefficient $\alpha_{\text{eff}} = \alpha - a$. If $\alpha < a$ (this happens at E/p less than a certain value for a given gas, see Sect. 7.2.5), multiplication becomes impossible.

4.4.3 Detachment

Experiments show that steady-state weakly ionized plasma is sustainable in electronegative gases at much lower values of E/p than short pulsed discharges require. This observation indicates that a discharge sustained for a long time accumulates active particles (in all likelihood, excited molecules) that release electrons upon collisions with negative ions. The detachment frequency ν_d and rate constant k_d are determined by the equations $(dn_e/dt)_d = -(dn_-/dt)_d = \nu_d n_- = k_d N n_-$. The constant per active molecule is $k_{d,act} \sim 10^{-10}\,\text{cm}^3/\text{s}$ (Table 4.6). Metastable molecules $N_2(A^3\Sigma_u^+)$, and also $O_2(b^1\Sigma_d^+)$ in air, are presumably efficient in air and laser mixtures of CO_2, N_2, and He (Table 4.2). The constants $k_{d,act}$ are unknown for them but are assumed to be of the same order of magnitude. Indirect estimates (Sect. 8.8.4) show that discharges are characterized by $k_d \sim 10^{-14}\,\text{cm}^3/\text{s}$ per any molecule. If $k_{d,act} \sim 10^{-10}$, the concentrations of active particles are about 10^{-4}.

The O^- ions formed in laser mixtures and in air in the reactions

$$O^- + CO_2 + M \to CO_3^- + M\,, \quad k = 1.1 \cdot 10^{-27}\,\text{cm}^6/\text{s} \quad \text{for} \quad M = CO_2\,,$$
$$O^- + O_2 + M \to O_3^- + M\,, \quad k = 1.05 \cdot 10^{-30}\,\text{cm}^6/\text{s} \quad \text{for} \quad M = O_2\,,$$

Table 4.6. Rate constants for the decay of negative ions at room temperature [4.8]

Reaction	Release of energy, eV	Rate constant k_d, 10^{-10} cm^3/s
$O^- + O \to O_2 + e$	3.6	2
$O^- + N \to NO + e$	5.1	2
$O^- + NO \to NO_2 + e$	1.4	1.6
$O^- + CO \to CO_2 + e$	4	4
$O^- + CO_2 \to CO_3 + e$	< 0	10^{-3}
$O^- + O_2(^1\Delta_g) \to O_3 + e$ [a]	0.5	3
$O_2^- + O_2 \to O_2 + O_2 + e$	-0.43	$2.2 \cdot 10^{-8}$; $3 \cdot 10^{-4}$ ($T = 600$ K)
$O_2^- + N_2 \to O_2 + N_2 + e$	-0.43	$1.8 \cdot 10^{-6}$ ($T = 600$ K)
$O_2^- + N \to NO_2 + e$	4.1	5
$O_2^+ + O_2(^1\Delta_g) \to O_2 + O_2 + e$	0.6	2
$H^- + H \to H_2 + e$	3.8	13
$H^- + O_2 \to H_2O + e$	1.25	12
$OH^- + O \to HO_2 + e$	0.9	2
$OH^- + H \to H_2O + e$	3.2	10

[a] $O_2(^1\Delta_g)$; see Table 4.2.

are transformed into more stable complexes, namely, O_3^- and CO_3^- *clusters*. It was established that CO molecules are efficient in destroying O^- ions but very inefficient in destroying CO_3^- (this is important for laser discharges).

4.5 Diffusional Loss of Charges

Breakdown and low-pressure discharges are usually greatly affected by electron losses due to the electron diffusion toward walls. These losses are irreversible: electrons go into the metal or attach to dielectrics and there recombine with ions. Obviously, the mean electron lifetime with respect to the diffusional loss is $\tau_{dif} = \Lambda^2/D$, where Λ is a length of the order of the minimal size of the vessel and D is the diffusion coefficient (for free or ambipolar processes). The value of Λ can be elaborated by solving the stationary diffusion equation of type (2.44) taking into account ionization sources, $D\Delta n_e + \nu_i n_e = 0$ (without attachment). Assume the field to be homogeneous, that is, $\nu_i(\boldsymbol{r}) = \text{const}$, with $n_e = 0$ at the walls. This gives us an eigenvalue problem, solvable by the method of separation of variables. For example, in a cylinder of radius R and length L, we have $n_e \propto J_0(2.4r/R)\cos(\pi z/L)$; J_0 is the Bessel function and z has its origin at the center of the cylinder's axis. The solution exists only if $\nu_i = \nu_{dif} \equiv D/\Lambda^2$, where

$$\begin{aligned}
&\text{(cylinder)} && (1/\Lambda)^2 = (2.4/R)^2 + (\pi/L)^2 \\
&\text{(sphere)} && (1/\Lambda)^2 = (\pi/R)^2\,,\quad \Lambda = R/\pi \\
&\text{(parallelepiped)} && (1/\Lambda)^2 = (\pi/L_1)^2 + (\pi/L_2)^2 + (\pi/L_3)^2
\end{aligned} \tag{4.12}$$

($L_{1,2,3}$ are the lengths of the sides). Obviously, $\nu_{dif} = \tau_{dif}^{-1}$ is the mean frequency of diffusional removal of electrons; Λ is known as the *characteristic diffusion length*.

For example, the ambipolar diffusion coefficient in the positive column of a nitrogen glow discharge at $p = 10\,\text{Torr}$ is $D_a = 200\,\text{cm}^2/\text{s}$. If the discharge is sustained in a long tube of radius $R = 1\,\text{cm}$, then $\Lambda = R/2.4 = 0.42\,\text{cm}$. The diffusion frequency is $\nu_{dif} = 1.1 \cdot 10^3\,\text{s}^{-1}$. Charges diffuse to the wall in a mean time $\tau_{dif} = 0.9 \cdot 10^{-3}\,\text{s}$. The rate of diffusional losses $(dn_e/dt)_{dif} = -\nu_{dif} n_e$ can be evaluated for more complicated and nonsteady cases as well, using the formula $\nu_{dif} = D/\Lambda^2$ with Λ given by (4.12). Thus, if the source is distributed uniformly on the axis of a long cylinder, the diffusion time is only $(2.4/2)^2 = 1.44$ times longer than the average value above.

4.6 Electron Emission from Solids

4.6.1 Work Function

A dc current in a gas discharge is sustained by the emission of electrons from the surface of the cathode. To extract an electron from a metal, it is necessary to spend a certain amount of energy; its minimum value is called the *work function* (Table 4.7). It is a function of the state of the surface, its contamination and roughness; on single crystals, it varies from face to face within 1 eV. The binding of electrons to a metal, $e\varphi$, can be interpreted as the work $e^2/4a$ against the attractive image force $e^2/4r^2$; this work is done to remove an electron from a distance a of the order of one interatomic spacing to infinity. If there is an external field E, the force applied to an electron is $F = e^2/4r^2 - eE$. The electron breaks loose of the metal if it is pulled to a distance r_K at which $F = 0$ and reverses its sign. The work function is reduced in comparison with $e\varphi = e^2/4a$ by a quantity

$$e\Delta\varphi = e\varphi - \int_a^{r_K} F dr = e^{3/2}E^{1/2} = 3.8 \cdot 10^{-4} \{E[\mathrm{V/cm}]\}^{1/2} \mathrm{eV} . \quad (4.13)$$

This is the so-called *Schottky effect* established in 1914.

Table 4.7. Work function of polycrystalline materials and the constant of thermionic emission. (The values of φ recommended in the handbook [4.15] on the basis of an analysis of measurements reported by numerous authors.)

Element	φ, eV	A_1, $\mathrm{A/(cm^2 \cdot K^2)}$
C	4.7	30–170
Al	4.25	
Fe	4.31	60–700
Ni	4.5	30– 50
Cu	4.4	60–100
Mo	4.3	60–150
Ba	2.49	60
W	4.54	40–100
Pt	5.32	10–170

4.6.2 Thermionic Emission

This occurs when a metal is heated: some electrons acquire sufficient energy to escape from the potential well that the metal represents for them. In the absence of an external field, the escaping electrons accumulate near the surface, and the field of this space charge prevents other electrons from escaping from the metal. The space charge is easily removed by a weak accelerating field. Unimpeded emission corresponds to the *saturation current*

$$j_T = A_0 D T^2 \exp\left(-\frac{e\varphi}{kT}\right) , \quad A_0 = \frac{4\pi m e k^2}{h^3} = 120\,\mathrm{A/(cm^2\,K^2)} . \quad (4.14)$$

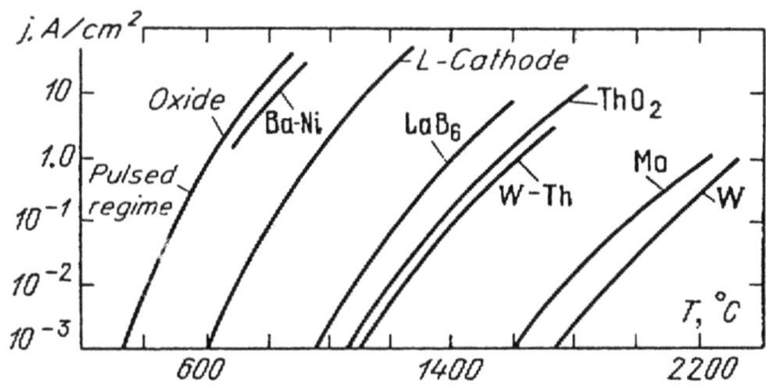

Fig. 4.14. Current density of thermionic emission as a function of cathode temperature for a number of materials. From [4.16]

The factor D in this Dushman-Richardson formula covers the quantum-mechanical effect of the reflection of electrons into the metal from the wall of the potential well, and $A_1 = A_0 D$ is in the range from 15 to 300 (Table 4.7, Fig. 4.14). Electrons leave the metal with a mean energy of $2kT$. The Schottky effect can greatly affect the ***thermionic current*** (Table 4.8). Thermionic emission is present in arc discharges.

Table 4.8. The currents of the thermionic (j_T), field electron (j_F), and thermionic field (j_{TF}) emission. (The following parameters were used in the computations: $T = 3000\,K$, $\varphi = 4\,V$, $A_1 = 80\,A/(cm^2 \cdot K^2)$, $\varepsilon_F = 7\,eV$.)

E, 10^7 V/cm	$\Delta\varphi$, V	j_T, A/cm²	j_F, A/cm²	j_{TF}, A/cm²
0	0	$1.3 \cdot 10^2$	0	0
0.8	1.07	$8.2 \cdot 10^3$	$2.0 \cdot 10^{-20}$	$1.2 \cdot 10^4$
1.7	1.56	$5.2 \cdot 10^4$	$2.2 \cdot 10^{-4}$	$1.0 \cdot 10^5$
2.3	1.81	$1.4 \cdot 10^5$	$1.3 \cdot 10^0$	$2.1 \cdot 10^5$
2.8	2.01	$3.0 \cdot 10^5$	$1.3 \cdot 10^2$	$0.8 \cdot 10^6$
3.3	2.18	$6.0 \cdot 10^5$	$4.7 \cdot 10^3$	$2.1 \cdot 10^6$

4.6.3 Field Electron Emission

The field that pulls the electrons away transforms the potential well into a potential barrier of finite width (Fig. 4.15); as a result, electrons can escape from the metal by tunneling. The result is ***field electron emission*** and the emission current is given by the Fowler-Nordheim formula. In numerical form,

$$j_F = 6.2 \cdot 10^{-6} \frac{(\varepsilon_F/\varphi)^{1/2} E^2}{\varepsilon_F + \varphi} \exp\left(\frac{-6.85 \cdot 10^7 \varphi^{3/2} \xi}{E}\right) \text{A/cm}^2 \,. \qquad (4.15)$$

Here ε_F[V] is the Fermi energy, φ[V]is the work function nonperturbed by the field, $\xi(\Delta\varphi/\varphi)$ is a correction factor for its reduction (Table 4.9), and E is measured in V/cm. In reality, appreciable current is obtained at $E \sim 10^6$ V/cm, which is less than implied by (4.15) and Table 4.8 by an order of magnitude. The

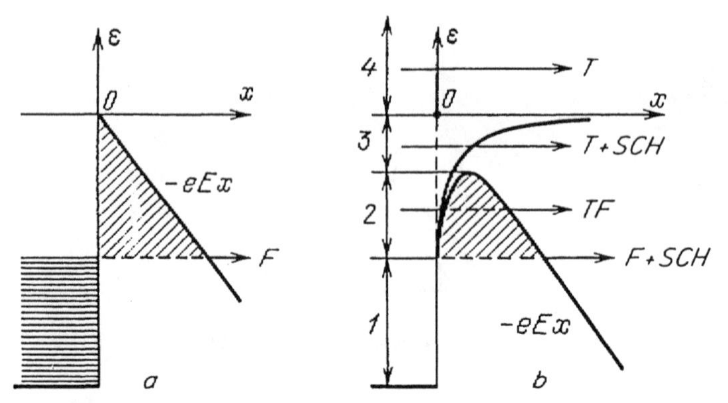

Fig. 4.15. Electron potential energy when an external field is applied to the metal. F – field emission, T – thermionic emission, T+SCH – Schottky-affected thermionic emission, TF – thermionic field emission, F+SCH – Schottky-affected field emission; (**a**) Mirror forces neglected; (**b**) mirror forces taken into account. The diagrams illustrate the nature of field-electron and thermionic-field emission, T+SCH – Schottky-affected thermionic emission, TF – thermionic field emission, F+SCH – Schottky-affected field emission

Table 4.9. Correction factor for the reduced work function in the Fowler-Nordheim formula

$\Delta\varphi/\varphi$	0	0.2	0.3	0.4	0.5	0.6	0.7	0.8	0.9	1
ξ	1	0.95	0.90	0.85	0.78	0.70	0.60	0.50	0.34	0

reason is a dramatic enhancement of the applied field at the microscopic protrusions that always exist on real metal surfaces (see Sect. 12.6.1). Field emission results in the breakdown of vacuum gaps.

4.6.4 Thermionic Field Emission

When a strong extracting field is applied to a heated metal surface, both factors (high temperature and field) affect the emission of electrons, in ways not restricted to the mechanisms discussed above. In Fig. 4.15, all allowed energy states of electrons in a metal are classified into four groups. At $T = 0$, electrons occupy states 1 with $\varepsilon \leq \varepsilon_F$. At $T > 0$, all four groups fill up, although the number of electrons falls off rapidly as the excess $\varepsilon - \varepsilon_F$ becomes greater. Electrons of group 1 undergo field emission as at $T = 0$. Electrons of group 4 would escape by thermionic emission even if the field were zero. Electrons of group 3 jump over the barrier, lowered thanks to the field, at the expense of thermal energy. As for the electrons of group 2, which exist only if the metal is heated, they face a narrower and lower barrier that they can cross by tunneling at a higher probability than group-1 electrons.

These are the electrons that generate the current of the *thermionic field emission* [4.17, 18] which is not expressible by simple formulas. The values of j_{TF} given in Table 4.8 are taken from the results of computer simulation [4.19] for the cathode spot of an arc discharge. At $T = 3000\,\mathrm{K}$ and $E > 0.8 \cdot 10^7\,\mathrm{V/cm}$, thermionic field emission is predominant, the more so at higher fields. If $E < 0.5 \cdot 10^7\,\mathrm{V/cm}$, mostly group-3 electrons are emitted; this is described by (4.14, 13).

4.6.5 Secondary Emission

This is caused by various particles: positive ions, excited atoms, electrons, and also photons. *Secondary emission* from a cold cathode produces breakdown of discharge gaps and also sustains small dc currents that are incapable of substantial heating of the cathode or of creating such a strong field at the cathode that thermionic field emission develops.

The most important among the various secondary mechanisms is the *ion-electron emission.* It is characterized by a coefficient γ_i: the number of electrons emitted per incident positive ion. The relatively small kinetic energies that ions acquire in discharges are ineffective for knocking out electrons, and the main mechanism, as established by *Penning* in 1928 is that the field of an ion approaching a surface to within a distance of atomic dimensions transforms a potential well on the surface into a potential barrier. The barrier is low and narrow because the field is tremendously strong, on the order of that around nuclei. An electron from the metal immediately tunnels into the ion and neutralizes it. If the energy released thereby, $I - e\varphi$, is greater than $e\varphi$, it may be spent on ejecting another (emission) electron. An empirical formula $\gamma_i \approx 0.016(I - 2e\varphi)\mathrm{eV}$ holds for clean surfaces (with an accuracy of about 50%). Thus $\gamma_i \approx 0.21$ for tungsten and He^+, 0.30 for Ne^+, 0.09 for Ar^+, and 0.02 for Xe^+; γ_i is almost independent of ε_i up to ion energies $\varepsilon_i \sim 1\,\mathrm{keV}$ [4.16]. For platinum and the ions H^+, H_2^+, we have $\gamma_i \approx 3 \cdot 10^{-3}$, for N^+, N_2^+ it is $5 \cdot 10^{-3}$, and for O^+, O_2^+ it is $5 \cdot 10^{-4}$ (for $\varepsilon_i \sim 0 - 10\,\mathrm{eV}$)[4.20].

Metastable atoms of inert gases are very efficient: $\gamma_m \approx 0.24$ for $He(2^3S)$ and Pt, 0.4 for $He(2^1S)$ and Pt, 0.4 for Ar^* and Cs. The difference $E^* - e\varphi$ goes to the released electron. In the case of Hg and Ni, $\gamma_m \sim 10^{-2}$. The photoeffect from the surface at $\hbar\omega > e\varphi$ is characterized by its *quantum yield,* that is, the number of electrons per photon γ_ν (Fig. 4.16). The yield in the visible and near-UV regions is $\sim 10^{-3}$, and in the far-UV region it is $\sim 10^{-2} - 10^{-1}$. In the first two regions, γ_ν is very sensitive to the state and contamination of the surface, and is considerably reduced by reflection. Photoemission often plays the crucial role in breakdown. *Secondary electron emission* is essential in the case of vacuum breakdown by high-frequency fields: the oscillating electron strikes the gap walls alternatingly, one after the other. The secondary emission coefficients γ_e for different metals with ε_e up to several keV vary from 0.4 to 1.6 [4.16]. Secondary electrons are also knocked out of dielectric surfaces. In glass and quartz, $\gamma_e \sim 1 - 3$ and the maxima are at $\varepsilon_e \approx 300$–$400\,\mathrm{eV}$. If $\varepsilon_e < 40$–$60\,\mathrm{eV}$, then $\gamma_e < 1$ [4.21]. The incident electrons attach to the dielectric, so that the surface is charged negatively if $\gamma_e < 1$ and positively if $\gamma_e > 1$.

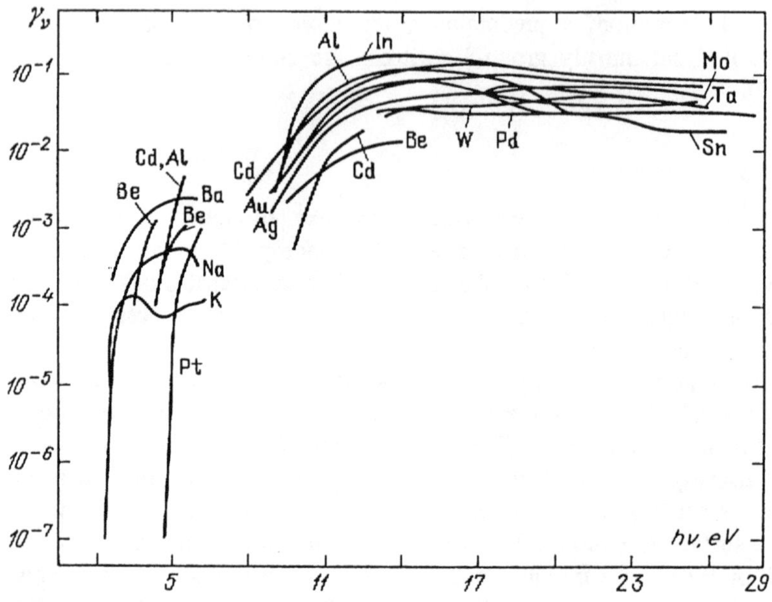

Fig. 4.16. Photoelectron emission coefficients (quantum yield) for various metals as functions of photon energy. From [4.2]

4.7 Multiplication of Charges in a Gas via Secondary Emission

4.7.1 Effect of Secondary Emission on the Enhancement of Primary Electron Current

Let us turn to the experiment for measuring the ionization coefficient α (Sect. 4.1.4). If the photocurrent from the cathode is $i_0 = e\mathcal{N}_0$, then the current recorded at the anode in the absence of secondary emission is $i = i_0 \exp(\alpha d)$. An electron leaving the cathode generates $\exp(\alpha d) - 1$ positive ions in the gap; all of them arrive at the cathode. The current at the cathode in the stationary state can be written $i = i_{\text{el}} + i_{\text{ion}} = i_0 + i_0[\exp(\alpha d) - 1]$. The dependence of $\ln i$ on d at p, E = const is linear. As E or d is increased, multiplication rapidly intensifies and the secondary electron emission from the cathode begins to affect the total current. Assume that this is the ion-electron emission. Each of the $\exp(\alpha d) - 1$ ions generated by a single electron leaving the cathode knocks out γ_{i} electrons from the cathode; this secondary electron current is added to the primary current i_0 of the external source. The electronic part of the cathode current i_1 is given by the equation $i_1 = i_0 + \gamma_{\text{i}} i_1[\exp(\alpha d) - 1]$. The current at the anode and in the external circuit is

$$i = i_1 \exp(\alpha d) = i_0 \exp(\alpha d)/\{1 - \gamma[\exp(\alpha d) - 1]\} \tag{4.16}$$

(we have dropped the subscript of γ).

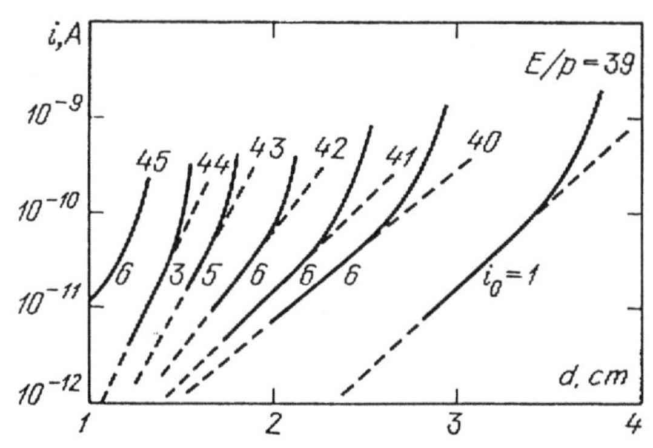

Fig. 4.17. Effect of secondary emission on current enhancement in a discharge gap of length d in air at $p = 200$ Torr. The *curves* are marked with the values of E/p, V/(cm·Torr), and cathode photocurrent i_0 in 10^{-15} A. From [4.22]

A formula of this type was first derived by *Townsend* in 1902 to explain the process of the ignition of a self-sustaining discharge. As a result of secondary emission, the region of linear growth of $\ln i$ with d turns steeply upward (Fig. 4.17). This process occurs when the denominator of (4.16), which is very close to unity at small enhancement coefficients αd, tends to zero as αd increases. When the denominator becomes zero, breakdown takes place and a self-sustaining discharge is formed: formally, $i = 0/0 \neq 0$ at $i_0 = 0$ (Sect. 7.2.2). Experimentally, this is achieved by raising the voltage between the electrodes. In order to follow a certain E = const curve in Fig. 4.17, one needs to increase V and d simultaneously and proportionally. By analyzing this plot using (4.16) and a known value of α (found from the slope of the linear segment of the curve), one can determine γ.[2]

4.7.2 Photoemission and Effective Coefficient γ

A cold cathode under discharge conditions may emit electrons in response to a large number of agents; it is not always possible to identify a specific agent. For this reason, one usually employs an effective secondary emission coefficient γ per ion. This coefficient characterizes the entire complex process and replaces the elementary-process characteristics γ_i, γ_ν, etc., that are found by bombarding a target with beams of particles or photons of a specific energy.

Let an electron excite along 1 cm of the field such a number of atoms that they later emit $\alpha_\nu \, \text{cm}^1$ photons capable of generating emission. The total number of

[2] Townsend was of the opinion that the secondary process that makes it possible for a non-self-sustaining discharge with exponential amplification of photocurrent i_0 to transform into self-sustaining discharge was the ionization of gas atoms by impact of positive ions. In addition to α, the theory had a second ionization coefficient β for ions. Numerous experiments later proved that ionization of a gas by ions is impossible in a discharge (for the reason given at the beginning of Sect. 4.2.2). The 'α, β'-theory was replaced with an 'α, γ'-theory.

such photons, which are in fact the progeny of one electron leaving the cathode, equals

$$\int_0^d \exp(\alpha x)\alpha_\nu \, dx = (\alpha_\nu/\alpha)[\exp(\alpha d) - 1] \, .$$

If emitted quanta are weakly absorbed by the gas (this happens if pressure is low) and ζ is the mean probability of their reaching the cathode, then the number of secondary electrons emitted from the cathode per primary electron is $(\gamma_\nu\zeta\alpha_\nu/\alpha)[\exp(\alpha d) - 1]$. Adding this quantity to the $\gamma_i[\exp(\alpha d) - 1]$ caused by positive ions, we arrive at the same formula, (4.16), with effective coefficient $\gamma = \gamma_i + \gamma_\nu\zeta\alpha_\nu/\alpha$. The contribution of metastable atoms can likewise be added to γ.

4.7.3 Results of γ Measurements. Positive Ions or Photons?

The results of determining the effective γ by the method outlined in Sect. 4.7.1 are summarized in [4.20, 23] and are illustrated in Fig. 4.18 and Table 4.10. They are not easy to interpret. The coefficient γ depends on E/p in an irregular manner and is very sensitive to the state of the cathode surface. In a number of cases, γ is found to be of the same order of magnitude as γ_i reported in experiments with

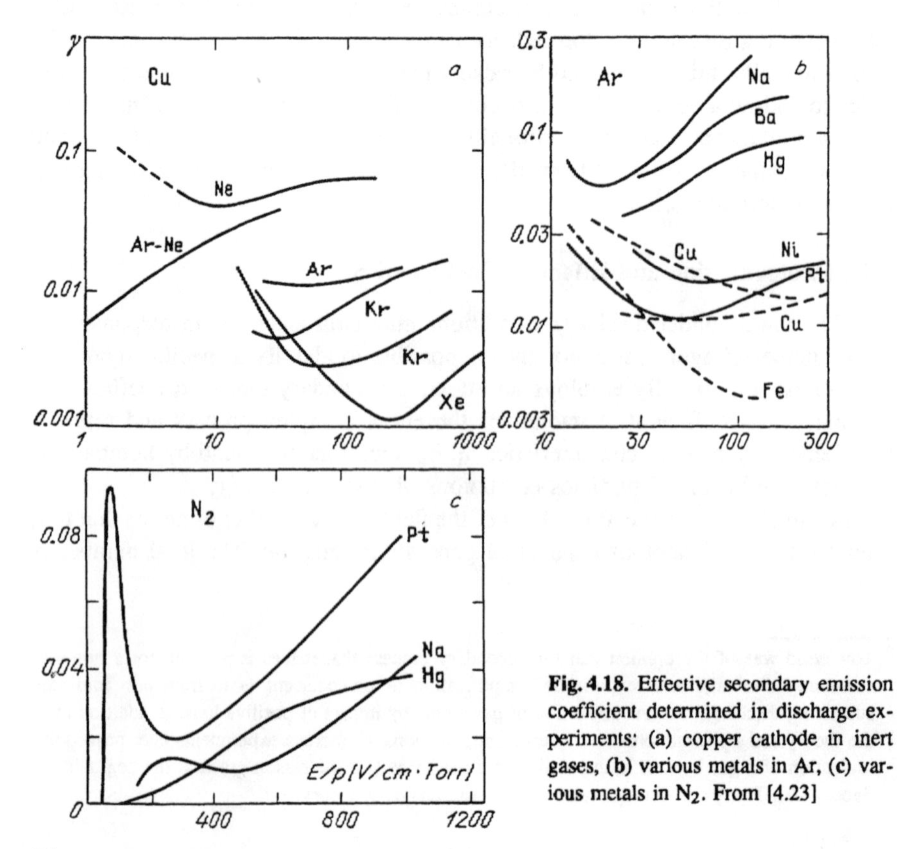

Fig. 4.18. Effective secondary emission coefficient determined in discharge experiments: (a) copper cathode in inert gases, (b) various metals in Ar, (c) various metals in N_2. From [4.23]

Table 4.10. Effective secondary emission coefficients at medium (~ 100 Torr) and high (~ 1 atm) pressures, see the references given in [4.21]

Gas	Cathode	State of surface	Conditions in the gas E/p V/(cm·Torr)	pd cm·Torr	γ	Mechanism
Air	Ni	cleaned	39–45		$8.0 \cdot 10^{-6} - 1.5 \cdot 10^{-4}$	
N_2	Ni	cleaned	39–45		$1.3 \cdot 10^{-4} - 3.7 \cdot 10^{-4}$	
N_2	Cu	cleaned		50	$1.5 \cdot 10^{-6}$	ions
N_2	Cu	oxidized		50	$> 10^{-3}$	ions
O_2	Ni		35.4		$4.5 \cdot 10^{-2}$	
O_2	Cu	cleaned		50	$\sim 10^{-7}$	
O_2	Cu	oxidized		50	10^{-6}	ions, photons
H_2	Ni	cleaned	20.3–25.1	$d = 2$	$1.0 \cdot 10^{-3} - 2.4 \cdot 10^{-3}$	
H_2	Cu	cleaned		50	10^{-6}	photons
H_2	Cu	oxidized		50	$5.0 \cdot 10^{-5}$	photons
organics: alcohol, methane, methylal					$\sim 10^{-9}$	

targets bombarded by ion beams. An indication of the emission mechanism may be obtained by studying the nonsteady-state process of generation of secondary, tertiary, etc., electrons (or avalanches, rather) after the primary ones have started, because the arrival times of ions and photons at the cathode are very different (Sect. 12.1.2).

The following preliminary conclusions can be drawn from the available data. Ion-electron emission with $\gamma \sim 10^{-1} - 10^{-3}$ seems to be predominant at $pd \sim 1 - 10$ cm Torr, where $E/p \gtrsim 100$–200 V/(cm Torr) are typical. This is the case for the breakdown of rarefied gases and for the cathode layer of glow discharges. In inert gases on a clean (annealed) cathode, $\gamma \approx \gamma_i$ at high pressures as well. On a contaminated surface, the γ_i are sharply reduced and photoemission is often predominant. Photoemission is dominant in most gases, except inert ones, at about atmospheric pressure and $E/p \approx 30$–40 V/(cm Torr), typical for the breakdown of dense gases. Secondary electrons may appear owing to photoprocesses in the gas itself (Sect. 12.1.3). The data on γ are incomplete and often contradictory. The uncertainty that usually mars the selection of γ for designing or analyzing an experiment is partly alleviated by the fact that γ is normally found in the formulas only within the logarithm (Sects. 7.2 and 8.3). As a rule, one assumes that $\gamma \sim 10^{-1} - 10^{-2}$.

5. Kinetic Equation for Electrons in a Weakly Ionized Gas Placed in an Electric Field

5.1 Description of Electron Processes in Terms of the Velocity Distribution Function

The behavior of electrons in an ionized gas placed in a constant or oscillating electric field was treated in Chaps. 2 and 3 in the framework of *elementary theory*. This concentrates all attention on a single electron, assuming that all electrons behave identically in the transition to macroscopic quantities. This approach allows an approximate calculation of a number of important characteristics of ionized gases: electric conductivity and dielectric permittivity, absorption coefficient for electromagnetic waves, and heating of electrons in the field. These results permit an analysis of various concrete processes: different types of discharge, propagation of radio and light waves in plasma, etc.; we will frequently resort to simple and illustrative notions of elementary theory. This approach is nevertheless quite imperfect, especially if one needs to analyze subtler and more complex effects: ionization and excitation of atoms by electron impact, excitation of molecular oscillations in molecular lasers, etc. These problems cannot be solved without knowing the *electron distribution function* which makes it possible to describe various effects of electron interaction, not only with atoms and molecules but also with the field, much more completely and in finer detail.

The velocity distribution function of electrons, $f(t, \boldsymbol{r}, \boldsymbol{v})$, is defined as follows. The quantity $f\,d\boldsymbol{r}\,d\boldsymbol{v}$ is the number of electrons at a moment t in an element of volume $d\boldsymbol{r} = dx\,dy\,dz$ around a point $\boldsymbol{r}$, with velocity components from v_x to $v_x + dv_x$, etc., so that $d\boldsymbol{v} \equiv dv_x dv_y dv_z$. The integral of f over all velocity components equals the electron density $n_e(t, \boldsymbol{r})$. Recalling that we have a preferred direction in space, defined by the electric vector $\boldsymbol{E}$, it is expedient to express velocity in terms of spherical, not Cartesian, coordinates. The vector $\boldsymbol{v}$ is characterized by its magnitude v, the angle ϑ it makes with the polar axis $\boldsymbol{E}$, and the azimuthal angle φ (Fig. 5.1). Besides, $d\boldsymbol{v} = v^2\,dv\,d\Omega$, where $d\Omega = \sin\vartheta\,d\vartheta\,d\varphi$ is an element of solid angle around the direction of $\boldsymbol{v}$.

It is easy to pass from the function $f(\boldsymbol{v})$ to distribution functions in absolute values of v, $\varphi(v)$, and in energies, $n(\varepsilon)$:

$$n(\varepsilon)d\varepsilon = \varphi(v)dv = v^2 dv \int f(\boldsymbol{v})d\Omega\,. \tag{5.1}$$

These are also normalized to the density n_e, and the relation between them follows from the equality $\varepsilon = mv^2/2$:

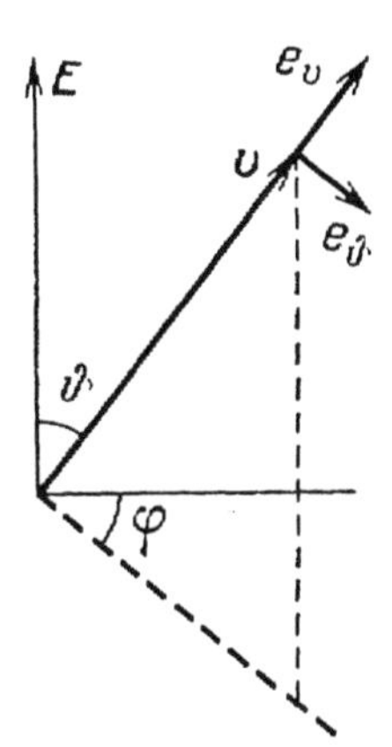

Fig. 5.1. Velocity vector in spherical coordinates

$$n(\varepsilon) = \varphi(v)/mv\;; \quad \varphi(v) = n(\varepsilon)\sqrt{2m\varepsilon}\;. \tag{5.2}$$

With the distribution function known, any quantity characterizing the electron gas can in principle be calculated. The frequency of ionization of atoms and molecules is given by (4.1). Frequencies of all inelastic collisions, of any reactions, are given by similar formulas. The density of the total electric current carried by electrons is

$$\boldsymbol{j}_t = -e \int \boldsymbol{v} f(\boldsymbol{v}) d\boldsymbol{v}\;. \tag{5.3}$$

The theory of kinetic equation gives expressions for the conductivity and the dielectric permittivity. They are free of the uncertainty in choosing the electron collision frequency inherent in formulas (3.23) and (3.24) of the elementary theory.

5.2 Formulation of the Kinetic Equation

The kinetic equation for electrons is a particular case of the general *kinetic Boltzmann equation* for the distribution function of particles in a gas. In fact, it gives the balancing of the number of particles in an elementary volume in *phase space*.

5.2.1 Balancing Equation

Let us take an elementary volume in the form of a cube around a fixed point in phase space $\boldsymbol{r}$, $\boldsymbol{v}$. One cannot draw a six-dimensional cube, so we trace an ordinary cube (Fig. 5.2) and, appealing to our imagination, think of a six-dimensional cube; one of its vertices has the coordinates x, y, z, v_x, v_y, v_z. At a moment t, the cubic volume $d\Gamma = dx\,dy\,dz\,dv_x dv_y dv_z$ contains $f\,d\Gamma$ particles, so that the distribution function f is interpreted as the *density* in phase space.

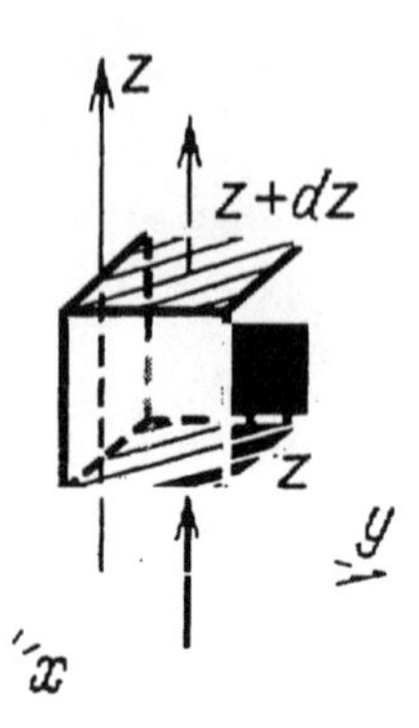

Fig. 5.2. Cube in three-dimensional space: Derivation of the balance equation for particle numbers in phase space

Even with no collisions, the number of particles in the cube changes. A particle moving at a velocity $\boldsymbol{v} = \dot{\boldsymbol{r}}$ changes its position $\boldsymbol{r}$; if subjected to a force $\boldsymbol{F}$, it undergoes acceleration $\boldsymbol{w} = \dot{\boldsymbol{v}}$ and changes its velocity $\boldsymbol{v}$. The particle moves in the phase space where the density f, in general, changes from point to point; hence, the number of particles entering the cube through one face may be greater or smaller than that leaving through the opposite face. Particles may thus accumulate in the volume or be depleted in it. Collisions produce the same effect. Some particles go out of $d\Gamma$ because their velocity vector sharply changes or because they disappear, others enter $d\Gamma$ after a collision or as a result of creation.

The number of particles entering the volume $d\Gamma$ per second through a specific face of the cube, say the lower one in Fig. 5.2 (there are 12 such faces), is $(fv_z)_z dx\, dy\, dv_x dv_y dv_z$. The subscript z with the z-component fv_z of the flux density signifies that the value of the flux is taken at a point z of the axis perpendicular to the face. The product of the five differentials is the area of the face (a face is *five*-dimensional). The number of particles per second leaving through the opposite (upper) face is $(fv_z)_{z+dz} dx\, dy\, dv_x dv_y dv_z$. The difference between the inflow and outflow,

$$\left[(fv_z)_z - (fv_z)_{z+dz}\right] dx\, dy\, dv_x dv_y dv_z = -\left[\partial(fv_z)/\partial z\right] d\Gamma \; ,$$

contributes to the rate of particle accumulation in the cube, $(\partial f/\partial t) d\Gamma$. A similar procedure is applied to the other five faces.

As for the collisions, their contribution to the rate of change of the number of particles in the volume $d\Gamma$ is proportional to the volume itself; we denote it by $(df/dt)_c d\Gamma$. Collecting the terms and cancelling the common factor $d\Gamma$, we arrive at the balancing equation for the number of particles:

$$\frac{\partial f}{\partial t} + \left[\frac{\partial}{\partial x}(fv_x) + \ldots + \frac{\partial}{\partial v_x}(fw_x) + \ldots\right] = \left(\frac{df}{dt}\right)_c \; . \tag{5.4}$$

This is quite similar to the ordinary continuity equation in the presence of sources (represented by the collision term). The sum in brackets is the six-dimensional divergence of the "flux density". Let us introduce into (5.4) the derivative df/dt *along the trajectory* of a specific group of particles in the phase space. This can be done by treating f as a composite function of time,

$$\frac{df}{dt} = \frac{\partial f}{\partial t} + \frac{\partial f}{\partial x}\frac{dx}{dt} + \ldots + \frac{\partial f}{\partial v_x}\frac{dv_x}{dt} + \ldots$$
$$= \frac{\partial f}{\partial t} + v_x\frac{\partial f}{\partial x} + \ldots + w_x\frac{\partial f}{\partial v_x} + \ldots \quad .$$

(The operation d/dt corresponds to the total derivative in hydrodynamics.) We obtain

$$\frac{df}{dt} + f\left[\frac{\partial v_x}{\partial x} + \ldots + \frac{\partial w_x}{\partial v_x} + \ldots\right] = \left(\frac{df}{dt}\right)_c . \tag{5.5}$$

The pairs of quantities v_x and x, and so on, are independent coordinates in phase space, v_x not being a function of x. If the ordinary space contains a field of force $\boldsymbol{F}(\boldsymbol{r})$, the accleration $\boldsymbol{w} = \boldsymbol{F}/m$ is a function of coordinates x, y, z. Even in the presence of a magnetic field, the Lorentz force $\boldsymbol{F} \propto \boldsymbol{v} \times \boldsymbol{H}$, and the component w_x depends on v_y and v_z but is independent of v_x, etc. Hence, the divergence of velocity vanishes: $[\partial v_x/\partial x + \ldots + \partial w_x/\partial v_x + \ldots] = 0$, so that (5.5) reduces to the equality

$$df/dt = (df/dt)_c . \tag{5.6}$$

In the absence of collisions, the number density in a specific group of particles does not change with time while the particles move along the trajectory in the phase space: $df/dt = 0$. The medium in the phase space is "incompressible".

5.2.2 Liouville's Theorem

Let us follow an ensemble of particles that occupy a small volume $\Delta\Gamma$ at a moment t. Without collisions, the number of particles in the group remains constant: $d(f\Delta\Gamma)/dt = 0$. However, $df/dt = 0$ and hence, $d\Delta\Gamma/dt = 0$. The phase volume occupied by a given set of particles travels through the phase space, undergoes deformation, but retains unchanged in volume. This statement is known as *Liouville's theorem.* It is clearly illustrated by Fig. 5.3, drawn for the one-dimensional case x, v_x in which the phase space is represented by the plane of the figure.

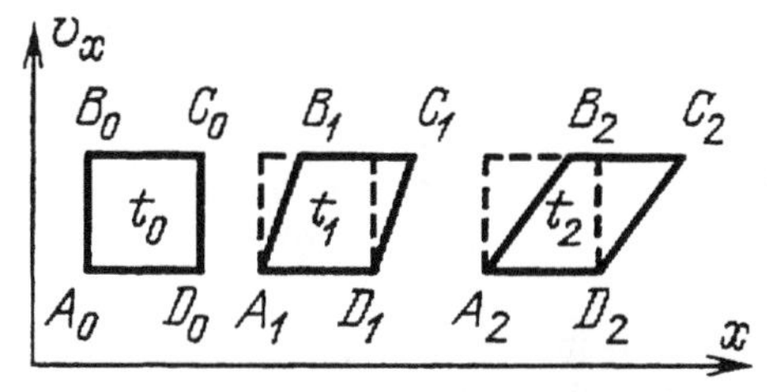

Fig. 5.3. Illustration of Liouville's theorem. A region that is rectangular at a moment t_0 transforms at subsequent moments t_1, t_2 into parallelograms of the same area

5.2.3 Application to Electrons in a Field

We shall not consider the cases with a strong magnetic field, even though this would not be difficult: The Lorentz force is small in the field of electromagnetic waves in comparison with the electric force (Sect. 3.1). Using $\boldsymbol{F} = -e\boldsymbol{E}$, we can rewrite the balance equation (5.6) in the form

$$\frac{\partial f}{\partial t} + \boldsymbol{v} \cdot \operatorname{grad} f - \frac{e\boldsymbol{E}}{m} \cdot \operatorname{grad}_v f = \left(\frac{df}{dt}\right)_c , \quad (5.7)$$

where the symbol "grad_v" denotes the gradient in velocity space. In spherical coordinates we have

$$\operatorname{grad}_v \equiv \boldsymbol{e}_v \frac{\partial}{\partial v} + \boldsymbol{e}_\vartheta \frac{1}{v} \frac{\partial}{\partial \vartheta} + \boldsymbol{e}_\varphi \frac{1}{v \sin \vartheta} \frac{\partial}{\partial \varphi} ;$$

here $\boldsymbol{e}_v$, $\boldsymbol{e}_\vartheta$, and $\boldsymbol{e}_\varphi$ are the unit vectors along the three directions (Fig. 5.1). Let us consider only spatially uniform fields, an approximation justified for electromagnetic waves because the amplitude of electronic oscillations is usually small in comparison with the wavelength (Sect. 3.1). The dependence of f on space coordinates in a uniform field can be caused only by the presence of walls and diffusion flows due to gradients. To avoid distracting attention from our main objective (to find the effect of field and collisions on the distribution function), we assume the entire space to be uniform (the effects of diffusion fluxes being taken into account later by simple techniques), which gives

$$\frac{\partial f}{\partial t} - \frac{eE}{m} \left[\cos \vartheta \frac{\partial f}{\partial v} + \frac{\sin^2 \vartheta}{v} \cdot \frac{\partial f}{\partial (\cos \vartheta)} \right] = \left(\frac{df}{dt}\right)_c . \quad (5.8)$$

The function $f(t, v, \vartheta)$ is independent of the angle φ because $\boldsymbol{E}$ defines an axis of symmetry.

5.2.4 Classification of Collisions into Elastic and Inelastic

Let us look at the right-hand side of (5.8). Assume the gas to be weakly ionized and neglect the collisions of electrons with other electrons and with ions, taking into account only those with neutrals. This is a very important assumption, greatly facilitating the problem of solving the kinetic equation by making it *linear*. The general Boltzmann equation for a gas is nonlinear because the right-hand side includes the collisions of particles of a given species with one another. These terms contain, of course, the products of distribution functions of the colliding particles. In our case, electrons collide with foreign particles, that is, heavy atoms "at rest", which are assumed here to have no distribution. The contributions of collisions of each type to the change in distribution function are simply added up. Let us divide all collisions into *elastic* and *inelastic*:

$$\left(\frac{df}{dt}\right)_c = \left(\frac{df}{dt}\right)_{el} + \left(\frac{df}{dt}\right)_{inel} = I(f) + Q(f) . \quad (5.9)$$

We subsume into the group of inelastic collisions, in addition to the processes of excitation of atoms and molecules, the creation of new electrons as a result of ionization and the possible annihilation processes. Inelastic collisions are important for the formation of the *energy spectrum* of electrons but, being much less frequent than elastic collisions, they have practically no influence on the field-electron interaction and on the change in electron velocity and energy caused by the field. Hence, inelastic processes do not affect the build-up of the *asymmetric* part of the distribution function that reflects the oriented action of the field and frequent elastic collisions. For these reasons, we do not yet specify the expression $Q(f) \equiv (df/dt)_{\text{inel}}$; this will be done only after we pass from the velocity vector distribution function to the distribution in electron energy.

5.2.5 Collision Integral

This name is applied to the term $I(f)$ representing the effect of elastic collisions. Assume that the atoms are at rest and, in addition, neglect the quantities of order m/M, assuming $M = \infty$. Under this approximation, the absolute value of the electron velocity v and its energy ε are exactly conserved in scattering. We will take elastic losses into account later, after having derived the final equation for the electron velocity distribution. It will be possible to use a simple line of reasoning to add to the equation an elastic loss term, and obtain an accurate result. If, however, the change in v due to scattering is introduced from the very beginning, this not too significant refinement makes the derivation of the collision integral considerably more complex.

The *collision integral* $I[f(\boldsymbol{v})]$ takes into account the change in the number of electrons with a given velocity vector $\boldsymbol{v}$, caused by the loss of electrons to the points of the phase space with a different vector $\boldsymbol{v}'$ as a result of scattering by atoms, and also by the arrival to $\boldsymbol{v}$ from all other points $\boldsymbol{v}'$. According to our assumption, the magnitude of velocity is conserved in scattering, so that it is sufficient to characterize $\boldsymbol{v}$ by a unit vector of direction, $\boldsymbol{\Omega}$. Since f has for argument the same absolute value of velocity v, we write simply $f(\boldsymbol{\Omega})$ instead of $f(\boldsymbol{v}) = f(v, \boldsymbol{\Omega})$. Owing to scattering, $f(\boldsymbol{\Omega})d\Omega\nu_c(v)$ electrons move in one second out of a given solid angle $d\Omega$ around the given direction $\boldsymbol{\Omega}$ of velocity, ν_c being the collision frequency. The electrons are lost to all other allowed directions $\boldsymbol{\Omega}'$. Let $q(v, \boldsymbol{\Omega}, \boldsymbol{\Omega}')d\Omega'$ be the probability for a colliding electron, moving in the direction $\boldsymbol{\Omega}$, to change its direction to $\boldsymbol{\Omega}'$ in the interval $d\Omega'$. The electron has to move in some direction, so that

$$\int q(\boldsymbol{\Omega}, \boldsymbol{\Omega}')d\Omega' = 1 .$$

The number of electrons leaving $d\Omega$ can be written in the form of a detailed expression:

$$f(\boldsymbol{\Omega})d\Omega\nu_c = \nu_c \int_{\boldsymbol{\Omega}'} f(\boldsymbol{\Omega})d\Omega q(\boldsymbol{\Omega}, \boldsymbol{\Omega}')d\Omega' .$$

The number of electrons arriving per second in the same volume $d\Omega$ and the given direction $\boldsymbol{\Omega}$ from other directions $\boldsymbol{\Omega}'$ is

$$\nu_c \int_{\Omega'} f(\boldsymbol{\Omega}')d\Omega' q(\boldsymbol{\Omega}', \boldsymbol{\Omega})d\Omega \, .$$

The difference between the gain and loss gives us $I[f(\boldsymbol{\Omega})]d\Omega$. Before we write out this expression, note that the probability of scattering from one direction to another depends not on the directions as such, but only on the angle between them: the scattering angle θ (Fig. 5.4). Hence, $q(\boldsymbol{\Omega}, \boldsymbol{\Omega}') = q(\boldsymbol{\Omega}', \boldsymbol{\Omega}) = q(\theta)$, and the probability can be integrated, with the same result, in final directions $\boldsymbol{\Omega}'$ or in initial ones, $\boldsymbol{\Omega}$. Then we can cancel the differential $d\Omega$ not involved in integration and finally obtain

$$I(f) = \nu_c(v) \int_{\Omega'} [f(\boldsymbol{\Omega}') - f(\boldsymbol{\Omega})]q(\theta)d\Omega' \, . \tag{5.10}$$

Integration is carried out here in all directions $\boldsymbol{\Omega}'$ at a fixed $\boldsymbol{\Omega}$. Equation (5.8) with the right-hand side (5.9), where the collision integral I is given by (5.10) and the inelastic collisions term Q is to be specified later, is the required kinetic equation.

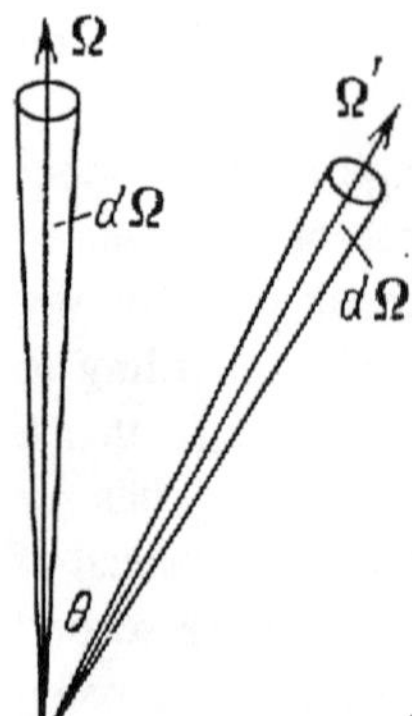

Fig. 5.4. Scattering angle θ

5.3 Approximation for the Angular Dependence of the Distribution Function

The kinetic equation is integro-differential in the angle ϑ and hence is mathematically very unwieldy. The factor that makes the distribution function depend on the direction of velocity, that is, on ϑ, is the field. In zero field, the distribution is isotropic. The field accelerates negative charges in the opposite direction to $\boldsymbol{E}$, therefore producing an excess of electrons moving in this direction and a shortage of those moving in the opposite direction.

5.3.1 Symmetric and Asymmetric Parts of the Distribution Function

Assume the field to be moderately strong, and the *anisotropy* caused by it to be small. (We take this into account approximately, as a correction to the main, *symmetric* part of the function.) If this operation is to be mathematically rigorous, the angular dependence $f(t, v, \vartheta)$ must be written as a series expansion, so as to describe in an accurate manner the detailed departure from symmetry. Only a system of orthogonal and normalized functions may be used for the expansion. Such a system satisfying the angular dependence is that of Legendre polynomials: $P_0 = 1$, $P_1 = \cos\vartheta$, etc. Our approximation limits the series to the first two terms,

$$f(t, v, \vartheta) = f_0(t, v) + f_1(t, v) \quad \cos\vartheta \,, \tag{5.11}$$

where f_0 and f_1 are the new sought-after functions; equations for finding these are to be formulated.

The new functions have a definite physical meaning, the first of them, the *symmetric* part, determines the electron *energy* spectrum. According to (5.1), we have

$$n(\varepsilon)d\varepsilon = \varphi(v)dv = 4\pi v^2 f_0(v)dv \,. \tag{5.12}$$

The *asymmetric* part $f_1 \cos\vartheta$ determines the *electric current*. In view of the axial symmetry $f(\boldsymbol{v})$ the current points along the field. Formulas (5.3) and (5.11) imply that its magnitude is

$$j_t = -e \int\int v(\cos^2\vartheta) f_1 2\pi v^2 \, dv \sin\vartheta \, d\vartheta = -\frac{4\pi}{3} e \int v^3 f_1 \, dv \,. \tag{5.13}$$

Approximation (5.11) is admissible only if the anisotropy of the distribution function is sufficiently small, that is, if the field is not too strong. The quantitative criterion of the notion "not too strong" will be clear after we find the correction f_1 to the main part of f and demand that $f_1 \ll f_0$ (Sect. 5.5.1). Approximation (5.11) is attributed to H. Lorentz, who formulated the kinetic equation for electrons in a dc field, and elaborated formula (2.7) for conductivity using approximation (5.11).

5.3.2 Equations for the Functions f_0 and f_1

The simplest way to derive these functions is to make use of the method of moments. The original equation for f is multiplied by a Legendre polynomial and integrated over the angles taking into account the properties of polynomials. In the case in question, it is sufficient to do it twice: first simply integrate (5.8) over the solid angle $d\Omega$ because the zeroth polynomial P_0 is equal to 1, and then multiply (5.8) by $P_1 = \cos\vartheta$ and integrate for the second time. As a result of the first integration (rather, of averaging, i.e., of the operation $\int d\Omega/4\pi$), taking into account that $\langle\cos\vartheta\rangle = 0$, $\langle\cos^2\vartheta\rangle = 1/3$, $\langle\sin^2\vartheta\rangle = 2/3$, we find

$$\frac{\partial f_0}{\partial t} - \frac{eE}{m}\left(\frac{1}{3}\frac{\partial f_1}{\partial v} + \frac{2f_1}{3v}\right) = Q(f_0) \,.$$

The integral $\int I d\Omega$ on the right-hand side vanishes automatically since it gives the change in the number of electrons moving in all directions as a result of elastic collisions and elastic collisions do not change the total number of electrons. Of course, the inelastic-collisions term is linear in f; in general, the effect of inelastic collisions is independent of the direction of velocity, being a function only of energy spectrum. Hence, Q becomes a function of the symmetric part of f. The obtained equation can be rewritten in the form

$$\frac{\partial f_0}{\partial t} = \frac{eE}{m} \frac{1}{3v^2} \frac{\partial (v^2 f_1)}{\partial v} + Q(f_0) . \qquad (5.14)$$

The second averaging of the kinetic equation with the weight $\cos\vartheta$ yields

$$\frac{1}{3} \frac{\partial f_1}{\partial t} - \frac{1}{3} \frac{eE}{m} \frac{\partial f_0}{\partial v} = \frac{\nu_c}{4\pi} \int \cos\vartheta \, d\Omega \int [f(\boldsymbol{\Omega}') - f(\boldsymbol{\Omega})] q(\theta) d\Omega' , \qquad (5.15)$$

where we have so far simply copied (5.10) without substituting (5.11) there, and neglected the contribution of inelastic collisions in comparison with that of elastic ones.

Let us take the right-hand side of (5.15). The inner integral in $d\Omega'$ is taken over all directions $\boldsymbol{\Omega}'$ at a fixed $\boldsymbol{\Omega}$. In fact, when integrating over the angle $\boldsymbol{\Omega}'$, one need not choose the vector $\boldsymbol{E}$ as the polar axis, as we did in constructing the original kinetic equation. Now it is more convenient to direct the polar axis along $\boldsymbol{\Omega}$ (Fig. 5.5) and describe the direction of $\boldsymbol{\Omega}'$ by the angles θ and φ', the azimuth φ' being measured from a fixed plane passing through the vectors $\boldsymbol{\Omega}$ and $\boldsymbol{E}$. In these coordinates, an element of the solid angle is $d\Omega' = d\varphi' \sin\theta \, d\theta$; this is very convenient because the factor q in the integrand is a function of θ. Now we substitute (5.11) and rewrite the inner integral, with a fixed angle ϑ:

$$J = \int [f(\boldsymbol{\Omega}') - f(\boldsymbol{\Omega})] q(\theta) \, d\Omega' = f_1 \int (\cos\vartheta' - \cos\vartheta) q(\theta) \times d\varphi' \sin\theta \, d\theta .$$

Expressing $\cos\vartheta'$ via the familiar formula of spherical trigonometry,

$$\cos\vartheta' = \cos\vartheta \cos\theta + \sin\vartheta \sin\theta \cos\varphi' ,$$

and taking into account that the term with $\cos\varphi'$ vanishes in the integration in φ', we find

$$J = f_1 \cos\vartheta \int (\cos\theta - 1) q(\theta) d\varphi' \sin\theta \, d\theta = f_1 \cos\vartheta (\overline{\cos\theta} - 1) .$$

Here $\overline{\cos\theta}$ is, by definition, the *mean cosine of the scattering angle* because $\cos\theta$ is averaged on the basis of the scattering probability $q(\theta)$ normalized to unity over all angles. The second of the sought equations is obtained by introducing the effective collision frequency $\nu_m = \nu_c(1 - \overline{\cos\theta})$ (note that we have just derived this expression rigorously), substituting the inner integral J into (5.15), and again integrating over $d\Omega$:

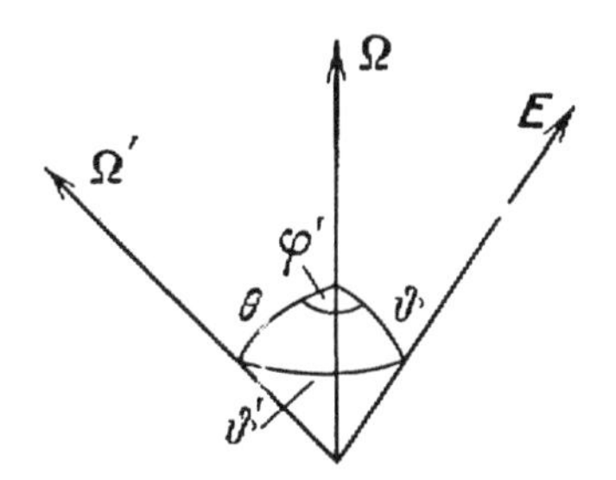

Fig. 5.5. Directions of field and velocities before and after scattering

$$\frac{\partial f_1}{\partial t} + \nu_m f_1 = \frac{eE}{m}\frac{\partial f_0}{\partial v} \,. \tag{5.16}$$

We have thus obtained, instead of the integro-differential equation (5.8), two differential equations for the functions f_0 and f_1 that approximate the true distribution function by (5.11). These equations hold for arbitrary dependence $E(t)$, that is, both for dc and for oscillating fields.

5.4 Equation of the Electron Energy Spectrum

5.4.1 Approximate Integration of the Equation for f_1

In order to proceed further, we have to specify the dependence of the field on time. Let us take a harmonic field $E = E_0 \sin \omega t$. It is obviously impossible to find the exact solution of the system (5.14, 16), so we make the following approximation. The correction f_1 to the symmetric part of the distribution function is caused by the field, which changes its direction periodically. The correction oscillates with the same frequency: first a greater number of electrons move along the field, and after half a cycle more electrons move in the opposite direction. As follows from (5.14), the dependence of the main (symmetric) function f_0 on time consists of two parts. On one hand, this is a relatively slow dependence due to the buildup of the electron energy spectrum as a result of various inelastic processes, creation and loss of electrons, and energy gain from the field. Since f_1 is also a harmonic function and is proportional to E_0, the first term in the right-hand side of (5.14) has a component, averaged over a period, that is proportional to E_0^2, and an oscillating component. The accumulation of energy is provided by the averaged component. On the other hand, f_0 contains an oscillating component due to the oscillating part of the first term in the right-hand side of (5.14). It resembles a ripple imposed on the slowly changing $f_0(t)$. Obviously, the variations of the spectrum during one period, that is, the ripple, need not bother us if we are interested in the electron energy spectrum at sufficiently high frequencies (see the criterion in Sect. 5.5.2). The object of physical interest is the smoothed function $\langle f_0 \rangle$ averaged over one period of field oscillations.

When integrating equation (5.16), we substitute the function $\langle \partial f_0/\partial v \rangle$ averaged over a period. The high-frequency component would contribute to f_1 a term of higher order in E. Neglecting the slow dependence of $\langle \partial f_0/\partial v \rangle$ on time in comparison with $\sin \omega t$, we integrate (5.16) as a linear equation, where $\langle \partial f_0/\partial v \rangle$ is independent of time. We find that

$$f_1 = -\frac{eE_0}{m(\omega^2 + \nu_m^2)} \left\langle \frac{\partial f_0}{\partial v} \right\rangle (\omega \cos \omega t - \nu_m \sin \omega t) , \tag{5.17}$$

and this expression can be rewritten in the form

$$f_1 = \frac{eE_0}{m\sqrt{\omega^2 + \nu_m^2}} \left\langle \frac{\partial f_0}{\partial v} \right\rangle \sin(\omega t - \alpha) , \tag{5.18}$$

where $\alpha = \arctan(\omega/\nu_m)$.

We conclude that $f_1 \propto E_0$; it oscillates at a frequency ω but its phase is shifted with respect to the field. In the limiting case of high frequencies ($\omega^2 \gg \nu_m^2$), the phase shift is $\alpha \approx \pi/2$, and

$$f_1 \approx -\frac{eE_0}{m\omega} \left\langle \frac{\partial f_0}{\partial v} \right\rangle \cos \omega t = -u \left\langle \frac{\partial f_0}{\partial v} \right\rangle \cos \omega t .$$

To an order of magnitude, $\partial f_0/\partial v \sim f_0/v$, where v is some characteristic, average velocity of random motion. According to (3.2), $eE_0/m\omega = u$ is the amplitude of the varying velocity of an electron in the oscillating field, so that to an order of magnitude,

$$f_1 \sim (u/v) f_0 . \tag{5.19}$$

In the opposite limiting case of low frequencies ($\omega^2 \ll \nu_m^2$), the phase shift α is small and

$$f_1 \approx \frac{eE_0}{m\nu_m} \left\langle \frac{\partial f_0}{\partial v} \right\rangle \sin \omega t \underset{\omega \to 0}{\longrightarrow} \frac{eE}{m\nu_m} \frac{\partial f_0}{\partial v} .$$

It is readily seen that the same result is implied directly by (5.16) if the field is assumed to be constant *ab initio*. The asymptotic constant value of f_1 that is reached after a time of about one collision period is

$$f_1 = \frac{eE}{m\nu_m} \frac{\partial f_0}{\partial v} \sim \frac{v_d}{v} f_0 , \tag{5.19'}$$

where, in view of (2.4), $v_d = eE/m\nu_m$ is the absolute value of the electron drift velocity. The smallness criterion f_1/f_0 is implied by (5.19, 19') (Sect. 5.5.1).

5.4.2 Equation for f_0

Now that we have expressed the correction f_1 in terms of the main function f_0, we may take the last step: to substitute it into (5.14) and thus obtain an equation for the symmetric part of the distribution function which is unambiguously related to the energy spectrum. If the field is harmonic, we substitute (5.17) into (5.14) and average the equation so obtained over one period of oscillation in order to eliminate the ripples and reveal the slow time-dependence of the spectrum, it being the only one of interest. Remarking that

$$\langle \cos\omega t \sin\omega t\rangle = 0\ , \quad \langle \sin^2\omega t\rangle = \tfrac{1}{2}\ ,$$

and dropping the averaging brackets $\langle\ \rangle$ with f_0, we arrive at an equation for the function $f_0(t,v)$:

$$\frac{\partial f_0}{\partial t} = \frac{1}{v^2}\frac{\partial}{\partial v}\left[\frac{e^2E^2}{3m^2}\left(\frac{\nu_{\mathrm{m}}(v)v^2}{\omega^2+\nu_{\mathrm{m}}^2}\right)\frac{\partial f_0}{\partial v}\right] + Q(f_0)\ . \tag{5.20}$$

Here we have substituted for E_0 the root-mean-square field $E = E_0/\sqrt{2}$. In the case of a dc field, the substitution of (5.19′) into (5.14) gives the same equation (5.20) but with $\omega = 0$. In other words, the exact limiting transition is possible from the case of the harmonic field to that of a dc field, provided we replace the root-mean-square field with a constant one as $\omega \to 0$; note that this appears to be intuitively natural.

5.4.3 Equation for $n(\varepsilon)$

If we recast (5.20) to a new independent variable $\varepsilon = mv^2/2$, $d\varepsilon = mv\,dv$, and replace the function $f_0(t,v)$ with the energy distribution function $n(t,\varepsilon)$ via (5.12), the result is an equation for the energy spectrum:

$$\frac{\partial n}{\partial t} = \frac{\partial}{\partial\varepsilon}\left(A\varepsilon^{3/2}\frac{\partial}{\partial\varepsilon}\frac{n}{\varepsilon^{1/2}}\right) + Q(n)\ ,$$

$$A = \frac{2e^2E^2}{3m}\frac{\nu_{\mathrm{m}}}{\omega^2+\nu_{\mathrm{m}}^2} = \frac{e^2E_0^2}{3m}\frac{\nu_{\mathrm{m}}}{\omega^2+\nu_{\mathrm{m}}^2}\ . \tag{5.21}$$

The case of a dc field is also obtained if we set $\omega = 0$ and replace the root-mean-square field by the constant field.

5.4.4 Diffusion Nature of the Equation

Carrying out the differentiation of $n\varepsilon^{-1/2}$ in (5.21) and introducing new notation for combinations of variables, we rewrite this equation:

$$\frac{\partial n}{\partial t} = -\frac{\partial J}{\partial\varepsilon} + Q\ , \quad J = -\mathcal{D}\frac{\partial n}{\partial\varepsilon} + nU\ ,$$
$$\mathcal{D} = A\varepsilon\ , \quad U = A/2\ . \tag{5.22}$$

The structure of (5.22) is quite similar to that of the equation of *one-dimensional diffusion* of particles. Indeed, ε is the coordinate, n is the density of particles, J is the flux, Q is the source, $\mathcal{D}$ is the diffusion coefficient that, in fact, is position dependent (this is also imaginable: say, the density of the main gas through which particles diffuse varies with the coordinate), and U is the velocity of the "kinematic" flow, that is, of the systematic motion in one direction, for instance, a drift that may be caused by the flow of the medium.

The physical meaning of the diffusion nature of energy accumulation in the field, that is, of the diffusive motion of electrons along the "energy axis," is very lucid. We have indicated in Sects. 2.3.3 and 3.2.5 that collisions can either increase or decrease the energy of electrons, in portions that equal, to an order of magnitude, mvu, where v is the velocity of random motion and u is the velocity of oriented motion caused by the field. In the high-frequency case, u is the amplitude of the oscillating electron velocity, and in the dc field, u equals the drift velocity v_{d}. Since on average, energy is gained or lost with almost equal probability, the change in electron energy resembles a random walk along the axis ε. The coefficient of ordinary diffusion, which manifests itself in treating the one-dimensional random walk of a particle, is approximately $D \approx \overline{\Delta x^2}/\tau$, where Δx is a step along the x axis and τ is the mean time interval between steps. In our case,

$$\mathcal{D} \approx (mvu)^2 \nu_{\mathrm{m}} \, .$$

Substituting here $u = eE_0/m\omega$ in the high-frequency range $\omega^2 \gg \nu_{\mathrm{m}}^2$, or $u = v_{\mathrm{d}} = eE/m\nu_{\mathrm{m}}$ (drift velocity) in the low-frequency range $\omega^2 \ll \nu_{\mathrm{m}}^2$, and recalling that $\varepsilon = mv^2/2$, we find for these limiting cases the diffusion coefficient defined by (5.22) and (5.21) (to within an unimportant numerical factor).

The kinematic velocity U also has a physical meaning. The positive kinematic flux directed towards increasing ε is mostly related to the predominant energy gain in collisions, in contrast to energy losses. We have seen in Sects. 2.3.3 and 3.2.5 that on the average the energy gained is greater than the energy lost by a quantity $\Delta\varepsilon_E \propto mu^2$, which is less by a factor v/u than a mean increment in any direction, mvu. The rate of systematic upward motion on the energy axis, $U \sim \Delta\varepsilon_E \nu_{\mathrm{m}}$, is indeed on the order of

$$U \sim mu^2 \nu_{\mathrm{m}} \sim (mvu)^2 \nu_{\mathrm{m}}/mv^2 \sim \mathcal{D}/\varepsilon \, ,$$

as we have formally obtained in deriving (5.22).

5.4.5 Elastic Loss Term

Following the remarks on the diffusion along the energy axis, it is not difficult to add to the energy spectrum equation a term describing elastic energy loss. Indeed, elastic losses also produce a flux along the energy axis, always pointing toward decreasing ε. The mean energy that an electron loses in an elastic collision, calculated in Sect. 2.3.4, is

$$\Delta\varepsilon_{\mathrm{el}} = (2m/M)(1 - \overline{\cos\theta})\varepsilon \, .$$

This is the quantity by which an electron slides "downward" on the ε axis after each collision. The time between collisions is $\tau_{\mathrm{c}} = \nu_{\mathrm{c}}^{-1}$. Hence, the corresponding velocity of downward motion is

$$U_{\mathrm{el}} = -\Delta\varepsilon_{\mathrm{el}}/\tau_{\mathrm{c}} = -(2m/M)\nu_{\mathrm{m}}\varepsilon \, ;$$

the kinetic flux corresponding to this velocity is nU_{el}. We add it to the flux J in (5.22). Returning now from (5.22) to the original equations (5.20, 21), we write them out, now taking into account the elastic loss term:

$$\frac{\partial f_0}{\partial t} = \frac{1}{v^2}\frac{\partial}{\partial v}\left[\frac{e^2E^2}{3m^2}\frac{\nu_{\text{m}}v^2}{\omega^2+\nu_{\text{m}}^2}\frac{\partial f_0}{\partial v} + \frac{m}{M}\nu_{\text{m}}v^3 f_0\right] + Q(f_0)\,, \tag{5.23}$$

$$\frac{\partial n}{\partial t} = \frac{\partial}{\partial \varepsilon}\left(A\varepsilon^{3/2}\frac{\partial}{\partial \varepsilon}\frac{n}{\varepsilon^{1/2}} + \frac{2m}{M}\varepsilon\nu_{\text{m}}n\right) + Q(n)\,,$$
$$A = \frac{2e^2E^2}{3m}\frac{\nu_{\text{m}}}{\omega^2+\nu_{\text{m}}^2}\,. \tag{5.24}$$

Note that our simple reasoning has led to the exact result covering elastic losses. If the finite mass of atoms has been introduced from the very beginning into the collision integral, we would have arrived at the same equations, (5.23) and (5.24).

5.4.6 Inelastic Collision Term

Let us specify the quantity Q that was to include all processes not related to the field and elastic collisions. The loss of electrons per second in 1 cm^3 from the energy interval $d\varepsilon$ due to the excitation and ionization of atoms equals $n(\varepsilon)d\varepsilon\nu^*(\varepsilon)$ and $n(\varepsilon)d\varepsilon\nu_{\text{i}}(\varepsilon)$, respectively, where $\nu^*(\varepsilon)$ and $\nu_{\text{i}}(\varepsilon)$ are, respectively, the frequencies of excitation of corresponding levels and of ionization at a given electron energy ε. The energy E^* lost by an electron in an act of excitation equals the excitation potential plus a small energy needed to impart to the atom the velocity necessary for the total momentum of the electron and the atom to remain unaltered. This additional energy loss is very small, as it is in elastic scattering; it is negligible in comparison with E^*. If the inelastic collision involved an electron with energy $\varepsilon' = \varepsilon + E^*$ in the interval $d\varepsilon' = d\varepsilon$, then the loss of energy shifts it into the interval $d\varepsilon$ at the point ε. Hence, the term in $Q(n)$ due to the excitation of a certain level can be expressed approximately in the form

$$Q^*(n) = -n(\varepsilon)\nu^*(\varepsilon) + n(\varepsilon+E^*)\nu^*(\varepsilon+E^*)\,, \tag{5.25}$$

where $\nu^*(\varepsilon) = 0$ if $\varepsilon \le E^*$. The excitation of vibrational levels in molecules is described by expressions of similar type. The total amount Q is the sum of this type of term over all relevant levels of atoms and molecules.

The term representing ionizing collisions, $Q_{\text{i}}(n)$, is more complicated. Let an electron have energy $\varepsilon' > I$. The electron spends energy I on liberating an electron from an atom; the remainder, $\varepsilon' - I$, is divided between the primary and the secondary electrons (the energy transferred to the ion being negligible). Let $\Phi(\varepsilon', \varepsilon)d\varepsilon$ be the probability for the energy of the electron knocked out of the atom to be between ε and $\varepsilon + d\varepsilon$ ($\int_0^{\varepsilon'-I}\Phi(\varepsilon',\varepsilon)d\varepsilon = 1$). The ionizing electron falls into the same interval if the new one acquires an energy from $\varepsilon' - I - \varepsilon - d\varepsilon$ to $\varepsilon' - I - \varepsilon$, and the probability of such an event is $\Phi(\varepsilon', \varepsilon' - I - \varepsilon)d\varepsilon$. Forming

$Q_\mathrm{i} d\varepsilon$ from the terms due to electrons arriving at and leaving the interval ε to $\varepsilon + d\varepsilon$ and then dividing by $d\varepsilon$, we obtain

$$Q_\mathrm{i} = -n(\varepsilon)\nu_\mathrm{i}(\varepsilon) + \int_{\varepsilon+I}^{\infty} n(\varepsilon')\nu_\mathrm{i}(\varepsilon')[\Phi(\varepsilon',\varepsilon) + \Phi(\varepsilon',\varepsilon' - I - \varepsilon)]d\varepsilon' . \tag{5.26}$$

Expression (5.26) transforms to (5.25) if we assume that all new electrons appear with identical energy ε_0 and $\varepsilon' > I + \varepsilon_0 \equiv I_1$. Then $\Phi(\varepsilon',\varepsilon) = \delta(\varepsilon - \varepsilon_0)$, where δ is the Dirac delta function. In this case

$$Q_\mathrm{i} = -n(\varepsilon)\nu_\mathrm{i}(\varepsilon) + n(\varepsilon + I_1)\nu_\mathrm{i}(\varepsilon + I_1) + \delta(\varepsilon - \varepsilon_0) \int_{I_1}^{\infty} n(\varepsilon')\nu_\mathrm{i}(\varepsilon')d\varepsilon' . \tag{5.26$'$}$$

The additional term when compared with (5.25) describes the source of new electrons.

The losses due to recombination or attachment of electrons are extremely easy to introduce into Q, for example, by a term $-n(\varepsilon)\nu_\mathrm{a}(\varepsilon)$, where $\nu_\mathrm{a}(\varepsilon)$ is the frequency of attachment to atoms or molecules.

5.4.7 Spatial Diffusion of Electrons

This can be rigorously calculated if the term $\boldsymbol{v}\,\mathrm{grad}\,f$ found in (5.7) is left in the left-hand side of the original equation (5.8). We have not done this in the preceding paragraphs in order to avoid complicating the manipulations and to focus the attention on the effects of interaction with the field. In the final equation for the spectrum, the diffusional losses of electrons can be taken into account approximately, adding to Q a term of type

$$Q_\mathrm{d} = -n(\varepsilon)\nu_\mathrm{d}(\varepsilon) ,$$

where $\nu_\mathrm{d} = D/\Lambda^2$ is the "diffusion frequency", that is, a quantity reciprocal to the characteristic time of diffusional removal of an electron from the chosen volume, $D = v^2/3\nu_\mathrm{m}$ is the diffusion coefficient (in ordinary space!), and Λ is the characteristic diffusion length (Sect. 4.5).

5.5 Validity Criteria for the Spectrum Equation

5.5.1 With Respect to Field Magnitude

The Lorentz approximation (5.11), on which the derivation of the equation was based, is valid if the asymmetry of the distribution function $f(\boldsymbol{v})$ is small: $f_1/f_0 \ll 1$. According to (5.19, 19′), this happens if the velocity of the electron directed along the field (amplitude u in the case of rapidly oscillating field or the drift velocity v_d in the case of dc field) is much less than the random velocity v. The terms with higher-order harmonics in the expansion of f in Legendre polynomials are proportional to the appropriate powers of the ratios u/v

and v_d/v. These ratios serve as the small parameters for series expansion of the distribution function f. The indicated conditions are satisfied in most practically interesting cases. Indeed, in a uniform not very high dc field in which an electron loses only a small fraction δ of its energy, $v_d/v \approx \sqrt{\delta} \ll 1$ [see (2.16)]. A quite similar relation, $u/v \approx \sqrt{\delta}$, is implied by (3.2) and (3.12) in the case of a rapidly oscillating field.

If a colliding electron loses a considerable fraction of its energy (formally, $\delta \sim 1$), the conditions are violated and the distribution function becomes essentially asymmetric (electrons move mostly along the field). This happens when an electron gains from the field in one free path length l (or in oscillations) the energy greater than that necesssary for the excitation of electron levels or for ionization of atoms, say, $eEl \gtrsim I$. Such situations occur mainly in exceptionally strong fields: in the cathode layer of the glow discharge, in focusing superpowerful optical pulses etc. Equations (5.23) and (5.24) have limited applicability on the side of small E, as well. The electron temperature in a very weak field may be comparable to the gas temperature T, which we assumed to be zero. The kinetic equation with $T \neq 0$ was discussed by *B.I. Davydov* in 1936.

5.5.2 With Respect to Field Frequency

The calculations of Sect. 5.4.1 ignored the modulation of the spectrum at the field frequency ω, since the symmetric part of the distribution function, f_0, was averaged over the oscillation period. This is admissible if the field oscillates rapidly in comparison with the time necessary for the buildup of the electron energy spectrum. Then only the root-mean-square field affects the spectrum and the gaining of energy from the field. In other words, the condition of applicability of the approximation used is the inequality $\omega \gg \nu_u = \nu_m\delta$, where $\tau_u = \nu_u^{-1}$ is the spectrum relaxation time, equal to the characteristic time of energy transfer from electrons to molecules (Sect. 2.3.7). In atomic gases this inequality is satisfied better than in molecular ones. It is satisfied practically always in the microwave and, of course, in the optical range of frequencies; almost always in the radiofrequency range in atomic gases; and sometimes also in the radiofrequency range in molecular gases. For example, in nitrogen $\nu_m \approx 4.2 \cdot 10^9 p[\text{Torr}]\,\text{s}^{-1}$, $\delta = 2.7 \cdot 10^{-3}$, $\nu_u \approx 1.1 \cdot 10^7 p[\text{Torr}]\,\text{s}^{-1}$; at a frequency $f = 13.7\,\text{MHz}$, $\omega = 0.86 \cdot 10^8\,\text{s}^{-1}$, the approximation holds only if $p \ll 10\,\text{Torr}$.

In the opposite limiting case, $\omega \ll \nu_m\delta$, the energy spectrum and the mean electron energy oscillate together with the field, tracking its relatively slow variations. The range $f \sim 10\,\text{kHz}$ is employed in discharge devices, for instance, in "ac" lasers (Sect. 14.4.6). The field is quasistationary in this case. The limiting transition from oscillating to dc field produced by imposing a weaker condition $\omega \ll \nu_m$ instead of $\omega \ll \nu_m\delta$ in Sect. 5.4.1 was a purely formal operation.

5.5.3 With Respect to Spatial Nonuniformity

When a group of electrons drift in a dc field, the energy spectrum builds up over one energy relaxation length $\Lambda_u \approx l/\sqrt{\sigma}$ (Sect. 2.3.7). The dc field must therefore be uniform over this length, otherwise the spectrum depends, not only on the field magnitude, but also on the distribution of potential in space (this *nonlocal* nature of the spectrum manifests itself in the cathode layer of the glow discharge). A field that is quasistationary in the sense $\omega \ll \nu_u$ is definitely uniform with respect to wavelength λ, because $\lambdabar = \lambda/2\pi \gg \Lambda_u(c\sqrt{\delta}/v) \gtrsim \Lambda_u$. The effect of field variation over one wavelength on the energy spectrum of electrons in electromagnetic waves of high frequencies is often negligible, but one must nevertheless be vigilant.

5.5.4 With Respect to the Degree of Ionization

One of the most important assumptions made in the analysis of the kinetic equation is that of neglecting collisions between electrons and thus making the equation linear. As in an ordinary gas, collisions between like particles lead to the Maxwellian distribution ("Maxwellization" of electrons). This is the situation in a sufficiently strongly ionized, dense low-temperature plasma. Colliding electrons exchange portions of energy that are of the order of the electron energies themselves. Electrons colliding inelastically with atoms and molecules also lose large amounts of energy, comparable with their energies. Hence, if inelastic collisions are possible, electron-electron collisions can be neglected if their frequency ν_{ee} (Sect. 2.2.2) is much less then ν_{inel}. At electron energies $\varepsilon \gtrsim$ 5–10 eV, sufficient for the impact excitation of atoms and molecules (and for their ionization), the condition $\nu_{ee} \ll \nu_{inel}$ in weakly ionized plasma is satisfied up to quite considerable degrees of ionization, $n_e/N \sim 10^{-4}$–10^{-3}. The same estimate is valid at smaller energies, $\varepsilon \sim$ 1–5 eV, in the case of molecular gases where electron impact excites molecular vibrations. Generally speaking, inelastic collisions distort the Maxwellian distribution by reducing the number of high-energy electrons.

The situation in atomic gases at energies below the excitation potential of atoms E^* ($E^* \approx$ 10 eV in inert gases) is different. Here only very weak elastic energy losses are efficient, especially if atoms are heavy, so that energy exchange in electron-electron collisions is active in spectrum formation process at much lower degrees of ionization. The condition of applicability of the linear kinetic equation in the energy range $\varepsilon < E^*$ is something like $\nu_{ee} \ll (m/M)\nu_m$, $n_e/N < 10^{-6}/A$, where A is the atomic mass. The spectrum in the range $\varepsilon < E^*$ may approach the Maxwellian form in steady-state conditions at not too weak ionization; only if $\varepsilon > E^*$, does it fall off much more steeply than the latter. If we wish the kinetic equation to describe these effects, it needs refinement, that is, the addition of electron-electron collisions.

5.6 Comparison of Some Conclusions Implied by the Kinetic Equation with the Results of Elementary Theory

5.6.1 Conductivity and Dielectric Permittivity

If we substitute f_1 of (5.17) into (5.13), the part of the current in phase with the field, that is, the part proportional to $\sin \omega t$ and E, gives the conduction current. The component proportional to $\cos \omega t$, that is, $\partial E/\partial t$, is the polarization current. Let us compare the result with the phenomenological formula (3.20) and equate separately the terms proportional to $\sin \omega t$ and $\cos \omega t$, as we did in Sect. 3.4.1. The rigorous expressions for the high-frequency conductivity σ_ω and dielectric permittivity ε_ω, derived in this way, are

$$\sigma_\omega = \frac{4\pi e^2}{3m} \int_0^\infty \frac{\nu_m(v) v^3}{\omega^2 + \nu_m^2} \left(-\frac{\partial f_0}{\partial v} \right) dv \, , \tag{5.27}$$

$$\varepsilon_\omega = 1 - \frac{16\pi^2 e^2}{3m} \int_0^\infty \frac{v^3}{\omega^2 + \nu_m^2} \left(-\frac{\partial f_0}{\partial v} \right) dv \, . \tag{5.28}$$

In the general case, these quantities depend on the electron energy spectrum. If, however, the collision frequency $\nu_m(v)$ is independent of velocity, the factors containing ν_m can be factored out from the integral. Integrating the resulting expression by parts and taking into account that there are no electrons with infinite energy, that is, $f_0 \to 0$ as $v \to \infty$, and the normalization condition for the function $f_0(v)$,

$$\int_0^\infty 4\pi v^2 dv \, f_0(v) = n_e \, , \tag{5.29}$$

we arrive at the formulas (3.23,24) of the elementary theory. We conclude that the latter formulas are valid for any spectrum provided $\nu_m(v) = \text{const}$. The dc conductivity is obtained automatically from (5.27) if we set $\omega = 0$:

$$\sigma = \frac{4\pi e^2}{3m} \int_0^\infty \frac{v^3}{\nu_m(v)} \left(-\frac{\partial f_0}{\partial v} \right) dv \, . \tag{5.30}$$

In practice, however, one normally resorts to elementary formulas (3.23,24, 27), selecting on the basis of pertinent arguments a value of the characteristic collision frequency that is the most plausible for the real spectrum. But once the spectrum is known, formulas (5.27,28,30) make it possible to choose this quantity correctly: They are employed in exact theories and for determining the correction coefficients to elementary formulas.

5.6.2 Rate of Change of the Mean Energy of the Spectrum

By definition, the mean electron energy is

$$\bar{\varepsilon} = \int_0^{\infty} \varepsilon n(\varepsilon) d\varepsilon \Big/ \int_0^{\infty} n(\varepsilon) d\varepsilon = n_e^{-1} \int_0^{\infty} \varepsilon n(\varepsilon) d\varepsilon \,. \tag{5.31}$$

Let us construct an equation for the rate of change of $\bar{\varepsilon}$, neglecting the effect of inelastic losses and taking into account only the effects of the field and elastic losses. Correspondingly, we multiply (5.24) [better still, (5.22) with an additional term nU_{el} in the expression for the flux J] by ε and integrate the product over the entire spectrum. The inelastic collisions term Q is omitted. Now we integrate twice by parts, recalling that $n \to 0$, $J \to 0$ as $\varepsilon \to \infty$, divide by n_e, and obtain

$$\frac{d\bar{\varepsilon}}{dt} = \frac{d\bar{D}}{d\varepsilon} + \bar{U} + \bar{U}_{el} = \frac{e^2 E^2}{m} \overline{\left(\frac{\nu_m}{\omega^2 + \nu_m^2}\right)} - \frac{2m}{M} \overline{(\nu_m \varepsilon)} \,, \tag{5.32}$$

where a bar means averaging over the spectrum. If $\nu_m(\varepsilon) = \text{const}$, this expression coincides exactly with the formula (3.11) of the elementary theory for the rate of change of the energy of the "mean" electron [in (3.11), $\delta = 2m/M$ because we include only elastic losses]. The condition of constancy of collision frequency again ensures the rigorousness of the elementary theory.

5.6.3 Similarity Laws

Similarity relations for drift velocity (Sect. 2.1.4), mean electron energy (Sects. 2.3.5 and 3.2.4), ionization coefficient (Sect. 4.1.4), etc., find confirmation and stringent justification in the kinetic equation. Consider the case of a dc field, $\omega = 0$. The frequencies of all inelastic, as well as elastic, collisions are proportional to the gas density N. Assume that there are no spatial gradients. If ionization is low, recombination is unimportant; then $Q \propto N$. Dividing (5.23) and (5.24) by N, we find that the distribution functions $f_0(v)$ and $n(\varepsilon)$ include E and N only as a combination E/N. In the non-steady-state case, $n(t, \varepsilon, E, N) = n(Nt, \varepsilon, E/N)$. As the gas density increases, the time of evolution is correspondingly reduced; this is natural because all processes are related to collisions.

In a high-frequency field $\omega^2 \gg \nu_m^2$ $n(t, \varepsilon, E, N, \omega) = n(Nt, \varepsilon, E/\omega)$, that is, the steady-state spectrum is completely independent of density and is determined by the ratio E/ω. If the collision frequency is assumed constant, $\nu_m(v) = \text{const}$, and $\omega \gg \nu_m \delta$ (Sect. 5.5.2), the kinetic equation and the spectrum for an oscillating field with root-mean-square magnitude E and amplitude $E_0 = \sqrt{2}E$ are identical to the equation and the spectrum in a dc field of effective strength

$$E_{eff} = E \left[\nu_m^2 / \left(\omega^2 + \nu_m^2\right)\right]^{1/2} < E \,. \tag{5.33}$$

Thus it is sometimes possible to make use of the richer store of computational and experimental dc data when studying discharge in rapidly varying fields. What is needed is an appropriate recalculation via (5.33).

5.7 Stationary Spectrum of Electrons in a Field in the Case of only Elastic Losses

On the whole, the results of Sects. 5.6.1 and 5.6.2 pointed to a high accuracy of the elementary theory. Now we will analyze an example revealing its imperfections and limited capabilities in comparison with those offered by the kinetic equation.

Imagine an ionized gas in an oscillating or dc field, and ignore the effects of inelastic collisions. Consider a stationary electron spectrum that is finally established as a result of exact balancing of energy gain in the field and elastic energy loss. This situation cannot be described as too abstract, since something like this is implemented if a monatomic weakly ionized gas occupying a large volume is placed in a weak field (the gas is monatomic in order to avoid the excitation of molecular vibrations). If the volume is large, the diffusion loss of electrons is small, especially because appreciable ionization leads to ambipolar electron diffusion (Sect. 2.6), which is much slower than free diffusion. In order to compensate for the small loss of electrons (so as to maintain the steady state), low-rate ionization by the relatively weak field is sufficient. Hence, electron energies are mostly low and very few electrons gain enough energy for the excitation or ionization of an atom. The effect of inelastic collisions on the spectrum is therefore not very significant.

5.7.1 What the Elementary Theory Has to Say

Let us look first at what the elementary theory predicts for this situation. This theory considers the behavior of a single, mean electron and assumes the states of all electrons to be identical. The electron energy ε varies in time as given by (2.12), for $\delta = 2m/M$:

$$\frac{d\varepsilon}{dt} = \left(\Delta\varepsilon_E - \Delta\varepsilon_{\text{el}}\right)\nu_{\text{m}} = \left[\frac{e^2E^2}{m(\omega^2+\nu_{\text{m}}^2)} - \frac{2m}{M}\varepsilon\right]\nu_{\text{m}} \,.$$

The energy $\Delta\varepsilon_E$ gained from the field in one collision is independent of ε (if $\nu_{\text{m}}(\varepsilon) = \text{const}$), while $\Delta\varepsilon_{\text{el}} \propto \varepsilon$. Hence, the electron energy reaches the value $\varepsilon_{\max}$ found from the equality $\Delta\varepsilon_E = \Delta\varepsilon_{\text{el}}$,

$$\varepsilon_{\max} = \frac{M}{2m}\,\frac{e^2E^2}{m(\omega^2+\nu_{\text{m}}^2)}\,, \tag{5.34}$$

and then remains constant. Indeed, if the energy of an electron drops by chance to less than $\varepsilon_{\max}$, it immediately starts to gain energy, $d\varepsilon/dt > 0$; if it grows above $\varepsilon_{\max}$, the electron starts losing energy, $d\varepsilon/dt < 0$. A steady (and stable) state corresponds to $d\varepsilon/dt = 0$ and to a delta distribution function: all electrons have identical energies $\varepsilon = \varepsilon_{\max}$. In view of the initial assumption, $\varepsilon_{\max}$ must be less than the excitation potential or, even more so, the ionization potential of atoms; otherwise inelastic losses will determine the behavior, instead of elastic ones. This leads to a question, however: how is ionization to occur if the gas has not a single electron with sufficient energy? Indeed, at least very low-rate ionization is necessary, otherwise unavoidable losses would gradually remove all electrons and the steady state would be impossible. Here the elementary theory is at a dead end.

5.7.2 Solution of Kinetic Equation (5.23)

In the stationary case, $\partial f_0/\partial t = 0$ and the expression in brackets (the flux) is constant if inelastic processes are neglected ($Q = 0$). However, as $v \to \infty$, f_0 and the flux vanish; hence, this constant equals zero. There is no flux at each point of the energy axis; the energy gained from the field is exactly balanced out by elastic losses at each energy ε. The second integration yields

$$f_0 = C \exp\left[-\frac{3m^3}{Me^2E^2}\int_0^v v\left(\omega^2 + \nu_{\rm m}^2\right) dv\right] , \tag{5.35}$$

where the integration constant C is determined by the normalization condition (5.29). The distribution function (5.35) is especially simple (Maxwellian) in the case $\nu_{\rm m}(v) = \text{const}$:

$$f_0 = C \exp - \left[\frac{3m^2(\omega^2 + \nu_{\rm m}^2)}{Me^2E^2}\frac{mv^2}{2}\right] = C \exp\left(-\frac{\varepsilon}{kT_{\rm e}}\right) , \tag{5.36}$$

with temperature $T_{\rm e}$ and mean energy $\bar{\varepsilon}$ equal to

$$\bar{\varepsilon} = \frac{3}{2}kT_{\rm e} = \frac{M}{2m}\frac{e^2E^2}{m(\omega^2 + \nu_{\rm m}^2)} = \varepsilon_{\max} . \tag{5.37}$$

The mean energy coincides with the single energy $\varepsilon_{\max}$ given by the elementary theory. The exact coincidence is more or less accidental. If different assumptions about the function $\nu_{\rm m}(v)$ are chosen, $\bar{\varepsilon}$ coincides with $\varepsilon_{\max}$ only to an order of magnitude.

The "true" spectrum is thus spread around $\varepsilon_{\max}$; it contains high-energy electrons (the tail of the Maxwell distribution) that produce ionization and sustain the stationary state. Electrons with $\varepsilon < \varepsilon_{\max}$ are also present. Electrons with energies not equal to $\varepsilon_{\max}$ appear in the stationary spectrum because the kinetic equation takes into account rigorously the force exerted by the field on the electrons; we have already seen in Sect. 3.2.5 that this field admits the possibility of gaining an energy in excess of $\Delta\varepsilon_E$ and of losing large portions of energy in collisions. Energetic electrons with $\varepsilon > \varepsilon_{\max}$ "survive" at the expense of gaining $\Delta\varepsilon \gg \Delta\varepsilon_E$ in this individual fashion, while slow electrons with $\varepsilon < \varepsilon_{\max}$ make use of high individual losses.

5.7.3 Margenau and Druyvesteyn's Distributions

The particular case of distribution (5.35) that is more often considered in gas discharge physics is the one where not the collision frequency but the free path

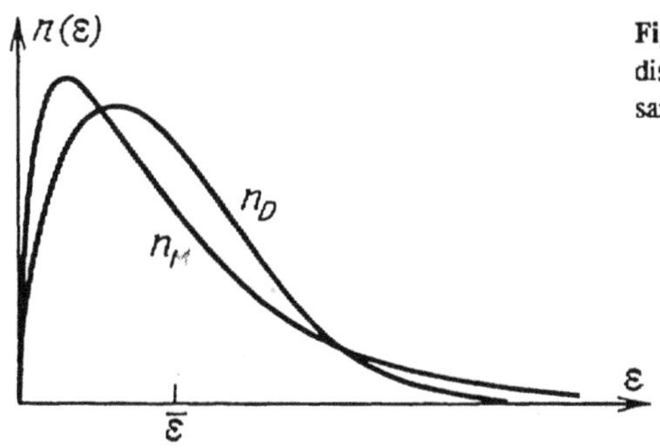

Fig. 5.6. Maxwell (n_M) and Druyvesteyn (n_D) distribution functions in energy, $n(\varepsilon)$, for the same mean energy $\bar{\varepsilon}$

length of electrons, $l = v/\nu_m$ (i.e., cross section for momentum transfer), is assumed to be independent of energy. Under this approximation, $\nu_m \propto v \propto \sqrt{\varepsilon}$ and the integration of (5.35) gives

$$f_0 = C \exp\left[-\frac{3m^3}{4Me^2E^2l^2}\left(v^4 + 2v^2\omega^2l^2\right)\right] . \tag{5.38}$$

This is the so-called *Margenau distribution* (1946); we shall not dwell on it here. In the dc case, the Margenau distribution transforms into *Druyvesteyn's distribution*

$$f_0 = C \exp\left[-\frac{3m}{M}\frac{\varepsilon^2}{\varepsilon_0^2}\right] \; ; \quad \varepsilon_0 = eEl \tag{5.39}$$

which was derived directly by Druyvesteyn in 1930. The parameter ε_0 is the energy gained by an electron from the field over a free path length. As follows from (5.39), the mean energy $\bar{\varepsilon}$ is gained over approximately $\sqrt{M/m}$ free paths [cf. (2.13)]. It is easy to see that $\bar{\varepsilon}$ also coincides, to an order of magnitude, with that unique energy $\varepsilon'_{\max}$ an electron is allowed to gain in the elementary theory. However, the quantity $\varepsilon'_{\max}$ is now different from (5.34) because $\Delta\varepsilon_E = e^2E^2/m\nu_m^2 \propto 1/\varepsilon$, in contrast to the high-frequency case where $\Delta\varepsilon_E = \text{const}$. Druyvesteyn's distribution is characterized by a considerably steeper decrease of the number of electrons in the "tail" than that of the Maxwell distribution (ε^2 in the exponential instead of ε); see Fig. 5.6.

5.7.4 Remark on Approximate Solution for Inert Gases in the Case of Predominant Inelastic Losses

The solution will be obtained in Sect. 7.5 in the analysis of electrical breakdown: it is unusual in that it employs the "infinite sink" approximation. In a certain sense, this case is the opposite of that treated in Sect. 5.7.

5.8 Numerical Results for Nitrogen and Air

Analytic solutions of the kinetic equation (5.23) are always obtained at the cost of considerable simplifications and assumptions, as in Sect. 5.7 (see also Sect. 5.8.1). In the case of molecular gases, any analytic solution would be hopeless, because of the need to take into account the vibrational and rotational excitations, in addition to the electron excitation. Advances in computer technology make it possible, however, to perform numerical integration, even though the computations require certain skill and are time-consuming. The most prominent physical aspect is the analysis of the processes to be taken into account and the choice of the most reliable data on cross sections. Considerable discrepancies exist between the results of various authors, which is connected with the complexity of the corresponding experiments. It is the shortage of cross section data that constitutes the main source of error in solving the kinetic equation.

The needs of modern molecular laser techniques provided a strong impetus for the computations. Numerical modeling was carried out for many mixtures of the type $CO_2 + N_2 + He$ (see Sect. 8.8). In addition to laser mixtures, such gases as nitrogen and air, widely used in discharge work, have also been studied. Here we will give the results of computations for N_2 [5.1] and for air [5.2] both as illustrations of the application of the kinetic equation and as material of interest for discharge research.

Equation (5.23) (slightly transformed for reasons of convenience) was integrated numerically for stationary conditions and a dc field. In Fig. 5.7, a function $\psi = \varepsilon^{-1/2} n(\varepsilon)/n_e$ for nitrogen is plotted on a semilogarithmic scale. The mean-

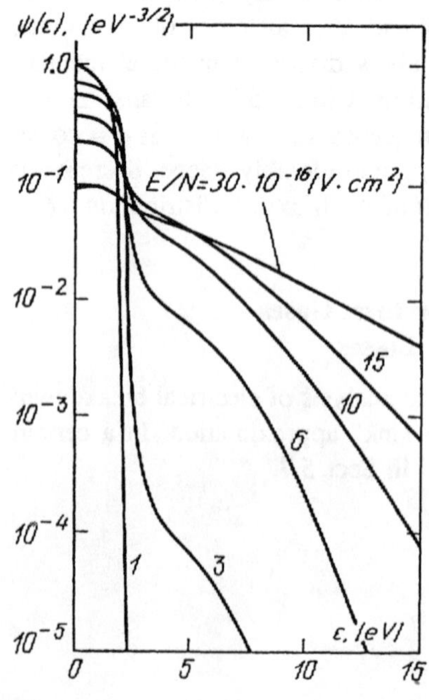

Fig. 5.7. Electron distribution function $\psi(\varepsilon) = n(\varepsilon)/n_e\sqrt{\varepsilon}$ in nitrogen

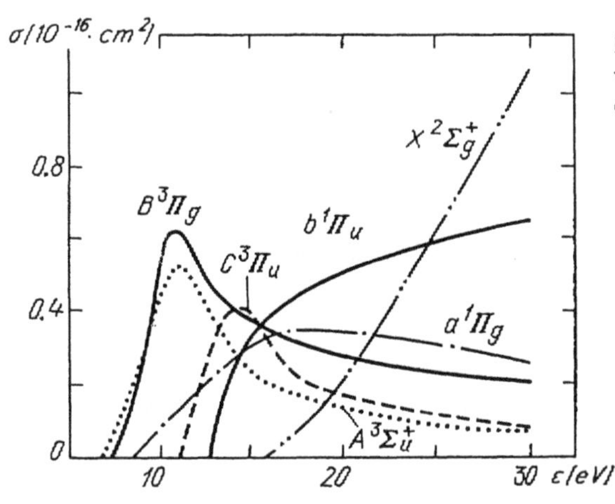

Fig. 5.8. Excitation cross sections of various energy levels, and ionization cross section of nitrogen molecule

ing of ψ becomes clearer if we take into account that in the case of the Maxwell distribution, $\psi_M(\varepsilon) = 2\pi^{-1/2}(kT_e)^{-3/2}\exp(-\varepsilon/kT_e)$ is the Boltzmann exponential up to the normalizing factor. Its semilogarithmic plot would be a straight line. The cross sections of excitation of a number of electron levels and ionization cross sections, employed in the computation, are shown in Fig. 5.8.

Figure 5.9 plots drift velocity, electron temperature, and characteristic energy given by (5.30) and (5.31) (see Sect. 2.4.2). (The diffusion coefficient was found by averaging $l_m v/3 = v^2/\nu_m$ over the spectrum; the mobility is found as v_d/E.) If the spectrum is Maxwellian, the characteristic energy coincides with temperature. The difference between T_e (given in volts) and D_e/μ_e is caused by the non-Maxwellian nature of the spectrum; this is seen in Fig. 5.7. Figure 5.10 plots the ionization rate constant k_i found from (4.1), and Fig. 5.11 shows the fractions of energy transferred from electrons to different degrees of freedom. The curves demonstrate that a predominant part of the work done by the field transforms into the energy of molecular vibrations in a wide range of E/N. This fact is

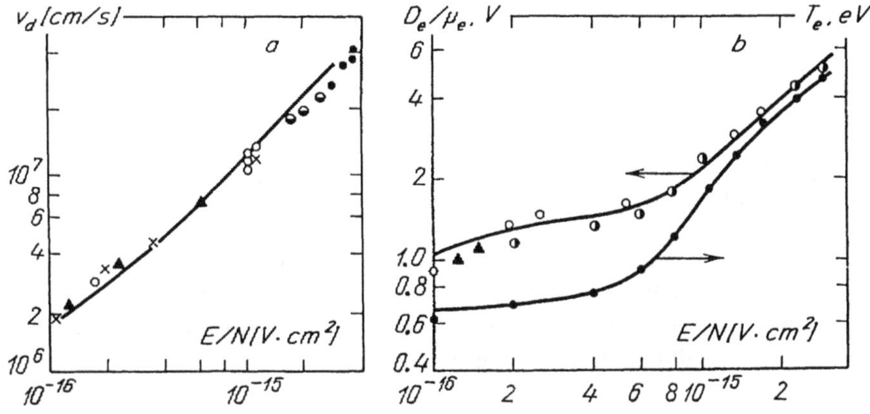

Fig. 5.9. (a) Drift velocity of electrons in nitrogen, (b) electron temperature as 2/3 of mean energy of the spectrum, and the characteristic energy D_e/μ_e in nitrogen

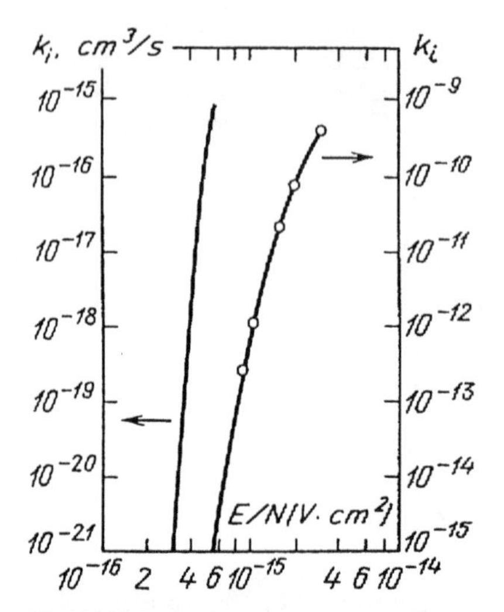

Fig. 5.10. Ionization rate constant $k_i \equiv \langle\sigma_{ion}v\rangle$ in nitrogen. *Dots* represent experimental data. Townsend's coefficient $\alpha/N = k_i/v_g$

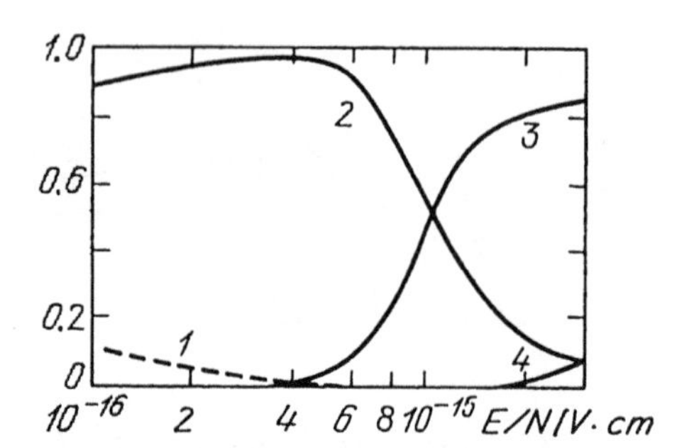

Fig. 5.11. Fraction of energy transferred by electrons to (*1*) rotation, (2) oscillations, (*3*) electronic excitation, and (*4*) ionization

of principal importance. This property of weakly ionized plasma in the field is essential for the operation of electric discharge molecular lasers. The calculated constants of the rate of ionization of nitrogen molecules from the ground and metastable $A^3\Sigma_u^+$ states, k_i, k_{iA}, and the rates of excitation of levels $A^3\Sigma_u^+$ and $B^3\Pi_g$ (Fig. 5.8), k_A^*, and k_B^* are approximated by the formulas [5.3]

$$\begin{aligned} &\log k_i = -8.3 - 34.8(N/E)\,, \quad \log k_{iA} = -6.1 - 27.5(N/E)\,, \\ &\log k_A^* = -8.35 - 14.9(N/E)\,, \quad \log k_B^* = -8.2 - 15.6(N/E)\,, \end{aligned} \tag{5.40}$$

where E/N is given in $10^{-16}\,\mathrm{V\,cm^2}$, and k in cm³/s. The rate constants increase in the case of strong vibrational excitation, which is typical of glow discharge in nitrogen, owing to the lower rate of vibrational relaxation. This is connected both with reduced reaction thresholds and with the enrichment of the spectrum with higher-energy electrons. If T_v is the vibrational temperature characterizing the ratio of populations of the levels with $v = 1$ and $v = 0$, and $Z = \exp(-\hbar\omega_k/kT)$, then we approximately have

$$\log[k(T_v)/k(0)] = 43.5(N/E)^2 Z\,, \quad Z = \exp(-3360/T)\,. \tag{5.41}$$

The formula is the same for all k and is valid up to $T_v = 5000\,\mathrm{K}$. Figures 5.12–14 plot the results of computations for dry air (in practice the water vapor content may vary from one experiment to another). The distribution functions qualitatively resemble those for nitrogen.

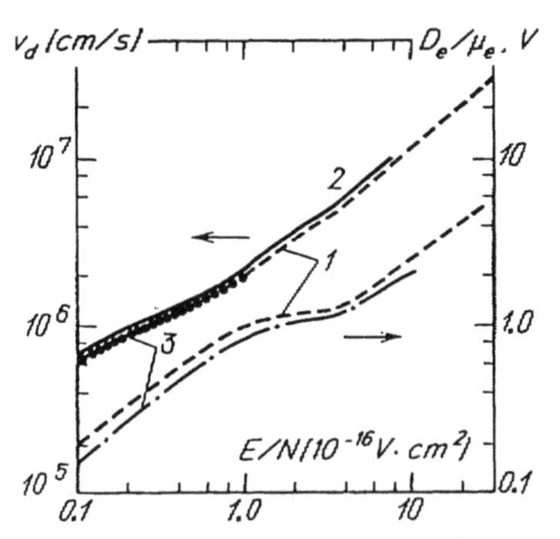

Fig. 5.12. Drift velocity (*upper curves*) and characteristic energy D_e/μ_e (*lower curves*) in air. (*1*) calculated values, (2) and (*3*) experimental data

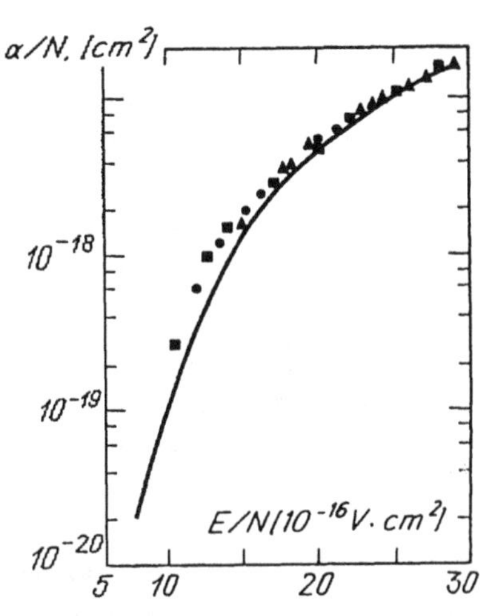

Fig. 5.13. Townsend's ionization coefficient α/N in air. *Dots* represent experimental data

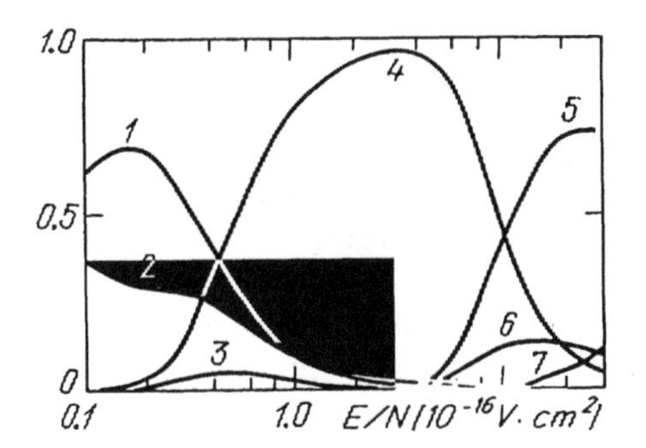

Fig. 5.14. Fraction of energy transferred by electrons in air to (*1*) vibrations of O_2, (2) rotations of O_2 and N_2, (*3*) elastic losses, (*4*) vibrations of N_2, electronic excitation of N_2 (*5*) and of O_2 (*6*), and (*7*) ionization of O_2 and N_2

5.9 Spatially Nonuniform Fields of Arbitrary Strength

Strongly nonuniform fields are produced in weakly ionized gases in the cathode layer of a glow discharge, in the electrode layers of a radio-frequency capacitively coupled discharge, in high-voltage pulsed discharges, and possibly in other situations. The Lorentz approximation fails in these cases (Sect. 5.5) and we have to return to the original kinetic equation. At present, the *Monte Carlo method* is regarded as the most efficient and accurate method of solving this equation. This is a computational procedure in which a computer simulates the process of random walk of a particle implied by the kinetic equation. These computations [5.4, 5] are very complicated, require considerable computer time, and are rarely worthwhile. As an alternative to Monte Carlo simulations, an approximate approach can be used in which the problem is reduced to equations that are *no more complicated* than the equations of the Lorentz approximation. The simplified method is a more manageable tool, in comparison with the Monte Carlo techniques, for studying discharge phenomena when moderate accuracy is acceptable. It is suitable for analyzing the behavior of electrons in any, strong or weak, uniform or nonuniform, fields, with all the qualitative features and corollaries of the exact equation retained.

5.9.1 "Forward-Backward" Approximation

We shall assume that electrons move only along the field. They are scattered in elastic collisions either forward or, with probability ξ, backwards. In inelastic collisions (with excitation or ionization) they lose energy but do not change the direction of motion. Let $\varphi_1(v,x,t)dv$ electrons move at velocities from v to $v+dv$ in the positive direction of the x axis ($E \equiv E_x < 0$), and $\varphi_2(v,x,t)dv$ move in the negative direction. Functions φ_1, φ_2 obey equations which are easily obtained from (5.7,9,10):

$$\begin{aligned}\frac{\partial\varphi_1}{\partial t}+v\frac{\partial\varphi_1}{\partial x}-\frac{eE}{m}\frac{\partial\varphi_1}{\partial v}&=\frac{1}{2}\nu_{\mathrm{m}}(\varphi_2-\varphi_1)+Q(\varphi_1)\\ \frac{\partial\varphi_2}{\partial t}-v\frac{\partial\varphi_2}{\partial x}+\frac{eE}{m}\frac{\partial\varphi_2}{\partial v}&=\frac{1}{2}\nu_{\mathrm{m}}(\varphi_1-\varphi_2)+Q(\varphi_2)\,.\end{aligned} \tag{5.42}$$

The effective frequency of elastic collisions is $\nu_{\mathrm{m}} = 2Nv\sigma_{\mathrm{c}}(v)\xi$, where $\xi = (1/2)(1-\cos\theta)$. Adding and subtracting equations (5.42), we now use the functions $\Phi_{0,1} = \varphi_1 \pm \varphi_2$. They characterize the spectral density and the flux of electrons and correspond to $f_{0,1}$ of the Lorentz approximation. In the weak field limit, equations for $\Phi_{0,1}$ become similar to (5.14, 16). If $\nu_{\mathrm{m}}(v) = \mathrm{const}$, the exact value of the drift velocity is obtained, while that of the diffusion coefficient is three times the true value, reflecting the "one-dimensionality" of the random motion. For details and the application to the *cathode layer of a glow discharge* see [5.6] and Sect. 8.4.10; see also [5.7].

5.9.2 Runaway Electrons

In very strong and sufficiently extended fields, electrons reach high energies. Inelastic collisions are then more frequent than elastic ones, and electrons are mostly scattered forward. Of the two equations in (5.42), only the first survives with ξ, $\nu_{\mathrm{m}} = 0$. The resulting equation for the mean energy of the spectrum corresponds to the equation for a "monoenergetic" beam,

$$\frac{d\varepsilon}{dx} = e|E(x)| - L(\varepsilon)\,, \quad L = N\sigma_{\mathrm{i}}I + N\sum\sigma_k^*E_k^*\,. \tag{5.43}$$

As with all cross sections $\sigma_{\mathrm{i}}(\varepsilon)$, $\sigma_k^*(\varepsilon)$, the inelastic loss function $L(\varepsilon)$ goes through a maximum, at $\varepsilon \sim 20$–$50\,\mathrm{eV}$. Therefore, an electron moving in the field $|E| > E_{\mathrm{crit}} = L_{\max}/e$ will be continuously accelerated (will "run away"), in spite of inelastic losses. In N_2, $(E/p)_{\mathrm{crit}} = 365$ V/(cm·Torr), and in He, 63 V/(cm·Torr).

6.1 Introduction. Electric Circuit

The main objectives in experimental determination of plasma parameters are to measure the density n_e of electrons, their temperature T_e (provided they have temperature), and, in the general case, the distribution function $f_0(v)$. The distributions of potential and electric field in space are of considerable interest in discharge research. The *probe method* developed by Langmuir in 1923 solves these problems if conditions are favourable. The probe method is unique among all diagnostic techniques in making it possible to determine directly the local plasma characteristics, that is, the spatial distribution of parameters; this is the reason for the special value of probe techniques.

To conduct a probe study, one introduces into a chosen place in the plasma an electrode and connects it to various potentials. The probe is a metal conductor coated with insulation almost to the tip. The naked surface of the probe, in contact with the plasma, may be given a plane, cylindrical, or spherical shape. The probe potential imposed by a power supply is determined with respect to a *reference* electrode: the anode, the cathode, or the grounded metal wall of the discharge chamber if one is present. The measurement circuit is shown in Fig. 6.1. In this circuit the current closes the probe circuit via the anode, so that the polarity of the probe current supply is chosen so as to have the probe potential lie between those of the anode and the cathode (as in the plasma). The probe potential is varied by a potentiometer. The experiment consists in measuring the current through the probe and the voltage applied to it, that is, in recording its *current-voltage* $(V - i)$ *characteristic*. Figure 6.2 shows several probe geometries. Probes are

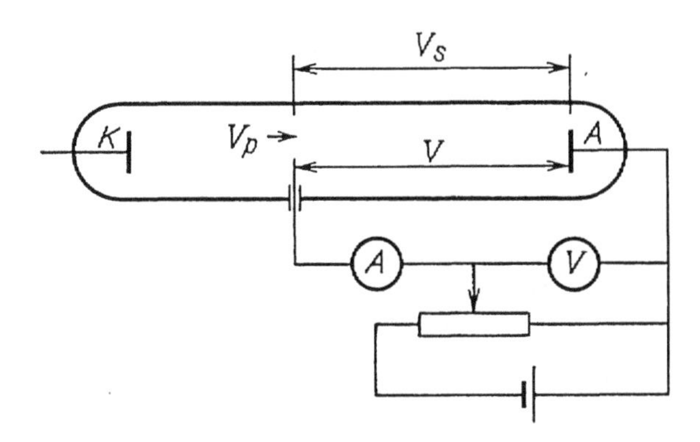

Fig. 6.1. Circuit for probe measurements

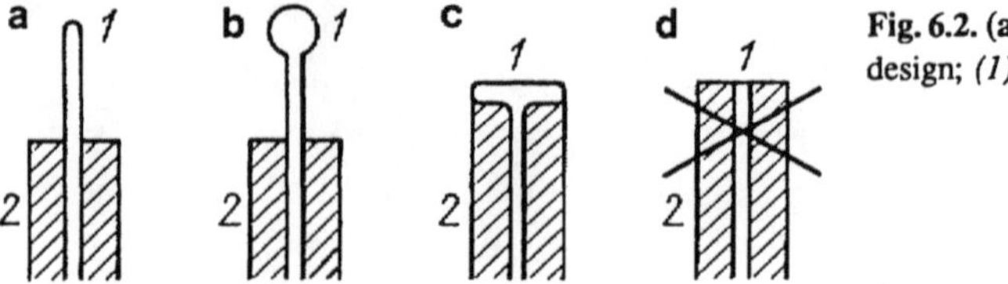

Fig. 6.2. (a–c) probe design, (d) wrong design; *(1)* probe, *(2)* insulator

usually made of refractory metals: tungsten, molybdenum, tantalum. Cylindrical probes are made of wire 0.5–0.05 mm in diameter. Special glass coatings of wire, ceramics, quartz, alumina (99 % Al_2O_3), etc. are used as probe insulators. In contrast to the wrong geometry in *d*, the probe geometry in *c* reduces edge effects. Indeed, data processing operates with the *current-collecting* surface area. The characteristic size of spherical and plane probes is about 1 mm.

The simplicity of equipment and experiment constitute the advantages of the probe method. The disadvantages lie in the complexities of the theory used to extract the plasma characteristics from probe measurement data. To put it more correctly: there is only a limited range of conditions under which the theory is only moderately complicated and thus does not lead to a considerable probability of obtaining erroneous results and faulty interpretations. In measuring the quantity, one must strive for a method based on a simple, reliable theory with a minimal number of assumptions and fuzzy constraints. In this respect, the probe method is sufficiently reliable only in rarefied gases where the free path length is greater than the characteristic size of the probe and the perturbed region around it.[1] In principle, though, probes can be used to study plasma in a very wide range of conditions: $p \sim 10^{-5} - 10^{2}$ Torr $n_e \sim 10^{6} - 10^{14}\,\text{cm}^{-3}$.

If the plasma is placed in a magnetic field, the theory becomes very complicated and measurements are only interpreted with considerable difficulty. The same is true if negative ions are present. If there is no reference electrode (e.g., in electrodeless high-frequency discharge or in a decaying plasma after the field has been switched off), a *single probe* is useless. In such cases, *double-probe* circuits are used (Sect. 6.6).

6.2 Current-Voltage Characteristic of a Single Probe

Figure 6.3 shows a somewhat idealized *probe characteristic,* that is, the electric current i through a plane probe is plotted as a function of its potential V. The choice of the reference point is immaterial as long as it is strictly fixed. This is

[1] This is the case for which Langmuir's theory is valid. Probes were introduced into discharge plasma much earlier, at the beginning of this century. The potential difference between the probe and the cathode was measured electrostatically. But it soon become clear that the probe and plasma potentials do not coincide, although the potential difference between neighbouring points can be determined in this way. Only with Langmuir's theory did probes turn into an efficient method of quantitative diagnostics

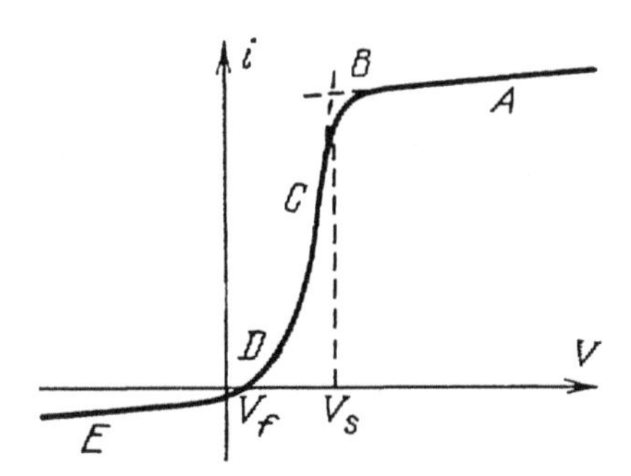

Fig. 6.3. Typical probe characteristic

why a reference electrode is introduced. Let us give a qualitative interpretation to the curve $i(V)$. Assume that the plasma is electroneutral in the absence of the probe: $n_e = n_+ = n_0$. We denote its potential at the point where the probe is located by V_S. The plasma (space) potential V_S is measured with respect to the reference electrode. Let V_S change only slightly within the region perturbed by the probe, that is, the potential of the nearby unperturbed plasma around the probe is V_S. The experimentally measured probe potential with respect to the reference electrode is $V = V_p + V_S$, where V_p is the probe potential with respect to the unperturbed plasma in the surrounding space (Fig. 6.1).

A probe does not emit charged particles, it only collects them from the plasma. We will operate with the absolute values of the electron and ion currents to the electrode, i_e and i_+, and use the following convention of probe current sign: $i = i_e - i_+$; it corresponds to the curve orientation in Fig. 6.3. If the probe and space potentials are identical, and the current-collecting surface is parallel to the direction of the external field between the anode and cathode, the charges reach the probe only owing to their thermal motion. In fact, electrons move much faster than ions, all the more so because their temperature in the weakly ionized plasma is much higher than the ionic (gas) temperature T. As a result, at $V = V_S$ the probe current is practically equal to the electronic one: $i \approx i_e$. Let us emphasize the fact that the conductor *collects* the electric current of electrons *in the absence* of potential difference between the conductor and the surrounding plasma.

When the potential applied to the probe is positive with respect to the plasma, $V > V_S$, ions are repelled by the probe so that the ionic current vanishes, and electrons are attracted. A *layer* of negative *space charge* (known as the sheath), screening the potential V_p, develops around the probe. The potential drop from V to V_S and the probe field are concentrated within the space charge layer, vanishing asymptotically in the transition to the unperturbed plasma. The effect is quite similar to plasma *polarization* around a charge and to the screening of the field of a charge by plasmas at distances greater than the *Debye radius* (2.38).

Let us introduce the exterior surface of the sheath, that is, the boundary beyond which the plasma can be considered approximately neutral and the field absent. Electrons are brought to the boundary from the outside and then transferred to the probe mostly via *thermal* motion; this factor determines their flux, which is only weakly dependent on probe potential. The probe current coincides with the more or less constant *saturation* electronic current $i_{e,sat}$. It corresponds to the *upper plateau* AB of the current-voltage characteristic. In the ideal case of an "infinite" plane, $i_{e,sat}$ = const, so that this part of $V - i$ curve would be horizontal. If the probe is small, the current grows with increasing positive potential, though more slowly than on the steep segment of the $V - i$ curve.

If the potential applied to the probe is negative with respect to the plasma, the electronic current drops sharply as $|V_p|$ increases, because the fraction of electrons with velocities sufficient for overcoming the decelerating field diminishes. Thus is the *steep* part C of the characteristic formed. The upper knee of the current-voltage characteristic, B, fixes the space potential V_S, corresponding to $V_p \approx 0$. It is in this way that it is determined experimentally. The probe is moved and the electric field is found from the difference between the potentials V_S at the neighbouring points.

The *current vanishes* at a certain negative potential $V_p = V_f$ (D in Fig. 6.3). The flux of a small number of energetic electrons, capable of overcoming the decelerating potential, to the probe is compensated in this state by the ion flux. This potential V_f (known as the *floating* potential) develops on an *insulated* body placed in the plasma. Returning to the footnote in Sect. 6.1, we can say that experiments with probes not connected to a current source yielded, not the plasma potential, but the more negative floating potential.

At still more negative potentials, the probe repels practically all the electrons but attracts ions. The probe is surrounded with an *ionic sheath* of positive space charge that screens the high negative potential V_p. The probe current is then purely ionic, determined by the flux of ions reaching the sheath boundary from the surrounding plasma. This flux is nearly independent of the probe potential, which is screened off, that is, the probe current varies slowly and coincides with the *ionic saturation current*. This current corresponds to the *lower plateau* of the current-voltage characteristic.

6.3 Theoretical Foundations of Electronic Current Diagnostics of Rarefied Plasmas

6.3.1 Electron Temperature

Consider the steep segment of the probe characteristic C, where the current is electronic and the potential decelerates electrons. Let the electrons cross the positive space-charge layer *without collisions*. When the probe voltage corresponds to the steep segment, the sheath thickness is of the order of the Debye radius. If $T_e \approx 1\,\text{eV}$, $n_e \approx 10^9\,\text{cm}^{-3}$, formula (2.38) yields $d \approx 10^{-2}\,\text{cm}$. If the free path length of electrons l_e is such that $l_e p \approx 0.03 - 0.01\,\text{cm}{\cdot}\text{Torr}$, then $l_e > d$ if $p < 10^{-1} - 1\,\text{Torr}$. We now calculate the electronic current to the probe, assuming for simplicity that the layer is thin in comparison with the curvature radius or size of the current-collecting surface. The problem can then be considered *plane*. In Sect. 6.3.4, it is shown that the result obtained below is true for any *convex* surface, for example, for a small spherical probe. Let us assume that the metal surface totally absorbs (does not reflect) the charges.

An electron impinging on the outer boundary of the sheath at a thermal velocity is decelerated by the field normal to the surface, $E_x = -d\varphi/dx$. As prescribed by the equation of motion, $m\dot{v}_x = d(mv_x^2/2)dx = e\,d\varphi/dx$, an electron can reach the probe surface only if at the beginning its velocity component v_x is such that $mv_x^2/2 \geq e|V_p|$ or $v_x \geq v_t = (2e|V_p|/m)^{1/2}$. If the electron distribution function at the outer boundary of the layer is $f(v_x, v_y, v_z)$, then the probe current density is

$$j_e = e\int_{-\infty}^{\infty} dv_y \int_{-\infty}^{\infty} dv_z \int_{v_t}^{\infty} f(v_x, v_y, v_z) v_x dv_x\,. \tag{6.1}$$

Integrating (6.1) for the Maxwell distribution (see Appendix) and multiplying the result by the surface area S, we find the probe current

$$i = S(en_0\bar{v}_e/4)\exp(eV_p/kT_e)\,,\quad \bar{v}_e = (8kT_e/\pi m)^{1/2}\,. \tag{6.2}$$

This formula, describing the steep fragment of the current-voltage characteristic, was derived by Langmuir and is widely employed in practical work. Having found the probe characteristic and then plotted $\ln i$ as a function of V, one can determine the electron temperature T_e from the slope of the obtained straight line. At the same time, the linearity of the $\ln i$ vs. V curve is evidence of the maxwellian distribution of electrons.

6.3.2 Saturation Current; the Potential and Charge Density in Plasmas

We have already mentioned that the space potential V_S is determined by the point of the upper knee on the $V - i$ curve. If $V > V_S$ (the field accelerating the electrons), formula (6.2) is invalidated because now one has to integrate v_x in (6.1) from zero, regardless of V_p. The probe current then coincides with the electronic saturation current corresponding to zero field, that is, $V_p = 0$ in (6.2):

$$i = i_{e,sat} = Sen_0\bar{v}_e/4\,. \tag{6.3}$$

This quantitity corresponds to the flux density of particles of the gas, crossing the area element from one side: $n\bar{v}/4$. One of the two factors of 1/2 in the 1/4 appears because only one half of the particles move in the necessary direction, and the other results from averaging over the hemisphere the cosine of the angle ϑ between the direction of the velocity $\boldsymbol{v}$ and the normal to the area element. Having found the thermal velocity $\bar{v}_e$ of electrons by measuring T_e on the steep segment of the current-voltage characteristic, and the current at the knee of the $V - i$ curve, one can find the electron density n_0 in the plasma from formula (6.3).

6.3.3 Criteria of "Rarefaction" of a Plasma

In order to justify the interpretation of n_0 as the electron density in nonperturbed plasmas, it is necessary that the presence of the probe not violate n_0 at the point of the last collision before electrons reach the probe. These points are at a distance of about one free path length from the probe. However, the density at a point is created by particles arriving from the sphere of a radius of the order of l. The surface of the probe, S, "eclipses" the fraction $S/4\pi l^2$ of this sphere and thus weakens the source of density formation. This fraction must be small, that is, $S/4\pi l^2 \ll 1$, for the free path length l to exceed the characteristic linear size of the probe, $\sqrt{S}$. Another condition of rerefaction is the smallness of the size of the space-charge layer in comparison with l.

6.3.4 How to Find the Electron Distribution Function

In the case of an arbitrary, nearly isotropic electron distribution in an unperturbed plasma, $f(v_x, v_y, v_z) \approx f_0(v)$, it is expedient to introduce into expression (6.1) for the electronic current to the probe surface the magnitude of velocity v and the angle between $\boldsymbol{v}$ and the inward normal to the surface, ϑ. Remarking that $v_x = v \cos\vartheta$, we rewrite (6.1) in the form

$$j_e = e\int_0^{\pi/2} \cos\vartheta\, 2\pi \sin\vartheta\, d\vartheta \int_{v_t/\cos\vartheta}^{\infty} v^3 f_0(v) dv \,. \tag{6.4}$$

Changing the sequence of integration but retaining the domain covered by the double integral, we integrate (6.4) over $\mu = \cos\vartheta$:

$$\begin{aligned} j_e &= 2\pi e \int_0^1 \mu\, d\mu \int_{v_t/\mu}^{\infty} v^3 f_0(v) dv = 2\pi e \int_{v_t}^{\infty} v^3 f_0(v) dv \int_{v_t/v}^{1} \mu\, d\mu \\ &= \frac{2\pi e}{m}\int_{v_t}^{\infty}\left(\frac{mv^2}{2} - e|V_p|\right) v f_0(v) dv \,. \end{aligned} \tag{6.5}$$

The Maxwell function, $f_0 = (m/2\pi kT_e)^{3/2}\exp(-mv^2/2kT_e)$, transforms (6.5) into (6.2). Both (6.2) and (6.5) are valid only if $V_p \leq 0$. If $V_p > 0$, the integration in v in (6.4) must begin from zero, regardless of V_p. This gives the saturation current (6.3) with non-maxwellian mean velocity $\bar{v}_e$.

Now we twice differentiate (6.5) with respect to V_p and obtain

$$\frac{d^2 i}{dV_p^2} = -S\frac{2\pi e^3}{m^2} f_0(v_t) \,, \quad v_t = \sqrt{\frac{2e|V_p|}{m}} \,. \tag{6.6}$$

In order to find the electron distribution function, one has to obtain the probe characteristic, differentiate it twice at each point of the steep segment, and assign to the point the potential V_p measured from the point B of the upper knee. The second derivative gives a number $f_0(v)$ for $v = (2e|V_p|/m)^{1/2}$. This method, first employed by *Druyvesteyn* in 1930, is still used nowadays, with certain improvements (Sect. 6.4).

6.3.5 The Applicability of the Theory of the Steep Segment of the $V - i$ Characteristic to Small-Sized Probes

Let us show that the fundamental diagnostics formula (6.5), derived above for the case of plane geometry, holds just as well for probes that are small in comparison with the size of the space-charge region. The only necessary condition is for the probe surface to be *convex*. The possibility of using small probes is extremely desirable, because such probes disturb the natural conditions in the plasma to be studied only slightly. The need to *intrude* into the plasma is one of the essential *drawbacks* of the probe technique.

Not every electron entering the space-charge layer around a finite-size probe reaches its surface. Some of them fly by the probe (Fig. 6.4). This is not a plane problem; therefore the derivation of formula (6.5) must be reconsidered. This shall do, following [6.1]. Denoting by $\boldsymbol{r}_\mathrm{p}$ the coordinates of a point on the probe surface, and by $\boldsymbol{v}_\mathrm{p}$ and $f(\boldsymbol{r}_\mathrm{p}, \boldsymbol{v}_\mathrm{p})$ the velocity and electron distribution function at the surface, the probe current density at a point $\boldsymbol{r}_\mathrm{p}$ is

$$j_\mathrm{e} = e \int_0^{\pi/2} \int_0^{\infty} v_\mathrm{p} \cos \vartheta f(\boldsymbol{r}_\mathrm{p}, \boldsymbol{v}_\mathrm{p}) 2\pi \sin \vartheta \, d\vartheta v_\mathrm{p}^2 dv_\mathrm{p} . \tag{6.7}$$

The surface being conves, electrons arrive at it from the entire hemisphere $0 \leq \vartheta \leq \pi/2$.[2]

For an electron to arrive at a surface point $\boldsymbol{r}_\mathrm{p}$ without collisions and with a velocity $\boldsymbol{v}_\mathrm{p}$, it has to leave a point $\boldsymbol{r}$ of nonperturbed plasma at a velocity $\boldsymbol{v}$. The force of deceleration has a potential, so that the change in the kinetic energy of electrons over this distance is independent of the spatial distribution of the potential and the shape of the trajectories:

$$mv_\mathrm{p}^2/2 = mv^2/2 - e|V_\mathrm{p}| . \tag{6.8}$$

According to (5.6), however, the distribution function cannot change along the trajectory of a particle in phase space. In nonperturbed plasma, it is isotropic and equal to $f_0(v)$. Hence, $f(\boldsymbol{r}_\mathrm{p}, \boldsymbol{v}_\mathrm{p}) = f(\boldsymbol{r}, \boldsymbol{v}) = f_0(v)$. After replacing f in (6.7) with $f_0(v)$, v_p^2 with the expression derived from (6.8), $v_\mathrm{p} dv_\mathrm{p}$ with $v\, dv$, and integrating over ϑ, we arrive at the last expression in (6.5). The current density

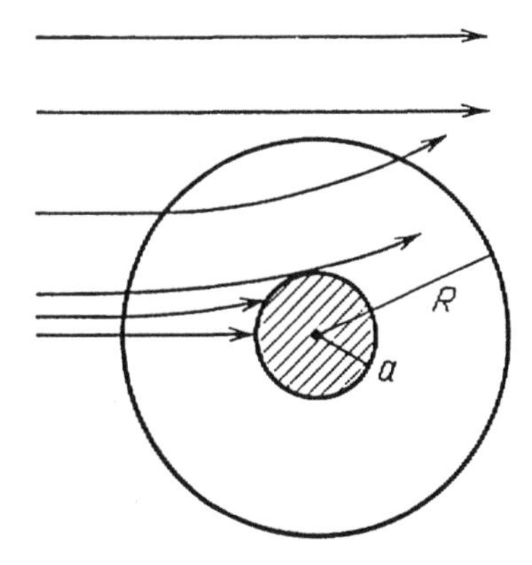

Fig. 6.4. Trajectories of particles near a repulsing spherical or cylindrical probe

[2] Some corners of concave areas are in the "shadow" of nearby protrusions, and thus have to be singled out of the integral.

is the same for all surface elements of a convex probe, that is, $i_e = Sj_e$, where S is the total probe surface area.

6.3.6 Why the Current to a Small Probe Does Not Saturate

For the electronic current to reach the potential-independent value (6.3) corresponding to saturation, it is sufficient to apply to the probe a small positive potential V_p both in the ideal plane case and in real situations, provided the space-charge thickness is small in comparison with the radius of curvature and the size of the current-collecting surface. If the probe size is small in comparison with the size (radius) of the space-charge region, the electronic current continues to increase with increasing positive potential, although it grows less steeply than on the steep part of the $V-i$ curve, which corresponds to decreasing deceleration potential. The reason for this behaviour of the $V-i$ curve is that not all electrons entering the space-charge layer (where they are in the attracting field) reach the probe. Some of them *pass by* the probe and escape from the layer without touching the probe. But it is obvious that the higher the accelerating potential V_p, the stronger the attractive force pulling electrons to the probe, the greater the fraction of electrons in the layer that are collected by the probe; hence, the current grows.

This situation is illustrated in Fig. 6.5, which represents a spherical and a long cylindrical probe. In the latter case, it shows projections of trajectories on the plane perpendicular to the axis. If a particle approaches the layer with an impact parameter ϱ greater than the layer radius R (the effective boundary beyond which the field vanishes), its straight path does not bend. If a particle enters the layer ($\varrho \leq R$), it can either pass or hit the probe. The outcome depends on the initial velocity v_0, the impact parameter ϱ, and the magnitude and radial distribution $V(r)$ of potential. The more energetic the electron, the smaller its impact parameter must be, or the higher the potential, for the electric force to be able to attract it to the probe. Very slow electrons with any $\varrho < R$ are collected by the probe.

This behavior is implied by the energy and angular momentum conservation laws:

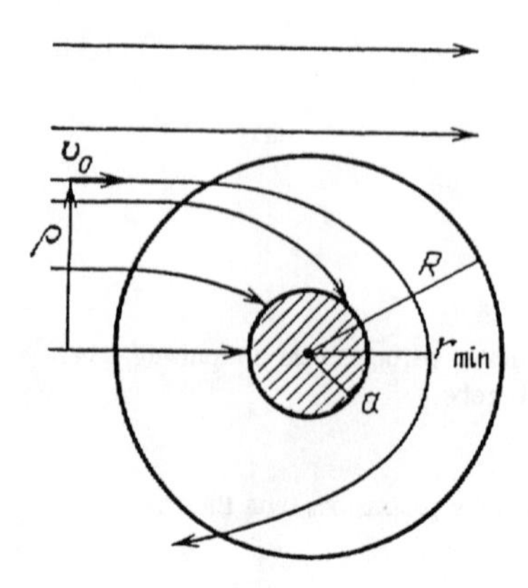

Fig. 6.5. Trajectories of particles near an attracting spherical or cylindrical probe

$$\frac{mv_0^2}{2} = \frac{mv^2}{2} - eV(r) \,, \quad mvr \sin\theta = mv_0\varrho \,, \tag{6.9}$$

where r is the distance from the center to a point moving along the trajectory, v is the velocity of the point, and θ is the angle between the direction of motion (tangent to the trajectory) and the radius vector $\boldsymbol{r}$ connecting the center to the moving piont. The minimum distance of approach $r_{\min}$ of the particle to the center satisfies (6.9) for $sin\,\theta = 1$, because the tangent to the trajectory is perpendicular to $\boldsymbol{r}_{\min}$. If $r_{\min}$ is less than the probe radius a, the particle inevitably strikes its surface, and if $r_{\min}$ is greater, it goes by the probe.

Note that these effects, characterizing charges of both signs, are of greater importance for ions. In practical work, the electronic part of the probe characteristic ($V_p > 0$), corresponding to attraction, is rarely used. In order to reduce the effect of a probe on the plasma, small probes are preferable; electronic current in the range $V > V_S$ is very high, melting the probe. Consequently, the *ionic* part of the $V - i$ curve is typically employed for measuring the charge density in plasma (Sect. 6.5).

6.4 Procedure for Measuring the Distribution Function

6.4.1 Application of a Low ac Voltage

Direct differentiation of an experimental $i(V)$ curve, let alone double differentiation, involves considerable errors. For this reason, d^2i/dV^2 is found by indirect means. Thus it is advisable to superpose on the constant probe voltage V_c a small ac component: $V = V_c + V_a \sin\omega t$. If $V_a \ll V_c$, then

$$i(t) \approx i(V_c) + \left(\frac{di}{dV}\right)_{V_c} V_a \sin\omega t + \frac{1}{2}\left(\frac{d^2i}{dV^2}\right)_{V_c} V_a^2 \sin^2\omega t$$

for each value of V_c. Averaging over time gives

$$\Delta i = \langle i \rangle - i(V_c) = \frac{1}{4}\left(\frac{d^2i}{dV^2}\right)_{V_c} V_a^2 \,.$$

To achieve greater accuracy, the main component $i(V_c)$ is cancelled out by a balancing circuit. Then the time-averaged increment to current, with known amplitude V_a, immediately yields the second derivative.

A small, constant increment in the current can be measured only if the discharge parameters are highly stable. The method can be improved by modulating the amplitude of a high-frequency (ω) the ac voltage with a low frequency ω_1, i.e. $V_a = V_{a0}(1 + \cos\omega_1 t)$. The second derivative is then related to the low-frequency oscillating component of current:

$$\Delta i = \frac{1}{2}\left(\frac{d^2i}{dV^2}\right)_{V_c} V_{a0}^2 \cos\omega_1 t \,.$$

This component is easier to measure. High-frequency-oscillations of current are automatically averaged by the instrument because it cannot resolve them. In the experiments to be described below, the carrier frequency was $\omega/2\pi = 8.4 \cdot 10^4$ Hz, and the modulation frequency was $\omega_1/2\pi = 500$ Hz.

6.4.2 A Typical Result

To illustrate the applications of probe techniques, we will give the results obtained for the electron distribution function in the positive column of a glow discharge. The discharge was produced in a glass tube 2.5 cm in diameter and 50 cm long. Cylindrical probes 0.03 and 0.06 mm in diameter and 6 mm long were sealed into the central part of the tube, parallel to the axis. Figure 6.6 shows a typical current-voltage characteristic (on a semilogarithmic scale) and gives the second derivative of current. At the knee of the $V-i$ curve, d^2i/dv^2 reverses its sign (see Fig. 6.3); this fact rather facilitates the determination of the bend point and of the space potential. Figure 6.7 plots energy distribution functions $n(\varepsilon)$ recalculated on the basis of $f_0(v)$. They are given in arbitrary units, the quantity 100 being assigned to the maximum. The corresponding Maxwell distributions are also shown. The distribution temperatures were found from straight segments in the range of low electron energies $\varepsilon = eV_p$ present on the plots of ln i vs. V. The contribution of energetic electrons to the spectrum is seen to drop in comparison with the maxwellian curve, as a result of electron energy loss to excitation.

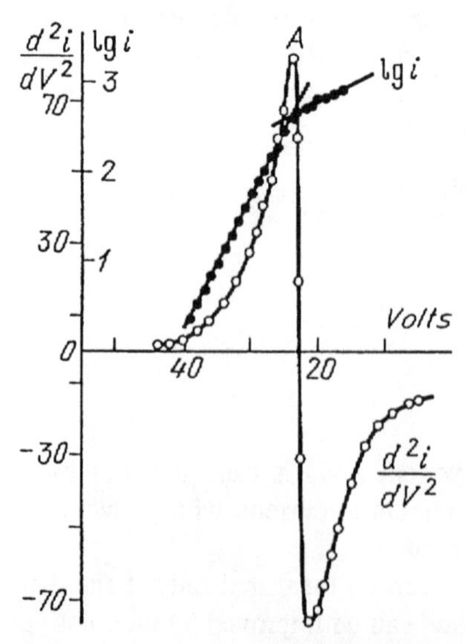

Fig. 6.6. Example of probe measurements. Discharge in mercury vapor. Wire probe on the tube axis, 0.03 mm in diameter, 6 mm in length [6.2]

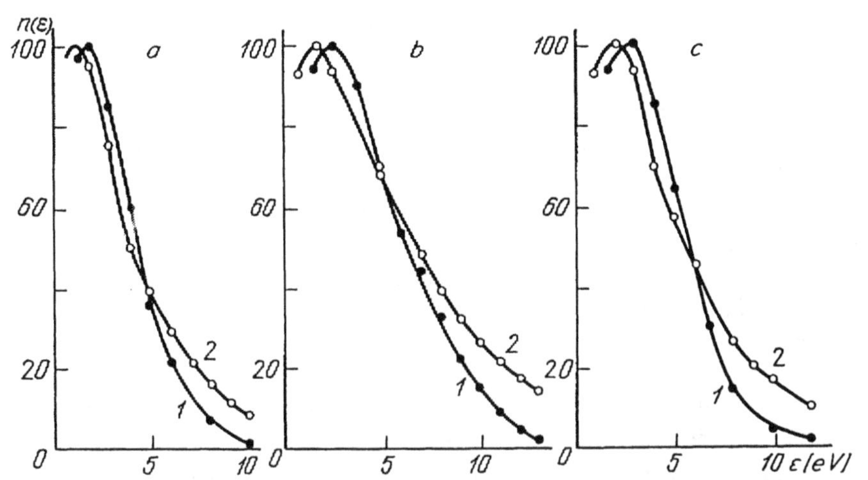

Fig. 6.7. Examples of distribution functions in neon, obtained by probe measurements. **(a)** $p = 1$ Torr, $i = 100$ mA; **(b)** $p = 1$ Torr, $i = 25$ mA; **(c)** $p = 1.6$ Torr, $i = 25$ mA. *(1)* Measured distribution, *(2)* Maxwell distribution of the same mean energy [6.2]

6.5 Ionic Current to a Probe in Rarefied Plasma

Let a probe be connected to a negative potential so much higher than the electron temperature (say, by an order of magnitude) that all electrons are repelled from the probe and make no contribution to the probe current. The probe is surrounded by a layer of positive space charge. Assume that the layer is thin, so that the surface area of its outer boundary differs only slightly from the probe area S. If the mean-free-path length of ions is much greater than the probe size (and hence, than the layer thickness), the surrounding plasma is only slightly perturbed and the ionic current can seemingly be evaluated using (6.3) after $\bar{v}_e$ is replaced with the thermal velocity of ions, $\bar{v}_+ = (8kT/\pi M)^{1/2}$. However, if the charge density n_0 of weakly ionized plasma is determined in this way from the measured ionic current at the lower low-slope part of the $V - i$ curve, it is found to be systematically *greater* than the value calculated by using (6.3) and the electronic saturation current (at the upper break point B).

Much theoretical work, including that of Langmuir himself and his coworkers, was devoted to clarifying this situation. The clear answer was given by *Bohm, Burhop,* and *Massey* in 1949. Varous versions of the detailed theory, which includes the analysis of ion trajectories in the space-charge layer (Sect. 6.3.6) and of the potential distribution in it, are very complicated [6.1, 6.3, 6.4]. We will describe the process in the simplest possible way, aiming only at clarifying the physical essence of the phenomenon and finding an evaluation formula for the ionic current.

6.5.1 Saturation Current

The gas and ion temperature T in a weakly ionized plasma is less, by more than an order of magnitude than the electron temperature $T_e \approx 1\,\mathrm{eV}$. The charge neutrality of plasma far from the probe begins to be violated where electrons are appreciably decelerated in the repulsive field, that is, where the negative potential with repsect to the nonperturbed plasma equals roughly $|V_b| \approx kT_e/e$. This is the external boundary of the space-charge layer (Fig. 6.8). The electron density at the boundary is given by the Boltzmann law: $n_b = n_0 \exp(-e|V_b|/kT_e) \approx n_0/\bar{e}$, where $\bar{e}$ is the base of natural logarithms. On the outside of this boundary, low-energy ions manage to "keep up" with more energetic electrons so as to maintain quasineutrality, that is, the ion density at the boundary is quite close to n_b. However, since $kT \ll kT_e \approx e|V_b|$, ions are relatively strongly influenced by the field outside the layer as well, namely, in the outer *pree-sheath*, where the potential V lies in the interval $kT < |eV| < kT_e$. Inside it the plasma is quasineutral but the field imparts to ions a velocity much higher than their thermal value. They enter the layer from the pre-layer with a velocity normal to the boundary surface given by

$$v_+ \approx (2e|V_b|/M)^{1/2} \approx (2kT_e/M)^{1/2} \approx (T_e/T)^{1/2}\bar{v}_+ \,.$$

As a result, the ionic saturation current in the idealized plane case is approximately equal to

$$i_{+\mathrm{sat}} \approx Sen_b v_+ \approx \left(\sqrt{2}/\bar{e}\right) Sen_0 \sqrt{kT_e/M} \,. \tag{6.10}$$

More detailed calculations give the same result, and even a very close value for the numerical coefficient that in (6.10) equals $\sqrt{2}/\bar{e} = 0.52$. In the case of a spherical or thin cylindrical probe, the ionic current grows with increasing negative potential, for reasons described in Sect. 6.3.6.

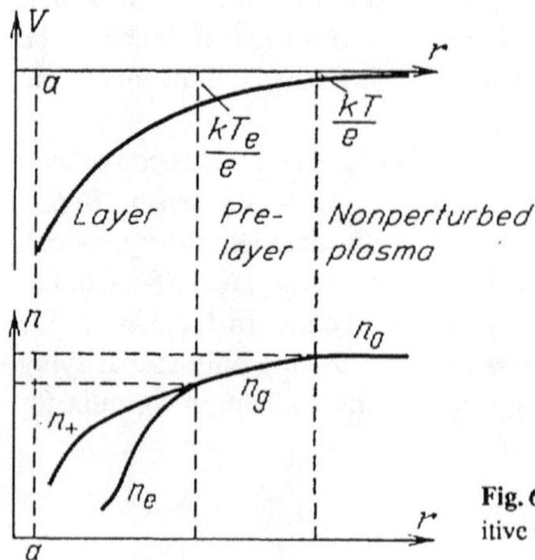

Fig. 6.8. Variation of potential, electron density, and positive ion density near a negative probe

6.5.2 Measurement of Charge Density in Plasma

This quantity is conveniently determined from T_e and the measured ionic current at the lower low-slope part of current-voltage characteristic, using (6.10) or corrected, more complex formulae provided by the small probe theory.[3] This is the method that is mostly used in practice for finding n_0, especially if the ionic current is weakly dependent on potential, the $V - i$ curve has a low-slope, and the simple formula (6.10) is valid. As a comparison of (6.3) and (6.10) shows, the electronic saturation current exceeds the ionic current by a factor of about $\sqrt{M/m} \sim 10^2$. We have mentioned already that the former produces greater plasma perturbations and melts small probes. Formula (6.10) is used for the rapid evaluation of spatial electron density distributions in plasmas. Typically, the electron temperature is uniform in space and need not be measured at each point.

6.5.3 Floating Potential

In the case of the Maxwell distribution of electrons, the electronic current at a negative probe potential is given by (6.2) and the ionic current by (6.10). The negative probe potential with respect to plasma, corresponding to zero probe current, is found by equating the two expressions:

$$e|V_f|/kT_e \approx \ln\left[\left(\bar{e}/\sqrt{4\pi}\right)\sqrt{M/m}\right] \approx \ln\left(0.77\sqrt{M/m}\right) . \qquad (6.11)$$

The floating potential V_f [V] is about $-3.3T_e$ [eV] in hydrogen and $-6.3T_e$ in argon. The simplest way to measure the spatial plasma potential distribution V_S is to determine at each point the probe potential V at which the probe current vanishes. The space potential is found by adding $|V_f|$ of (6.11) to the measured potential. If the electron temperature (in a more general case, the mean energy) is spatially uniform, that is, $V_f \approx$ const, a constant potential bias does not affect the electric field distribution, determined by the differences of V between neighbouring points.

6.6 Vacuum Diode Current and Space-Charge Layer Close to a Charged Body

Estimates of the possibility of working with the simplest formulae of the probe theory obtained for the plane case are based on comparing the space-charge layer thickness with the probe size. We will evaluate the layer thickness in the simplest situations, but those of main interest to probe diagnostics, in the extreme situations in which the sign of particles flowing to the probe coincides with that

[3] In fact, difficulties are encountered in estimating the correspondence of the experimental conditions to the limits of applicability of a specific approximate formula; these limits are not always clearly defined.

of the space charge. These are the cases of positive or sufficiently high negative potentials, that is, of saturation currents. In the range of the steep part of the $V-i$ curve, where the space charge at the probe repelling electrons is positive and the current is electronic, this current is independent of layer geometry (Sect. 6.3.5). The classical problem of current in the *vacuum diode* can be used as a suitable model fo considering a layer at the probe and a number of other situations, such as the cathode layer of arc discharges (Sect. 10.5.3).

6.6.1 Space-Charge-Limited Current in the Vacuum Gap

Let a voltage V be applied to plane electrodes separated by a distance h. External factors cause the emission of charges from one electrode, so that dc current flows through the circuit. This model represents an actual device, namely, the vacuum diode with a heated cathode emitting thermionic current. If the heater current is low and so is the emission current, the number of electrons in the surrounding space is also small. They do not produce an apprciable field, so that the potential φ is distributed in the gap exactly as in the absence of charges: $\varphi = -Ex$, where $E = -V/h$ is the field; φ and x are measured from the cathode (curve 1 in Fig. 6.9). If the voltage is not too low, the field transports all the electrons to the anode. The current coincides with the emission current (of density $j_{\rm em}$) and is independent of V.

If the emission current is considerable, the gap fills up with a large number of charges that produce their own field. The potential distribution $\varphi(x)$ is affected by the space charge. The electron density distribution, in its turn, depends on $\varphi(x)$. This *self-consistent* picture is described by Poisson's equation, current continuity equations, and the equation for electron energy, $mv^2/2 = mv_0^2 + e\varphi$, that determines their velocity $v(x)$. An electron leaves the cathode at a velocity v_0. If one assumes for simplicity that all electrons have identical velocities, the absolute value of current density is $j = en_e v = \text{const}$. Therefore

$$\frac{d^2\varphi}{dx^2} = 4\pi e n_e = \frac{4\pi j}{v} = \frac{4\pi j}{v_0(1 + 2e\varphi/mv_0^2)^{1/2}}\,; \tag{6.12}$$

note that $\varphi(0) = 0$, which has already been used in writing the equality for $v(x)$, and $\varphi(h) = V$.

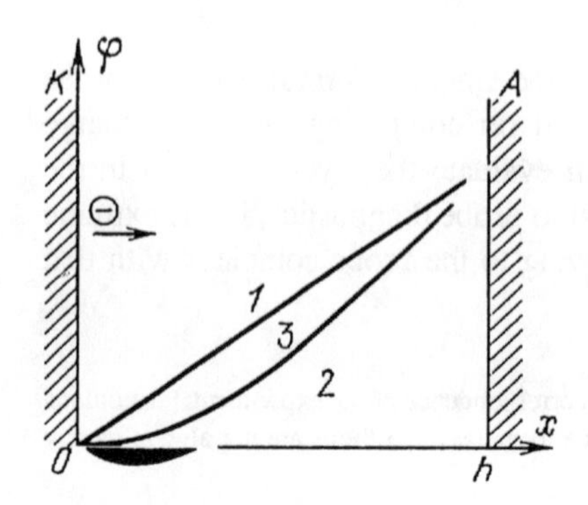

Fig. 6.9. Potential distribution in plane vacuum gap when electrons are emitted by the cathode. *(1)* very low emission current, the field is not distorted by space charge; *(2)* potential barrier is formed for electrons when they leave the cathode at a finite velocity; *(3)* electrons are ejected at zero velocity

Let us look at the behaviour of $\varphi(x)$. Electrons accumulate in the vicinity of the cathode, while the field has not yet accelerated them. This negative charge repels the electrons of the metal away from the surface, forming a *double layer*: plus on the metal and minus in its vicinity. A reverse field $E > 0$ appears at the cathode, decelerating the emerging electrons, so that the potential φ falls below the potential $\varphi = 0$ of the cathode itself. However, $\varphi > 0$ at the anode, and hence, $\varphi(x)$ passes through a minimum $\varphi_{\mathrm{m}} < 0$ (curve 2 in Fig. 6.9). Electrons face a potential barrier, $-e\varphi_{\mathrm{m}}$, that they have to overcome at the expense of the initial energy $mv_0^2/2$.

How high can this barrier be? Assume that $|e\varphi_{\mathrm{m}}| > mv_0^2/2$. Then none of the electrons could cross it and the current is zero. This is possible only at very low V; this case holds no interest for us. Let $|e\varphi_{\mathrm{m}}| < mv_0^2/2$. Then all (assumed to be monoenergetic) emerging electrons cross the barrier; the current is that of saturation, j_{em}. The case of interest is intermediate: the space charge limits, but does not cut off, the current, $0 < j < j_{\mathrm{em}}$; a case that is realized if the barrier height (assuming electron energies to be identical) is exactly equal to the initial energy of the electrons. Formally, electrons pass across the barrier peak at zero velocity, their density at this point being infinite although j ="$0 \cdot \infty$" is finite. The solution of (6.12) that satisfies the condition $d\varphi/dx = 0$ for $\varphi = \varphi_{\mathrm{m}} = -mv_0^2/2e$ selects the value of current that the gap can let through at the voltage V when the space charge does not let all the emitted electrons reach the anode.[4]

Let us introduce a dimensionless potential $\psi = 2e\varphi/mv_0^2$ into (6.12):

$$\frac{d^2\psi}{dx^2} = \frac{4}{9x_0^2}\frac{1}{\sqrt{1+\psi}}\,, \quad x_0 = \left(\frac{mv_0^3}{18\pi ej}\right)^{1/2}. \tag{6.13}$$

The physical meaning of the length scale x_0 will be clear immediately. Note that $d^2\psi/dx^2 = (1/2)d(d\psi/dx)^2/d\psi$ and integrate (6.13). To find the arbitrary constant, we make use of the condition $d\psi/dx = 0$ for $\psi = -1$. We obtain

$$\frac{d\psi}{dx} = \mp\frac{4}{3x_0}(1+\psi)^{1/4}\,, \quad (\mp) \text{ for } x \lessgtr x_{(\mathrm{min})}\,.$$

Let us integrate these two equations. The constant in the equation with the sign (−) is found from the condition $\psi = 0$ at $x = 0$. The result shows that the minimum $\psi_{\mathrm{m}} = -1$ is attained at a distance $x_{(\mathrm{min})} = x_0$ from the cathode. The constant in the solution of the equation with (+) is found by stipulating that the solution pass through the same point x_0, $\psi = -1$. This gives the potential distribution plotted by curve 2 in Fig. 6.9:

$$\frac{2e\varphi}{mv_0^2} = \left[\pm\left(\frac{x}{x_0} - 1\right)\right]^{4/3} - 1\,, \quad (\pm) \text{ for } x \gtrless x_0\,. \tag{6.14}$$

[4] Real emitted electrons possess various velocities and a quite definite barrier is formed: $|e\varphi_{\mathrm{m}}|$. The electrons with higher initial energy overcome the barrier while the slower ones are turned back. The correct solution is obtainable only by numerical integration; actually, the same result is obtained in the limit $V \gg |\varphi_{\mathrm{m}}|$ [6.5].

Assigning (6.14), with the sign (+), to the anode coordinate $x = h$, where $\varphi = V$, we find the relation defining the current $j(V, h, v_0)$. Nearly always $eV \gg mv_0^2/2 \sim kT$, where T is the cathode temperature (several tenths of a volt). Hence, $h \gg x_0$ and the "1"s in (6.14) can be neglected everwhere except in the immediate vicinity of the cathode. Hence, $\varphi \propto x^{4/3}$, $E \propto x^{1/3}$, $v \propto x^{2/3}$, $n_e \propto x^{-2/3}$, and the current is

$$j = \frac{1}{9\pi}\sqrt{\frac{2e}{m}} \cdot \frac{V^{3/2}}{h^2} = 2.34 \cdot 10^{-6} \frac{\{V[\mathrm{V}]\}^{3/2}}{\{h[\mathrm{cm}]\}^2} \frac{\mathrm{A}}{\mathrm{cm}^2} . \qquad (6.15)$$

This relation, giving the space-charge-limited current in a planar vacuum diode, is known as the Child-Langmuir equation derived in 1913 or the *law of three-halves power*: the current is proportional to $V^{3/2}$. In this approximation, the vanishingly small minimum of φ sits on the cathode. Law (6.15) can be derived in an elementary manner, by integrating (6.12) under an additional condition $E = 0$ for $x = 0$, which is equivalent to assuming that electrons leave the cathode at zero velocity. The independence of j from j_{em} must be interpreted as the unlimited emission capability of the cathode. On the side of high voltages and currents, the validity of (6.15) is limited by the condition $j < j_{\mathrm{em}}$, and on the side of low values, by the condition $V \gg kT/e$ and the stipulation that space charge strongly affects the current.

We will illustrate this with a numerical example. If $T \approx 1000\,\mathrm{K} \approx 0.1\,\mathrm{eV}$, an oxide cathode emits $j_{\mathrm{em}} \sim 1\,\mathrm{A/cm^2}$. If a voltage $V = 100\,\mathrm{V}$ is applied to a gap $h = 1\,\mathrm{cm}$, the current will be only $j \approx 2.3 \cdot 10^{-3}\,\mathrm{A/cm^2}$; only two electrons out of 1000 will break through the barrier raised by the electrons themselves.

The $i \propto V^{3/2}$ equation (6.15) holds for spherical and cylindrical diodes, but h^2 is replaced with the product of one radius squared by a function of the ratio of the radii; these functions are tabulated in [6.5]. Formula (6.15) also holds for ion emitters after m is replaced with M, and V with $|V|$.

6.6.2 Evaluation of Plane Layer Thickness

This can be evaluated using the result above (and the relations for the appropriate diodes in the cases of the spherical and cylindrical probes). The probe acts as a charge-collecting electrode. The boundary of the quasineutral plasma, from which particles are injected into the space-charge region, acts as the emitting electrode. Note that the dependent variable changes. In a diode, the gap width is fixed and the current "tunes up" to the applied voltage. The layer at the probe receives a certain saturation current, imposed by the thermal gas-kinetic electron flux in the case of positive probe potential, and by the V_{p}-independent ion flux from the pre-layer to the layer in the case of high negative potentials. As for the layer, its thickness h adjusts itself to the probe potential.

For $V_{\mathrm{p}} > 0$ and the Maxwell electron distribution, formulas (6.3) and (6.15) yield

$$h \approx (8/9)^{1/2}(eV_{\mathrm{p}}/kT_{\mathrm{e}})^{3/4} d\,, \quad d = (kT_{\mathrm{e}}/4\pi e^2 n_0)^{1/2} .$$

The thickness scale for the negative space-charge layer is the *Debye radius* d, but the thickness increases with increasing potential. If the negative potential is high, so that the current is due only to ions, (6.10) and (6.15) (with m replaced by M) give the same thickness of the positive charge layer:

$$h \approx (4\bar{\epsilon}/9)^{1/2}(e|V_p|/kT_e)^{3/4}d\,.$$

But in this case the potential $|V_p|$ is much higher. It exceeds the electron temperature by about an order of magnitude, so as to have the electron current smaller than the ion current. The ionic layer thickness is therefore an order of magnitude greater than the Debye radius. Qualitatively similar results are obtained for the sphere and cylinder, but the theory gets very complicated when these cases deviate very much from the plane case, that is, when h is much greater than the probe radius [6.3, 4].

6.7 Double Probe

A double probe designed for plasma diagnostics in the absence of a reference electrode was first used in [6.6, 7]. Two Langmuir probes are introduced into the plasma and connected via a potentiometer to a dc supply unit so as to vary not only the voltage V between the probes but its polarity as well (Fig. 6.10).

6.7.1 Probe Characteristic

The current-voltage characteristic of a double probe in an electrodeless high-frequency discharge is shown in Fig. 6.11. Let us discuss its physical meaning under the assumption of *identical probes* and *identical plasma parameters* at the points where the probes are located. The characteristic in the figure is *symmetric,* which demonstrates that in this particular experiment the above conditions were

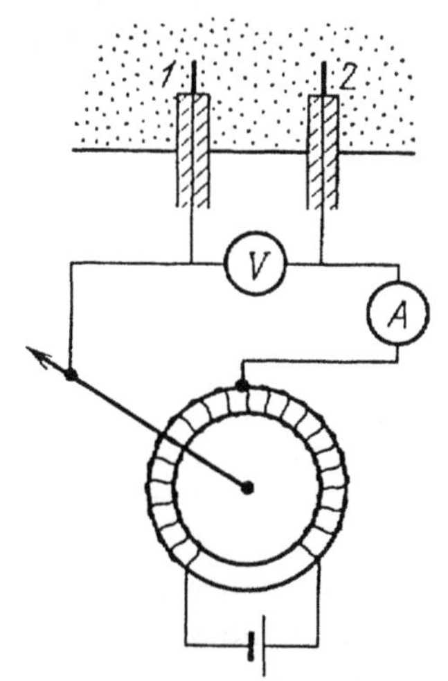

Fig. 6.10. Double probe circuit with circular potentiometer [6.4]

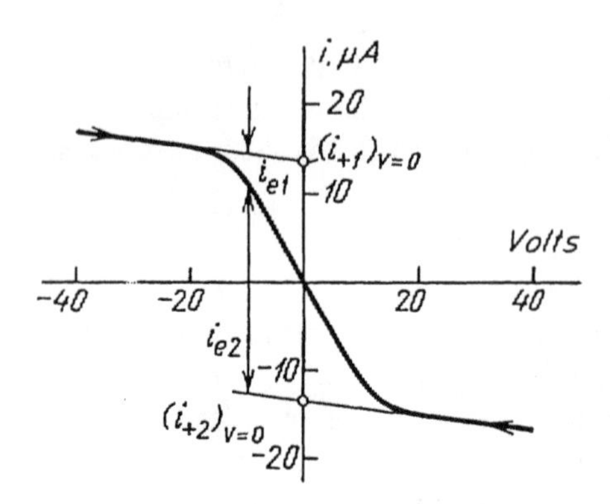

Fig. 6.11. Characteristic of symmetric double probe. Recorded by the circuit of Fig. 6.10 in electrodeless rf discharge [6.4]

satisfied with high accuracy. If the plasma potentials V_S at the two probes are equal, then no current flows through the probe circuit at zero voltage, $V = 0$: $i = 0$. This condition has also been met in this experiment: apparently, the probes were placed very close, and the potential gradient in the plasma was small. The two probes are at the same floating potential $V_f < 0$ (Fig. 6.12a).

Let us denote the potential of the left-hand probe with respect to the plasma by V_{p1}, and that of the right-hand probe by V_{p2}. For the direction of the voltage axis, we set: $V = V_{p1} - V_{p2}$. The electric current i is assumed positive if it flows from the plasma into the left-hand probe; as before, i_e and i_+ are the magnitudes of the electronic and ionic probe currents (the signs of i and V are in accord with the orientation of the curve in Fig. 6.11). The amount of positive charge flowing from the plasma into one of the probes equals the amount flowing into the plasma from the other probe; hence,

$$i = i_{+1} - i_{e1} = -(i_{+2} - i_{e2}) \,, \quad i_{+1} + i_{+2} = i_{e1} + i_{e2} \,. \tag{6.16}$$

The potential of neither of the probes can be positive. Indeed, if V_p were positive the probe would receive the electronic saturation current. According to (6.16), the electric circuit must be closed at the other probe where a substantially smaller ionic current flows. Therefore, not only is the entire probe system floating, that is, negatively charged with respect to the plasma, but each of the probes is necessarily negative. Let the negative side of the power supply be connected to the left-hand probe and the positive terminal to the right-hand one ($V < 0$). The current in the plasma flows from (+) to (−). Therefore, the ionic curent dominates at the left-hand probe and the electronic current at the right-hand one. If the voltage $|V|$ is high (in comparison with kT_e/e), the left-hand probe is strongly negative, while the right-hand one is less negative than the floating potential (Fig. 6.12b). The left-hand probe receives the purely ionic saturation current. The dependence $i(V)$ is then weak, corresponding to the left-hand low-slope part of $V - i$ curve. If V is small and negative, the ionic current to the left-hand probe is partly compensated for by the electronic current. However, the latter strongly depends on the probe potential, [namely, by Boltzmann's law (6.2)]. This is the steep segment of the $V - i$ curve; the current drops sharply to

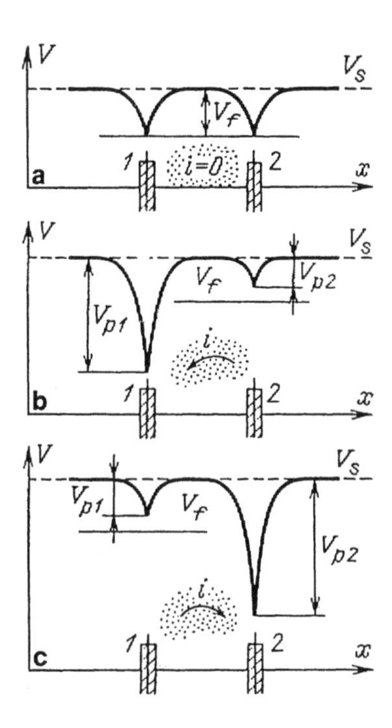

Fig. 6.12. Double probe potential: (a) Floating potential, no current through the probes; (b) Left-hand probe is strongly negative, and receives ionic saturation current; (c) Polarity reversal; the right-hand probe is strongly negative

zero as $V \to 0$. The right-hand part of the $V - i$ curve exactly retraces the left-hand one, corresponding to polarity reversal: "plus" connected to the left-hand probe and "minus" to the right-hand probe (Fig. 6.12c).

Langmuir's formula (6.2) is applicable to electronic currents i_{e1} and i_{e2}, because the potentials of both probes are negative with respect to the plasma. In view of (6.16),

$$i = i_{+1}(V_{p1}) - i_{e,sat} \exp(eV_{p1}/kT_e) = -i_{+2}(V_{p2}) + i_{e,sat} \exp(eV_{p2}/kT_e) , \tag{6.17}$$

where ionic currents weakly depend on the negative probe potential that accelerates ions.

6.7.2 Measurement of Plasma Parameters

Differentiating the first equation of (6.17) with respect to V, and then considering the symmetry point of the $V - i$ characteristic, where $i = 0$ by setting $i_{e1}(V_f) = i_{+1}(V_f) \equiv (i_+)_0$ results in

$$\left(\frac{di}{dV}\right)_0 = \left[\left(\frac{di_{+1}}{dV_{p1}}\right)_0 - \frac{e}{kT_e}(i_+)_0\right]\left(\frac{dV_{p1}}{dV}\right)_0 . \tag{6.18}$$

If the probe current vanishes at zero potential difference V, polarity reversal simply changes the roles of the two probes. As a result, in addition to the relation $V = V_{p1} - V_{p2}$, the probe potentials are related by the equation $V_{p1}(V) = V_{p2}(-V)$.

Differentiating these equations with respect to V, we obtain

$$1 = (dV_{p1}/dV)_V - (dV_{p2}/dV)_V \ , \quad (dV_{p1}/dV)_V = -(dV_{p2}/dV)_{(-V)} \ .$$

Hence, at the symmetry point, $(dV_{p1}/dV)_0 = 1/2$. Now (6.18) implies

$$e/kT_e = (i_{+1})_0^{-1} \left[(di_{+1}/dV_{p1})_0 - 2(di/dV)_0 \right] \ . \tag{6.19}$$

This formula is used to measure electron temperature. The derivative $(di/dV)_0$ is found from the slope of the measured $V - i$ curve, at a point where $i = 0$. The ionic current and its derivative (note that the latter is much lower than the derivative of the total current) can be found by extrapolating the low-slope part of the $V - i$ curve linearly to the symmetry point (Fig. 6.11). We assume here that $V \approx V_{p1}$ in the range of high negative voltages, that is, the function $i_{+1}(V)$ measured in this range can be treated as $i_{+1}(V_{p1})$, which justifies the extrapolation.

Another method of measuring T_e resembles the technique employed in the single-probe method. Let us divide the second equation of (6.16) by i_{e1}, use Langmuir's formula (6.2) for electronic currents, and take the logarithm of the result:

$$\ln \left(\frac{i_{+1} - i_{+2}}{i - i_{+1}} - 1 \right) = \frac{eV}{kT_e} \ . \tag{6.20}$$

Now T_e can be found from the slope of the straight line in the neighbourhood of the symmetry point after ionic currents have been determined by extrapolation and the semilogarithmic curve (6.20) has been plotted as a function of V.

The electron density in the plasma is found as in the single-probe technique, using (6.10) for the ionic saturation current, after T_e has been found and the current $i \approx i_{+\mathrm{sat}}$ at the low-slope part of the $V - i$ curve measured.

6.7.3 Measurement of Electric Field

In the case of a potential gradient in a plasma, the space potentials at the pints where the probes are placed, V_{S1} and V_{S2}, are different. For both probes to float and probe currents to vanish, they must be placed at a potential difference $\Delta V = V_{p1} - V_{p2} = V_{S1} - V_{S2}$. By varying the voltage to bring the probe current to zero, and knowing the distance Δx between probes, we find the field $E_x = \Delta V/\Delta x$. However, the plasma potential cannot be found by the double-probe technique in prinicple.

If $\Delta V \neq 0$, the methods of determining T_e and n_0 remain valid, but the symmetry point on the $V - i$ curve at which $i = 0$ shifts along the V axis by ΔV. Formulas (6.19) and (6.20) must operate with this new point. The current in a double probe being much lower than the electronic current of a single probe (which can be used for positive ion currents only), the former produces a much smaller disturbance in the plasma. This is why it is used also for studying nonstationary processes that are especially sensitive to perturbation, since the entire picture of plasma evolution may be disrupted otherwise.

6.8 Probe in a High-Pressure Plasma

When we speak of high pressure, we mean the situation in which the free path lengths of electrons and ions, l_e and l_+, are very small compared with the probe size. The probe is assumed to be spherical, of radius a. For instance, for $p =$ 30 Torr we find $l_e \sim 10^{-3}$ cm, $l_+ \sim 10^{-4}$ cm, while $a \sim 10^{-1} - 1$ cm. Under typical discharge conditions ($T_e \approx 10^4$ K, $n_0 \sim 10^9 - 10^{10}$ cm^{-3}), the Debye radius is $d = (kT_e/4\pi e^2 n_0)^{1/2} \sim 10^{-2}$ cm, so that $l \ll d \ll a$. In a dense gas, fluxes of charged particles to a probe are formed as a result of *diffusion* and *drift*. In this case the theory seems to be even more complicated than for rarefied plasma, and has not been fully developed. Furthermore, quantitative results can be obtained only using the data on free-path lengths or mobilities; this introduces additional errors into the final results. Without going into the details of the theory, let us dwell on several general features that will facilitate our understanding of the physical processes.

6.8.1 Approximate Equilibrium in Electron Gas

Fluxes of charged particles can be treated as sums of the diffusion and drift components [see (2.20)]. The electric field $\boldsymbol{E}$ is produced by the voltage applied to the probe and by plasma polarization. Let a negative potential, decelerating the electrons, be applied to the probe. If the potential equals the floating value, the electron and the ion fluxes to the probe are exactly equal. If the potential is more negative, the electron flux is less than the ionic flux. If it is more positive, but only slightly, than the floating value, the electron flux exceeds the ionic flux but has a comparable magnitude. In such cases (when the probe potential is neither positive nor weakly negative), the ionic layer of positive space charge around the probe transforms into a layer of perturbed but now quasineutral palsma that gradually changes into nonperturbed plasma of charge density n_0. The perturbation due to the probe manifests itself in the quasineutral layer as a gradient of the densities $n_e \approx n_+ \approx n$. The densities decrease towards the probe, which ensures the diffusion flux of charges from a remote region to the probe surface absorbing them. The quasineutrality as such follows from the smallness of the Debye radius in comparison with the characteristic length of density variation, which is of the order of the probe size.

As we have mentioned, if the probe potential is sufficiently negative, the electronic current is comparable to the ionic one or is smaller. But the diffusion coefficient D_e and mobility μ_e of electrons are much higher than D_+ and μ_+ of ions. Therefore, oppositely directed high diffusion and drift electronic fluxes balance each other out to yield a relatively small resultant electron flux Γ_e comparable with (or less than) the ionic flux Γ_+. In view of the Einstein relation (2.24) between D_e and μ_e, and assuming in (2.20) that $\Gamma_e \approx 0$, we find that

$$\frac{\nabla n_e}{n_e} \approx -\frac{\mu_e}{D_e}\boldsymbol{E} = \frac{e}{kT_e}\nabla V , \quad n_e \approx n_0 \exp\frac{eV}{kT_e} , \tag{6.21}$$

where $V(r) < 0$ is the potential measured with respect to the unperturbed plasma. The electron density distribution in a remote neighbourhood of a negative probe is thus approximately described by the Boltzmann law.

6.8.2 Spatial Distributions of Charge Density and Potential in the Quasineutral Region Around a Negative Probe

The electric field in the quasineutral region (quasi-equilibrium for electrons) is related to the gradient of n_e by the first formula of (6.21). Since $n_+ \approx n_e \approx n$, we introduce this value of $\boldsymbol{E}$ into equation (2.20) for the ionic flux. The total ionic current across any spherical surface is

$$i_+ \approx 4\pi r^2 e \left[D_+ + \mu_+ \frac{kT_e}{e} \right] \frac{dn}{dr} \, . \tag{6.22}$$

The expression in brackets is the familiar *coefficient of ambipolar diffusion* (2.25) in which $\mu_e \gg \mu_+$. If the probe is at floating potential and the electronic and ionic fluxes to the probe are equal, the flux of ions to the absorbing body is in fact the result of ambipolar diffusion (Sect. 2.6). At a different potential, the electron and ion fluxes differ by a quantity comparable with the ambipolar flux. In any case, the ionic flux in the quasineutral zone coincides in magnitude and direction with the ambipolar diffusion flux. In a nonequilibrium flux with $T_e \gg T$, the drift component of the ionic flux, which is proportional to the second term of (6.22), is greater than the diffusion component by a factor of $T_e/T \gg 1$; one is justified then to speak of the pure drift of ions in the total field $\boldsymbol{E}$ of the probe and polarization fields. We neglect the small term depending on D_+ in (6.22).

Under stationary conditions and neglecting relatively slow processes of creation and removal of charges in the plasma region perturbed by the probe (assumed to be small), we have $i_+(r) = \text{const}$. Integration of (6.22) yields the charge distribution in th quasineutral zone which transforms asymptotically into the nonperturbed plasma as $r \to \infty$:

$$n = n_0 \left(1 - \frac{R}{r}\right) \, , \quad R = \frac{i_+}{4\pi e n_0 \mu_+ (kT_e/e)} \, . \tag{6.23}$$

The physical meaning of R will now be clarified. According to (6.21) and (6.23), and since $n_e \approx n$, the potential and field distributions in the quasineutral region are

$$V = -\frac{kT_e}{e} \ln \frac{1}{1 - R/r} \, , \quad E = -\frac{kT_e}{eR} \frac{R^2}{r^2} \frac{1}{1 - R/r} \, . \tag{6.24}$$

As we move from infinity to the probe, V and E grow from zero, with scales kT_e/e and kT_e/eR, and tend formally to $-\infty$ as $r \to R$, when formally $n \to 0$.

The radius R corresponds to the boundary of the space charge that separates the probe from the quasineutral region: as $r \to R$, (6.23) and (6.24) cease to be valid. Indeed, quasineutrality is violated where the potential $|V(r)|$, decelerating electrons, reaches several times the electron temperature kT_e/e. At this distance, however, $1 - R/r \ll 1$, so that $r \approx R$. Therefore, R can be interpreted as the effective layer radius; the second relation of (6.23) gives the ionic probe current as a function of R.

6.8.3 Ionic Saturation Current and Evaluation of Charge Density in Plasma

The higher the negative probe potential $|V_p|$, the greater the positive space-charge layer thickness $h = R-a$ and the greater R. According to (6.23), the ionic current grows with increasing R. However, if the layer is thin, R is approximately equal to the probe radius a and the ionic current is independent of potential. The corresponding quantity i_+ is the ionic saturation current. This occurs either when the probe is large and the layer thickness h (in general, it is characterized by the Debye radius d) is small in comparison with a, or if the Debye radius is small, that is, the electron density in the plasma is sufficiently high.

Expressing the ion mobility in (6.23) in terms of free path length, $\mu_+ = el_+/M\bar{v}_+$, where $\bar{v}_+ = (8kT/\pi M)^{1/2}$, and introducing the probe surface area $S = 4\pi a^2$, we can rewrite the ionic saturation current in the form

$$i_{+,\text{sat}} = S\left(\frac{\pi}{8}\right)^{1/2} en_0 \left(\frac{kT_e}{M}\right)^{1/2} \left(\frac{T_e}{T}\right)^{1/2} \frac{l_+}{a} . \tag{6.25}$$

This differs from (6.10) for rarefied plasmas in the last two factors. The first one, $(T_e/T)^{1/2}$, does not exceed 10; the second one is, according to the initial assumption, a very small quantity, much less than 10^{-1}. The ionic saturation current is, therefore, much lower in dense ionized gas than in rarefied gas, and the higher the pressure the lower it is.

The probe characteristic at high pressures is qualitatively similar to the $V-i$ curve in a rarefied plasma; its lower low-slope part corresponds to the ionic current. The charge density n_0 can be estimated on the basis of the measured ionic current via (6.25). Even if the electron temperature in (6.25) cannot be reliably measured, we know that it lies in a much narrower range of values ($T_e \approx 1\,\text{eV}$) than the electron density, which may vary by orders of magnitude in different discharge conditions.

6.8.4 On the Use of Electronic Probe Current for Evaluating Electron Temperature

If the negative probe potential is higher than the floating value, the electronic current is greater than the ionic and increases rapidly as the decelerating potential $|V_p|$ decreases. As in the case of rarefied plasmas, $i = i_e - i_+ \approx i_e$. This situation corresponds to the steep part of the $V-i$ curve, present at high pressure as well. Electrons reach the probe immediately after the last collision, that is, from the sphere of radius $r_1 \approx a + l_e$ that lies at a small distance of order $l_e \ll a$ from the probe surface. Let us denote the density and potential at the radius r_1 by n_{e1}

and V_1, repsectively. The electron probe current can be expressed by a formula similar to (6.2),

$$i_e \approx S(n_{e1}\bar{v}_e/4) \exp\left[e(V_p - V_1)/kT_e\right] , \tag{6.26}$$

because electrons from the sphere r_1 reach the probe without colisions. The electron temperature and $\bar{v}_e(T_e)$ are much less perturbed than the density.

If (6.21) for equilibrium electron density $n_e(V)$ is extrapolated up to the sphere of radius r_1, where $V = V_1$, we define n_{e1} and obtain from (6.26) exactly the expression (6.2) for the probe current, $i \propto \exp(eV_p/kT_e)$. There is some hope, therefore, that the steep part of the $V - i$ characteristic can be used to evaluate the electron temperature (using the semilogarithmic plot ln i vs. V_p). A more detailed analysis [6.8] taking into account that the electron flux in (2.20) is nonzero demonstrated that the electronic current is reduced in comparison with (6.2) by a factor of $\gamma \approx 1 + \alpha(h/l_e)(kT_e/eV_p)$, where α is a numerical coefficient $\approx 1/2$. The reduction may be substantial, of the order of 10, but it depends on V_p much less than the Boltzmann exponential in (6.2). That is why T_e can be estimated on the basis of the slope of the electronic part of the probe characteristic.

6.8.5 Layer of Positive Space Charge

In order to evaluate the thickness or radius of the layer separating the probe from the quasineutral plasma, we have to integrate Poisson's equation. Neglecting the small electron density in the ionic layer, we can write

$$\frac{1}{r^2}\frac{d}{dr}r^2 E = 4\pi e n_+ , \quad E = -\frac{dV}{dr} < 0 . \tag{6.27}$$

As in Sect. 6.6, we give the ion density in terms of the ionic flux (drift flux in this case): $n_+ = \Gamma_+/\mu_+|E|$.[5] Introducing the total ionic current (which is independent of r), according to (6.23), we obtain

$$n_+ = \frac{i_+}{4\pi r^2 e\mu_+(-E)} = -n_0\frac{kT_e}{eE}\frac{R}{r^2} . \tag{6.28}$$

For boundary conditions to (6.27) and (6.28), we can set that the field and potential are zero at the boundary between the layer and the quasineutral region at $r = R$. This approximation reflects the fact that the field is small compared with the average field in the layer.

Even after these simplifications, a compact formula for layer thickness, $h = R - a$, is obtained only if the layer is thin, that is, "plane"; correspondingly, $r^2 \approx \text{const} \approx a^2$. In this case,

---

[5] At very high negative potentials V_p, the field in the layer may be so high that the drift of ions is of the *anomalous* type: $v_{+d} \propto \sqrt{|E|}$ (Sect. 2.5.4).

$$h/a = \left[(3/2\sqrt{2})(d/a)(e|V_p|/kT_e)\right]^{2/3} . \tag{6.29}$$

For a collisionless layer, formula (6.15) gives $h \propto |V_p|^{3/4}$. The plane layer approximation is valid as long as $h < a$. For exmaple: if $d \sim 10^{-2}$ cm, $a \sim 10^{-1}$ cm, and $a/d = 10$, the layer thickness increasing with potential becomes comparable with probe radius, $h/a \approx 1$ for $e|V_p| \approx 10kT_e$. With sphericity taken into account the formulae are very unwieldy, even though the integration does not involve any fundamental problems. If the layer is thick, its radius R and the probe potential obey the crude approximate relation:

$$eV_p/kT_e \approx -(2/3)^{1/2}(a/d)(R/a)(R/a-1) . \tag{6.30}$$

For example, if $a/d = 12.3$, $R/a = 1.5$ for $e|V_p|/kT_e = 7.5$ and $R/a = 2$ for 20. In the limit of large $|V_p|$, $R/a \sim \sqrt{|V_p|}$, and hence (6.23) implies the probe current $i \approx i_+ \sim \sqrt{|V_p|}$.

6.8.6 Floating Potential and Determination of Potential Distribution

Equating the expressions for the ionic (6.25), and the electronic current (6.2), and taking into account the decreasing factor γ mentioned in Sect. 6.8.4, we arrive at a relation similar to (6.11),

$$\frac{e|V_f|}{kT_e} \approx \ln\left[\frac{2}{\pi}\left(\frac{M}{m}\right)^{1/2}\left(\frac{T}{T_e}\right)^{1/2}\frac{a}{l_+}\gamma^{-1}\right] , \tag{6.31}$$

that defines the floating potential. It is of the order of $10(kT_e/e)$. Measuring the absolute value of potential in a dense plasma is not a simple problem, but the spatial potential distribution is readily obtainable. This can be accomplished, for example, by connecting the probe without any dc supply to the reference electrode in series with a very high resistor Ω. The almost insulated probe is then at the floating potential. The potential difference between the probe and the reference electrode is found by measuring on the weak probe current: $i\Omega = V_S + V_f$. If the floating potential is the same everywhere, we move the probe and find the plasma potential distribution with respect to the reference electrode potential V_S up to the constant V_f. The field distribution is found from potential differences between nearby points.

7. Breakdown of Gases in Fields of Various Frequency Ranges

7.1 Essential Characteristics of the Phenomenon

In the most general sense, *electric breakdown* is the process of transformation of a nonconducting material into a conductor as a result of applying to it a sufficiently strong field. The ionized state produced in the gas by breakdown builds up in a time which varies from 10^{-9} to several seconds, although usually it is between 10^{-8} to 10^{-4} s. Ionization reaches appreciable values, so much so that, as a rule, breakdown is accompanied with a light flash visible to the naked eye. Some modes of flash are commonly known as "sparks". If the external field is applied for a sufficiently long time, the breakdown may start a discharge, sustained as long as the field is there. This occurs in any electric field: constant, pulsed, periodic, or produced by electromagnetic waves, including light waves. Concrete conditions dictate to what limit the degree of ionization will grow. It may reach 10^{-8}, as in the glow discharge where the current is limited by a high resistance in the external circuit, or it may be the total single ionization of all atoms, as occurs in the breakdown by high-intensity laser pulse.

The primary element of the often very complicated breakdown process is the *electron avalanche,* which develops in the gas when a strong enough electric field is applied to it. An avalanche begins with a small number of "seed" electrons that appear accidentally, say, due to cosmic rays. It can even be triggered by a single electron. An artificial source of primary electrons is employed to facilitate breakdown build up in experimental studies, in order to start up the avalanche reliably. For example, the cathode or the gas may be irradiated with UV light to produce photoelectrons. An electron picks up energy in the electric field. Having reached energy somewhat greater than the ionization potential, the electron ionizes a molecule, therebgy losing its energy. The result is the production of two slow electrons. They are again accelerated in the field, ionize molecules, thereby producing four electrons, and so forth. In principle, it is unimportant whether this occurs in an avalanche that drifts systematically in a constant field, or by electrons which are "marking time" executing oscillatory motion in a rapidly oscillating field, although the details and the outward manifestations of the process may be very different.

Gas breakdown is essentially a *threshold* process. This means that breakdown sets in only if the field exceeds a value characterizing a specific set of conditions. Thus no changes in the state of the medium are noticeable for some time while the voltage across a discharge gap or the intensity of electromagnetic radiation is gradually increased. Suddenly, ionization rises dramatically at a certain value of voltage or intensity, instruments detect a current, and a flash is observed. The threshold is a consequence of the steep dependence of the rate of atomic ionization by electron impact on field strength and by the fact that ionization, producing electron *multiplication,* is accompanied by mechanisms that create obstacles to the development of the avalanche.

The avalanche is *slowed down* by electron energy losses and by the loss of electrons themselves. The former losses slow down the accumulation of energy sufficient for ionization. The latter terminate chains in the multiplication *chain reaction.* Electrons lose energy to excite electron states of atoms and molecules, molecular vibration, and rotation; energy is also lost in elastic collisions. Electron impact chains are also terminated as a result of diffusion leading to the removal of electrons from the field (e.g., precipitation on the walls), and of the attachment in electronegative gases. When gas breakdown occurs between electrodes, the field applied to them removes electrons to the anode. Recombination is not amongst the mechanisms of electron removal that appreciably influence the breakdown threshold. The fate of an avalanche (whether it will grow or die out) is decided at its *early* stage, when the numbers of electrons and ions are so small that their encounters have a very low probability. The recombination rate is proportional to the electron density squared. At low densities, recombination is much less effective than removal mechanisms that are linear in the electron density: transport to the anode, diffusion to the walls, and attachment. However, recombination intensifies after a large number of *generations* of secondary electrons and may set the limit to further ionization, thereby finalizing the level reached by ionization in the breakdown.

Electron energy losses must rather be treated as a factor reducing the ionization frequency. Formally, they do not eliminate the possibility of ionization, only slow it down; practically, though, these losses in insufficiently strong fields suppress the ionization rate. The disappearance of an electron breaks a chain, setting a limit to the possibility of sustaining the chain reaction of mulitplication. The creation and removal of electrons are competing processes. The rate of creation of new electrons is determined by the ionization frequency and is extremely sensitive to field strength. The rate of removal is much less dependent on the field. Even if the field is slightly lower than the *threshold value,* the ionization rate is considerably smaller than the rate of removal, and no multiplication occurs. If the field exceeds the threshold, the ionization process is speeded up catastrophically. The higher the field is above the threshold, the easier and swifter the breakdown develops.

The breakdown threshold is determined by the relation between creation and removal of electrons only if the field is maintained for a sufficiently long time, adequate for producing numerous generations of electrons. If a pulse is very short, the field must be so high that a sufficient, "macroscopic" number of electrons be born during the pulse, even if losses are absent. This is known to happen, for example, in gas breakdown by focussed "giant" laser pulses that last only

$(2-4)\cdot 10^{-8}$ s. A visible flash appears when about 10^{13} electrons are produced in the focal region.

This chapter treats the effects of gas discharges in all frequency ranges, from dc fields to optical frequencies. However, the principal phenomena in discharge-gap breakdown by voltage applied to electrodes are discussed here only with regard to the breakdown of the entire gap volume and the triggering of a self-sustained discharge in moderate-pressure gases. The breakdown in relatively long gaps filled with high (atmospheric) pressure gas, known as the *streamer, leader* or *spark,* discharge in which a thin ionized channel grows from one electrode to the other, will be treated in Chap. 12.

7.2 Breakdown and Triggering of Self-Sustained Discharge in a Constant Homogeneous Field at Moderately Large Product of Pressure and Discharge Gap Width

7.2.1 Non-Self-Sustaining Current in a Discharge Gap

Consider what happens in a plane gap connected to a circuit with a dc power supply if the voltage V on the electrodes is gradually raised. The applied electric field is assumed to be homogeneous, $E = V/d$, where d is the electrode separation. Electrons appear at the cathode occasionally, and the field transports them towards the anode. An electron may not reach the anode: it may stop on the side wall of the discharge chamber, or attach itself to an electronegative molecule. Then ions may recombine. The fraction of electrons lost on the way is smaller, the faster they cross the gap, that is, the stronger the field. As a result, the electric current i in the circuit, determined by the number of charged particles that reach the electrodes in 1 s, increases (at first) with increasing V. Beginning at a certain voltage, practically all the charged particles (electrons and ions) randomly created in the gas reach the electrodes. The current reaches *saturation* and ceases to depend on V. It is determined by the rate of charge generation due to outside sources, cosmic rays, or an artificial ionizer. This discharge is *not self-sustaining*. Its static current-voltage characteristic is shown in Fig. 7.1. It is *static* since it corresponds to a steady state. The voltage is assumed to be raised so slowly that a stationary state is attained at each value of V.

At still higher voltages, the electron impact ionization of gas molecules starts, amplifying the current due to outside sources. Assume, for example, that the cathode is irradiated with the light of a UV lamp producing a photocurrent i_0; attachment is absent (for its effects, see Sect. 7.2.5). The electronic current at the anode and the circuit current i are enhanced in comparison with the current of electrons leaving the cathode by a factor $\exp(\alpha d)$, where α is Townsend's coefficient for ionization (Sect. 4.1.2): $i = i_0 \exp(\alpha d)$. The total cathode current in the steady state also equals i. It consists of the electron current i_0 and the current of ions generated in the course of ionization and pulled by the field to

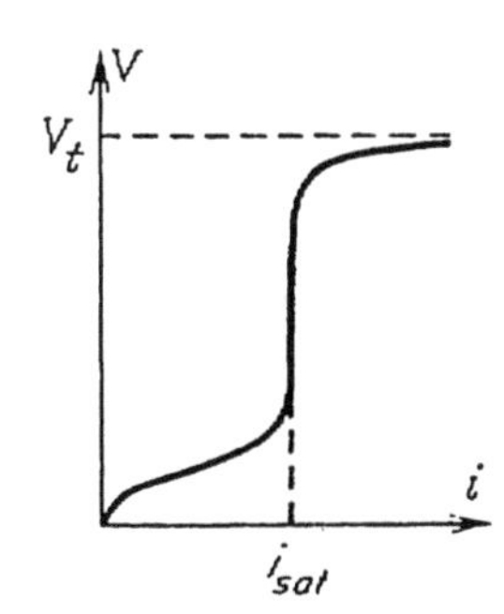

Fig. 7.1. $V - i$ characteristic of non-self-sustaining discharge between plane electrodes

the cathode, $i_0[\exp(\alpha d) - 1]$ (Sect. 4.7.1). The vertically rising curve in Fig. 7.1 becomes less steep. As voltage grows further, *secondary* processes come into play: creation of electrons by particles that appear as a result of the *primary* process of electron impact ionization. Secondary processes affect amplification more strongly if they produce electron *emission* from the cathode. An emitted electron covers the entire path from cathode to anode and therefore produces more ionization than an electron "born" halfway. With secondary emission taken into account, the steady discharge current is given by (4.16):

$$i = i_0 \exp(\alpha d) / \{1 - \gamma[\exp(\alpha d) - 1]\} ,$$

where γ is the effective *secondary emission coefficient for the cathode*; the emission is caused by positive ions, photons, and metastable atoms produced in the gas as a result of ionization and excitation of atoms by electrons. As long as the denominator is positive, the current remains non-self-sustained. As V increases, the current grows even steeper than in the range of simple amplification, owing to a decrease in the denominator which equals unity at small values of the amplification coefficient αd.

7.2.2 Condition for Initiating a Self-Sustaining Discharge

If the voltage between the electrodes $V > V_t$ is such that $\mu = \gamma[\exp(\alpha d) - 1] > 1$, the denominator in the last formula is negative and the expression becomes meaningless. This signifies that the current cannot be steady at this voltage. On the other hand, the current at $V < V_t$, with $\mu < 1$, is steady and non-self-sustained. The transition condition is $\mu = 1$, or

$$\gamma\left[\exp(\alpha d) - 1\right] = 1 , \quad \alpha d = \ln(1/\gamma + 1) ; \tag{7.1}$$

this represents a *steady self-sustained current in a homogeneous* field $E_t = V_t/d$, where the threshold voltage V_t is found from equality (7.1).

Indeed, formally at $V = V_t$, $i = 0/0 \neq 0$ for $i_0 = 0$, that is, current flows even in the absence of an outside source of electrons. Processes in the discharge gap ensure the *reproduction* of electrons removed by the field, without outside help. One electron emitted by the cathode produces $\exp(\alpha d) - 1$ ions which, hitting the cathode, knock out γ electrons each (if this is the ion-electron emission). One primary electron is replaced with one secondary electron ($\mu = \gamma(e^{\alpha d} - 1) = 1$), etc. The *transition* of non-self-sustaining to self-sustaining discharge can also be interpreted as the *onset of breakdown*. The breakdown voltage V_t is defined by condition (7.1) as a function of gap width d, in terms of γ and the known function $\alpha(E)$. If $\gamma \sim 10^{-1} - 10^{-3}$ (Sect. 4.7.3), an electron triggers a self-sustained discharge if it takes part in $\alpha d / \ln 2 \approx \ln \gamma^{-1} / \ln 2$ (3 to 10) ionizing collisions along the path d.

7.2.3 Formative Time of Breakdown

Strictly speaking, breakdown cannot be sustained if the voltage applied to electrodes is exactly V_t, as it ensures only the primitive *reproduction* of electrons, $\mu = 1$. A negligible seed current at the cathode must grow to a macroscopic value, otherwise we cannot speak of a self-sustained discharge. It will happen if there is at least a small *overvoltage* $\Delta V = V - V_t > 0$ ensuring *expanded* reproduction of electrons, $\mu > 1$. In this case, if, say, a single electron has left the cathode at the initial moment, $\mu > 1$ electrons will be emitted in the next cycle, then μ^2, etc. The current and ionization in the gas will increase until the growth is stopped by recombination or the ohmic resistance Ω of the circuit. As the current increases, this resistance accepts a progressively greater part of the power supply voltage, $i\Omega$, and the voltage across the electrodes decreases. When V drops to V_t, i ceases to grow and the self-sustaining current becomes stationary. Thus starts the so-called dark (Townsend) discharge (Sects. 8.2.2 and 8.3.1). For this to happen, the circuit resistance must be very high, limiting the discharge current to a very small value at which the positive space charge accumulating in the gap does not distort the external field. Otherwise the field becomes spatially inhomogeneous and a glow discharge develops (Sects. 8.3, 8.4). The outlined breakdown is also known as *Townsend process* (to distinguish it from the spark breakdown mechanism).

Let us find the law of current growth at the stage when the overvoltage can be regarded as constant. Assume that emission is caused by positive ions. Ions are created mostly close to the anode, where multiplication results in the maximal number of electrons. By τ we denote the time required to pull an ion from the anode to the cathode. Electron emission from the cathode at a moment t is caused by ions produced by electrons emitted at the time $t - \tau$. The electronic current i_1 from the cathode obeys an approximate equation:

$$i_1(t) \approx i_0 + \mu i_1(t - \tau) \approx i_0 + \mu \left[i_1(t) - \tau \frac{di_1}{dt} \right] ,$$

where i_0 is the seed current due to the external ionizer. Integrating this equation with the initial condition $i_1(0) = i_0$ at the moment of switching on the field, we obtain the following expression for the discharge current (Schade, 1937)

$$i(t) = i_1(t) e^{\alpha d} = i_0 e^{\alpha d} \left[\frac{\mu}{\mu - 1} \exp\left(\frac{\mu - 1}{\mu} \frac{t}{\tau} \right) - \frac{1}{\mu - 1} \right] . \qquad (7.2)$$

The current grows with time exponentially, and the faster, the higher the overvoltage and $\mu - 1$. The time scale of current increase is $\mu\tau/(\mu - 1)$. If the emission is caused by photons, τ is of the order of the drift time of electrons (not ions) that is, breakdown develops two orders of magnitude faster. The ionization coefficient α is a steep function of field, while the amplification $\exp(\alpha d)$ and reproduction coefficient μ depend on α exponentially. Therefore, several percent overvoltage is already sufficient for μ to be appreciably greater than unity and for the breakdown to develop rapidly. For this reason, the critical value V_t, found from the condition $\mu = 1$, is a sufficiently definite characteristic of the breakdown threshold. The real time of breakdown buildup after the voltage has been applied consists of two parts: that discussed above, with a scale $\mu\tau/(\mu - 1)$, and the time until the first seed electron appears (unless an artificial source of sufficient intensity is used). The latter time has a *statistical spread*. The *retardation* time of the Townsend breakdown is of the order of 10^{-5}–10^{-3} s.

The evolution of the Townsend breakdown is best thought of as the multiplication of *avalanches*. Each cycle, from the moment an individual electron leaves the cathode until all $\exp(\alpha d)$ electrons that are its descendents reach the anode, can be treated as a single avalanche. If the breakdown has started with a single spurious electron, then the second cycle following the first avalanche involves, on the average, $\mu > 1$ avalanches, the third cycle involves μ^2 avalanches, and so on. Each avalanche spreads transversally somewhat owing to electron diffusion, so that a new avalanche starts at a different spot on the cathode (which may be quite far in the case of photoemission). Furthermore, a process is not necessarily started by a single electron: several may be emitted simultaneously from different points. As a result, the Townsend breakdown most often involves in a diffuse manner the *entire volume* of the gap. This is a clear external difference to the spark discharge.

7.2.4 Ignition Potential

This is an equivalent term for the break-down voltage V_t. This quantity, and the corresponding breakdown field E_t, depend on the gas, the material of the cathode, the pressure, and the discharge gap width. To arrive at explicit expressions, we make use of (4.5) for α. Substituting it into (7.1), we obtain

$$V_t = \frac{B(pd)}{C + \ln pd}\,, \quad \frac{E_t}{p} = \frac{B}{C + \ln pd}\,, \quad C = \ln \frac{A}{\ln(1/\gamma + 1)}\,. \tag{7.3}$$

The *ignition potential* V_t and E_t/p depend only on the product pd. This is a manifestation of a *similarity law*. The calculation of V_t by (7.3), with experimentally determined constants A and B (Table 4.1), gives a satisfactory agreement with experiment. The experimental curves $V_t(pd)$, the so-called *Paschen curves*, are plotted in Figs. 7.2, 7.3. There exists the minimal breakdown voltage for a discharge gap, and according to (7.3), the parameters of this minimum point are ($\bar{e} = 2.72$) :

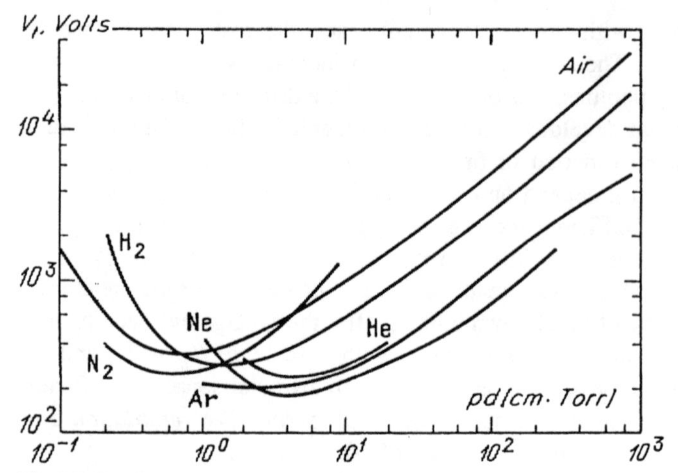

Fig. 7.2. Breakdown potentials in various gases over a wide range of pd values (Paschen curves) on the basis of data given in [7.1,2]

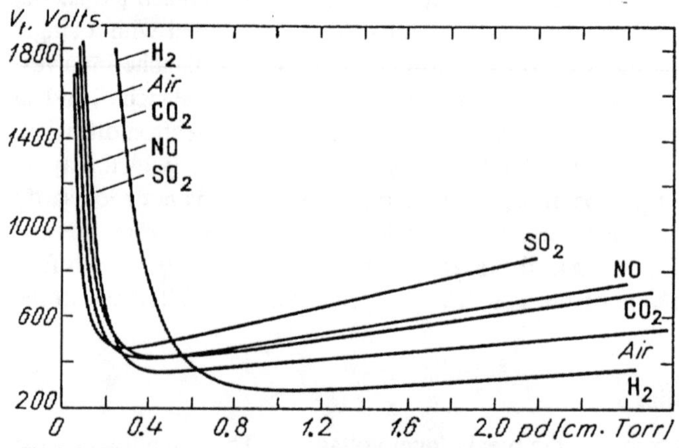

Fig. 7.3. Paschen curves on an enlarged scale [7.3]

$$(pd)_{\min} = \frac{\bar{e}}{A}\ln\left(\frac{1}{\gamma}+1\right), \quad \left(\frac{E}{p}\right)_{\min} = B, \quad V_{\min} = \frac{\bar{e}B}{A}\ln\left(\frac{1}{\gamma}+1\right). \tag{7.4}$$

The value of E/p at the minimum corresponds to Stoletov's point (Sect. 4.1.6), where the ionization capability of electrons, $\eta = \alpha/E = A/B\bar{e}$, is at a maximum. The conditions for breakdown are the easiest because the conditions for multiplication are optimal. In contrast to $V_{\min}$ and $(pd)_{\min}$, the product $(E/p)_{\min}$ is independent of the cathode material (of γ), as demonstrated by (7.4) and experimental data [7.1,2]. Let us compare the estimate given by (7.4) with measurements. In air, $A = 15$, $B = 365$. For $\gamma = 10^{-2}$, formulas (7.4) give $C = 1.18$, $(pd)_{\min} = 0.83$ Torr·cm, $(E/p)_{\min} = 365$ V/(cm·Torr), $V_{\min} = 300$ V. Experiments with an iron cathode give: $(pd)_{\min} = 0.57$ Torr·cm, $V_{\min} = 330$ V,

$(E/p)_{\min} = 580\,\text{V/(cm·Torr)}$. In inert gases, $(pd)_{\min}$ is greater but $V_{\min}$ is smaller. Thus in argon with an iron cathode, $(pd)_{\min} = 1.5$, $V_{\min} = 265$, $(E/p)_{\min} = 176$.

In the range of relatively large pd on the *right-hand branch* of the Paschen curve, the threshold value E_t/p decreases rather slowly (logarithmically) as pd increases. Correspondingly, the breakdown voltage increases almost proportionally to pd (slightly slower). This behaviour of threshold values arises because in the case of elevated pressures and large gaps an electron can produce numerous ionizing collisions even at not very high E/p. In this case, however, α depends sharply on E/p, and the condition of the necessary amplification (7.1) fixes the value of E/p rather rigidly.

On the other hand, the possibilities for collisions are very limited on the *left-hand branch* at low pd. A very high value of α/p, that is, a very strong field is required to achieve the necessary amplification. The breakdown voltage grows rapidly as pd decreases. Hence, the voltage has a minimum. The effective ionization cross section being limited, the ionization coefficient is also limited (by Ap). As a result, the necessary amplification cannot be obtained at sufficiently low pd, regardless of the field. In the framework of this approximation, as pd is reduced to its limiting value

$$(pd)_{\lim} = A^{-1} \ln(1/\gamma + 1) = (pd)_{\min}/\bar{e}\,, \tag{7.5}$$

$V_t \to \infty$. In fact, the growth of E_t/p and V_t on the left-hand branches of the Paschen curves is not as steep, nor does it tend to infinity. Very different mechanisms come into play "to the left" of the left-hand branches (Sect. 7.2.6).

7.2.5 Breakdown Fields in Moderately Large Gaps in Air and Other Electronegative Gases at Atmospheric Pressure. Limiting Values of *pd* for the Townsend Breakdown Mechanism

This mechanism is characterized by low pressure and not too large $pd(\lesssim 1000\,\text{Torr cm})$. If the gap is not too large (and the field is homogeneous), the mechanism of avalanche multiplication is predominant at atmospheric pressure. In room-temperature air in plane gaps, it is realized roughly for $d < 5\,\text{cm}$ ($pd < 4000\,\text{Torr·cm}$). At such high pd, the breakdown voltage is more or less proportional to pd, that is, it is only slightly dependent on pd; more or less definite values of breakdown voltage or $(E/p)_t$ are characteristic for atmospheric-pressure gases. Figure 7.4 plots the results of measurements in room-temperature air in the range of d where the Townsend mechanism still acts. The typical figure for centimetres-wide gaps is $(E/p)_t \approx 32\,\text{kV/(cm·atm)} = 42\,\text{V/(cm·Torr)}$. In large gaps (tens of centimetres wide), the breakdown field in room-temperature air reduces to a limit, $E_t \approx 26\,\text{kV/cm}$. In general, the spark mechanism sets in if $d > 6\,\text{cm}$ (Chap. 12).

The limiting values of threshold fields, $E \approx 32 - 26\,\text{kV/cm}$, observed at sufficiently high pd, are not accidental. They are clearly related to the possibility of electron multiplication in a gas with *attachment* of electrons. The *attachment coefficient* a, defined by analogy to the ionization coefficient α (Sect. 4.4.2), is

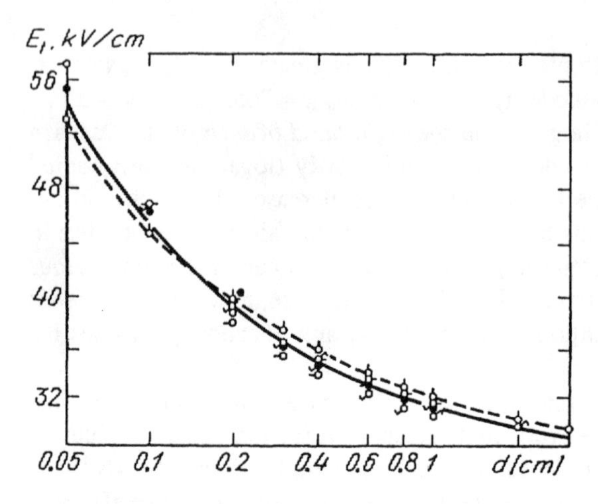

Fig. 7.4. Breakdown fields in a plane gap of length d in air at p = 1 atm. From [7.1]

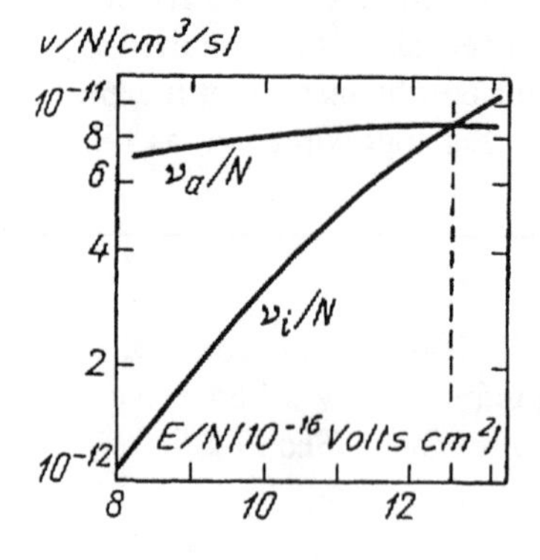

Fig. 7.5. Ionization and attachment frequencies in air, calculated using the solution of the kinetic equation. Intersection at E/p = 41 V/cm·Torr

also known to grow appreciably with E/p but much slower than α. The curves of α/p and a/p as functions of E/p intersect at a certain value $(E/p)_1$. Figure 7.5 shows the results of calculation of the coefficients using the kinetic equation; these calculations are similar to that described in Sect. 5.8 (cf. similar plots in Figs. 8.18 and 8.19 in Chap. 8 for laser mixtures with an electronegative component, CO_2). The intersection point lies at $(E/p)_1 \approx 41$ V/(cm·Torr). This figure is close to the limit for the breakdown potential of air; in fact, it is somewhat exagerated, perhaps because of the imperfect data, used in the evaluation, on cross sections of a number of processes. The avalanche equation (4.4) with the effective ionization coefficient $\alpha_{\text{eff}} = \alpha - a$ is

$$dN_e/dx = (\alpha - a)N_e \ , \quad N_e \propto \exp(\alpha - a)x \ ;$$

the measurements of α_{eff} show that $\alpha_{\text{eff}} \to 0$ at $(E/p)_1 \approx 35$ V/(cm·Torr), in good agreement with $(E/p)_{\text{lim}} \approx 26$ kV/(cm·atm). If $E/p < (E/p)_1$, the multiplication of electrons is obviously impossible; this fact affects the limits of breakdown.

Measurements show that breakdown thresholds in strongly electronegative halogen-containing gases at atmospheric pressure are very high. This is shown in Table 7.1, which also gives the data for gases manifesting no attachment. At low

Table 7.1. Approximate values of breakdown threshold at high pressure

Gas	Constant field, gap width less than several cm, $p \sim 1$ atm		Microwaves, $p \sim 100$–300 Torr
	E/p kV/(cm·atm)	E/p V/(cm·Torr)	E/p V/(cm·Torr)
He	10	13	3
Ne	1.4	1.9	3–5
Ar	2.7	3.6	5–10
H_2	20	26	10–15
N_2	35	46	~ 25
O_2	30	40	35
Air	32	42	~ 30
Cl_2	76	100	
$CCl_2F_2^*$	76	100	
CSF_8	150	200	
CCl_4	180	230	
SF_6	89	117	

* Freon

pressure (small pd), the values of $(E/p)_t$ are much greater than $(E/p)_1$ (Fig. 7.4), and α is appreciably greater than a, so that the electronegative properties of gases are not manifested as clearly as at high pd. The high electric strength of electronegative gases has important practical applications.

7.2.6 Breakdown of Vacuum Gaps

If $pd < 10^{-3}$ Torr·cm, an electron crosses the gap practically without collisions, so that there is no multiplication in the volume. This does not mean, however, that a vacuum gap can be an ideal insulator (Figs. 7.6). If high voltage is applied to a narrow gap, a high field is generated, capable of causing field emission from the cathode (Sect. 4.6.3). The field is additionally enhanced in the vicinity of microscopic protrusions. Breakdown occurs in wider gaps at fields insufficient for ejecting an electron from the metal. A spurious electron is accelerated in the field, knocks an ion from the anode, or emits a bremsstrahlung photon. The ion or the photon knock out, in turn, an electron from the cathode, etc. This multiplication proceeds without the residual gas. A process is also possible in which electrodes are sputtered by the particles accelerated in the field, so that the gap gets filled with metal vapor in which gas enhancement then occurs.

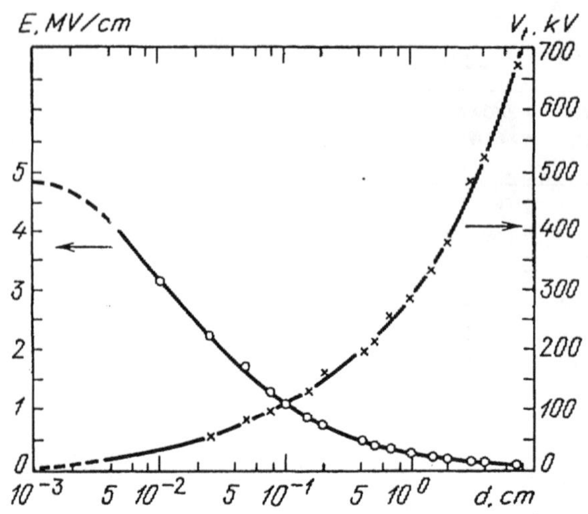

Fig. 7.6. Breakdown voltage and field in the vacuum gap between a steel sphere 2.5 cm in diameter and a steel disk 5 cm in diameter as functions of gap width [7.4]

7.3 Breakdown in Microwave Fields and Interpretation of Experimental Data Using the Elementary Theory

For the time being, let us postpone the case of *radio frequency* fields, since they are more complex and diversified. The *microwave* range is characterized by a small amplitude of electron vibration compared with the size of the discharge volume (which is comparable with the wavelength $\lambda \sim 1$–10 cm, see a numerical example at the end of Sect. 3.1.2). As a result, the evolution of an electron avalanche is localized and almost independent everywhere, the field does not push particles towards the walls, and the emission of the walls is insignificant. The process is of bulk nature and relatively simple. It has been studied quite thoroughly both experimentally and theoretically in [7.2, 5].

7.3.1 Measurements

When the breakdown threshold is measured experimentally, the controlled power of a cw or pulsed magnetron is fed to a *resonator cavity* through a waveguide. The threshold field of a given frequency f depends on the size of discharge volume. This effect is caused by the diffusion leakage of electrons to the walls. On the other hand, the resonator size is related to λ, that is, $f = c/\lambda$. As a result, not just any geometry allows changes of cavity size at unaltered frequency. Actually, such changes are necessary to establish the dependence of the threshold field on size, other conditions being equal. This difficulty can be avoided by using a cylindrical resonator excited at such a *mode* that the resonance frequency is a function of cylinder radius but is independent of its height (Fig. 7.7). The diffusion length Λ (Sect. 4.5) can be varied by changing the cylinder height at unaltered radius and field frequency. Increasing the field by bringing up the power fed into the

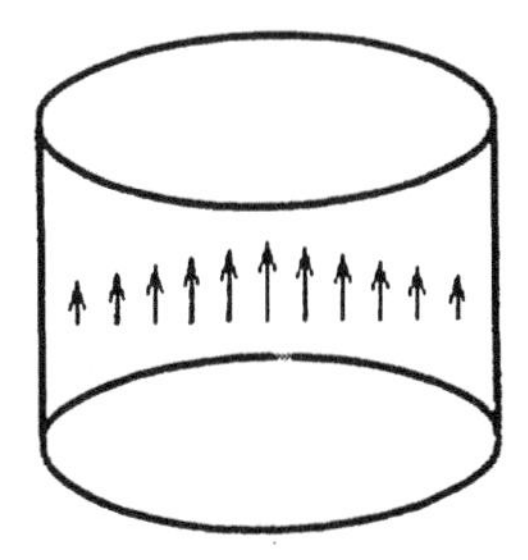

Fig. 7.7. Electric field distribution in the resonator of an experimental device for measuring microwave breakdown thresholds

cavity, one fixes the parameters at which the transmitted power drops abruptly. This is a sign of breakdown followed by the dissipation of electromagnetic energy in the plasma. The breakdown occurs first of all at the central part of the cavity, where the field amplitude is maximal; this field is assumed to be the threshold.

The threshold field (root-mean-square value E_t in Fig. 7.8) as a function of pressure always has a minimum. On the left-hand branch, the threshold decreases with increasing pressure. This threshold is the lower, the greater the discharge volume and the lower the field frequency. The same is true for the minimum value. The minimum at lower frequencies lies at lower pressures. On the right-hand branch, where the threshold increases with increasing pressure, the depen-

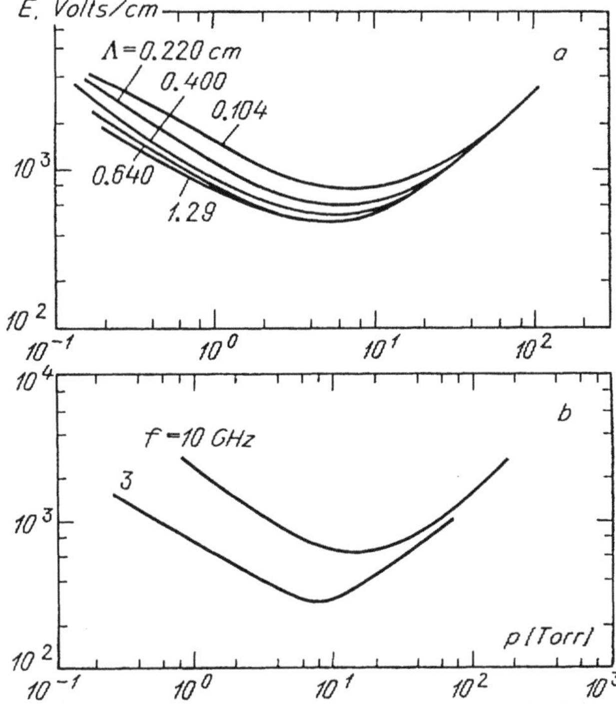

Fig. 7.8. Measured thresholds of microwave breakdown [7.6] (a) air, f = 9.4 GHz, diffusion length Λ is indicated for each curve; (b) Heg gas (He with an admixture of Hg vapor), Λ = 0.6 cm

dence of E_t on size and frequency becomes less and less pronounced, and almost vanishes in the limit of high pressures: all curves asymptotically merge.

The point of interest is the unusual facility of breakdown in mixtures of helium and neon with a small admixture of argon. The reason is the Penning effect (Sect. 4.2.2); owing to it, electron energy loss to the excitation of He and Ne impedes only slightly, or not at all, the progress of an avalanche. The excited He* and Ne* atoms ionize Ar atoms. The rate of this two-stage process is proportional to the densities of the main and admixture gases, so that the effect is better pronounced at higher pressures. Ionization in the so-called Heg gas (a mixture of He with some mercury) occurs similarly. The cross section of ionization of Hg by metastable He* atoms is anomalously high (Sect. 4.2.2). It can be said that each event of He excitation immediately produces a new electron. The frequency of ionization of a gas mixture by electrons, ν_i, then coincides with the excitation frequency of the main gas atoms (He). Inelastic losses are as if absent. One can thus resort to the *elementary theory* and understand quite a few of the essential features of breakdown without addressing the *kinetic equation*. Qualitatively, they still hold with inelastic losses present.

7.3.2 Ionization Kinetics Equation

When oscillation displacements are small, electron densities obey an equation of type (2.44):

$$\partial n_e/\partial t = D\nabla^2 n_e + (\nu_i - \nu_a)n_e \,, \quad D \equiv D_e \tag{7.6}$$

(electrons diffuse freely in breakdown). If the condition $\omega \gg \nu_m\delta$ (Sect. 5.5.2) holds (it is satisfied for microwave frequencies), the electron energy distribution is quasisteady and the ionization and attachment frequencies, ν_i and ν_a, are determined by the root-mean-square field E. The dependencies $\nu_i(E)$, $\nu_a(E)$ are much stronger than $D_e(E)$, so that $D_e(E) \approx$ const. For simplification, assume that the field is spatially homogeneous, and hence, ν_i and ν_a are independent of coordinates. Averaging (7.6) over the volume, we obtain, in accord with the results of Sect. 4.5, an equation for the mean density, or (which is equivalent) for the total number of electrons, N_e, in the discharge volume:

$$dN_e/dt = (\nu_i - \nu_a - \nu_d)N_e \,, \quad \nu_d = D/\Lambda^2 \,, \tag{7.7}$$

where ν_d is the frequency of diffusion losses of electrons. This equation describes the ionization kinetics of the gas.

7.3.3 Steady-State Background Criterion

Assume that the external field is switched on in a time small in comparison with the characteristic time of multiplication and remains constant during the avalanche buildup. This constraint covers not only stationary, but also pulsed fields with not too short pulses and sufficiently small rise time. Under this assumption, $\nu_i(t)$, $\nu_a(t) = \text{const}$ after the moment $t = 0$ at which the field is switched on, and (7.7) has an exponential solution typical of an avalanche process:

$$N_e = N_{e0} \exp[(\nu_i - \nu_a - \nu_d)] = N_{e0} \exp(t/\Theta) , \qquad (7.8)$$

where Θ is the *avalanche time constant,* and N_{e0} is the number of seed electrons that start the avalanche.[1] Breakdown is impeded in experiments with short pulses, since the probability of an electron appearing in the region of the field at the necessary moment is quite low and the avalanche has to be initiated by injecting a small number of electrons. For this purpose, a weak radioactive source is used.

According to (7.8), an avalanche develops if $\nu_i - \nu_a - \nu_d > 0$; this condition is met if the field exceeds a threshold E_t determined by the *steady-state breakdown criterion*:

$$\nu_i(E_t) = \nu_d + \nu_a(E_t) . \qquad (7.9)$$

As an example, consider breakdown in helium, for $p = 1$ Torr, $\lambda = 3$ cm, diffusion length $\Lambda = 1$ cm, $D = 2 \cdot 10^6\,\text{cm}^2/\text{s}$, time of diffusion to the walls $\nu_d^{-1} \approx 5 \cdot 10^{-7}$ s, diffusion frequency $\nu_d \approx 2 \cdot 10^6\,\text{s}^{-1}$, and no attachment. The avalanche develops if $\nu_i > \nu_d \approx 2 \cdot 10^6\,\text{s}^{-1}$. We will show a little later that the ionization frequency $\nu_i \propto E^2$ under the most favorable conditions for multiplication (zero electron energy losses). If losses, especially inelastic, are nonzero, the ν_i vs. E curve is much steeper. Hence, if the field increases by 10 % in comparison with E_t, then $\Theta^{-1} = \nu_i - \nu_d \geq 0.2\nu_d \approx 4 \cdot 10^5\,\text{s}^{-1}$. The number of electrons is doubled every $\Theta/\ln 2 \leq 1.7\,\mu$s, which is a very high rate. In many cases, it is sufficient for a reliable realization of breakdown. As a result, stationary criterion (7.9) determines with good accuracy [like criterion (7.1)] the breakdown threshold of gases for "not too short" pulses.

7.3.4 Low Pressure

Let us evaluate threshold fields for the Heg gas. At low pressure the diffusion coefficient $D \propto 1/p$ is high and electron diffusion losses are substantial. The losses can be compensated for if the ionization rate is large, that is, if the field is strong. Recall that the role of elastic energy losses in high fields is negligible. Indeed, electron energies ε are not greater than a quantity of the order of the helium excitation energy $E^*_{\text{He}} = 19.8$ eV, because at $\varepsilon > E^*_{\text{He}}$ an electron enters an inelastic collision with high probability and dissipates energy. The elastic-collision energy transfer to an atom is limited by the quantity $(\Delta\varepsilon_{\text{el}})_{\max} = (2m/M)E^*_{\text{He}}$. Formula (3.10) shows that the energy $\Delta\varepsilon_E$ gained in a collision is proportional to E^2. In sufficiently strong fields necessary to balance out high diffusion losses, an electron gains energy from the field at a rate $d\varepsilon/dt = \Delta\varepsilon_E \nu_m$, where ν_m is defined

[1] The rate of generation of electrons in the atmosphere by cosmic rays at sea level is on the order of $10\,\text{cm}^{-3}\text{s}^{-1}$. Usually about 10–$10^2$ electrons per cm^3 are present in non-electronegative gases. In air, they immediately attach to oxygen, so that charges exist as ions; their density is on the order of $10^3\,\text{cm}^{-3}$.

as in Chap. 2. The energy $E^*_{\rm He}$ is reached over a time $\tau_E = (E^*_{\rm He}/\Delta\varepsilon_E)\nu_{\rm m}^{-1}$. The ionization frequency is determined simply by τ_E because excitation and Penning ionization follow immediately:

$$\nu_{\rm i} = \tau_E^{-1} = (\Delta\varepsilon_E/E^*_{\rm He})\nu_{\rm m} = e^2E^2\nu_{\rm m}/m\omega^2E^*_{\rm He}\,. \tag{7.10}$$

We have taken into account here that at low pressure, $\nu_{\rm m}^2 \ll \omega^2$. As implied by (7.9), the root-mean-square breakdown field is

$$E_{\rm t} = \left(\frac{Dm\omega^2E^*_{\rm He}}{e^2\nu_{\rm m}\Lambda^2}\right)^{1/2} = \left(\frac{mE^*_{\rm He}}{3}\right)^{1/2}\frac{\omega}{e\sigma_{\rm m}N\Lambda} \propto \frac{\omega}{p\Lambda}\,. \tag{7.11}$$

This calculation operates with the expressions for the diffusion coefficient, $D = lv/3 = l^2\nu_{\rm m}/3$,and the mean-free-path length, $l = 1/N\sigma_{\rm m}$, of electrons. The threshold field is proportional to frequency and inversely proportional to gas density (pressure) and to discharge volume size; this is in complete agreement with experiments. Moreover, if the collision cross section $\sigma_{\rm m} = 4\cdot10^{-16}\,{\rm cm}^{-2}$, corresponding to the midpoint of the electron spectrum, $\varepsilon \approx E^*_{\rm He}/2 \approx 10\,{\rm eV}$, is substituted into the formula, a satisfactory fit with Fig. 7.8 is obtained. Formula (7.11) gives a correct description of the asymptotic lines on the logarithmic plot which the left-hand branches of the function $E_{\rm t}(p)$ tends to for different values of ω and Λ.

7.3.5 High Pressure

In this case the diffusion losses of electrons are negligible, and breakdown occurs even at not very high ionization rate. The most important factor now is the energy dissipation, and specifically, purely elastic losses in the Heg gas. These losses limit the ionization frequency. In terms of elementary theory, an electron cannot gain more energy than the limit (3.12) imposed by electronic losses. In the case of high pressures, when $\nu_{\rm m}^2 \gg \omega^2$, we have

$$\varepsilon_{\rm max} = (M/2m)e^2E^2/m\nu_{\rm m}^2 \propto (E/p)^2\,. \tag{7.12}$$

If this energy is less than $E^*_{\rm He}$, electrons cannot excite helium atoms and no avalanche can develop. Hence, the possibility of breakdown is determined by the condition $\varepsilon_{\rm max}(E) \geq E^*_{\rm He}$; the breakdown field $E_{\rm t}$ calculated by this equality is

$$E_{\rm t} = (m\nu_{\rm m}/e)(2E^*_{\rm He}/M)^{1/2} \propto p\,. \tag{7.13}$$

The breakdown field is proportional to p and is independent of both the volume size Λ (in the framework of the approximation chosen here; see Sect. 7.3.8) and the frequency; this is also found to be in qualitative agreement with experiments. The quantitative fit for the Heg gas is also satisfactory. The threshold is frequency-independent because if $\omega^2 \ll \nu_{\rm m}^2$, the effect of an oscillating field on electrons is indistinguishable from that of a constant field.

7.3.6 The Position of the Minimum

Under a crude approximation, the position of the minimum on the threshold curve $E_t(p)$ can be found using the condition that separates to some extent the limiting cases of low and high pressure, that is, of $\nu_m^2 \ll \omega^2$ and $\nu_m^2 \gg \omega^2$. This condition states that the collision and field frequencies are of the same order of magnitude: $\nu_m \sim \omega$. At this value of ν_m, $d\varepsilon/dt$ as a function of p has a maximum [see (3.11)]. The frequency at the minimum of threshold field is proportional to pressure. This result is qualitatively supported by experimental data. The breakdown of gases by microwave fields is easiest at $p \sim 1$–10 Torr.

7.3.7 Inelastic Losses; Molecular and Electronegative Gases

Inelastic losses are important in most gases; they affect threshold fields in almost the same qualitative manner that elastic losses do. This is demonstrated even by the perfect apparent similarity of $E_t(p)$ curves in Heg and other gases. At low pressure, the threshold field is determined mostly by diffusion. Threshold fields are high, so that electrons gaining energy rush rapidly thruogh the "danger zone" of energies between the excitation and ionization potentials. As a result the probability of energy loss due to atomic excitations is not too high. The threshold field is then given by a formula of type (7.10) in which E_{He}^* must be replaced with the ionization potential.

Diffusion at high pressure being slow, the threshold is mostly determined by energy losses. The rate of both inelastic and elastic energy losses is proportional to pressure. The condition of balancing of energy losses by energy gained from the field implies that if $\omega^2 \ll \nu_m^2$, the mean electron energy as a function of E/p [see (3.12)]. For ionization to proceed, the mean energy cannot be too small compared with the ionization potential; this fact somewhat fixes the ratio E/p. Hence, the threshold field $E_t \propto p$; if $\omega^2 \ll \nu_m^2$, it is independent of ω. In the framework of the approximation chosen here, it is also independent of Λ, as it is in the case of purely elastic loss.

Molecular gases undergo breakdown at higher fields than atomic gases because an electron spends much of its energy on the excitation of vibrational and lower electronic levels of molecules; the rate of energy buildup is thereby slowed down. Thresholds in electronegative gases are also high because of additionial electron losses to attachment.

7.3.8 Similarity of Threshold Values of E/p in Constant and Oscillating Fields at High Pressure and Their Independence of Size

If the effect of oscillating fields on electrons at high pressure is nearly the same as that of constant fields, and if the threshold values of E_t/p in the microwave range are almost independent of p and Λ, there is every reason to expect that the values of E_t/p are close to those for constant field and high pd (where they are also almost independent of pd). On the whole, experiments bear out this conclusion. The right-hand branches of microwave breakdown correspond to the right-hand branches of Paschen curves far from the minima. The E_t/p values differ by a factor of 1.5–2 (Table 7.1).

It must be emphasized (this is important for understanding the process) that the independence of E_t/p of the diffusion length Λ in (7.13) is a result solely of the elementary theory. The absence of Λ in the formula is equivalent to neglecting electron losses: $\Lambda = \infty$, $\nu_d = 0$. In fact, ionization, even if very weak, takes place in any (no matter how weak) field because the spectrum always contains some energetic electrons. Therefore, the breakdown would have zero threshold if electrons suffered absolutely no losses. The number of electrons with energies $\varepsilon \approx I$ sufficient for ionization is exponentially small: it is mainly proportional to $\exp(-\varepsilon/\varepsilon_0)$ in the case of the Maxwell distribution, and to $\exp(-\varepsilon^2/\varepsilon_0^2)$ in the case of Druyvesteyn's distribution. The scales ε_0 approximately coincide with the mean energies of the spectra or with ε_{max} of the elementary theory, and increase with increasing field. As a result, the real ionization frequency is a characteristic exponential function of E, of type $\exp[-\text{const}/f(E/p)]$.

As an example, take Townsend's law $\nu_i \sim p\exp(-\text{const}\cdot p/E)$. The results of Sect. 7.4.7, where the kinetic equation was solved approximately, imply that this law is not far from the thruth. The breakdown condition $\nu_i = \nu_d$ yields

$$E_t/p = \frac{\text{const}}{\text{const}' + \ln(p\Lambda)} \,. \tag{7.14}$$

This logarithmic dependence (weak for large $p\Lambda$) is quite similar to the dependence on pd in (7.3); formally, it ensures that the threshold vanishes as $\Lambda, d \to \infty$.

The lengths Λ and d play essentially identical qualitative roles, characterizing the rate of removal of electrons from the discharge volume. The field pulls an electron out over a time from 0 to d/v_d, depending on where the motion started. The inverse, v_d/d, is the scale of the "removal" frequency (loss frequency). The Townsend criterion (7.1), $\alpha d = k$, where k lies between one and ten, and $\alpha = \nu_i/v_d$, can be interpreted, by analogy to (7.9) for $\nu_a = 0$, as the condition of equality of the ionization and loss frequencies: $\nu_i = kv_d/d = \nu_d'$. The mean removal frequency ν_d' is greater than the minimum value v_d/d by a factor of k, because the majority of electrons in an avalanche are created close to the anode.

7.4 Calculation of Ionization Frequencies and Breakdown Thresholds Using the Kinetic Equation

7.4.1 Derivation of the Equation of Ionization Kinetics from the Kinetic Equation

Using (5.24–26), one can recast the kinetic equation for the electron energy distribution function $n(\varepsilon, t)$ in the following convenient form for analysis:

$$\frac{\partial n}{\partial t} = -\frac{\partial J}{\partial \varepsilon} + Q^* + Q_i - \nu_a(\varepsilon)n - \nu_d(\varepsilon)n \,,$$

$$J = -A\varepsilon \frac{\partial n}{\partial \varepsilon} + \frac{A}{2} n + n V_{el} , \tag{7.15}$$

$$A = \frac{2}{3} \frac{e^2 E^2 \nu_m}{\omega^2 + \nu_m^2} , \quad V_{el} = -\frac{2m}{M} \varepsilon \nu_m .$$

The flux J along the energy axis reflects energy gains from the field and elastic losses. The term Q^* describes the excitation of atoms, Q_i represents ionization, attachment is given by $\nu_a(\varepsilon)n$, and diffusion losses by $\nu_d(\varepsilon)n$.

As a result of inevitable excitation and ionization events at energies above the corresponding potentials, electrons cannot reach very high energies. As $\varepsilon \to \infty$, the distribution function falls off very rapidly. The flux also vanishes: $J(\infty) = 0$. Particle sources in equation (7.15) are distributed along the ε axis. There are no electron sources with zero energy, and negative kinetic energy is impossible. Hence, $J(0) = 0$. If we turn to an analogy with one-dimensional diffusion of particles in ordinary space, $x \equiv \varepsilon$, the situation is found to correspond to an impenetrable and nonemitting wall at $x = 0$.

In view of this, we integrate (7.15) over the entire spectrum from 0 to ∞. The integral of Q^* vanishes automatically, since excitation events do not change the number of electrons. Integration in ε of the first term in formula (5.26) for Q_i yields $-\nu_i n_e$, where ν_i is the ionization frequency averaged over the spectrum. The second term gives $2\nu_i n_e$; this is readily verified if the order of integration in the double integral is reversed. It is as if one electron disappears in each ionization event, while two new ones appear. This gives the equation of kinetics for electron density,

$$dn_e/dt = (\nu_i - \nu_a - \nu_d) n_e , \tag{7.16}$$

equivalent to (7.7); ν_a and ν_d are also frequencies averaged over the spectrum.

7.4.2 Separation of Variables

Strictly speaking, the initial electron distribution function $n(\varepsilon, 0)$ must be specified as the initial condition to (7.15). It is physically clear, however, that the initial spectrum is forgotten after one or two new generations of electrons are generated, and a new spectrum forms, corresponding to the effects of the field and collisions. Indeed, the build-up (relaxation) time of the spectrum is characterized by the mean time during which an electron covers the entire path along the ε axis from $\varepsilon \approx 0$ to the highest realizable energies. In fact, this is the time necessary for ionization and multiplication. If the point of interest is any well-developed stage of the avalanche, there is no sense in going into details of the relaxation process; rather, one should directly search for the stationary spectrum, that is, seek the solution of the nonstationary equation in the form $n(\varepsilon, t) = n(\varepsilon)\Phi(t)$.

The substitution into (7.15) gives

$$n(\varepsilon, t) = n(\varepsilon) \exp(t/\Theta) , \tag{7.17}$$

where the separation constant Θ has the meaning of the time constant of the avalanche, and the spectral function $n(\varepsilon)$ is normalized to the initial density n_e^0. A solution of type (7.17) would always be exact if the spectrum of initial electrons coincided with the one that is established by the end. According to (7.16, 17), the constant Θ is related to spectrum-averaged frequencies by an obvious equality

$$\Theta^{-1} = \nu_i - \nu_a - \nu_d \,. \tag{7.18}$$

7.4.3 Equation for the Spectral Function

We now substitute (7.17) into (7.15) and replace Θ via (7.18). The diffusive removal of electrons to the walls was taken into account in (7.15) in an approximate manner, in order not to add complications to the already complex dependence of the distribution function on spatial coordinates. It would hardly be advisable, therefore, to retain the rather weak and largely unimportant dependence of the spatial diffusion coefficient and ν_d on energy. Let us replace $\nu_d(\varepsilon)$ in (7.15) with the mean value of ν_d. Then, if (7.17, 18) are substituted into (7.15), the diffusive loss term vanishes completely from the equation for the spectral function $n(\varepsilon)$, so that the equation takes the form

$$(\nu_i - \nu_a)n = -dJ/d\varepsilon + Q^* + Q_i - \nu_a(\varepsilon)n \,. \tag{7.19}$$

In this approximation, the spectrum and frequency of ionization are independent of geometry and volume size, and do not differ from those that would be obtained for infinite space and homogeneous field. This is the standard procedure in solving the kinetic equation.

The solution of (7.19) gives the spectrum $n(\varepsilon)$ and ionization frequency ν_i as functions of field and gas characteristics. The rate of multiplication (or removal) of electrons as a function of field and diffusive loss is defined by (7.18). The condition $\Theta^{-1} = 0$ corresponds to the steady state and the steady-state breakdown criterion (7.9). If this condition is imposed on the solution, one can find the threshold field E_t. In fact, the dependence $\nu_i(E)$ obtained in this way has a more general significance. It can be employed for studying some other processes, such as the positive column of a glow discharge (however, one should bear in mind that if ionization is strong, diffusion is *ambipolar* and ν_d is considerably smaller).

7.4.4 Similarity Laws

As we have mentioned in Sect. 5.6.3 and as (7.15) and (7.19) directly demonstrate, the stationary spectrum in the low-frequency limit $\omega^2 \ll \nu_m^2$ (and in constant field) is described by a function $n(\varepsilon, E/p)$, and in the high-frequency limit $\omega^2 \gg \nu_m^2$, by $n(\varepsilon, E/\omega)$. Correspondingly, the ionization frequency is a function of the type $\nu_i = pf_1(E/p)$ in the former case, and $\nu_i = pf_2(E/\omega)$ in the latter case. If the collision frequency is assumed to be constant, $\nu_m(\varepsilon) = \text{const}$, the ionization frequency in an oscillating field, $\nu_{i\omega}$, is expressed in terms of the ionization frequency at constant ν_{i0} by introducing an effective field E_{eff} using (5.33):

$$\nu_{i\omega}(\omega, p, E) = \nu_{i0}(p, E_{\text{eff}}) = pf_1(E_{\text{eff}}/p) \,. \tag{7.20}$$

In a gas without attachment, we make use of the stationary breakdown criterion (7.9) and of the dependence $\nu_d \propto 1/p\Lambda^2$, and find that the threshold field in the low-pressure (high frequency) limit is $E_t = \omega F_1(p\Lambda)$, and in the high-pressure (low frequency) limit is $E_t = pF_2(p\Lambda)$. The asymptotic form of the functions F_1 and F_2 is given by (7.11), (7.14), or (7.3). Similarity laws for the attachment frequency are the same as for the ionization frequency. Therefore, the breakdown value of E/p in electronegative gases, with attachment dominating diffusion at high pressures, is constant.

7.4.5 Formulation of a Simplified Problem of the Effect of Inelastic Losses on Ionization Frequency

If the complexity of (7.19) is not ignored, only numerical solutions are possible. Cumbersome and time-consuming computations of this type became feasible only with the advent of computers. Numerical solutions do supply information valuable for practical work, but an *analytic* solution, even if crude, is frequently a greater help in understanding the nature and identifying the mechanisms which are important. In view of this, we will construct a simplified solution that clearly demonstrates the effect of inelastic losses on the ionization rate, which is beyond the reach of the elementary theory.

Let us consider heavy inert gases (argon, xenon) which manifest no attachment, have no low energy levels, and where the role of elastic losses is negligible. Assume that the frequency ν_m of elastic collisionss is constant. Assume also that the atomic excitation frequency ν^* is equally energy-independent if the energy exceeds the level E_1^* that is slightly higher than the excitation potential. Next, assume that electrons gaining energy I_1 a little higher than the ionization potential enter in inelastic colision instantaneously, ionizing an atom with a probability β or exciting it with a probability $1-\beta$. The quantities E_1^*, I_1, ν^*, β can be adjusted in a reasonable way after an analysis of cross section curves $\sigma^*(\varepsilon)$ and $\sigma_i(\varepsilon)$. In inert gases, E_1^* and I_1 are 1 to 2 eV higher than the corresponding potentials E and I; $\beta \approx 0.2$.

The assumption of *instantaneous* inelastic collisions at $\varepsilon > I_1$ (which is by no means too crude, because the corresponding frequencies are high) makes it possible to *exclude* the region $\varepsilon > I_1$ from consideration, after replacing the effect of negative sources $Q^* + Q_i$ in this region by an adequate *boundary condition*. Electrons moving along the energy axis then have an infinite capacity sink at the point $\varepsilon = I_1$, so that $n(I_1) = 0$ at this point. In a simple model we "collect" the real positive sources $Q^* + Q_i$ located in the low-energy region and assign them to the point $\varepsilon = 0$. Now the flux $J(0)$ is *nonzero*. It is connected with the flux $J(I_1)$ or ionization frequency, which equals, by definition

$$\nu_i = \beta J(I_1) \Big/ \int_0^{I_1} n(\varepsilon) d\varepsilon \,. \tag{7.21}$$

Indeed, $J(I_1)\,\mathrm{cm}^{-3}\mathrm{s}^{-1}$ electrons reach the sink at $\varepsilon = I_1$. They immediately enter into ionizing and exciting inelastic collisions, so that $2\beta J(I_1) + (1-\beta)J(I_1)$

electrons with "zero" energy are created. In addition, electrons that produce excitations in the $E_1^* < \varepsilon < I_1$ zone also emerge with "zero" energy. Therefore,

$$J(0) = (1+\beta)J(I_1) + \nu^* \int_{E_1^*}^{I_1} n(\varepsilon)d\varepsilon \ \mathrm{cm^{-3}s^{-1}} . \tag{7.22}$$

This is the second boundary condition. The terms for elastic collisions, attachment, ionization, and distributed sources in the region $0 < \varepsilon < E_1^*$ are dropped from (7.19), which takes the form

$$\begin{aligned} \nu_i n &= -dJ/d\varepsilon & 0 < \varepsilon < E_1^* , \\ \nu_i n &= -dJ/d\varepsilon - \nu^* n & E_1^* < \varepsilon < I_1 , \\ J &= A\varepsilon\, dn/d\varepsilon + An/2 , \quad & A = 2e^2E^2\nu_m/3m(\omega^2+\nu_m^2) . \end{aligned} \tag{7.23}$$

It is not difficult to verify by integrating (7.23) in ε from 0 to I_1 that only one of relations (7.21), (7.22) is independent. The other is implied by the result of integration.

7.4.6 Results Obtained from the Model

Equations (7.23) may be integrated for functions of type $\exp(\pm \mathrm{const}\sqrt{\varepsilon})$. When the general solution is subject to the boundary conditions $n(I_1) = 0$ and (7.22) and to the continuity of n and J at the boundary between the regions, $\varepsilon = E_1^*$, the outcome is a rather unwieldy transcendental equation for the ionization frequency $\nu_i(E)$ [7.7]. It is successfully solved, however, in two limiting cases; the resulting expressions have a very lucid physical meaning that greatly clarifies the nature of the process.

Let us refer to the quantity

$$\nu_E = \tau_E^{-1} = \frac{1}{I_1}\left(\frac{d\varepsilon}{dt}\right)_E = \frac{e^2E^2\nu_m}{m(\omega^2+\nu_m^2)I_1} = \frac{3}{2}\frac{A}{I_1} \tag{7.24}$$

as the ***energy gaining frequency***. A slow electron would need a time $\tau_E = \nu_E^{-1}$ for gaining, in the absence of energy losses, the energy I_1 required for ionizing an atom. The inequality $\nu^* \ll \nu_E$ corresponds to a low probability of inelastic losses in the course of accumulating the ionization energy. In this limiting case, we find $\nu_i \approx \beta\nu_E \propto E^2$. An electron crosses the "dangerous" stretch $E_1^* < \varepsilon < I_1$, where its motion along the ε axis toward the energy I_1 may be impeded, and then produces ionization with a probability β. Multiplication takes a time $\tau_i = \nu_i^{-1} \approx \beta^{-1}\tau_E$, as expected.

In the opposite case of high inelastic loss, $\nu^* \gg \nu_E$,

$$\nu_i \approx a^2\beta\xi\nu_E , \quad \xi = \frac{J(I_1)}{J(E_1^*)} \approx 2a \exp\left[-\frac{a-1}{a}\left(\frac{6\nu^*}{\nu_E}\right)^{1/2}\right] , \tag{7.25}$$

where $a = (I_1/E_1^*)^{1/2}$ is a number, equal to approximately 1.2 for all inert gases. The factor $\xi \ll 1$ is the ratio of fluxes at the end and beginning of the dangerous zone; it is the probability for an electron to cross this zone without dissipating energy on the excitation of an atom. On the average, an electron goes through the almost complete cycle of accumulating energy ξ^{-1} times, dissipating it each time "uselessly" on excitation before the barrier of inelastic loss is broken and the electron accomplishes ionization with the specified probability β. The time necessary for multiplication, up to an unimportant factor a^2 of order unity, is $\tau_i = \nu_i^{-1} \approx \tau_E/\beta\xi$. Of course, the result (7.25) is significant not because it states this equally obvious fact, but because it leads to the calculation of the probability ξ. Note that the expression for ξ can be transformed to a form typical for a stochastic process [(7.15) implies that the motion of electrons along the energy axis is indeed stochastic [7.7]].

In the general case, the transcendental equation for ν_i has to be solved numerically. However, this need be done only once, by using similarity laws, because most gases have nearly equal values of a and β (Fig. 7.9). In the figure, $a = 1.2$, $\beta = 0.2$. The situation when $\beta = 1$, useful for some cases of optical breakdown, is also represented (Sect. 7.5).

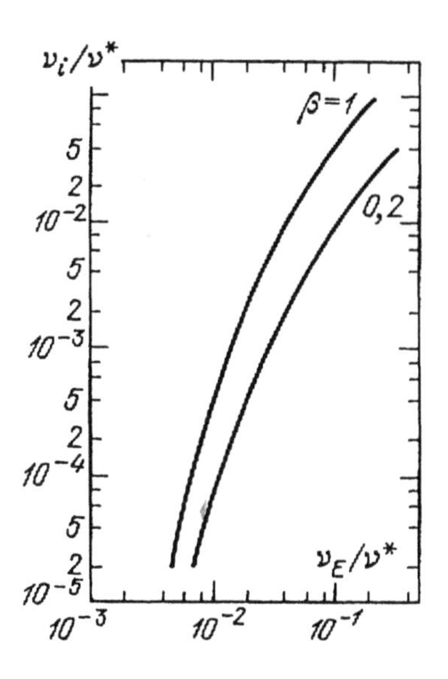

Fig. 7.9. Universal dimensionless function for calculating the ionization frequency as a function of the energy increase frequency which is proportional to E^2 [7.7.]

7.4.7 Comparison of Calculated Ionization Frequencies and Breakdown Thresholds with Experimental Data

Let us compare the theoretical formula (7.25) with the available experimental data on α in view of the relation of Townsend's ionization coefficient α for constant field to ionization frequency and drift velocity, $\alpha = \nu_i/v_d = \nu_i m\nu_m/eE$. Assuming $\omega = 0$ and factoring out pressure via formulae $\nu_m^* = \nu_{m1}p$, $\nu^* = \nu_1^* p$, using (7.25), we find the ionization coefficient:

$$\frac{\alpha}{p} = A_1 \frac{E}{p} \exp(-B_1 p/E) ,$$

$$A_1 = \frac{e}{I_1} 2a^3\beta = \frac{2a^3\beta}{I_1[\mathrm{eV}]} \approx \frac{0.68}{I_1[\mathrm{eV}]} V^{-1}., \tag{7.26}$$

$$B_1 = \frac{a-1}{a}\left(\frac{6I_1 m\nu_{m1}\nu_1^*}{e^2}\right)^{1/2} \approx 1\cdot 10^{-8}\sqrt{I_1[\mathrm{eV}]\nu_{m1}\nu_1^*}$$

$$\mathrm{V\cdot cm^{-1}Torr^{-1}} .$$

Formulas (7.25, 26), describing the case of strong inelastic losses, correspond to E/p being only 5–20 $\mathrm{V{\cdot}cm^{-1}Torr^{-1}}$. Although the exponential factor in (7.26) has the form identical to that in Townsend's formula (4.5), the constant B_1 differs essentially from the data of Table 4.1, which are valid for high values of $E/p(> 100)$. Moreover, the meaning of B_1, connected first of all with the excitation cross section, has nothing in common with the semi-quantitative interpretation given to Townsend's constant B (Sect. 4.1.5).

If the experimental curve of α for argon in the range of $E/p \approx 5$–20 is approximated by (7.26) (Sects. 4.2.6), we obtain $B_1 = 31$, $A_1 = 0.01$. For argon, $I_1 = I + 1 = 16.8\,\mathrm{eV}$, $\nu_m = 7\cdot 10^9 \cdot p[\mathrm{Torr}]\,\mathrm{s}^{-1}$, $\nu^* \approx 2.6\cdot 10^8 p[\mathrm{Torr}]\,\mathrm{s}^{-1}$, which gives $B_1 = 53$ and $A_1 = 0.04$. This agreement between a very simplified theory and experimental data should be regarded as satisfactory, especially in view of the fact that no "adjustment" parameters were used in the calculation. The agreement for xenon is even better. Choosing from the cross section data the values $I_1 = 13.1\,\mathrm{eV}$, $\nu_m = 1.5\cdot 10^{10} p\,\mathrm{s}^{-1}$, $\nu^* = 4\cdot 10^8 p\,\mathrm{s}^{-1}$, we find $B_1 = 85$ and $A_1 = 0.05$. The approximation of the experimental α curve gives $B_1 = 85$, $A_1 = 0.1$.

Figure 7.10 compares the calculated and measured microwave breakdown thresholds of argon and xenon. The calculations were based on the theory presented above of ionization frequency and the stationary criterion $\nu_i = \nu_d$. In the high-pressure range, at the ends of the right-hand branches, $\nu_m^2 \gg \omega^2$, $E/p \sim 10\,\mathrm{V{\cdot}cm^{-1}Torr^{-1}}$. Here, the asymptotic formula (7.25) is valid. The calculation fits the experimental data quite well. The description of the minimum region, where Fig. 7.9 has to be employed, is also good. Discrepancies are greater at low pressures, where inelastic losses are small and another asymptotic formula holds: $\nu_i = \beta\nu_E$. However, at $p < 10^{-1}$ Torr the free path length of electrons reaches the diffusion length (the cavity size), so that the theory of diffusion losses becomes invalid and collisions of electrons with walls become significant when they move in a transverse direction to the applied field; hence, good agreement cannot be expected here.

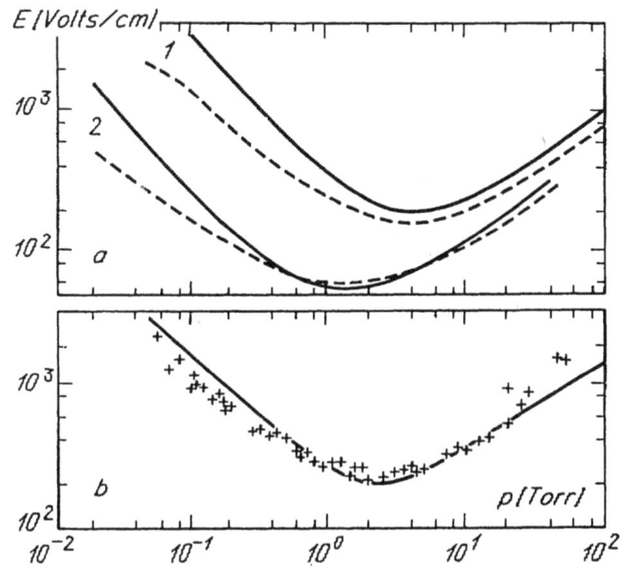

Fig. 7.10. Thresholds of microwave breakdown: (**a**) Ar, *(1)* f = 2.8 GHz, Λ = 0.15 cm; *(2)* f = 0.99 GHz, Λ = 0.63 cm; (**b**) Xe, f = 2.8 GHz, Λ = 0.10 cm. *Solid curves,* results of calculations [7.7]; *dashed curves* and *crosses* give experimental data [7.5]

7.5 Optical Breakdown

The discovery of the *optical breakdown* effect, in 1963 [7.8], became possible only after the development of Q-switched lasers that produce light pulses of tremendous power, called "giant pulses". When the light of such a (ruby) laser was passed through a focusing lens, a spark flashed in the air, in the focal region, as in the electrical breakdown of a discharge gap. The discovery was a complete surprise for physicists and produced a sensation at the time, though the element of surprise has worn off by now. Gas breakdown at optical frequencies requires a tremendous field strength, 10^6–10^7 V/cm, in the light wave; this was unthinkable before the advent of the laser. Furthermore, the necessary light intensity, about 10^5 MW/cm^2, could only be reached by focusing the light of not just an ordinary laser, but one operating in th giant pulse regime. The new effect caused unparalleled interest among physicists. In a short time, it was experimentally and theoretically investigated to such a degree [7.7], that by now we know at least as much about it as about its closest analogue, the microwave field breakdown.

7.5.1 Experimental Arrangement

The arrangement of the pioneer experiments on measuring optical breakdown thresholds [7.9] is typical for much later work. Giant pulses of a ruby laser with the following parameters were employed: power output 1 J, pulse length 30 ns= $3 \cdot 10^{-8}$ s, the maximum (peak) power 30 MW= $3 \cdot 10^{14}$ erg/s. These parameters are typical for modern moderate-power systems.[2] The pulse energy was measured by a calorimeter. The pulse shape was roughly triangular, the rise time being shorter than the decay time.

In order to increase the radiation flux density, the light beam is focused. The diameter of the focal spot d is determined by the divergence angle of the original light beam, θ, and the focal length f of the lens, $d = f\theta$. (Normally, $\theta \sim 10^{-3} - 10^{-2}$ rad, $f \sim$ 3–10 cm.) *Meyerand* and *Haught* [7.9] had $d \approx 2 \cdot 10^{-2}$ cm. The spot diameter was measured by the size of the hole burnt in very thin gold foil (0.05 μ thick). At the peak power of 30 MW, the radiant power density at the focal spot was $S \approx 10^5$ MW/cm^2 = 10^{18} erg/(cm^{2}s), the rms electric field in the light wave was

$$E = \sqrt{4\pi S/c} = 19\sqrt{S[\mathrm{W/cm^2}]}\,\mathrm{V/cm} \approx 6 \cdot 10^6\,\mathrm{V/cm}\ ,$$

and the photon flux density of the ruby laser, $\hbar\omega = 1.78$ eV, was

$$F = 3.4 \cdot 10^{18} S[\mathrm{W/cm^2}]\,\mathrm{cm^{-2}s^{-1}} \approx 3.4 \cdot 10^{29}\,\mathrm{cm^{-2}s^{-1}}\ .$$

The light beam was focused into a chamber filled with a gas to be investigated, at various pressures. The breakdown was deduced from the appearance of a visible light flash (lasting about 50 μs). Moreover, the focal spot was placed between a pair of electrodes to which a voltage of $\approx$ 100 V was applied. About 10^{13} electron charges were extracted by this field from the focal region when breakdown occurred.[3] The radiant powe was varied by an attenuator in order to determine the breakdown threshold, which was found to be very abrupt.

7.5.2 Results of Experiments

The threshold field decreases monotonically as the pressure increases (Fig. 7.11). However, this is observed only in a limited range of pressure. When a wider range from several atm. to two thousand atm. was scanned [7.10] it was found that the threshold has a minimum, as in the case of microwave breakdown. In this case the minima are not at 1–10 Torr, but at hundreds of atmospheres (Fig. 7.12). An interesting fact is that the position of the minimum approximately satisfies the same relation $\nu_m \approx \omega$ (Sect. 7.3.6). For example, for an Ar and ruby laser, $\omega = 2.7 \cdot 10^{15}\,\mathrm{s}^{-1}$, this formula gives $p \approx 225$ atm., while experiments give 190 atm.; for He, $p \approx 1450$ atm., while experiments give 700 atm. The rason for a minimum is the same as in the microwave field: the rate of energy accumulation by an electron in the field of a given frequency is maximal at $\nu_m \approx \omega$ (ν as a function of p).

[2] Solid state lasers have been developed recently that produce thousands of megawatts for several nanoseconds.

[3] Breakdown also reduces the power of the radiation that passes through the focal spot because the resulting plasma absorbs some radiation.

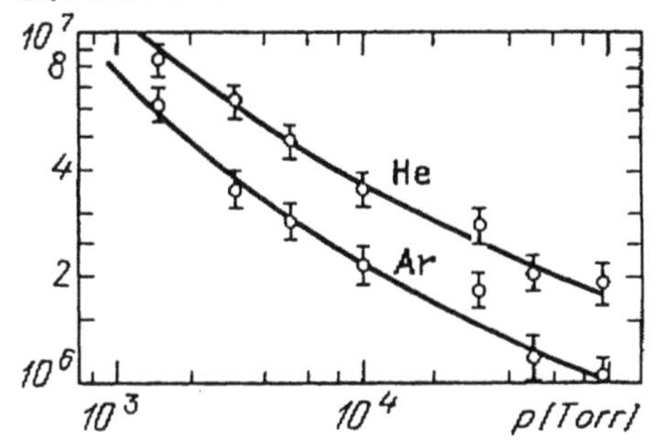

Fig. 7.11. Measured threshold fields for the breakdown of Ar and He by ruby laser radiation; pulse length 30 ns, diameter of focal spot $2 \cdot 10^{-2}$ cm [7.9]

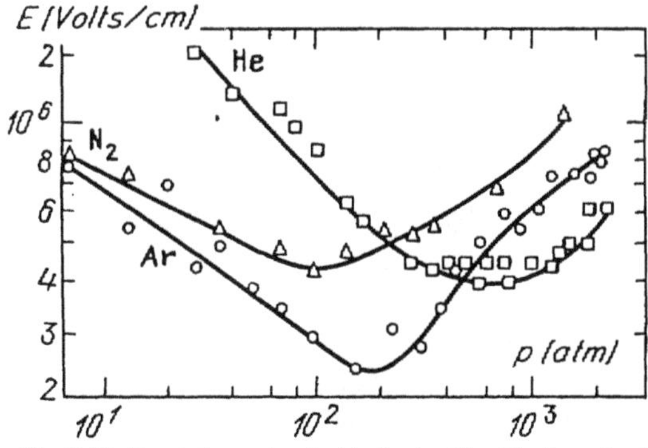

Fig. 7.12. Breakdown thresholds in Ar, He, N_2 for ruby laser radiation over a wide pressure range [7.10]. Pulse length 50 ns, focal spot diameter 10^{-2} cm

When the radiation from a pulsed CO_2 gas laser is focused into a gas, breakdown is also observed. This radiation ($\lambda = 10.6\mu \approx 10^{-3}$ cm) occupies an intermediate position between that emitted by ruby and neodymium lasers ($\lambda \sim 10^{-4}$ cm) and that of microwave radiation ($\lambda \sim 1$ cm), although it is closer to the visible part of the spectrum. As a rule, CO_2 lasers give longer pulses, of about 1 μs. The relation $\nu_m \approx \omega$ for the position of the maximum holds quite well (Fig. 7.13). For example, in Xe $\nu_m \approx 9 \cdot 10^{12} p\,[\text{atm.}]\,\text{s}^{-1}$; $\omega_{CO_2} = 1.78 \cdot 10^{14}\,\text{s}^{-1}$. We find $p \approx 20$ atm., while experiments give about 15 atm.

The breakdown threshold decreases when the focal spot, that is, the size of the region subjected to the field, is increased. Measurements were conducted in the range of diameters from 10^{-2} to 10^{-1} cm, by using lenses with different focal lengths ($d = f \cdot \theta$). This result is qualitatively fairly clear: the greater the region where the field is high, the lower the importance of the loss of electrons due to their diffusive escape from this region. In fact, the situation is more complicated. Estimates show that in some cases the pulse length is too short for an electron to diffuse across the distance d. The effect may be caused by the diffusion, not from the entire focal spot, but from the "hot" points with high local fields that appear in the focal spot as a result of cross-sectional inhomogeneity of the laser beam. The electron avalanche mostly develops at these points.

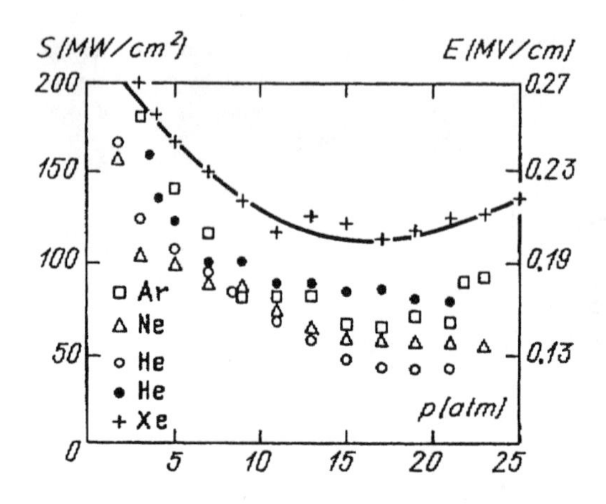

Fig. 7.13. Breakdown thresholds of inert gases in the radiation of a CO_2 laser [7.7]. The *black dots* represent data for helium of a higher purity

7.5.3 Breakdown Thresholds of Atmospheric Air

These data are very important. Quite a few physical experiments employ high-intensity laser beams. Electrical breakdown of air on the beam path to the target is an obstacle for light propagation because of absorption in the plasma. For example, in such experiments with high-power beams as target irradiation for fusion experiments one has to send the beam to the target through vacuum. The threshold intensity for the giant pulse of a ruby laser and an ordinary focal spot diameter of 10^{-2} cm is $S_t \approx 10^{11}$ W/cm^2, and the field is $E_t \approx 6 \cdot 10^6$ V/cm.

The breakdown threshold of nonfiltered air by focused CO_2 laser radiation is roughtly $2 \cdot 10^9$ W/cm^2, and that of dust-free air is not lower than 10^{10} W/cm^2. The tiniest dust particles floating in the air greatly facilitate the breakdown by CO_2 laser radiation, while their effect is negligible for the neodymium and, in particular, ruby lasers. This difference appears because the short-wave radiation of solidstate lasers "supplies itself" with the seed electrons required for starting an avalanche. The long-wave radiation of CO_2 lasers cannot do this in a pure gas.

7.5.4 Multiphoton Photoelectric Effect

Two quite different mechanisms of gas ionization by high-intensity light can be proposed. The first one, the development of an *electron avalanche*, is essentially the same as in fields of other frequencies. The differences are in the detailes the process of energy-gain in the field, which is essentially quantum in nature.

The second mechanism of ionization is characteristic of photons: it has a purely quantum nature. Electrons may be detached from atoms as a result of the *multiphoton photoelectric effect*, that is, in response to the simultaneous absorption of several photons. In the visible light range, the single-photon photoelectric effect is impossible, because atomic ionization potentials are several times greater than the quantum energy. For instance, the ruby laser photon energy is $\hbar\omega = 1.78$ eV, while the argon ionization potential is $I_{Ar} = 15.8$ eV, that is, nine photons are required to detach an electron. Multiphoton processes usually have a low probability, but the process rate increases sharply when the photon density (the light intensity) increases; at the extremely high intensities that result in optical breakdown, the probability may become substantial.

Both calculations (see below) and experiments show that nanosecond (and longer) laser pulses at pressures above several tenths of one atmosphere always produce *avalanche ionization*. The rate of avalanche ionization is sufficient for breakdown at fields that are insufficient for intensive multi-photon ionization. However, the latter plays an important role as a source of the initial, seed electrons required to ignite the avalanche. The arrival of a spurious electron at a small focal region during a very short pulse is a highly improbable event.

We have mentioned several times that the excitation of atoms by electron impact slows down the avalanche because an electron dissipates its energy and has to accumulate it again and again before it manages to overcome the excitation zone and reach the ionization potential. This process characterizes all fields except that of light. If the photon energy is high, a small number of quanta may be sufficient to eject an electron from an excited atom by the multi-quantum photoelectric effect. In this case excitation even accelerates the growth of the avalanche because it is enough for electrons to reach the excitation, not ionization, potential.

For example, in argon $I_{\mathrm{Ar}} = 15.8\,\mathrm{eV}$ and the potential of the first excitation level $E^*_{\mathrm{Ar}} = 11.5\,\mathrm{eV}$. The energy necessary to detach an electron from an excited atom is 4.3 eV, that is, what is required is the simultaneous absorption of three photons of a ruby laser or four photons of a neodymium laser ($\hbar\omega = 1.17\,\mathrm{eV}$). A four-photon process has very low probability, but a three-photon process may occur under certain conditions at the ruby laser frequency. There exists both experimental and theoretical evidence supporting the reality of this mechanism.

7.5.5 Nonstationary Breakdown Criterion

If light pulses are very short (as in our case), the field prescribed by the stationary criterion (7.9) may prove insufficient for appreciable ionization of the gas. Indeed, the situation cannot be classified as a breakdown if only two to three generations of electrons are produced during one pulse. Ionization must reach a substantial level. In fact, one typically associates breakdown with a visible light flash. Thus the experments described above recorded that a flash and the breakdown correspond to about 10^{13} avalanche electrons. Assuming an avalanche to be started by a single electron, we find that $\log_2 10^{13} \approx 43$ generations are created by breakdown during one pulse. The field must be so strong that the avalanche time constant Θ be $\ln 10^{13} \approx 30$ times shorter than the pulse duration $t_1 \approx 3 \cdot 10^{-8}\,\mathrm{s} : \Theta \approx 1\,\mathrm{ns}$.

The threshold field E_t is found from the condition that an avalanche initiated by N_0 electrons multiplies during the time t_1 to N_1 electrons:

$$t_1/\Theta(E_t) = (\nu_i - \nu_a - \nu_d)t_1 = \ln(N_1/N_0)\,. \qquad (7.27)$$

E_t is insensitive to the rather uncertain quantity N_1/N_0 (due to the logarithmic relationship). For calculations, one can use $t_1/\Theta \approx 30$. The nonstationary criterion (7.27) generalizes the stationary criterion (7.9) and tends to it as $t_1 \to \infty$. If the pulses are very short, the threshold is found to be very high and electron losses become unimportant: $\nu_a, \nu_d \ll \nu_i$. The nonstationary criterion makes the threshold nature of the breakdown even more pronounced. If the ionization frequency ν_i is reduced by half (this requires only a slight field reduction), only 21 generations are produced instead of 43, that is, the number of electrons decreases by 6 to 7 orders of magnitude.

7.5.6 Classical or Quantum Picture?

We all remember from school lessons that light is emitted and absorbed as discrete quanta. Radiation and matter cannot exchange energy if $\Delta\varepsilon < \hbar\omega$: this is forbidden. The classical description of the interaction between electromagnetic waves and electrons is valid if $\Delta\varepsilon \gg \hbar\omega$ in each elementary event. Otherwise quantum theory must be used. It is easy to verify via (3.10) that in microwave fields the mean energy gained by an electron from the field in a collision with an atom $\Delta\varepsilon_E \gg \hbar\omega$. The true amounts of energy, $\Delta\varepsilon_\pm$, exchanged by an electron and the field in individual collisions (Sect. 3.2.5) exceed $\hbar\omega$ by even more. There are no doubts as to the applicability of the classical theory that we have worked with so far. However, for optical frequencies, even at tremendous breakdown fields ($E \sim 10^7$ V/cm), (3.10) gives $\Delta\varepsilon_E \sim 10^{-2}$ eV$\ll \hbar\omega \sim 1$ eV. The true increments $\Delta\varepsilon_\pm$ are also smaller than $\hbar\omega$, or at least comparable to $\hbar\omega$. This means that the interaction of light radiation with the electrons of an ionized gas is of a quantum nature.

The interaction proceeds as follows. An electron colliding with an atom may absorb a quantum (photon), $\hbar\omega$, or produce stimulated emission of a photon $\hbar\omega$, provided its energy is sufficiently high.[4] All these events are random, so that the changes in the energy of the electron are of the random walk type (one-dimensional diffusion), by $\hbar\omega$ jumps along the energy axis ε. However, on average the energy of the electron increases with time, just as a particle diffusing away from an impenetrable wall is on the average receding (cf. Sect. 3.2.5). This goes on until the energy reaches the ionization potential and the electron produces a new electron, as in any other field. This is how an avalanche evolves.

It can be shown, in the calculation of $d\varepsilon/dt$ as a difference between the mean rates of quantum absorption and stimulated emission of radiation, that the quantum-mechanical expression reduces to the classical formula (3.11) under the condition $\hbar\omega \ll \varepsilon$, which is much less severe than the classical-physics condition $\hbar\omega \ll \Delta\varepsilon$ [7.7, 11]. The resulting light absorption coefficient, determined by the difference between the rates of absorption and stimulated emission of photons under the same condition $\hbar\omega \ll \varepsilon$, is reduced to the classical coefficient μ_ω

[4] In the field of high-intensity laser radiation, spontaneous emission takes place much less frequently than stimulated emission, so that the former can be ignored.

(3.28). Under the same condition, the quantum kinetic equation scrutinized in [7.7] transforms into the classical (5.24).

These arguments do not mean that our understanding of the relationship between the classical and quantum approaches to the interaction has changed. Actually, when (3.11) is applied to the quantum case, it must be given a statistical meaning. For instance, if this formula gives $\Delta\varepsilon_E = 0.01\hbar\omega$ but $\hbar\omega \ll \varepsilon$, this means that an electron did not interact with radiation at all in, say, 99 collisions but in the 100th collision it suddenly gained $\hbar\omega$ in one portion, 100 times that of the symbolic averaged value $\Delta\varepsilon_E$.

7.5.7 Calculation of Threshold Fields

When evaluating whether the condition $\hbar\omega \ll \varepsilon$ is satisfied, one obviously has to compare $\hbar\omega$ with the mean energy of the electron spectrum; under breakdown conditions, the latter equals about one half of the ionization potential, that is, $\bar{\varepsilon} \sim$ 8–13 eV in Ar and He. Hence, the condition is very well satisfied in the case of the CO_2 laser ($\hbar\omega = 0.117$ eV), is only passably met for the radiation of the neodymium laser ($\hbar\omega = 1.17$ eV), and can be considered to be satisfied for the ruby laser radiation, with a number of qualifications. Consequently, one can use the theory presented in Sect. 7.4 and the non-stationary criterion (7.27). In the case of ruby laser radiation, a plausible assumption is that of the fast *three-photon ionization of excited atoms*. The breakdown process is then regardes as similar to that in the Heg gas: inelastic losses are absent and $\beta = 1$. With neodymium and, of course, CO_2 lasers, inelastic losses are important. The CO_2 laser breakdown obeys the stationary criterion (7.9). Experimental data fit the calculated curves quite satisfactorily (Figs. 7.14, 15). We conclude that breakdown processes in the optical and microwave ranges are very similar.

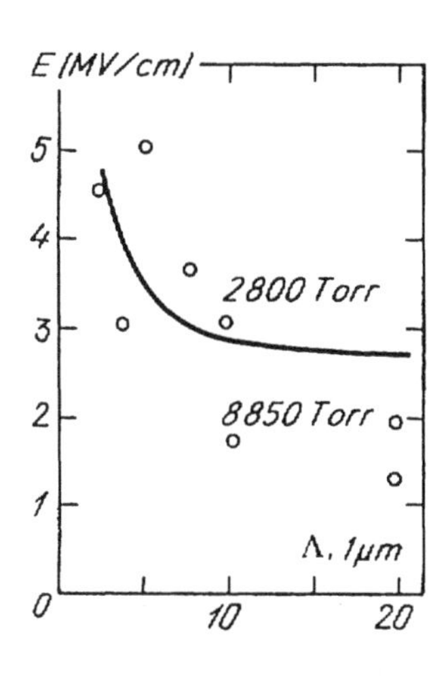

Fig. 7.14. Threshold fields in Ar calculated on the assumption that excited atoms are ionized instantaneously by the incident ruby laser radiation [7.7]. The *circles* are the experimental results obtained using a single-mode laser

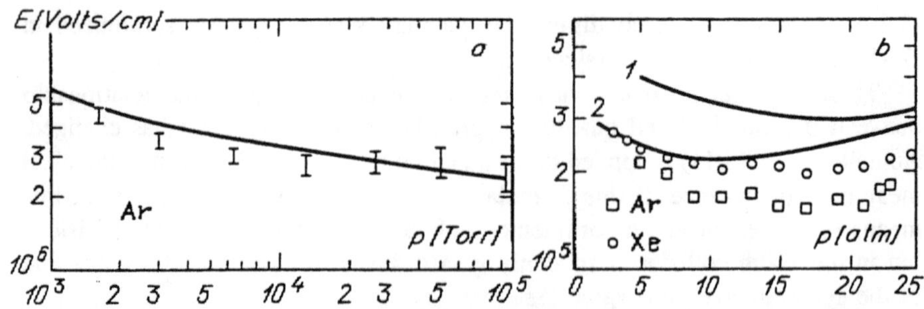

Fig. 7.15. Calculation of breakdown thresholds [7.7]: (**a**) Ar with neodymium laser, experimental data are shown by *error bars*, $\Lambda = 1.64 \cdot 10^{-3}$ cm; (**b**) Ar, Xe with CO_2 laser; *circles* and *squares* show experimental data; focal spot radius $4 \cdot 10^{-3}$ cm, pulse length 1 μs

7.5.8 A Bridge Between Microwave and Light Ranges

The experimental observation that the classical similarity law $E_t^2 \sim S_t \propto \omega^2$ holds for threshold values in a wide range of optical frequencies right down to the microwave range is an especially conclusive confirmation of the last statement in the preceding subsection. In the microwave range, the law $S_t \propto \omega^2$ holds only at low pressure ($p < 10$ Torr), corresponding to the left-hand branch of the threshold curves. To generalize, we have to assume that $S_t \propto (\omega^2 + \nu_m^2)$. As for the optical frequency range, here even tens of atmospheres is a low pressure; for example, the entire plot in Fig. 7.11 represents the left-hand curve.

Figure 7.16 shows the curve $S_t \propto (\omega^2 + \nu_m^2)$, which on a logarithmic scale degenerates into a straight line if $\nu_m^2 \ll \omega^2$; experimental pints for air are also shown. In addition to numerous data on the breakdown of atmospheric air by the radiation of ruby, neodymium, and CO_2 lasers [7.7, 11], data for D_2O [7.12] ($\lambda = 385$, $\mu = 0.38$ mm, the last "mastered" range), HF, and DF [7.13] ($\lambda = 2.7\mu$, 3.8μ) lasers were reported recently. Note that the points group well around the theoretical straight line, although the law cannot be expected to hold strictly since different experiments were conducted in nonidentical conditions. Note also that

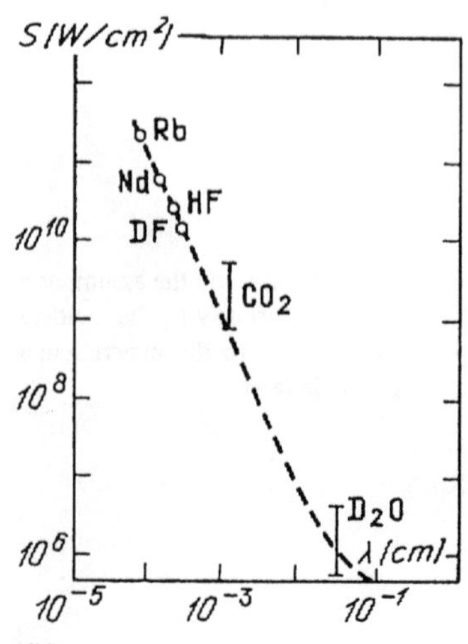

Fig. 7.16. Breakdown thresholds of air for radiation of different lasers. *Dashed curve* shows the classical dependence

the D_2O point lies at the limit of the range where the condition $\omega^2 \gg \nu_m^2$ is satisfied: $\omega = 4.9 \cdot 10^{12}\,s^{-1}$, $\nu_m \approx 3.8 \cdot 10^{12}\,s^{-1}$.

The law $S_t \propto \omega^2$ is obviously violated in the middle of the visible and UV ranges of the spectrum, for secondharmonics of the neodymium ($\hbar\omega = 2.34\,eV$) and especially ruby ($\hbar\omega = 3.56\,eV$) lasers, where quantum effects become important [7.7].

7.5.9 Long Spark

When the suprathreshold power is moderate, optical breakdown is obtained by focusing the laser beam with a short-focus lens. If a laser is very powerful, however, the intensity is sufficient for the breakdown on a long path along the caustic of a long-focus lens. The result is a very impressive optical breakdown: "long spark". A record length of spark – more than 60 meters – was produced in 1976 [7.14] using a neodymium laser pulse of 160 Joules power output, 5 GW mean power and a lens of $f = 40$ m. The beam was directed out of the laboratory window into the courtyard outside. Long sparks are not continuous: ionized stretches alternate with nonionized gaps (Fig. 7.17). This may be connected with the statistical behavior of seed electrons and also with the space-time and angular inhomogeneity of the beam. A long spark is also obtainable with CO_2 lasers, at intensities of $1\text{–}2 \cdot 10^8\,W/cm^2$; the threshold increases to $3 \cdot 10^9\,W/cm^2$ in dust-free air. Long laser sparks are effective in initiating the breakdown in wide gaps between electrodes. The threshold field of breakdown by dc voltage is then reduced to 250 V/cm. In fact, the breakdown is considerably facilitated by the joint action of the laser radiation and microwave or constant field. In this way one can produce a directed or even zigzag breakdown channel between electrodes (some relevant details and references can be found in the review [7.11]).

Fig. 7.17. Photograph of a long spark obtained by neodymium laser. Spark length 8 m, focal length of the lens 10 m [7.14]

7.6 Methods of Exciting an RF Field in a Discharge Volume

These methods are classified into two main groups depending on whether the lines of force of the electric field in the discharge plasma reconnect or not, in other words, whether this is a *vortex* (rotational) or a *potential* field. The first group comprises *induction* methods based on using electromagnetic induction.

A typical – and most frequent – approach to implementing this principle is as follows (Fig. 7.18). A high-frequency current is passed through a solenoid "coil" (in fact, the coil may consist of only one or several turns). The oscillating magnetic field of this current within the coil is directed along its axis and induces a vortex electric field, whose lines of force are closed circles concentric with the turns of the coil. This electric field can ignite and sustain a discharge, its currents also being closed and flowing along the closed circular lines of force of the electric field. In actual experiments, a dielectric tube filled with a gas to be studied is inserted into the coil so that breakdown occurs under certain conditions and the discharge can be sustained after breakdown. Pulsed discharges can be produced if a sufficiently strong current pulse is fed into the coil. This type of discharge is known as the inductively coupled, or H-type, rf discharge, with the latter H pointing to the decisive role of the magnetic field. Inductively coupled discharges are apparently *electrodeless*.

In the methods belonging to the second group, the high-frequency (or any other waveform) voltage is applied to the electrodes. In the simplest (and the most widespread) geometry, two parallel plane electrodes are employed. The electrodes may be *bare* and be in direct contact with the discharge plasma, or they may be *insulated* by a dielectric (Fig. 7.18b,c). A system of two electrodes behaves with respect to a variable voltage as a capacitor, so that in contrast to iniduction discharges, those in this category are known as *capacitively coupled*, or E-type, rf discharges (ccrf). The letter E symbolizes the decisive role of the electric field. A capacitively coupled discharge can be ignited in a tube via a pair of ring electrodes fixed on the outside surface at the ends of the tube, creating the longitudinal field. As a result, the discharge can be observed through the end faces.[5]

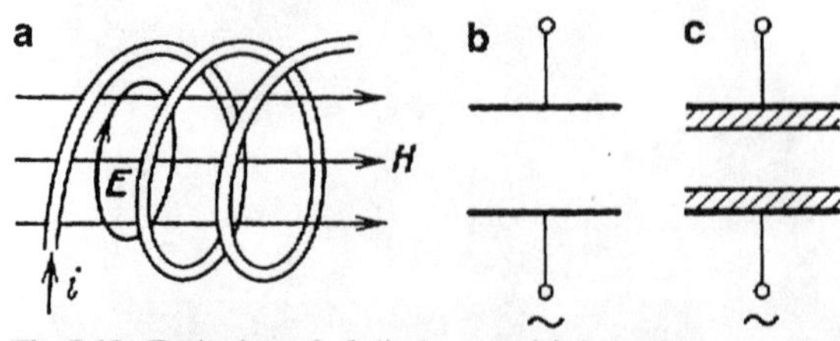

Fig. 7.18. Excitation of rf discharges: (a) inductively coupled through a solenoid coil; (b) voltage applied to electrodes in contact with plasma; (c) electrodes insulated from plasma (electrodeless, capacitively coupled rf discharge)

[5] The ccrf discharge may also arise in the case of inductive coupling, as a result of gas breakdown by voltage between turns.

The induction electric field increases with increasing frequency; it is proportional to frequency in the absence of plasma. Such fields are difficult to create at low frequencies since dangerously high currents have to flow through the inductor. As a result, the induction technique is typically used in the range of $f \sim 10^{-1} - 10^2$ MHz. Electrodeless capacitively coupled discharges also present a problem at low frequencies because an insulating dielectric layer has a high impedance, consuming most of the applied voltage. Electrodes used at frequencies $f \leq 10$ kHz are mostly bare. Actually, electrodes can be insulated with a thin layer of high dielectric constant (e.g., barium titanate) but this is not always practicable. As a rule, difficulties arise only if a voltage of many kV is to be applied to a discharge gap, that is, under conditions of wide gaps and high pressures. As for hundreds of volts or a kilovolt, serious problems are not encountered. Practical work in radio-frequency discharge is such that induction techniques are mostly used to sustain plasmas at high pressures about one atmosphere (Sect. 11.3), while capacitor techniques are preferable at medium and low pressures (Chap. 13).

As in the case of discharge analysis in a constant field, the electrode configuration may be arbitrary: two spheres, a sphere and a plane, a wire and a concentric cylinder, etc. In all these geometries, the field non-uniform and additional difficulties appear in the interpretation of results. Plane, parallel electrodes with a relatively narrow gap, that is, planar geometry, are easier and more convenient for experimental data processing, estimates, and theoretical evaluations.

7.7 Breakdown in RF and Low-Frequency Ranges

Many factors may drastically affect the characteristics of the process; these include frequency, pressure, method of introducing the field into the discharge volume, geometry and size of this volume, orientation of the electric vector.

The phenomena can be classified to some extent by comparing their characteristic lengths. Three such lengths are: volume size d, electron free path length l, and the oscillation amplitude of free electrons a, or of drift electrons A (Sect. 3.1.3). The choice between a and A is determined by which of the frequencies is higher: field frequency ω or electron collision frequency ν_m.[6] If pressures are so low that $l \gg d$, an electron suffers no collisions with the atoms of the gas. In typical cases, $d \sim 1$ cm and the upper boundary of such low pressure ($l \sim d$) in most gases is $p \sim 10^{-2}$ Torr. Let us very briefly consider several sets of conditions that produce more or less clearly pronounced effects.

[6] The field oscillation wavelength can be used to characterize frequency but few factors are indeed determined by it. At rf and lower frequencies, $\lambda > 1$ m, $\lambda \gg d \sim 1$ cm, that is, oscillations do not introduce additional spatial inhomogeneity in the field distribution.

7.7.1 Electron Oscillation Amplitude Is Small; Collisions Are Numerous

This is the case of $a, A \ll d, l \ll d$. It is realized at sufficiently high frequencies but not too low pressure, and differs but little from the microwave-field breakdown discussed earlier (Fig. 7.19).

The dependence of V_t on p at a fixed distance, that is, $E_t = V_t/d$ as a function of p, is very similar to Fig. 7.8 for microwaves. The interpretation is straight forward. The amplitude of drift oscillations at the ends of right-hand branches is $A = eE/m\nu_m\omega \approx 2.6 \cdot 10^{-3}\,\mathrm{cm} \ll d$ ($\omega = 10^9\,\mathrm{s}^{-1}$; $E/p \approx 3\,\mathrm{V/cm{\cdot}Torr}$; $\nu_m \approx 2 \cdot 10^9 p\,[\mathrm{Torr}]\,\mathrm{s}^{-1}$ for such E/p). In the region of the minima ($E/p \approx$ 50–100, $\nu_m \approx 4 \cdot 10^9 p$), the amplitude is greater by an order of magnitude, $2.5 \cdot 10^{-2}$, but is nevertheless small in comparison with $d \approx 1\,\mathrm{cm}$. The positions of the minima fit rather well the earlier-mentioned relation $\nu_m \approx \omega$ that yields $(p)_{min} \approx 2.5\,\mathrm{Torr}$. At elevated pressures $p \approx 50\,\mathrm{Torr}$ at the ends of right-hand branches, the threshold values are $E_t/p = 2.2\,\mathrm{V/cm}$ for $d = 2\,\mathrm{cm}$ and $E_t/p =$ 4.5 V/cm·Torr for $d = 0.5\,\mathrm{cm}$, being quite close to those in the microwave range and in constant field (Table 7.1). We are apparently trying to interpret a typical picture of bulk breakdown in which electrons are removed by diffusion.

Figure 7.19 shows plots of the sustaining voltages, measured in the same experiments, of the already ignited discharge. Ionization in the stationary discharge process, determined by the same frequency $\nu_i(E)$, also compensates for the diffusion loss of electrons but charge densities are now considerable and the diffusion, being ambipolar, proceeds much more slowly.[7]

7.7.2 Oscillation Amplitude Is Comparable with Volume Size; Collisions Are Frequent

This situation arises at lower frequencies. It has been known for quite some time that $V_t(p)$ curves similar to those of Fig. 7.19 sometimes display an additional minimum. Measurements plotted in Fig. 7.20 were carried out in a long (30 cm) cylindrical tube with outer electrodes (capacitively coupled rf discharge). With electrodes placed at the end faces (the field along the axis), the pattern was the same as in Fig. 7.19. The second minimum (right) observed on $V_t(p)$ curves at not too high frequencies was found only if the field was applied transversely to the tube. In this case the electron oscillation amplitude is comparable with the distance to the walls in the direction of charge motion, that is, with the tube diameter $d = 2\,\mathrm{cm}$.

[7] On the weaker field required to sustain a discharge (than to initiate the breakdown), see also Sect. 8.6.3.An example of this is a simple but extremely demonstrative experiment. If ever you have undergone uhf therapy (the field frequency is 40 MHz), you know how the nurse tests the normal functioning of the equipment. A tiny lamp probe (a low-pressure neon-filled bulb) is introduced into the field. If everything is in order, the lamp lights up in bright red. On bringing the probe slowly to an electrode or a lead wire, the lamp flares up. By moving, it slowly out of the field along the same route, the discharge shrinks and then dies out; this occurs quite far from the point of lighting up, in a considerably weaker field.

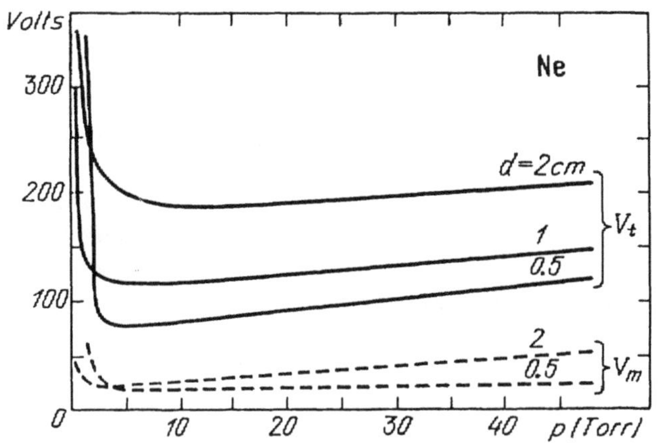

Fig. 7.19. Ignition potential of capacitively coupled rf discharge in neon (V_t), f = 158 MHz; d is the distance between the planar electrodes (which are covered by glass). *Dashed curves* show the burning voltage of a steady discharge, V_m [7.15]

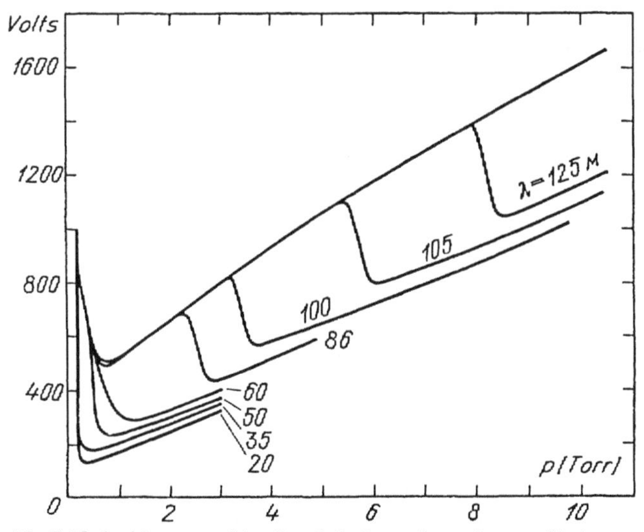

Fig. 7.20. Ignition potentials of ccrf discharge for various oscillation wavelengths [7.15]

Moving along the pressure scale from right to left, we first trace the ordinary right-hand branch, of the type shown in Fig. 7.19, corresponding to diffusion losses. As the pressure is reduced, the amplitude of the electron drift oscillations increases to $d/2$ so that electrons strike the walls on each swing. The loss is sharply increased because diffusion transports electrons to the wall at a slow rate. Discharge initiation requires a higher field; the breakdown potential increases. The positions of the jump and of the minimum are given by the obvious relation $A = eE_0/m\nu_m\omega \approx d/2$, whence $E_0/\omega \approx$ const. According to the data of Fig. 7.20, this relation is indeed valid. The numerical agreement is also satisfactory. For instance, at a frequency f = 3 MHz (λ = 100 m) we have

$\omega = 1.9 \cdot 10^7\,\text{s}^{-1}$, $E/p \approx 100$, $\nu_m \approx 4 \cdot 10^9 p$, $A \approx 1$ cm. To the left of the jump, we see the right-hand branch corresponding to enhanced losses. The main minimum (left) corresponds to the ordinary condition $\nu_m \approx \omega$; it is of the same nature as at higher frequencies (Sects. 7.3.6, 7.5.2). At sufficiently high frequencies, $A < d$ for any p, and the second minimum vanishes.

If the field frequency is reduced, with other conditions remaining unchanged, the field required for breakdown abruptly increases at a certain *boundary frequency*. This is observed when drift oscillations bring electrons to the walls. The threshold frequency satisfies the same approximate formula

$$\frac{2eE_0}{m\nu_m\omega d} \approx 0.14\frac{(E_0/p)[\text{V/cm}\cdot\text{Torr}]}{f[\text{MHz}]\cdot d[\text{cm}]} \approx 1\,, \tag{7.28}$$

where d is the distance between the opposite walls normal to the direction of the field, and E_0/p refers to the lower part of the jump. The frequency used in the numerical formula is $\nu_m \approx 4 \times 10^9 p$.

7.7.3 Wide Frequency Range, Including Low Frequencies; Collisions are Frequent

In hydrogen, the threshold field jumps at the boundary frequency are clearly pronounced (Fig. 7.21). This frequency agrees rather well with estimate (7.28). At frequencies below the boundary value, the threshold field is almost constant in a very wide frequency range, from 1 MHz to 50 Hz. This result is puzzling. Indeed, if a frequency is much lower than the boundary value, electrons are "herded" rather rapidly to one wall and kept there for a relatively long time, after which the field rushes them just as rapidly to the opposite wall, and so forth. Most of the time, there are no electrons in the discharge volume. If ionization stopped during these periods, breakdown would be appreciably impeded. It seems, therefore, that some other ionizing agent has not been identified in the volume. In neon, the threshold somewhat increases with λ over a wide range of frequencies (Fig. 7.22). It changes by a factor of 1.5–2 in response to a change in frequency by 3–5 orders of magnitude after a gentle "jump" at the boundary frequency (first rise on the left). Effects due to photons, excited atoms, photoemission from the walls, etc.

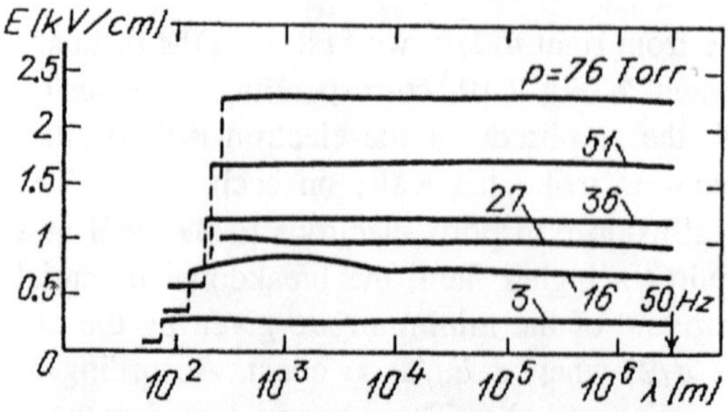

Fig. 7.21. Ignition field of ccrf discharge in hydrogen in a 2 cm long glass cylinder at different pressures over a wide range of frequencies [7.15]

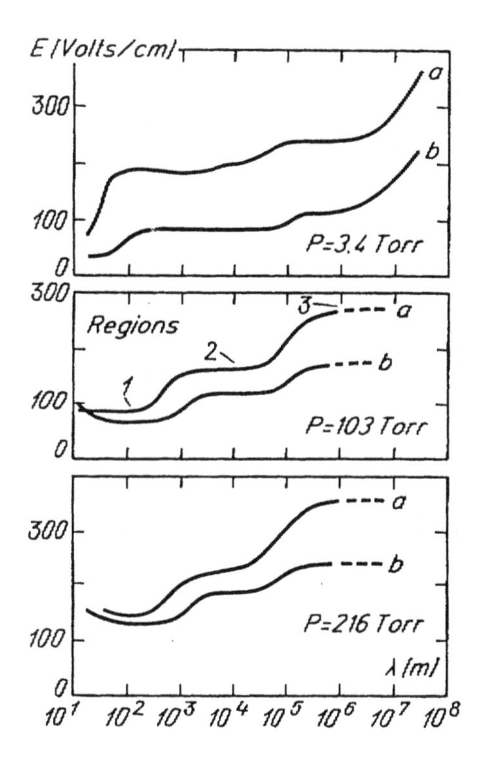

Fig. 7.22. Ignition field of ccrf discharge in neon in glass vessels *(a)* 1 cm in diameter and *(b)* 2.2 cm in diameter over a wide frequency range [7.15]

[7.15] are tentatively employed to explain these and a number of other, complex and tangled circumstances characterizing low-frequency discharges, but much remains to be done before clarity is achieved.

At low frequencies, of the order of 1 kHz and less, especially if electrodes are bare, breakdown develops anew in each half-period and evolves almost as in a constant field. In this respect, it is important that the rms threshold field in hydrogen at elevated pressure $p = 76$ Torr, $E_t/p \approx 20$ V/(cm·Torr), is quite close, according to the data of Fig. 7.21, to the corresponding value for the constant field (Table 7.1). In general, ignition potentials at elevated pressure depend on pd, by analogy to the right-hand branches of the Paschen curves, and assume similar values.

7.7.4 Breakdown of "Vacuum"

In the case of a highly rarefied gas, with electrons undergoing very few collisions ($l \gg d$), multiplication proceeds through secondary electron emission from the walls. Dielectric materials (glasses) produce quite strong emission capable of triggering breakdown in volumes insulated from electrodes. As a result, a discharge takes place in the residual gas in response to breakdown, a light flash appears, and a change in current is measured in the rf voltage generator. Nevertheless, breakdown occurs only at frequencies above the boundary value frequency f_b (Fig. 7.23).

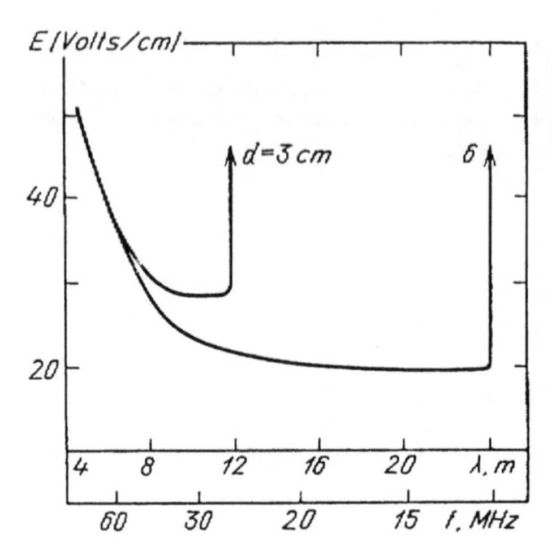

Fig. 7.23. High-frequency vacuum breakdown. Discharge ignition field in glass tubes of length d. The field is parallel to the axis: Hydrogen, $p = 10^{-3}$ Torr [7.15]

This can be explained [7.16]. For multiplication to occur, the energy of the incoming electron, ε, must suffice to detach more than one electron per projectile electron which penetrates the body and stays there (reflection of incident electrons does not lead to multiplication and can be ignored). Efficient emission requires $\varepsilon_m \approx 100\,\text{eV}$ (Sect. 4.5.6). Furthermore, electrons need to cross the gap in synchrony with the field. When an electron leaving one wall is accelerated and reaches the opposite wall, the field must be reversed for the emitted electron to be accelerated to the former.

Let an electron be emitted from the wall at $x = 0$ with velocity $\dot{x} = 0$ at a moment $t_0 = \pi/\omega$ at which the field $E_x = E_0 \sin \omega t$ reverses its direction and begins accelerating the emitted electron. According to (3.1), it reaches the opposite wall $x = d$ with a velocity $v_m = (2\varepsilon_m/m)^{1/2}$ at the moment of the next field reversal, $t_1 = 2\pi/\omega$, if $2eE_0/m\omega = v_m$, $\omega d = \pi(eE_0/m\omega)$ [the second equality corresponds to (7.28), only for free motion]. For $\varepsilon_m = 100\,\text{eV}$, $v_m = 5.8 \cdot 10^8$ cm/s, this gives the cut-off frequency $f_b \approx 140/d\,[\text{cm}]$ MHz and the threshold field $E_{0_t} \approx 120/d\,[\text{cm}]$ V/cm, in reasonable agreement with Fig. 7.23 and the empirical relation $f_b \approx 80/d$ MHz.

If $f < f_b$, the electron reaches the opposite wall at a moment when the accelerating force still points along its line of motion. The emitted electron is hopelessly "trapped" and multiplication possibilities are severely limited. If, however, $f > f_b$, the electron begins decelerating before reaching the wall. Now, something can be done: by transferring more energy to the emitted electron via the increasing field. Therefore, breakdown is possible for $f > f_b$, but the threshold increases as the difference $f - f_b$ grows (see Fig. 7.23). For details concerning radio- and low-frequency fields, see [7.15].

8. Stable Glow Discharge

8.1 General Structure and Observable Features

8.1.1 Distinctive Features

The glow discharge is a *self-stustaining* discharge with a *cold cathode* emitting electrons due to secondary emission mostly due to positive ion bombardment. A distinctive feature of this discharge is a layer of large positive space charge at the cathode, with a strong field at the surface and considerable potential drop of 100–400 V (or more). This drop is known as *cathode fall,* and the thickness of the cathode fall layer is inversely proportional to the density (pressure) of the gas. If the interelectrode separation is sufficiently large, an *electrically neutral* plasma region with fairly weak field is formed between the cathode layer and the anode. Its relatively homogeneous middle part is called the *positive column.* It is separated from the anode by the *anode layer.* The positive column of a dc glow discharge is the best pronounced and most widespread example of a *weakly ionized nonequilibrium plasma* sustained by an electric field. In contrast to the cathode layer, whose existence is vital for the glow discharge, the positive column is not an essential part. No such column is formed if the cathode layer fills the interelectrode gap. If, however, the distance is insufficient for the formation of the required cathode layer, the glow discharge cannot be ignited.

8.1.2 Discharge Devices

The glow discharge is one of the most studied and widely applied types of gas discharge. The *discharge tube* is a device that has been employed for decades for discharge generation and analysis (Fig. 1.1). The glow discharge in tubes of radius $R \sim 1\,\text{cm}$ and length $L \sim 10$–$100\,\text{cm}$, at typical pressure $p \sim 10^{-2}$–10^{2} Torr, is characterized by an electrode voltage $V \sim 10^{2}$–10^{3} V and a current $i \sim 10^{-4}$–10^{-1} A. In a number of modern laser systems, the discharge volume is a plane channel through which a gas is pumped (the flow is not essential for the discharge process as such). Electrodes may be placed along the larger surfaces of the channel or at its narrow ends (Fig. 8.1). The configuration of electrodes is quite different from the parallel disks typical for discharge tubes. An electrode may consist of several segments of various shapes, distributed over a plane, or it may be a long tube. A large discharge channel may allow reaching very high currents and voltages as does elevated pressure, resulting in kilovolts of voltage and amperes of current. Nevertheless, the main attributes of glow

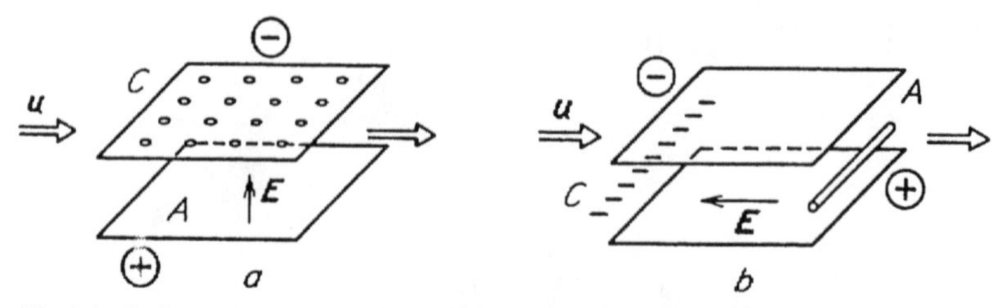

Fig. 8.1. Typical geometry of glow discharge in modern electric discharge CO_2 lasers; u is the direction of gas flow. (a) Transverse discharge (current is perpendicular to the gas flow). The upper plate is covered with cathode elements C, the lower plate is the anode A. (b) Longitudinal discharge. Cathode elements C are upstream of the gap, the anode A is a tube

discharge are retained, that is, a cathode layer with its inherent structure, and the region of electrically neutral, weakly ionized nonequilibrium plasma filling the space between anode and cathode layers. The familiar term, "positive column", is still applied to the homogeneous electroneutral region. This column forms the *lasing medium* of lasers (see Chap. 14). The main properties of glow discharges being insensitive to specific conditions, our discussion of this discharge will mainly refer to the classical discharge in a tube.

8.1.3 Pattern of Light Emission

The glow discharge normally manifests a stratification into dark and bright luminous layers; a name being ascribed to each (Fig. 8.2). The pattern is easily discernible at low pressure, when laysers are extended along the tube. Indeed, all discharge processes are connected with electrons. The distances from the cathode to characteristic points are dictated by the number of electron free paths, $l \propto p^{-1}$, within these distances. Hence, the coordinate at the boundary of a layer, x_1, corresponds to a specific value px_1. A layered pattern extends to centimetres if $p \sim 10^{-1}$ Torr. Sometimes a positive column has a *periodic* layered structure composed of *striations*. The formation of striations is not inevitable, or they may be not resolvable; in such cases the positive column emits light homogeneously up to the anode region.

If the pressure is low, $p \sim 10^{-2}$ Torr, and the separation between electrodes is moderate, the positive column has no space in which to form and what is seen is mostly the region of *negative (glow)* emission that gave the name to the discharge mode as a whole.

The quiet, sometimes slightly trembling light of a glow discharge has an enchanting beauty. As a rule, the positive column is less bright than the negative glow and is differently coloured. Helium manifests a red cathode layer, a gree negative glow, and a reddish-purple positive column; the respective combination in neon is yellow, orange and red, and in nitrogen, pink, blue and red. Each gas has a characteristic set of colours reflecting its spectrum; this is employed in coloured advertisement tubes. If the pressure in a long tube is not too low, we mostly observe the positive column. In very wide tubes and spherical vessels, the glow of the positive column is weak and often invisible.

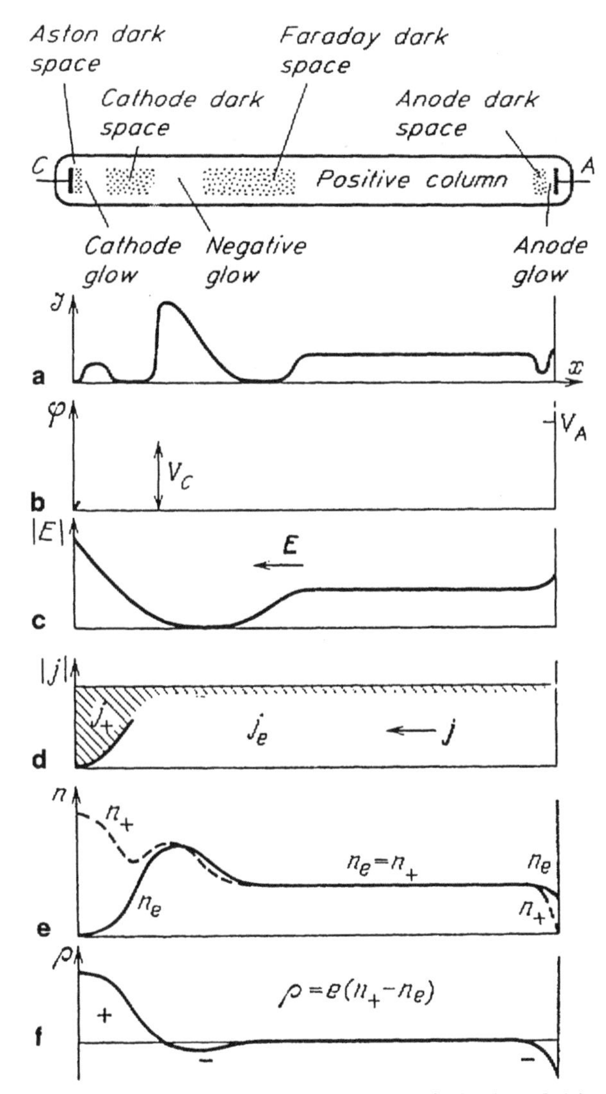

Fig. 8.2. Glow discharge in a tube and the distribution of: (a) glow intensity, (b) potential φ, (c) longitudinal field E, (d) electronic and ionic current densities j_e and j_+, (e) charge densities n_e and n_+, and (f) space charge $\varrho = e(n_+ - n_e)$

8.1.4 Variation of Conditions

As the pressure increases, all the layers become thinner and shift closer to the cathode. At $p \sim 100\,\text{Torr}$, it looks as if the cathode is burning, although we are merely observing the luminous gas. A more extended Faraday dark space (as in Fig. 8.2) can be distinguished; the rest of the tube or channel is occupied by the positive column. An elevated pressure *causes* the column to *contract* to the axis, while at low pressures the cross section of the tube is filled with the column in a *diffuse* manner. If the electrodes are moved closer at constant pressure, the positive column is shortened. Intermediate regions between the column and cathode (known as *negative* regions) remain unaltered for some time. If the cathode is shifted, these regions move together with it. Furthermore, if the cathode disk is rotated in a wide vessel with a fixed anode, all negative layers rotate together as if glued to the cathode surface, while the positive column bends so as to reach the anode. This situation is also realized in vessles of complex shape. *Negative layers* are "glued" to the cathode surface and the positive column finds a path connecting the end of the Faraday space with the anode. As the electrodes get closer, the column disappears, then the Faraday space is "eaten up", and finally the negative glow vanishes. When there is no space even for the cathode edge of this glow, the discharge goes out unless the voltage is increased. This discharge is sometimes said to be *obstructed.*

8.1.5 Distribution of Parameters over Length

The sequence of layers and the distribution of brightness along the discharge tube are compared in Fig. 8.2 with the distributions of the main discharge parameters. This is a qualitative picture but a fairly reliable one. It is supported by probe measurements and theoretical arguments. Among the principal features of this picture are a *large space charge* and high field at the cathode, which decreases almost linearly to a very low level at the cathode boundary of the negative glow. This region is known as the *cathode layer;* it is defined not by a visually apparent attribute (light emission), but by an "objective" characteristic, namely, electric field distribution.

This region is followed by a zone of very weak field which sometimes may even be slightly negative, that is, directed to the anode. The longitudinal field in the Faraday space increases and then stays constant over the entire length of the positive column. This column can be arbitrarily long provided the power supply circuit is adequate for maintaining the necessary potential difference across the column. The constancy of the axial potential gradient in the column has been confirmed by probe measurements; it proves the *electric neutrality* of plasma. There is a small region of slight *anode fall of potential* by the anode.

8.1.6 Qualitative Interpretation of the Light Emission Pattern

Electrons are ejected from the cathode at energies less than 1 eV. This is not enough for exciting an atom. The result is the formation of the *Aston dark space*. The field accelerates these electrons to an energy sufficient for excitation, and the *cathode glow* appears. Two, even three layers of cathode glow may be formed. They correspond to the excitation of different atomic levels, lower ones closer to the cathode and higher ones further out. These layers have different colours. The energy of accelerated eelectrons then grows above the *excitation function* maxima, where cross sections fall off (Fig. 5.8). Electrons cease to excite atoms and the *cathode dark space* is formed. This is the region where ionization of atoms predominantly takes place, where most electrons are multiplied. The newborn ions move much slowly and a large *positive space charge* builds up. The current is transferred mostly by *ions*.

By the end of the cathode layer, the electron flux gets fairly large; as a result of the avalanche process of multiplication, most electrons are generated at the very end of the layer, where the field is not strong any more and continues to fall off. These electrons have moderately high energies, in the region of the maxima of the excitation function. The *negative glow* appears. Close to the cathode, the electron energy increases with the distance from the cathode and more easily excited spectral lines appear (first and second cathode glows), but after the cathode layer, the electron energies decrease with increasing distance from the cathode. Correspondingly, the negative glow first reveals the lines that are emitted from higher atomic levels and then the lines from lower levels, in an order reversed with respect to the cathode glow (*Seeliger's rule*). As electrons dissipate their energy, acts of excitation become less and less frequent because electrons do not gain new energy in the weak field. The negative glow gives way to the *Faraday dark space*.

Most but not all of the electrons in the negative glow region have moderate energies. Some electrons here are energetic ones liberated deep inside the cathode layer or at the cathode, having traversed the cathode layer with only a few inelastic collisions. They ionize atoms; as a result, the electron density immediately after the cathode layer is higher than in the positive column.

In the Faraday space, the longitudinal field gradually increases to the value characterizing the positive column. The column has a random velocity distribution typical of nonequilibrium weakly ionized plasma, with slight asymmetry introduced by the drift towards the anode. The electron energy averaged over the spectrum in the positive column is 1–2 eV. However, the spectrum contains some energetic electrons as well. They excite atoms and generate the luminescence of the column. The anode repels ions but pulls out electrons from the column. Thus a region of *negative space charge* is formed; its higher field accelerates electrons. The result is the *anode glow*.

8.1.7 Guiding Effect of Charges Precipitated on the Walls

Observations show that a discharge can be maintained in tubes of very complicated shapes. Electrons (and ions) that transfer electric current are bound to move along the gas channel but have to follow the lines of force of the electric field. In fact, the lines of force of the applied external field trace their own paths from the anode to the cathode, intersecting quite often the walls of the discharge tube (Fig. 8.3a). How then can current flow?

Actually, charges (mostly electrons) are first transported along a line of force of the external field to a dielectric wall; there they stick and accumulate until they start repelling subsequently arriving charges of like sign away from the wall. The electrostatic field of the precipitated charges adds up vectorially with the external field and redirects a part of the lines of force of the resultant field along a path through the tube that is accessible for charges (Fig. 8.3b, c). Owing to this effect,

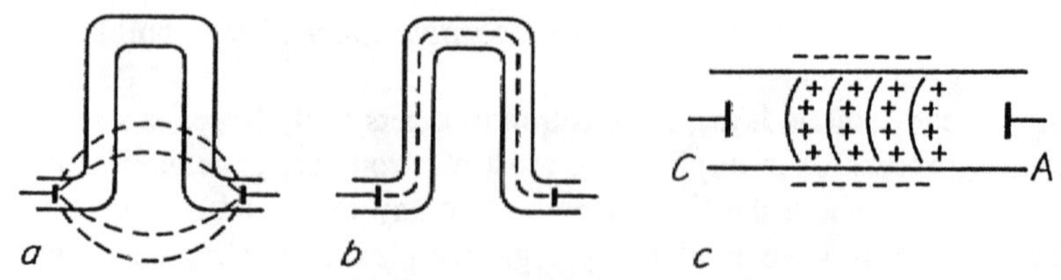

Fig. 8.3. Glow discharge in tubes of complicated shape: (a) Lines of force of the applied field, (b) lines of force of the resulting field (applied field plus that of charged deposited on the walls), (c) equipotential surfaces in a straight tube curved by the field of negative charge deposited on the walls

the longitudinal field in a straight tube or plane channel becomes more uniform in cross section. Nevertheless, a transverse (radial) field component is present in the discharge. Thus in a long positive column, it is uniform along the length and is directed from the axis to the negatively charged wall. Equipotential surfaces in the tube are convex, the convexity pointing to the cathode. Sometimes it is possible to see that the boundary between the positive column and the Faraday dark space, and the striations, are indeed convex. At a flat cathode, the boundaries of negative layers are usually plane; presumably, this indicates the absence of a transverse field component.

8.2 Current-Voltage Characteristic of Discharge Between Electrodes

Let us continue an analysis of the dc current-voltage characteristic ($V-i$ curve) begun in Sect. 7.2, and move to higher currents. As the breakdown voltage is reached across the electrodes, $V = V_t$, a *self-sustaining* discharge begins to burn in the gas. In the framework of the idealized scheme we used in Sect. 7.2.2, the current at $V = V_t$ tends to infinity. Any real circuit with a discharge gap always has an ohmic resistance Ω (a specially introduced resistance or the resistance of the lead wires, and power supply) which sets an *absolute limit* to the current achievable for a given electromotive force $\mathcal{E}$ of the power supply unit. As the discharge current scale largely determines the discharge type (the value of current dictates the degree of gas ionization), the resistance Ω imposes the type of discharge that is produced after breakdown.

8.2.1 Load Line

The equation for the voltage of a closed circuit with a discharge gap is

$$\mathcal{E} = V + i\Omega . \tag{8.1}$$

This equation is plotted as a straight line in V vs. i coordinates (Fig. 8.4); it is known as the *load line*. The line is the steeper, the larger the external resistance; the intercept on the abscissa axis is the limiting current $\mathcal{E}/\Omega$. The circuit realizes

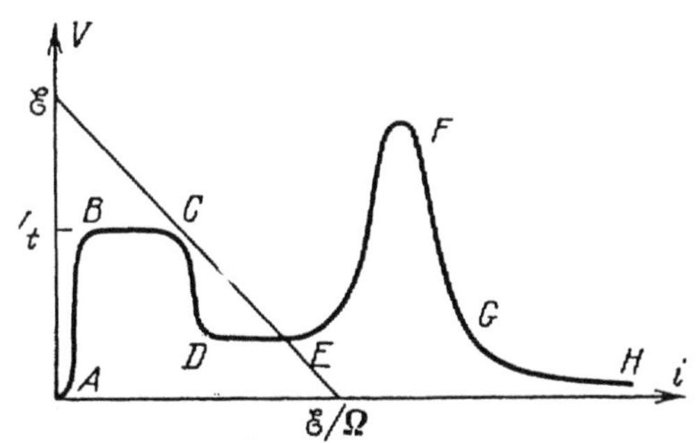

Fig. 8.4. $V - i$ characteristic of discharge between electrodes for a wide range of currents, and the loading line: *(A)* region of non-self-stustaining discharge, *(BC)* Townsend dark discharge, *(DE)* normal glow discharge, *(EF)* abnormal glow discharge, *(FG)* transition to arcing, *(GH)* arc

those values of i and V that correspond to the intersection of load line and the $V - i$ curve, $V(i)$.

8.2.2 Townsend Dark Discharge

Assume the resistance Ω to be so high that the circuit can supply only an extremely weak current. The densities n_e and n_+ are then negligible and the space charge is so small that the external field is not distorted. Thus if the distance L between plane electrodes is small in comparison with the transverse size of electrodes, the field is the same as in the absence of ionization: $E(x) \approx \text{const} = V/L$. This discharge is made self-sustained by applying to the electrodes the voltage equal to the *ignition potential* V_t. This voltage ensures the stationary reproduction of electrons ejected from the cathode and pulled to the anode (Sect. 7.2.2). As long as the field $E(x)$ is independent of charge (and current) densities, the $V - i$ curve of discharge is $V(i) = \text{const} = V_t$. This situation corresponds to the segment BC in Fig. 8.4.[1]

This *self-sustaining* discharge mode is indeed observed experimentally in ordinary tubes at currents of $i \sim 10^{-10}$–10^{5} A. This mode is called the *Townsend dark* discharge. Ionization is so small that the gas emits no appreciable light. The current is measured by high-sensitivity instruments.

8.2.3 Glow Discharge

Let us gradually increase the current. This can be realized by reducing the load resistance Ω or by increasing the e.m.f. $\mathcal{E}$. The voltage across the electrodes begins to decrease after a certain current is reached. The fall then stops and the current remains almost constant over a fairly wide range of values (sometimes of

[1] Note that we are now discussing the $V - i$ curve of a steady-state, stationary process. There must be no overvoltage (see Sect. 7.2.3) in comparison with the ignition potential. Overvoltage is needed for the development of the breakdown, that is, for the implementation of the nonstationary process of current buildup. The current increases in the course of breakdown (when $V > V_t$) to a value necessary to eliminate overvoltage.

several orders of magnitude). This segment of the $V-i$ curve, DE, corresponds to the so-called *normal glow discharge*. The lower part of the transition region CD corresponds to a below normal glow discharge.

The normal discharge has one remarkable property. As the discharge current is varied, its density at the cathode remains unchanged. What changes is the area through which the current flows. When Ω or $\mathcal{E}$ is varied, the luminous current spot on the cathode surface expands or contracts.

When no more free surface is left on the cathode, the current is increased by increasing the voltage, hence extracting more electrons from unit surface area. Indeed, the cathode current density must grow. This discharge is said to be *abnormal*. It corresponds to the climbing section EF of the $V-i$ curve. The transition to the abnormal mode is interesting to observe. The glow first covers the entire cathode surface facing the anode, then reaches every spot unprotected by dielectric on the lateral and inner surfaces and on the support pin, and only having exhausted these possibilities does it become more extended and intense to a degree typical of the abnormal discharge. When $i \sim 1$ A, the glow discharge cascades down to an *arc*. The segment FG describes the transition, and GH represents the *arc discharge*.

We have followed the $V-i$ curve as if "turning the handle" that varies Ω or $\mathcal{E}$. In experiments, a certain resistance is in the circuit at the moment of switching on an e.m.f.; if $\mathcal{E}$ is greater than the ignition potential, the discharge mode that sets in immediately after the breakdown corresponds to the point of intersection of the $V-i$ curve and the loading curve. In contrast to the schematic Fig. 8.4, Fig. 8.5 gives the actual $V-i$ curves [8.1]. The curves cover the dark, normal, and partially abnormal modes. The higher the pressure, the wider the current range in which the normal mode is realized (the reason for this will be clear in Sect. 8.4.4). The picture observed in H_2, N_2, and Ar is almost the same as in Ne.

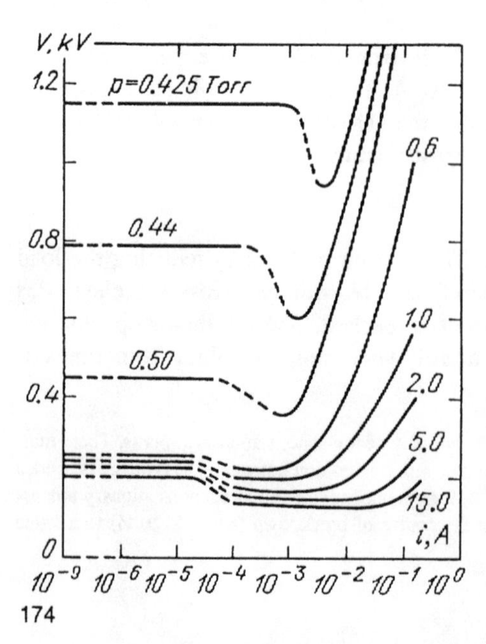

Fig. 8.5. Measured $V-i$ characteristics of discharge in neon bewteen copper disks 9.3 cm in diameter, gap width 1.6 cm. Plateau on the *left*: dark discharge; *lower* plateau or region of minimum: normal glow discharge; rising curve on the *right*: abnormal discharge [8.1]

8.3 Dark Discharge and the Role Played by Space Charge in the Formation of the Cathode Layer

The main distinction of the glow discharge from the dark discharge (with its extremely weak current) lies in a sharply *nonuniform* distribution of the potential difference applied across the gap to the electrodes. In order to find out why and by what mechanism the field is redistributed in the interelectrode gap, let us begin with the dark discharge that does not disturb the external field.

8.3.1 Charge Distribution in Weak-Current Dark Discharges

Assume the gap width to be small in comparison with the transverse electrode size. We choose the x axis to point from the cathode to the anode. The analysis is based on continuity equations (2.22), (2.20) for charge densities. Diffusion fluxes are small in comparison with drift fluxes, the diffusion in the lateral directions also being insignificant; even more so is the recombination. Bulk charge sources are related only to gas ionization: $q = \nu_i n_e = \alpha v_{ed} n_e$; fluxes are due to drift only. Let us work in terms of current densities $j_e = -e n_e v_{ed}$, $j_+ = e n_+ \cdot v_{+d}$. In the steady-state case,

$$\frac{dj_e}{dx} = \alpha j_e , \quad \frac{dj_+}{dx} = -\alpha j_e , \quad j_e + j_+ = j = \text{const} . \tag{8.2}$$

The third equality stating the constancy of the total current is implied by the first two. The boundary condition at the cathode ($x = 0$) describes the secondary emission, and that at the anode ($x = L$) describes the absence of ionic emission:

$$j_{eC} = \gamma j_{+C} = \frac{\gamma}{1+\gamma} j , \quad j_{+A} = 0 , \quad j_{eA} = j . \tag{8.3}$$

If equation (8.2) for j_e is integrated, starting at the cathode, for the first condition of (8.3) and $\alpha[E(x)] = \text{const}$, we obtain

$$j_e = \frac{\gamma}{1+\gamma} j e^{\alpha x} , \quad j_+ = j \left(1 - \frac{\gamma}{1+\gamma} e^{\alpha x}\right) . \tag{8.4}$$

Condition (8.3) can be satisfied at the anode only if the *criterion of ignition (self-sustainment)* is met (see Sect. 7.2.2):

$$e^{\alpha L} - 1 = 1/\gamma , \quad \alpha L = \ln(1 + 1/\gamma) . \tag{8.5}$$

We recast (8.4) using (8.5):

$$j_e/j = \exp[-\alpha(L - x)] , \quad j_+/j = 1 - \exp[-\alpha(L - x)] ;$$

$$j_+/j_e = \exp[\alpha(L - x)] - 1 .$$

The ionic current much exceeds the electronic current over a large part of the gap, beginning with the cathode (Fig. 8.6). For example, for $\gamma = 10^{-2}$ and

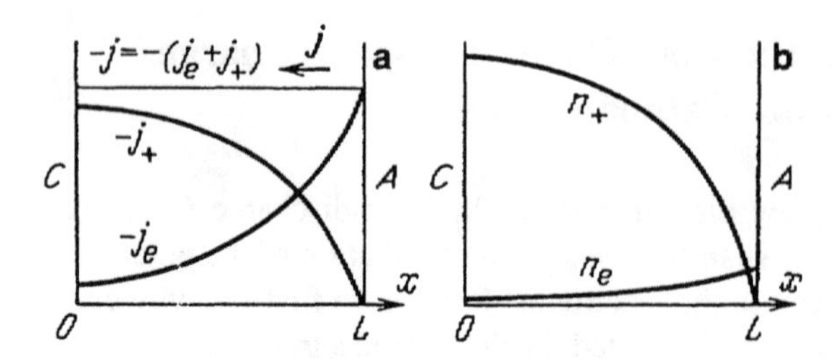

Fig. 8.6. Distributions of electronic and ionic currents (a), and charge density distribution (b) when the field in the gap is not distorted by space charge

$\alpha L = 4.6$, j_e reaches j_+ only at $x = 0.85L$. The difference in charge densities is even greater. Thus if $\mu_e/\mu_+ = 100$, we find that $n_+/n_e = (\mu_e/\mu_+)(j_+/j_e) = 1$ only if $x = 0.998L$. In most of the gap, $n_+ \gg n_e$ (Fig. 8.6). Practically the entire space is positively charged in dark discharges, but the space charge is small because j and n_+ are low. These quantities are arbitrary, being determined by the current allowed by the circuit and the area of electrodes (see Sect. 8.2).

8.3.2 Distortion of an External Field

This is produced by space charge. Let us evaluate the effect, taking for the zeroth approximation the charge density distributions obtained in the assumption $E(x) = \text{const}$. The spatial field distribution is determined by the equation

$$dE/dx = 4\pi e(n_+ - n_e)\ , \quad E \equiv E_x\ . \tag{8.6}$$

Assuming approximate equalities $n_+ \gg n_e$, $|j_+| \gg |j_e|$, $n_+ \approx j/ev_{+d} \approx j/e\mu_+E$ and denoting by E_C the field at the cathode, we find

$$E = E_C\sqrt{1 - x/d}\ , \quad d = \mu_+E_C^2/8\pi j\ . \tag{8.7}$$

The field decreases in the vicinity of the anode and increases in the vicinity of the cathode, the more the higher the current density (Fig. 8.7). The plane $x = d$ where the extrapolated quantity $E(d)$ vanishes lies far beyond the discharge gap (if the current is low). As j increases, it moves closer to the anode and coincides with the anode surface for $j_L = \mu_+E_C^2/8\pi L$ $(d = L)$. As j further increases, at $d < L$, the field implied by (8.7) tends formally to zero inside the gap, and the closer to the cathode the higher the current (Fig. 8.7). However, in this case distribution (8.7) becomes meaningless in the interval $d < x < L$ because the original assumptions become invalidated. Actually, the distributions n_{ge}, n_+, and E take the form given in Fig. 8.2.

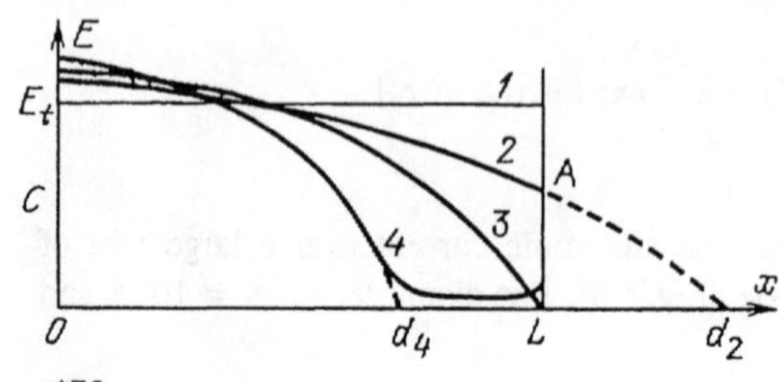

Fig. 8.7. Field evolution due to space charge: *(1)* undisturbed fields as $j \to 0$; *(2)* weak current, $j < j_L$; *(3)* $j = j_L$; *(4)* $j > j_L$, transition to glow discharge

8.3.3 Limiting Current for the Existence of Dark Discharge

When the currents are so small that the field is very little distorted, the field at the cathode, E_C, is quite close to the nonperturbed breakdown field E_t of this gap. As the current increases, the field E_C deviates from E_t more and more; but as long as $d > L$, E_C retains the same order of magnitude (see footnote in Sect. 8.3.5). Hence, E_C in the expression for j_L at $d = L$ can be replaced with E_t. The current density at which the field and discharge structure are considerably modified and which manifests the beginning of dark-to-glow transition of discharge, is given within an order of magnitude, by the formula

$$\frac{j_L}{p^2} \approx \frac{(\mu_+ \cdot p)(E_t/p)^2}{8\pi(pL)} = \frac{(\mu_+ \cdot p)V_t^2}{8\pi(pL)^3} . \tag{8.8}$$

It is given a form corresponding to similarity laws. For instance, (8.3) gives the values $E_t/p = 62\,(\text{V/(cm·Torr)})$, $V_t = 6200\,\text{V}$ for nitrogen, with α defined by (4.5); $A = 12\,\text{cm}^{-1}\text{Torr}^{-1}$ and $B = 342\,\text{V/(cm·Torr)}$ taken from Table 4.1; and $\gamma = 10^{-2}$ at $pL = 100\,\text{cm Torr}$. For $(\mu_+ p) = 1.5 \cdot 10^3\,\text{cm}^2\text{Torr/(V·s)}$, we find $j_L/p^2 = 2.5 \cdot 10^{-9}\,\text{A/(cm·Torr)}^2$. If, say, $p = 10\,\text{Torr}$ and $L = 10\,\text{cm}$, then $j_L = 2.5 \cdot 10^{-7}\,\text{A/cm}^2$. If the electrode surface area is $100\,\text{cm}^2$, the limiting dark discharge current is $i = 2.5 \cdot 10^{-5}\,\text{A}$.

8.3.4 The Condition of Self-Sustainment of Discharge in a Plane Gap in the Case of Inhomogeneous Fields

It is immediately implied by (8.2, 3) that

$$\int_0^L \alpha\,[E(x)]\,dx = \ln(1 + 1/\gamma) . \tag{8.9}$$

Equality (8.9) generalizes (8.5) and expresses the same fact. A specific number of generations have to be produced in the electron avalanche propagating from the cathode to the anode. This number is determined only by the secondary emission coefficient and is independent of whether the field is homogeneous or not. The integral in (8.9) is exactly equal to the value of $\alpha(E_t)L$ corresponding to the breakdown of a given gap in a homogeneous field. Note that $|E_C| > E_t$ at the cathode and $|E_A| < E_t$ at the anode, because $\alpha(E)$ is an increasing function of E, but that $|E(x)|$ is a decreasing function of x if E is distorted by the space charge.

8.3.5 What Happens to Voltage When Space Charge Builds Up

A qualitative tendency is clear from an analysis of the integral in (8.9), of the voltage integral $V = |\int_0^L E\,dx|$, and of the dependence of α on E according to (4.5). If the field is not too strong, the function $\alpha \propto \exp(-Bp/E)$ increases with increasing E at an increasing slope $d^2\alpha/dE^2 > 0$; if the field is very high, the slope decreases: $\alpha \to \text{const}$. The point of inflection of $\alpha(E)$ lies at $E = Bp/2$. For a nonperturbed breakdown field $E_t < Bp/2$, the conditions for multiplication are facilitated by a redistribution of potential: the enhanced field contributes to integral (8.9) more than the weakened field takes out. Conversely, integral (8.9) is conserved if $E_t - |E_A|$ of the decreasing field is greater than the increment $|E_C| - E_t$. However, if at one place the field is enhanced less than it is weakened at another place, the potential difference $|\int E\,dx|$ decreases. If $E_t > Bp/2$, the situation is reversed: redistribution either impedes multiplication or increases voltage.

The correctness of these qualitative arguments can be verified by considering a weak inhomogeneity, using formulas (8.9, 7) and (4.5) and expanding in a small parameter $j/j_L = L/d \ll 1$ (the integral cannot be calculated in the general case).[2] A small increment to the gap voltage in comparison with the breakdown potential $V_t = E_t L$ for a homogeneous field is found to be

$$V - V_t = -(1/48)(Bp/2E_t - 1)(j/j_L)^2 . \tag{8.10}$$

According to (7.3), the condition $E_t < Bp/2$ at which the voltage of sustained discharge falls below the ignition potential is satisfied for gaps with $(pL) > e(pd)_{min}$, where $(pd)_{min}$ corresponds to the minimum breakdown voltage. If $(pL) < e(pd)_{min}$, then $V > V_t$. In actual experiments, one usually has to deal with gaps that are long in the sense defined here; therefore the voltage decreases in the transition from the dark to the glow discharge (Figs. 8.4, 5).

8.4 Cathode Layer

8.4.1 What Is Its Purpose

In a plane gap undergoing breakdown, and in weak-current dark discharges, the loss of charged particles due to field-pulling onto electrodes is made up for by avalanche ionization over the entire length up to the anode. However, if a gap is wide in the sense $pL \gg (pd)_{min}$, this situation (implied by the homogeneity of the field) is obviously not optimal. Unjustifiably high voltage is required to meet the self-sustainment condition (8.9). A lower voltage would be sufficient if the potential drop were concentrated. Indeed, the multiplication efficiency is higher in strong fields.

The potential distribution would be ideal if the potential difference equal to the minimal breakdown voltage, V_{min}, were concentrated over the corresponding length $(pd)_{min}$ at the cathode. This would ensure reproduction at minimum applied voltage. To sustain further flow of electronic current generated in this cathode layer through the remaining (even if long) part of the gap, the additional voltage need only compensate, via weak ionization, for the losses of electrons caused by ambipolar diffusion to the walls, recombination, and attachment. Na-

[2] The field at the cathode is found by setting integral (8.9) equal to $\alpha(E_t)L$; $|E_C| = E_t[1 + L/4d + O(L/d)]$. Second-order terms must be retained. The problem was originally analyzed in [8.2, 3].

ture reveals unparalleled wisdom in its organization. The normal glow discharge comes quite close to the optimum. One of the main mechanisms of optimizing the potential distribution across the gap is the effect of space charge that is generated automatically at the cathode, creating there an enhanced field and a potential drop; this was described in Sect. 8.3.

8.4.2 Current-Voltage Characteristic

The theory of the cathode potential drop (or cathode fall) was developed by *von Engel* and *Steenbeck* in 1934 [8.2]; it has considerable significance for the physics of the glow discharge. Subsequent elaboration and more profound understanding of the process do not nullify the essential aspects of the theory. We will present it here in the simplest and most lucid form that brings forth the fundamental features of the phenomena. Let us take a stationary cathode-fall layer. Assume that pressures and currents are not too low, so that the current spot in the cathode is large and the layer is thin. It can then be assumed plane and one-dimensional. The field at the anode end of the layer, at $x = d$, is substantially less than that at the cathode: $E(d) \ll E(0) \equiv E_C$. Assume $E(d) \approx 0$. Assume also that even if some ionic current enters the cathode layer on the side of the anode, it is very weak (in the electrically neutral part of the gap, $j_+/j_e = v_{+d}/v_{ed} = \mu_+/\mu_e \sim 10^{-2}$). The layer is then an autonomous system satisfying the condition of self-sustainment of current, (8.9). We only have to replace the distance between electrodes L in (8.9) with the layer thickness d. The electronic current generated in this system reaches the anode. It coincides with the total discharge current up to a small quantity of order μ_+/μ_e.[3] The cathode fall is

$$V_C = \int_0^d E\,dx\ , \quad E \equiv |E|\ . \tag{8.11}$$

Von Engel and Steenbeck solved the system of equations (8.9, 11, 6) assuming (4.5) for $\alpha(E)$ and prescribing, in view of the results of probe measurements, a linear field distribution:

$$E(x) = E_C(1 - x/d)\ , \quad 0 \le x \le d\ . \tag{8.12}$$

The integral (8.9) with the field (8.12) is not expressible in terms of elementary functions. Essentially the same results that differ only by numerical factors of order unity but manifest a clear analytic form can be obtained assuming $E(x) =$ const $\approx E_C$ in (8.9) for $x < d$. Then (8.9) transforms into (8.5), with d for L. However, since we assumed the field in the layer to be homogeneous, (4.5) and a trivial relation $V_C = E_C d$ implied by (8.11) yield (7.3) for the breakdown of the gap d in the homogeneous field:

[3] According to (2.39), a stationary process has $\operatorname{div} \boldsymbol{j} = 0$, whence $i =$ const. In the one-dimensinal case, $j(x) =$ const as well.

$$\frac{E_C}{p} = \frac{B}{C + \ln pd}, \quad V_C = \frac{Bpd}{C + \ln pd}, \quad C = \ln \frac{A}{\ln(1 + 1/\gamma)}. \tag{8.13}$$

These formulas relate the cathode fall V_C to the cathode layer thickness pd.

Let us now find the relation of these quantities to the current density at the cathode, j. As shown in Sect. 8.3.1, the region of self-sustainment through multiplication is characterized by $n_+ \gg n_e$, $j_+ \gg j_e$. By virtue of (8.6), the ion density in the layer is approximately equal to

$$n_+ \approx (4\pi e)^{-1} |dE/dx| \sim E_C/4\pi ed,$$

where we have already taken into account that the actual field is not constant but decreases from E_C to zero. Hence,

$$j = (1 + \gamma)en_+\mu_+E \approx (1 + \gamma)\mu_+E_C^2/4\pi d \approx (1 + \gamma)\mu_+V_C^2/4\pi d^3. \tag{8.13'}$$

Together with (8.13), this formula defines a parametric dependence of the cathode fall V_C and the field at the cathode, E_C, on the current density j. The parameter is the layer thickness d. The function $V_C(pd)$ has a minimum [see (8.13)]. In this approximation, it describes a Paschen curve (Sect. 7.2.3) with V_{min} equal to the minimum gap breakdown voltage.

By virtue of (8.13, 13'), V_C as a function of j has a minimum, reaching the same value V_{min}. It will be convenient to rewrite these formulas in dimensionless form, using the quantities corresponding to the minimum potential difference as dimensional scales. We mark them with subscript "n" for "normal" instead of "min" (they are indeed realized in normal discharge) and denote dimensionless quantities by a tilde:

$$\tilde{V} = \frac{V_C}{V_n}, \quad \tilde{E} = \frac{E_C/p}{E_n/p}, \quad \tilde{d} = \frac{pd}{(pd)_n}, \quad \tilde{j} = \frac{j}{j_n}.$$

The scales V_n, E_n/p, $(pd)_n$ are defined by formulae (7.4); the current density scale, taking into account similarity laws, is

$$\frac{j_n}{p^2} = \frac{(1 + \gamma)(\mu_+p)V_n^2}{4\pi(pd)_n^3} = \frac{(1 + \gamma)}{9 \cdot 10^{11}} \frac{(\mu_+p)V_n^2}{4\pi(pd)_n^3} \text{ A/(cm} \cdot \text{Torr)}^2. \tag{8.14}$$

The parametric relations of dimensionless quantities via $\tilde{d}$ are

$$\tilde{V} = \frac{\tilde{d}}{1 + \ln \tilde{d}}, \quad \tilde{E} = \frac{1}{1 + \ln \tilde{d}}, \quad \tilde{j} = \frac{1}{\tilde{d}(1 + \ln \tilde{d})^2}. \tag{8.15}$$

Figure 8.8 plots $\tilde{V}$, $\tilde{E}$, and $\tilde{d}$ as functions of $\tilde{j}$ according to (8.15). The curve $\tilde{V}(\tilde{j})$ gives the "current-voltage" characteristic of the cathode layer, the inverted commas signifying that the argument is current density, not current.

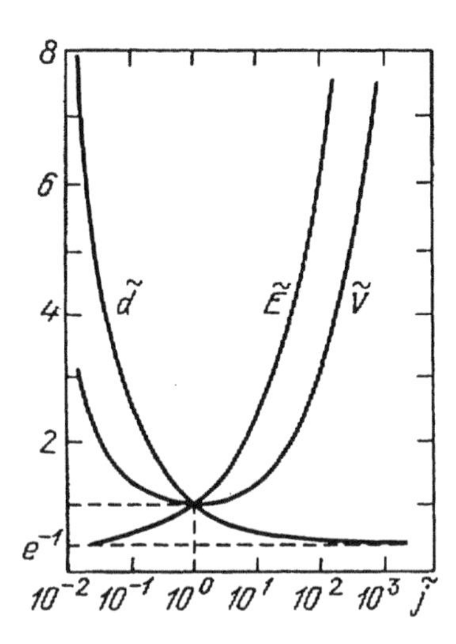

Fig. 8.8. Cathode potential fall ($\tilde{V}$), field at the cathode ($\tilde{E}$) and cathode layer thickness ($\tilde{V}$) as functions of current density (in dimensionless coordinates)

8.4.3 Normal Cathode Fall and Current Density

Formally, (8.15) and Fig. 8.8 imply that as j decreases from j_n, V_C and d increase and E_C decreases. When d grows to the gap width L, solution (8.15) becomes equal to that of Sect. 8.3, and j given by (8.13′) is nearly equal to j_L of (8.8). It may seem that the evolution of dark to glow discharge has been traced.

Experiments definitely show that nothing of the sort takes place. The falling branch of the curve $V(j)$ to the left of $\tilde{j} = 1$, $j = j_n$ is not realized. The mode that sets in at currents i less than $S_C j_n$, where S_C is the cathode surface area, corresponds to the minimum point of the "$V - i$ curve" of the cathode layer. The same mode is realized when current is varied and when $\mathcal{E}$ and Ω are such that, after ignition, the state falls into the region DE of Fig. 8.4. A current spot lights up automatically on the cathode, with area S such that the current density is $j_n \approx i/S$ and the cathode fall is V_n. A $V - i$ curve of a real discharge has nothing in common with the left-hand branch of Fig. 8.8 (this would be the case if the current i passed through the entire cathode, $i = S_C j$). The discharge voltage when the cathode is not totally covered is current-independent, exceeding V_n by the potential drop on the positive column. If this fall is negligible (low pressure, short tube), the voltage on the electrodes is almost equal to V_n. This glow discharge is known as *normal*, and the corresponding values of cathode fall and current density are also said to be *normal*.

The theoretical values V_n, j_n, and $(pd)_n$ somewhat depend on the assumptions about the $E(x)$ profile that were used in the calculation. In the simplest approximation, introduced in Sect. 8.4.2, V_n and $(pd)_n$ coincide exactly with the parameters of the minimum on the Paschen curve, V_{min} and $(pd)_{min}$, and j_n is given by (8.14). If approximation (8.12) is used, we find $V_n = 1.1 V_{min}$, $(pd)_n = 1.4(pd)_{min}$,[4] and j_n is 1.8 times the value given by (8.14). Any reasonable theory, as well as

[4] Formula (137) for $(pd)_n$ given in the monograph [8.2], and later in [8.4], contains a wrong numerical coefficient: 0.82 instead of 3.78. The correct coefficient in the above formula is 1.4.

experiment, give for the normal cathode fall and layer thickness values that are close to $V_{\min}$ and $(pd)_{\min}$ for the breakdown of a plane discharge gap in the same gas and for the same cathode material. Calculations with the mentioned numerical coefficients derived by von Engel and Steenbeck on the basis of approximation (8.12), give satisfactory agreement with the experimental parameters of normal discharges (see Tables 8.1, 2, 3) if the coefficients A and B are taken from Table 4.1 and $\gamma \sim 10^{-2}$–10^{-1} (the dependence on γ is logarithmic).

Table 8.1. Normal cathode fall V_n [V] [8.2, 4]

gas / cathode	air	Ar	He	H_2	Hg	Ne	N_2	O_2	CO	CO_2
Al	229	100	140	170	245	120	180	311	–	–
Ag	280	130	162	216	318	150	233	–	–	–
Au	285	130	165	247	–	158	233	–	–	–
Bi	272	136	137	140	–	–	210	–	–	–
C	–	–	–	240	475	–	–	–	526	–
Cu	370	130	177	214	447	220	208	–	484	460
Fe	269	165	150	250	298	150	215	290	–	–
Hg	–	–	142	–	340	–	226	–	–	–
K	180	64	59	94	–	68	170	–	484	460
Mg	224	119	125	153	–	94	188	310	–	–
Na	200	–	80	185	–	75	178	–	–	–
Ni	226	131	158	211	275	140	197	–	–	–
Pb	207	124	177	223	–	172	210	–	–	–
Pt	277	131	165	276	340	152	216	364	490	475
W	–	–	–	–	305	125	–	–	–	–
Zn	277	119	143	184	–	–	216	354	480	410
glass [a]	310	–	–	260	–	–	–	–	–	–

[a] Thin soft glass disk heated to 300° C. The same holds for Tables 8.2 and 8.3

Table 8.2. Normal cathode layer thickness $(pd)_n$ [cm·Torr] at room temperature [8.2, 4]

gas / cathode	air	Ar	H_2	He	Hg	N_2	Ne	O_2
Al	0.25	0.29	0.72	1.32	0.33	0.31	0.64	0.24
C	–	–	0.90	–	0.69	–	–	–
Cu	0.23	–	0.80	–	0.60	–	–	–
Fe	0.52	0.33	0.90	1.30	0.34	0.42	0.72	0.31
Mg	–	–	0.61	1.45	–	0.35	–	0.25
Hg	–	–	0.90	–	–	–	–	–
Ni	–	–	0.90	–	–	–	–	–
Pb	–	–	0.84	–	–	–	–	–
Pt	–	–	1.00	–	–	–	–	–
Zn	–	–	0.80	–	–	–	–	–
glass [a]	0.30	–	0.80	–	–	–	–	–

[a] see Table 8.1

Table 8.3. Normal current density j_n/p^2 [μA/(cm^{2}Torr^2)] at room temperature [8.2, 4]

gas cathode	air	Ar	H_2	He	Hg	N_2	O_2	Ne
Al	330	–	90	–	4	–	–	–
Au	570	–	110	–	–	–	–	–
Cu	240	–	64	–	15	–	–	–
Fe, Ni	–	160	72	2.2	8	400	–	6
Mg	–	20	–	3	–	–	–	5
Pt	–	150	90	5	–	380	550	18
glass [a]	40	–	80	–	–	–	–	–

[a] see Table 8.1

8.4.4 Abnormal Discharge

After the entire cathode has been covered with discharge, any further increase of current inevitably increases the density at the cathode in comparison with the normal value. This *abnormal* discharge corresponds to the right-hand branch of the $\tilde{V}(\tilde{j})$ curve of Fig. 8.8; now it actually describes the $V-i$ characteristic of the cathode layer and a discharge without a positive column, sine $i = \text{const} \cdot j = S_C j$. The theoretical curve fits the experimental data (region EF in Fig. 8.4). As $\tilde{j} \to \infty$, equation (8.15) implies that the cathode layer thickness decreases asymptotically to a finite value $\tilde{d} = e^{-1} = 0.37$; $\tilde{V}$ and $\tilde{E}$ grow as $\tilde{j}^{1/2}$. In actual situations, a cathode fall of more than several kV and current densities of order 10–10^2 A/cm^2 result in intense heating of the cathode and transition to an *arc* discharge. Experimental $V-i$ curves of abnormal discharges can be seen in Fig. 8.5.

8.4.5 The Current Range in Which Normal Discharge Is Possible

A dark discharge occupies the entire cathode. The same is true for the normal discharge at the upper limit of its existence. Correspondingly, the current increases from the transition of dark current to normal to the transition of normal current to abnormal by a factor of about j_n/j_L. By virtue of (8.8, 14), $j_n/j_L \approx \tilde{L}(1+\ln \tilde{L})^2$, where $\tilde{L} = (pL)/(pd)_n$. Therefore, the current range in the normal mode (in the region DE of Fig. 8.4) is the greater, the higher the pressure and the longer the tube (Fig. 8.5). For example, one of the experimental versions shown in Fig. 8.5 is characterized by $p = 15$ Torr, $L = 1.6$ cm, $(pd)_n \approx 0.7$ cm·Torr, $\tilde{L} = 34$, and the ratio of currents equals 700, in agreement with experimental data.

8.4.6 Subnormal Discharge

This is a transition region between the glow and dark discharge regions (rather nearer to the normal region) that corresponds to currents so weak that the size of the "quasinormal" cathode spot is found to be comparable to the cathode layer thickness. The loss of charges in the lateral direction is harmful for multiplication, so that the voltage across the layer required for self-sustainment of the discharge is found to be higher than for the normal regime.

8.4.7 Obstructed Discharge

This mode arises at very low pressure in narrow gaps of widths L, such that the product pL is less than the normal layer thickness $(pd)_n$. Roughly speaking, these conditions correspond to the left-hand branch of the Paschen curve, where $V > V_{min}$. The interelectrode separation is insufficient for "normal" multiplication, so that voltage has to be raised in comparison with the normal value. If this is not possible the discharge is extinguished.

8.4.8 Normal Discharge and Minimum Power Principle

Why is it that the discharge current to a partially covered cathode occupies precisely the area preserving the current density? The problem is equivalent to the task of justifying the postulate of the von Engel–Steenbeck theory on the realization of the minimum possible cathode fall by the normal discharge. It is this assumption that provides a good explanation of the experimental facts. The creators of the theory were already able to explain why states with cathode current density below the normal are not observed [8.2]. These states are unstable because they refer to the falling branch of the $V_C(j)$ curve of Fig. 8.8 (in general, discharges with falling $V - i$ characteristics typically produce unstable states; see Sect. 8.7.5). Indeed, if a fluctuation $\delta j > 0$ appears at some point on the cathode layer surface, a lower voltage is required to sustain the current $j + \delta j$ than is actually there, so that j increases. If $\delta j < 0$, the actual voltage is lower than the necessary level and j has to drop still lower. In this sense, the states with $j > j_n$ on the rising branch of the $V - i$ curve, $V_C(j)$, are quite stable. Nevertheless, "abnormal" cathode spots never appear on a partially covered cathode. Obviously, the neighbouring zero-current region proves to be unstable. How does it happen and what does stabilize the boundary of a normal cathode spot?

Twenty years later *von Engel* [8.3] discussed the incomprehensibility of the phenomenon, without resorting to the stability arguments, and appealed to the "minimum power principle". This principle left an important trace in discharge physics; sometimes it is resorted to even now and thus deserves being mentioned. The power released in the cathode layer volume is

$$P_C = S \int_0^d jE\,dx = SjV_C(j) = iV_C(j)\,.$$

If the area S is varied at constant total current i, the power is found to be minimal precisely at the normal current density, such that $V_C(j) = \min$. Gaseous discharges also manifest some other phenomena that realize just those states that require minimal voltages and (or) power, for example, *striations* (Sect. 9.7). Steenbeck proposed in 1932 the above minimum principle on the basis of such facts, demonstrating the spectacular expediency in the organization of nature. It may have been due to the great prestige of Steenbeck, however, that this principle was later employed not only for a better illustration of the observations, but also as a missing condition for completing a theoretical model. This proved to be fraught with errors (masked by an apparent agreement with experimental data), because the principle is not implied by the fundamental laws of physics. In fact, it is not necessary if the mechanism of the phenomenon has been understood and its theory has been constructed in the usual manner. Such was the case with striations, and with the channels of the arc (Sect. 10.10), induction (Sect. 11.3), and microwave (Sect. 11.4) discharges. The approach to *normal current density* must also follow these lines.

8.4.9 Mechanism by Which Normal Current Density Is Reached

This has only been clarified rather recently [8.5–8].[5] The effect of interest, $i/S \approx$ const, is also revealed by numerical modeling of glow discharge using equations (2.22, 20, 33) for n_e, n_+, $E = -\nabla\varphi$ with boundary conditions of type (8.3) and with α given by (4.5) [8.6, 8]. When the current and cathode layer surface area are sufficiently great, as in Fig. 8.9, the middle part of the layer behaves as a quasi-one-dimensional system whose parameters are described rather well by the formulae of the one-dimensional theory [8.2].[6]

Let us turn to the map of equipotentials in Fig. 8.9 and to the schematic dependence of the voltage V_C across the cathode layer on its thickness d (Fig. 8.10). This dependence follows from an equality of type (8.9) that we write in the form

$$\mu \equiv \gamma \left\{ \exp\left(\int_0^{d(r)} \alpha\,[E(l)]\,dl \right) - 1 \right\} = 1\,, \tag{8.16}$$

where μ is the charge reproduction coefficient (Sect. 7.2.2). We integrate here along a line of current that sinks into some point r of the cathode surface. If we choose to ignore the diffusion of charges, the lines of current coincide with the lines of force of the field. If $E(l)$ = const, the dependence $V_C(d)$ coincides with the Paschen function (8.13). Note that the linear law (8.12) is fairly well supported by both one-dimensional [8.10] and two-dimensional [8.6, 8] calculations; this law gives the curve [8.2], which is not very different from (8.13). The stationary current mode corresponds to the curve of Fig. 8.10 for $\mu = 1$. Above this curve $\mu > 1$, and below it $\mu < 1$. This is a corollary of the boundedness of the ionization coefficient $\alpha(E)$ as $E \to \infty$, so that the function $\mu(d)$ at V = const has a maximum. This is the root of the effect.

For a qualitative analysis, the curve $V_C(d)$ can be treated as the current-voltage characteristic $V_C(j)$, provided the axis j is directed counter to d. If $j < j_n$, that is, $d > d_n$, the stationary states on the curve are unstable (Sect. 8.4.8). Fluctuations may destroy the cathode layer in its middle part as well, but at the edges the layer decays even without fluctuations. At these edges, where the space charge decreases, the equipotentials shift away from the cathode (Fig. 8.9). When moving away from the midpoint of the spot, we shift to the right of point 1 in Fig. 8.10, where $\mu < 1$. Hence, the current at the edge vanishes with time. According to

[5] The problem was discussed in [8.9], but clarity had not yet been achieved at that time; see [8.8].

[6] Taking into account the correction mentioned in footnote 4 to Sect. 8.4.3.

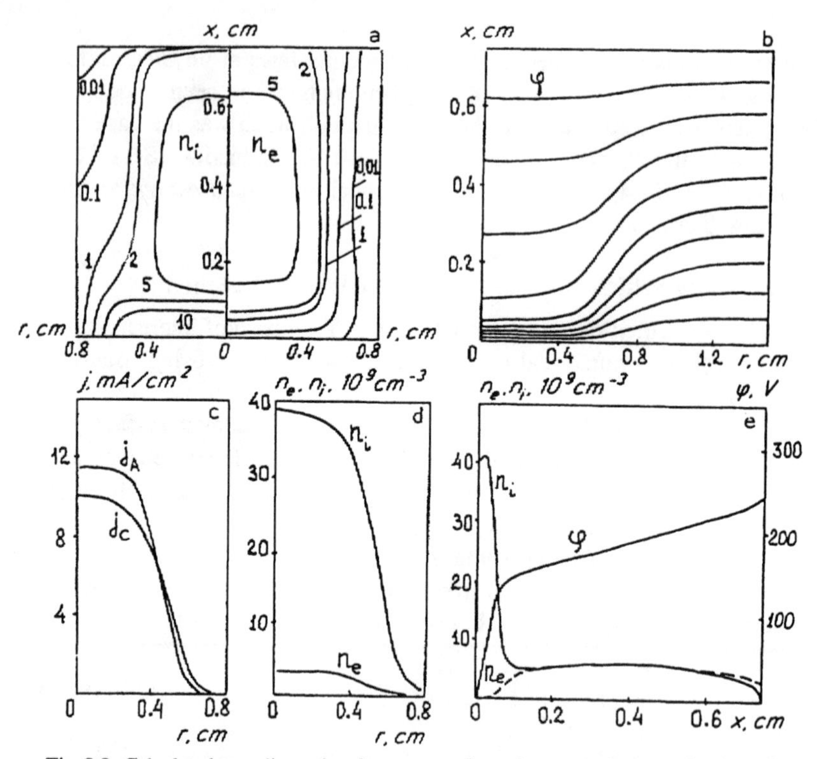

Fig. 8.9. Calculated two-dimensional structure of steady-state axisymmetric glow-discharge column between plane electrodes (nitrogen, p = 15 Torr, interelectrode spacing 0.75 cm, e.m.f. 2500 V, external resistance 300 kOhm). The cathode is placed below, the anode on top and on the electrodes V = 250 V, i = 7.5 mA. **(a)** Lines of equal densities of electrons and ions are plotted (n_e and n_i in $10^9\,cm^{-3}$, **(b)** equipotentials for each $V/10$ = 25 V, **(c)** radial current density distributions j_C on the cathode and j_A on the anode, **(d)** radial distributions n_i on the cathode and n_e on the anode, and **(e)** the distributions n_e, n_i, φ along the discharge axis [8.8]

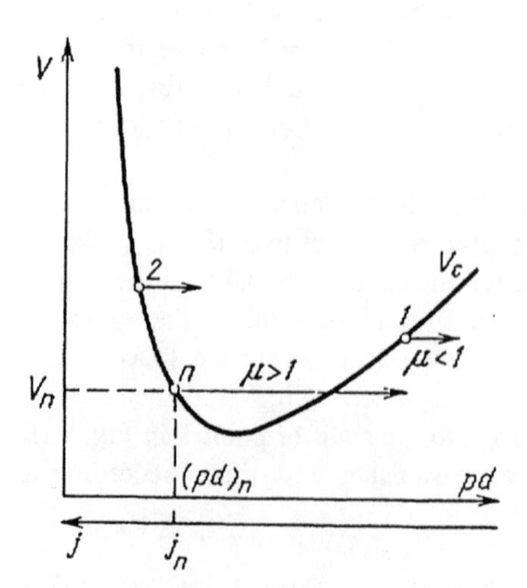

Fig. 8.10. Dependence of cathode fall V_C on layer thickness d and current density j

(8.1), the discharge voltage then increases due to the external resistance Ω, so that j in the homogeneous part of the layer increases until it reaches j_n.

Assume that a strongly supernormal layer has developed (point 2 in Fig. 8.10). The state in its central, quasihomogeneous part is stable. However, when we move to the right of point 2 (edgeward), where equipotential surfaces deviate from the cathode, we enter a region of increasing reproduction of charges, $\mu > 1$. Breakdown occurs at the edge of the cathode spot. A growth in the spot size increases the total current. The voltage across the electrodes decreases and j in the homogeneouss part of the layer drops. Point 2 slides downward until the state stabilizes at the edge, as well as in the middle of the layer.

The normal, that is, completely stable, state corresponds to point n, slightly above the minimum of the supernormal branch [8.5]. This is connected with the diffusion transport of charges to the edge zone, where $\mu < 1$; this transport sustains there a nondecaying non-self-sustaining current. This sustainment requires, however, that charge generation be enhanced ($\mu > 1$) in the region that lies closer to the midpoint. In the absence of diffusion, point 2 would move downward to the bottom, the normal state would coincide with the minimum V_C, and the edge of the cathode spot would become sharply defined. However, this state is incompatible with electrostatics. Equipotentials cannot be parallel up to the edge of the space charge zone and start deviating only beyond this zone.

8.4.10 Nonlocal Nature of Electron Spectrum and of the Ionization Coefficient in the Cathode Layer

So far, when establishing the conditions of self-sustainment of the discharge and the integral characteristics of the layer (V_C, d, j), we regarded the ionization coefficient α as a function of the local field $E(x)$. The (Townsend) dependence $\alpha(E)$ was taken from experimental data on ionization in homogeneous fields. This approximation is sufficient for obtaining integral characteristics, but it gives a severe distortion to the pattern of ionization produced by electrons at the end of the cathode layer and in the adjacent region; the understanding of the processes in this region is therefore seriously thwarted (Sect. 8.5). The point is that within the cathode layer, the field varies by a factor of 10^2–10^3; the thickness of this layer does not exceed 10 free path lengths for inelastic collisions, or $\ln(1 + 1/\gamma) \approx 3$ ionization length (α^{-1}). This inhomogeneity of E is too sharp for the equilibrium energy spectrum [corresponding to the local field $E(x)$] to set up, as it would in the case of a weak inhomogeneity. The actual ionization coefficient $\alpha(x)$ is also different from the Townsend one, $\alpha[E(x)]$.

Electrons move in the direction of weaker field; their spectrum is *harder* (i.e. of higher energy) than the equilibrium spectrum, and the ionization coefficient is greater because before arriving at a given point, electrons had gained energy in a stronger field and did not "forget" this fact. In the limiting case of a very small number of inelastic collisions, the energy of electrons is determined not by the field as such, but by the *potential difference* traversed. As a result of *nonlocal effects,* electrons with substantial energies are present among those emerging from the cathode layer; some of them have been created at the cathode and crossed the entire layer without a single inelastic collision. Their energies are eV_C, that is, hundreds of electron volts. These electrons, discovered a long time ago, are known as the "beam".

Current efforts in the cathode-layer theory are aimed at taking into account the nonlocal effects.[7] The rigorous approach is possible only on the basis of the solution of the kinetic equation for electrons in an inhomogeneous field. One of the approximations is to replace the local field $E(x)$ in the Townsend coefficient α with the mean field over the preceding path Δx on which the electron gains the energy $\int_{x-\Delta x}^{x} eE\,dx$ equal to th ionization potential [8.11]. The most complete and reliable information is obtained by *Monte Carlo simulation* of the stochastic process in the cathode layer; this has so much advanced in recent years [8.12] that it has become essentially a method of numerical solution of the kinetic equation. The solution is obtained by extensive, time-consuming computations that are only feasible if powerful computers are employed.

The results of computations, even though involvinig a number of simplifications, are impressive (Figs. 8.11, 8.12) and enable us to look at a sequence of stages of the process. Electrons with energies from small values up to 10–

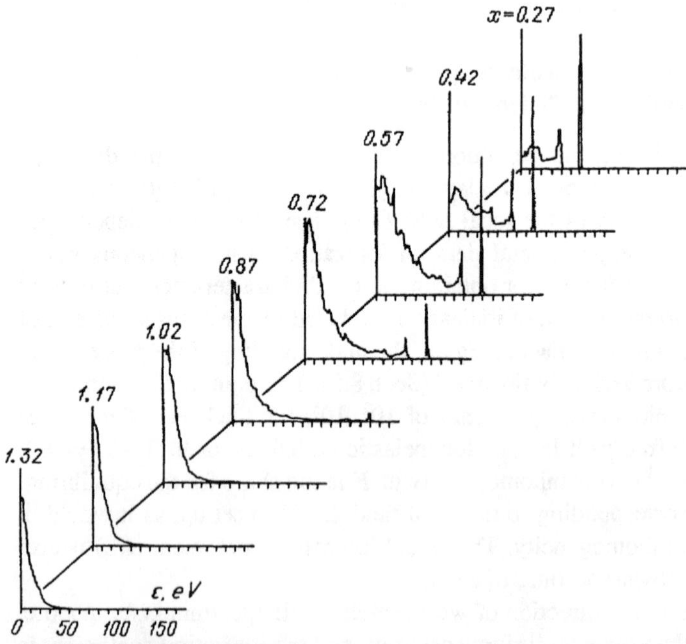

Fig. 8.11. Electron energy spectra at various distances x from the cathode in the cathode layer of normal glow discharge in He at p = 1 Torr. Calculations assume V_n = 150 V, the field $E = 230(1-x/1.3)+1$ V/cm in the cathode layer if $0 < x < 1.3$ cm, and $E = 1$ V/cm if $1.3 < x < 1.5$ cm [8.12]

[7] This aspect is unrelated to the problem of normal current density.

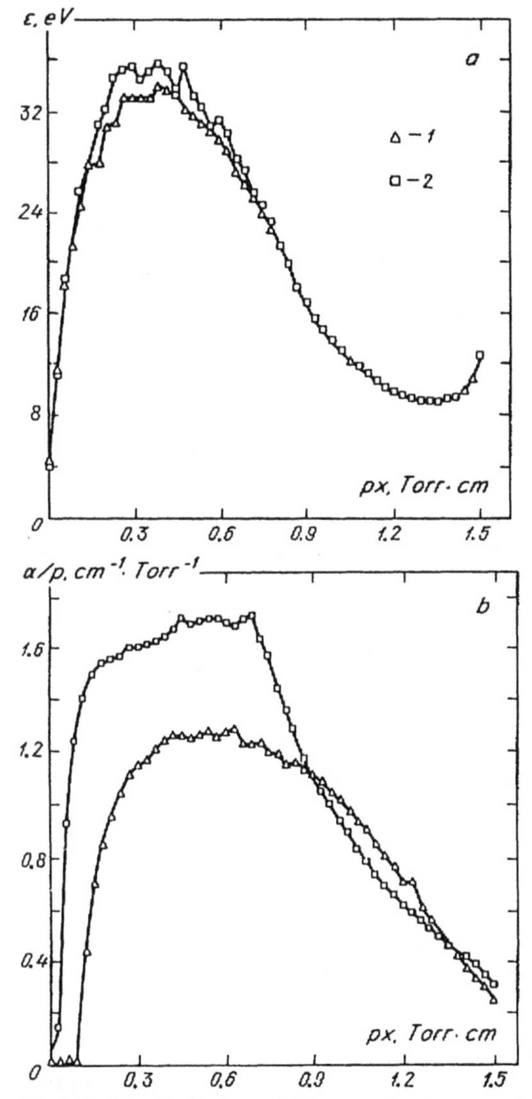

Fig. 8.12. The distributions of (a) mean electric energy $\bar{\varepsilon}$ and (b) the ionization coefficient α as a function of the coordinate x of the cathode (p = 1 Torr); calculations used the spectrum of Fig. 8.11 [8.12]:
△ – isotropic elastic scattering
□ – anisotropic elastic scattering, with forward scattering dominating

20 eV are detected on emergence from the layer, and a low-intensity beam is also found. Away from the cathode, the peak of the "beam" rapidly diminishes and disappears on the scale of Fig. 8.11. On the emergence from the cathode layer, it constitutes 10^{-3} of the distribution function for $\varepsilon = 0$; the mean energy on the emergence from the layer at $x = 1.3$ cm is $\bar{\varepsilon} \approx 10$ eV. If nonlocal effects are neglected, one gets $\bar{\varepsilon} \approx 2.9$ eV ($E/p = 1$ V/(cm·Torr); Sect. 2.3.5) and $\alpha \approx 0$.

The real ionization coefficient at the farther layer boundary is half the maximum value at the middle; on the other hand, the equilibrium coefficient $\alpha[E(x)]$ has its maximum at the cathode ($1.7\,\mathrm{cm}^{-1}$), and is many orders of magnitude smaller (practically zero) at the boundary. However, the details of the output spectrum do not agree with experimental data. Three groups of electrons were detected at the emergence from the cathode layer in helium: electrons with the mean energy of 2 eV (they are predominant), those with 22.5 eV (their number is two orders of magnitude less), and a weak beam with $\varepsilon = 150\,\mathrm{eV} \approx eV_n$ [8.13].

Section 5.9 outlined the "forward-backward" approximation for solving the kinetic equation in strong nonuniform fields. The approximation was tested under conditions assumed in the calculations of [8.12]. Satisfactory agreement with Figs. 8.11 and 8.12 was obtained. The method was used to take into account nonlocal effects in the von Engel–Steenbeck self-consistent calculation of the current-voltage characteristics of the cathode layer in helium (Sect. 8.4.2). The normal-discharge parameters found in [8.14], $V_n = 142\,\mathrm{V}$, $j_n = 3 \cdot 10^{-6}\,\mathrm{A/cm}^2$, $d_n = 2.2\,\mathrm{cm}$ for $p = 1\,\mathrm{Torr}$, and also the abnormal branch, are in much better agreement with the experiments than those using the Townsend coefficient $\alpha[E(x)]$ and A and B of Table 4.1. The rate of electron production increases away from the cathode, reaches a maximum (in normal discharge) of $q \approx 8 \cdot 10^{12}\,\mathrm{cm}^{-3}\mathrm{s}^{-1}$ at the end of the cathode layer, for $x \approx 2\,\mathrm{cm}$, and vanishes only at a distance of $x \approx 4$ to $5\,\mathrm{cm}$. Roughly a half of the electrons are produced byond the cathode layer, in the weak field assumed to be 0.1 V/cm. A self-consistent (in contrast to [8.12]) cathode layer calculation using Monte Carlo techniques has recently appeared in [8.15]. Like the calculation of [8.10] with $\alpha(E)$, it mostly confirmed the linear law (8.12) of field decrease with distance from the cathode.

8.5 Transition Region Between the Cathode Layer and the Homogeneous Positive Column

The section heading refers to the regions of negative glow and Faraday dark space, terms that reflect the visual attributes. We are interested in the processes responsible for the longitudinal structure of glow discharges: the field and charge density distributions along the x axis and the current transport. In this respect, the two regions are a coherent whole (Fig. 8.2).

8.5.1 The Decisive Role of Energetic Electrons Supplied by the Cathode Layer

These electrons dissipate energy in the excitation and ionization of the gas, thereby causing intensive light emission (and sharply increased ionization in the region where the field is insufficient for such processes). On the other hand, intense ionization is a factor causing a drop in field strength, because the current density along the direction of current remains constant in one-dimensional stationary discharges (this is also valid for discharge in a tube). Therefore, in the framework of the frequently justified assumption of constant mobility μ_e, we have $n_e E$ = const; hence, the field E at the point $n_{e,max}$ is weaker than in the positive column, where n_e is lower; this is indeed supported by Fig. 8.2, which schematically represents the results of experiments.

The rôle of nonlocal effects that produce a powerful electron source not attributable to the effect of the local field is especially well pronounced if we ask what would happen if the ionization decreased everywhere monotonously as the field decreased (as in the case of ionization with equilibrium Townsend coefficient). The distributions of n_e, n_+, and E along the x axis within the gap are then described by (8.2,6,3). Actually, (8.2) must be complemented with terms for charge loss in order to allow for the formation of a homogeneous lpositive column, with ionization compensating for losses (Sect. 8.6). Both a qualitative analysis and a numerical integration of the system show that the field monotonically decreases from the cathode value to that corresponding to the positive column, and n_e also increases in a monotonic manner. We thus come to a natural transition from the cathode layer to the positive column, without a region of field drop [8.9]. We can say that the Faraday space would not form without a flux of energetic electrons outside the layer in which current is self-sustained, that is, without nonlocal effects.

8.5.2 Probe Measurements

The relevant data are plotted in Fig. 8.13. The measured electron temperature $T_e \approx 0.12$ eV is practically constant across the region and does not depend on the current. The potential is almost unchanged within the investigated length, and E/p does not exceed 0.01 V/(cm·Torr). The latter result is likely to be beyond the accuracy of the measurements. It is easy to evaluate that this E/p is too small for supporting the drift transport of the current at $n_{e,max}$, when both the gradient of n_e and diffusion current vanish. Slow electrons are maxwellian because the frequency of electron-electron collisions at such low temperatures substantially exceeds the energy loss frequency (2.17) in collisions with atoms. These are the electrons that were created at the very end of the cathode layer, where the field that could accelerate them is almost zero, and also electrons, created by high-energy electrons, that have dissipated their energy. They are known as "final" electrons. Probe measurements of the distribution function pointed to the exis-

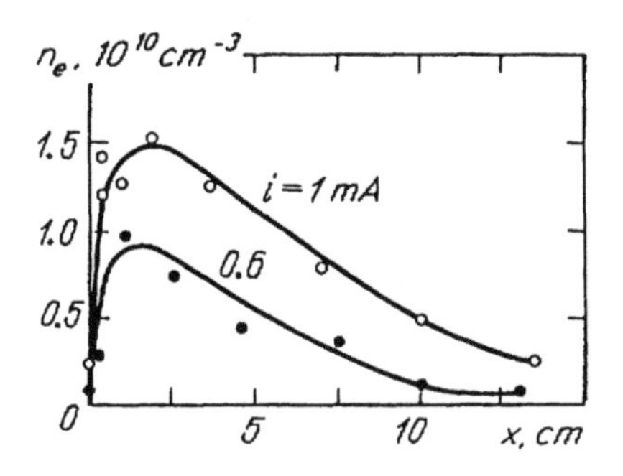

Fig. 8.13. Measured density distributions of slow electrons in the negative glow and Faraday space on the tube axis. Discharge in helium, p = 1.5 Torr. Coordinate x is measured from the beginning of the negative glow, towards the anode [8.1, 16]

tence of a second group of electrons with $T_e \approx 3$–$4\,\mathrm{eV}$. Their density is smaller by a factor of 100–200. These electrons are sometimes called "secondary". They are produced somewhat deeper in the cathode layer and are slightly accelerated in the field. Not too energetic electrons having undergone an inelastic collision and retaining this energy are likely to belong to this group too. The third group comprises a small number of electrons; this is the beam (the three groups were also mentined in Sect. 8.4.10).

Recent measurements in helium under similar conditions [8.17] ($p = 1$ Torr, $R = 1.5\,\mathrm{cm}$) but at a higher current ($i = 15\,\mathrm{mA}$) yielded greater values of E/p and $T \sim 1\,\mathrm{eV}$ and a shorter Faraday space (about 3 cm). It is of interest that the diffusional current is almost balanced out by the oppositely directed thermodiffusional current, so that only the drift component is present.

8.5.3 Role Played by Electron Diffusion

As a result of a sharp maximum of electron density that falls off steeply towards the anode, the current in the region of density drop may be sustained by electron diffusion, as described in Sect. 2.7.2.[8] The field is thereby cancelled. Owing to the drop in electron density, the diffusion flux gradually decreases, the field gets restored [see (2.42)], the diffusion is gradually replaced with drift, and the Faraday space is transformed into the positive column (provided the anode is still far removed; see Fig. 8.2). This situation also arises in a low-voltage arc. It has been approximately modelled in [8.18, 19, 1].

The drop in n_e from the maximum towards the anode is caused by electron losses not replenished by ionization (which is absent). Under the conditions of Fig. 8.13, electrons are removed by ambipolar diffusion to the walls, where subsequent neutralization takes place. *Bulk recombination* (and attachment in electronegative gases) becomes predominant at high pressures when diffusion is impeded.

8.5.4 Main Factors Determining the Longitudinal Structure of Discharge in a Gap

Three factors are thus important for the formation of the transition from the cathode layer to the homogeneous positive column via the negative glow and Faraday space. (1) The presence of a powerful ionization source at the end of the cathode layer; the source is not dependent on the local field strength. (2) Local charge loss (bulk recombination, ambipolar diffusion to the wall) not related to its current to the electrodes. (3) Diffusion of electrons along the current direction. No adequate theory of the transition region is possible if anyone of these factors

[8] It is sometimes mentioned (see, e.g. [8.1]) that the beam participates in charge transfer. In fact, only a small fraction of the electrons emitted from the cathode is transformed into the beam; besides, the electronic current from the cathode is a small fraction, $\gamma/(1+\gamma)$, of the total. The current is not increased by accelerating the beam to a high velocity, but the electron density is reduced.

is ignored (advances in this theory are very moderate). Early work in this field is described in [8.1], and the problem is discussed in [8.9].[9]

8.5.5 Hollow Cathode Discharge

If the cathode is arranged as two parallel plates (the anode being shifted to the side) and the cathode plates are brought closer and closer, the current increases hundreds and thousands of times after a certain distance is reached. This takes place when two formerly nonoverlapping regions of negative glow merge: the glow becomes considerably more intensive, and the voltage changes slightly. A similar effect can be obtained if the cathode is a hollow cylinder and the anode lies far along the axis. The pressure must be such that the cathode layer thickness is comparable with the cylinder diameter. In a hollow cathode, electron streams converge to the axis and produce intensive ionization and excitation of the gas. Photoemission excited on the cathode by UV radiation produced in this region also plays a role here.

8.6 Positive Column

8.6.1 The Function It Serves; Causal Relationships

The positive column closes the electric circuit in the space between the cathode layer and the anode; this is its only function. The state of the plasma in a sufficiently long column is completely independent of the situation in the *regions adjacent to the electrodes*. It is determined by *local* processes and by the electric current. The inevitable loss of charge carriers (electrons) in the column must be

[9] *Note added in proof.* The author and M.N. Shneider have recently computed the glow discharge longitudinal structure. Distributions of n_e, n_+, E, $\bar{\varepsilon}$ (or T_e) are obtained along the x-axis from the cathode to the uniform positive column (PC) inclusively. The patterns look like those in Fig. 8.2. Non-local effects in the cathode layer (CL) and the negative glow region (NG) are described by the method discussed at the end of Sects. 8.4.10 and 5.9. Besides the three necessary factors mentioned in Sect. 8.5.4, electron heat conductivity and thermodiffusion are also taken into account. In helium for $p = 1$ Torr and $j = 2.7\ 10^{-5}$ A/cm^2 (this is an order of magnitude higher than normal j_n) the CL stops at $x = 1.3$ cm, and NG at $x = 2$ cm. The Faraday space (FS) stops at $x \approx 4$–6 cm for a tube radius of $R = 1.35$ cm and $x \approx 8$–10 cm for $R = 5$ cm (the length of the FS for low pressure is about a few R). At the beginning of the FS the field falls to a very small magnitude and changes direction, remaining very weak. The temperature T_e falls to a few tenths of an eV in the middle of the FS. The maximum of n_e at the end of the NG is $n_{e\,\max} \approx 3 \cdot 10^9\ \mathrm{cm}^{-3}$ while in the PC $n_e \approx 10^8\ \mathrm{cm}^{-3}$ (for $R = 1.35$ cm). From a physical point of view the magnitude of $n_{e\,\max}$ is determined by the condition that the current in the region of falling n_e, after the maximum is carried by diffusion $j \approx eD_e dn_e/dx \approx eD_e n_{e\,\max}/X$, where the scale X of n_e decrease is determined by ambipolar diffusion of electrons to the wall, with the longitudinal drift as a background $X \approx v_d(R/2.4)^2/D_a$. Such an estimation of $n_{e\,\max}$ agrees well with calculations. The transition from the FS to the PC is found to be non-monotonic (n_e passes through a minimum, E and $\bar{\varepsilon}$ through a maximum). This is connected with the time lag of $\bar{\varepsilon}$ and ionization growth, while the restoring E (as in the case of striations; Sect. 9.7). These results are published in [8.14].

compensated for by ionization. The field strength E necessary for sustaining a stationary plasma is fixed because the ionization rate depends, and quite sharply, on the field, through the dependence of the electron energy distribution. This determines the longitudinal potential gradient and the voltage difference across a column of a given length. If the spectrum is maxwellian, the relationship can be separated into two causally linked parts: (1) The requirement of loss compensation by ionization shows what the electron *temperature* T_e must be; (2) The *field* must supply the necessary energy to electrons. The relation between E and T_e follows from the balancing of the electron energy (Sect. 2.3). The *gas temperature* T is determined by the balance of the gas energy as a whole. In the positive column of glow discharge, we have $T_e \gg T$.

The creation and removal of electrons in the column proceed against a steady background of unceasing electron replacement due to the drift motion from the cathode to the anode. It cannot be said that a considerable fraction of charge carriers are generated in the glow discharge column. Rather, the majority of electrons reaching the anode enter the column from the outside (from the cathode region). The probability for them to be lost on the way is not high, except for cases of exceptionally long interelectrode separations.

8.6.2 Balance of Charge Numbers in Cases Without Attachment

Consider a long positive column in a tube (or plane chanel) so long that it can be treated as homogeneous along the current direction x. According to (3.14), curl $\boldsymbol{E} = 0$ in stationary conditions; hence, a longitudinal field homogeneous in x is independent of transversal coordinates (the transverse polarization field is neglectigible in comparison with the longitudinal one). The charge density in the quasineutral plasma of the column is described by (2.44), where q includes ionization and bulk recombination. Denoting the transverse part of the Laplacian by the subscript $\perp$, we arrive at the equation

$$D_a \nabla_\perp^2 n + \nu_i(E) n - \beta n^2 = 0 . \tag{8.17}$$

Assume that the precipitation of charges on the walls is more intensive than bulk losses. It is said in such cases that the discharge is *controlled by diffusion or by recombination at the walls*. Without the term βn^2 and with the boundary condition $n = 0$ at $r = R$, (8.17) results in the Bessel radial profile $n \propto J_0(2.4r/R)$, (see Sect. 4.8) and in the condition of equality of ionization frequency and the effective frequency of diffusional loss:

$$\nu_i(E) = D_a/\Lambda^2 \equiv \nu_{da} , \quad \Lambda = R/2.4 \tag{8.18}$$

[*Schottky*, (1924)].[10] The case $\beta n \ll \nu_{da}$ is realized at low pressure and small transverse dimensions ($\nu_{da} \propto 1/p\Lambda^2$), at not too high currents, so that n is moderate; it is facilitated in monatomic gases where the bulk recombination proceeds slower than in molecular gases.

[10] In plane geometry, the profile is cosine-shaped and Λ is given by (4.12).

Let us look at an example. In nitrogen, $\mu_+ p \approx 1.5 \cdot 10^3\,\mathrm{cm^2{\cdot}Torr/(V{\cdot}s)}$, $T_e \approx 1\,\mathrm{eV}$, $D_a p = (\mu_+ p)T_e \approx 1.5 \cdot 10^3\,\mathrm{cm^2 Torr/s}$. If $p = 10\,\mathrm{Torr}$, $R = 1\,\mathrm{cm}$, and $D_a \approx 150\,\mathrm{cm^2/s}$, we find $\nu_{da} \approx 900\,\mathrm{s^{-1}}$. If the recombination coefficient (dissociative: see Sect. 4.3.2) $\beta = 1.6 \cdot 10^{-7}\,\mathrm{cm^3/s}$, the condition $\beta n < \nu_{da}$ holds up to $n \approx 6 \cdot 10^9\,\mathrm{cm^{-3}}$, which corresponds to current density $j = e(\mu_e p)n_e(E/p) \approx 1.2\,\mathrm{mA/cm^2}$ [by Table 2.1, $\mu_e p \approx 4.2 \cdot 10^5\,\mathrm{cm^2 Torr/(V{\cdot}s)}$; $E/p \approx 3\,\mathrm{V/(cm{\cdot}Torr)}$]. The total current $i \approx j\pi R^2 \approx 3.5\,\mathrm{mA}$. if the current is smaller than this value, the discharge is controlled by diffusion, otherwise it is controlled by *bulk recombination.*

The probability for an electron drifting along the column of length L to attach to the wall is $\nu_{da}t$, where $t = L/v_d$ is the drift time. The probability can be expressed in terms of gas characteristics and size using (2.36, 24, 21, 16):

$$\frac{\nu_{da}L}{v_d} = \frac{D_e L}{\Lambda^2 v_d}\frac{\mu_+}{\mu_e} = \frac{1}{3}\frac{lL}{\Lambda^2}\frac{\bar{v}}{v_d}\frac{\mu_+}{\mu_e} \approx \frac{0.4lL}{\Lambda^2\sqrt{\delta}}\frac{\mu_+}{\mu_e}\,. \tag{8.19}$$

In our case the probability is $7 \cdot 10^{-4}$ per 1 cm of length. Even a metre-long column transmits 93 % of electrons (provided the bulk recombination is low). The estimate assumes $lp = 0.03\,\mathrm{cm{\cdot}Torr}$, $\delta = 1.2 \cdot 10^{-3}$.

If $\beta n \gg \nu_{da}$, the recombination of charges in the bulk dominates over their diffusion to the walls. If the diffusion term in (8.17) is dropped, we find $\nu_i(E) = \beta n$, that is, the density is constant over the cross section. In fact, a large gradient of n appears at the absorbing walls, and diffusion there cannot be neglected. Density changes only slightly in the main part of the section but drops sharply near the walls. Using (8.17), one can write an interpolated balance equation covering both limiting cases and ensuring a smooth transition between the two:

$$\nu_i(E) - \nu_{da} - \beta n = 0\,. \tag{8.20}$$

8.6.3 Field Strength and Current-Voltage Characteristic

If the column is diffusion controlled, E is found from (8.18) and is independent of electron density, and hence, of current. This occurs because the rates of creation and removal of electrons are proportional to n. In this approximation, the $V - i$ curve is represented by a horizontal line both for the column and for the discharge as a whole, provided it is normal: $V(i) = V_n + EL = \mathrm{const}$. The field in the column (Fig. 8.14) obeys the similarity law $E/p = f(p\Lambda)$, which follows from the dependences $\nu_i = pf_1(E/p)$, $\nu_{da} \propto 1/p\Lambda^2$. As we see from Fig. 8.14, some decrease in E/p as the current increases by an order of magnitude is caused by a rise in gas temperature (Sect. 8.7.4). It is worthy of note that the discharge is sustained in air, in which ionization dominates attachment only if $E/p \gtrsim 35$ (Sect. 7.2.5), at substantially lower values of E/p. The reason is that many molecules that are active with respect to detachment accumulated under stationary conditions to high concentration. Attachment is then partially balanced out by detachment, with electron losses being lower than in breakdown or in short transient discharges (Sect. 8.8). The effect of two-stage ionization (ionization of excited molecules) may also be involved.

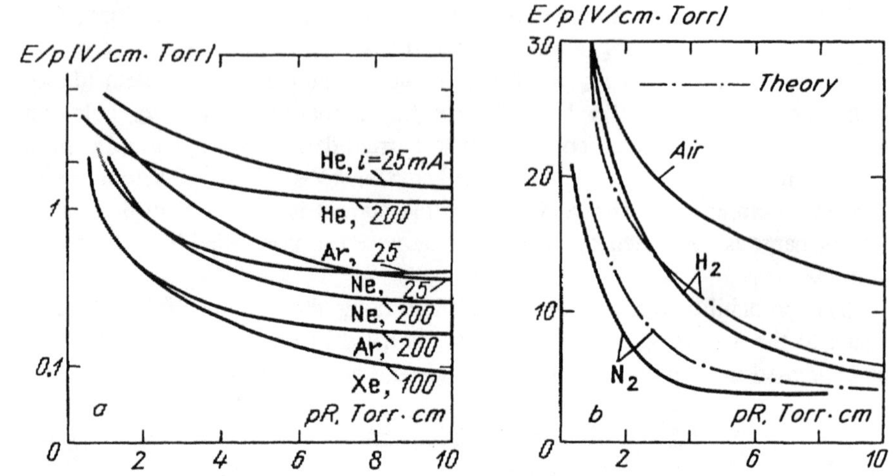

Fig. 8.14. Measured values of E/p for the positive column in tubes: (a) inert gases, (b) molecular gases [8.3]

It must be emphasized that the field required for plasma sustainment in the column is always lower than that required for gaseous breakdown under the same conditions. In breakdown, electrons are freely diffusing towards the walls. The diffusion in a well-developed discharge is ambipolar and substantially slower.

If discharge currents are considerable, the degree of ionization increases and bulk recombination becomes important. The current-voltage characteristic of the column is then given by the expression

$$n_e = \frac{\nu_i(E) - \nu_{da}}{\beta}, \quad j = \frac{e(\mu_e p)}{\beta}\frac{E}{p}[\nu_i(E) - \nu_{da}] \tag{8.21}$$

that follows from (8.20). If the discharge is controlled by bulk recombination ($\nu_i \gg \nu_{da}$), E increases with j, albeit slowly. This mode is more likely to manifest itself in molecular gases, where recombination is strong. In inert gases the *contraction* of the positive column into the *current filament* occurs earlier (Sect. 9.8).

8.6.4 Electron Temperature and Its Relation to Field Strength

Consider a diffusion-controlled discharge. Substituting (4.2) for ν_i in (8.18) if the spectrum is maxwellian, and (2.36) for D_a, gives an equation for T_e:

$$\left(\frac{kT_e}{I}\right)^{1/2} \exp\frac{I}{kT_e} = \frac{C_i}{\mu_+ p}\left(\frac{8I}{\pi m}\right)^{1/2}\frac{N}{p}(p\Lambda)^2 = \text{const}\,(pR)^2\,. \tag{8.22}$$

In this way von Engel and Steenbeck [8.2] determined the dependence, universal for all gases, of kT_e/I on cpR (Fig. 8.15); here c is a constant, specific for each

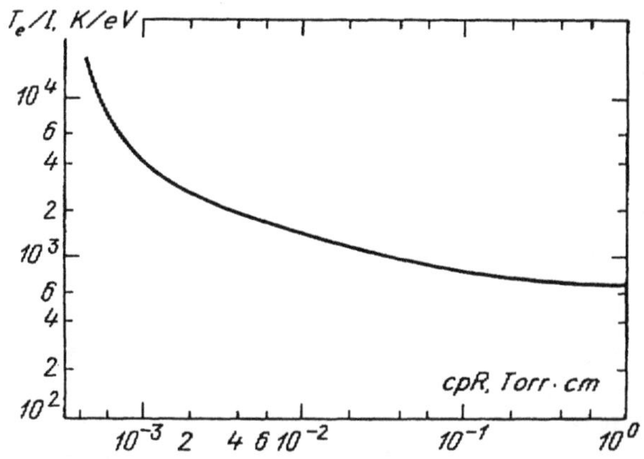

Fig. 8.15. Universal curve for calculating T_e in a positive column as a function of cpR. The constants c for several gases are given in the text [8.3]

gas, and is calculated using the data on μ_+, C_i, and I. The constant c in different gases is:[11]

$$\text{He} - 4 \cdot 10^{-3}\,, \quad \text{Ne} - 6 \cdot 10^{-3}\,, \quad \text{Ar} - 4 \cdot 10^{-2}\,,$$
$$\text{H}_2 - 1 \cdot 10^{-2}\,, \quad \text{N}_2 - 4 \cdot 10^{-2}\,.$$

For instance, for nitrogen and $R = 1\,\text{cm}$, $p = 10\,\text{Torr}$, we find $T_e \approx 0.9\,\text{eV} = 10{,}400\,\text{K}$. The electron temperature decreases with increasing tube radius and pressure: diffusion losses are reduced and a lower ionization rate is sufficient. It is clear why the positive column is dark in wide vessels: the electron temperature is too low.

The relation between the mean electron energy $\bar{\varepsilon} = 3kT_e/2$ and the field was discussed in Sect. 2.3.5. Assuming the electron path length to be independent of energy, we have $T_e \propto E/N$; if the collision frequency is assumed constant, $T_e \propto (E/N)^2$. The former assumption is preferable. It was shown in Sect. 2.3.6 that except at very low pressures, the electron drift velocity under the conditions exactly corresponding to the glow discharge positive column is much lower than the random one. Roughly speaking, the electron spectrum is said to be maxwellian if the frequency of electron-electron collisions ν_{ee} is appreciably higher than the energy loss frequency $\nu_u = \delta\nu_m$ (Sect. 2.3.7).

8.6.5 Why Is the Degree of Ionization in Weakly Ionized Gas Discharge Plasma Strongly Nonequilibrium?

The electron density in a *diffuse* positive column of a glow discharge (a *noncontracted* column filling the entire cross section of a tube is said to be diffuse) is of order $n_e \sim 10^8$–10^{11}, or 10^{12} at the most. If $p \sim 1$–$10\,\text{Torr}$, $N \sim 3 \cdot 10^{16}$–$3 \cdot 10^{17}\,\text{cm}^{-3}$, the degree of ionization of the gas is 10^{-8}–10^{-7}. At the same

[11] The proportionality factor in (8.22) is equal to $1.2 \cdot 10^7 c^2$.

time, the actual electron temperatures $T_e \approx 1$–$3\,\mathrm{eV}$ correspond to thermodynamical equilibrium ionizations of 10^{-2} to 1. This discrepancy results from the violation of the foremost requirement that ensures the attainment of thermodynamic equilibrium: the direct and reversed processes in the main reactions must be balanced.

In a glow discharge, atoms are ionized by electron impact, often from the ground state. Charge losses in cold rarefied gases occur at the walls and via dissociative recombination. The charge loss frequency in the example considered in Sect. 8.6.2 was $\nu_{da} \sim 10^3\,\mathrm{s}^{-1}$ for $n_e \lesssim 10^{10}\,\mathrm{cm}^{-3}$. The field and the electron temperature accomodated to these losses and produced the required ionization rate. However, three-body recombination with an electron trapped in the ground state of the atom [this process is reversed with respect to ionization described by (4.2)] gives a loss frequency of $10^{-10}\,\mathrm{s}^{-1}$. This is less by 13 orders of mangitude! There are no reverse processes of ionization that oppose the fast diffusional and dissociative losses.

The situation with strongly ionized equilibrium plasma is different. Diffusion and dissociative recombination are not important there because the gas is hot and dense, and molecular ions are rare. The charge density is high, and collision-radiative recombination is predominant, with electrons captured to upper levels (Sect. 4.3.4). Atoms are ionized as a result of a reverse process to this recombination: stepwise ionization from excited states. This is how the thermodynamically equilibrium ionization is achieved.

8.6.6 Inhomogeneous Plasma Column

When the geometry of the discharge volume and the electrode configuration are complicated, and there is a rapid transverse gas flow that offsets and bends the current channel, the plasma column between electrodes may be quite inhomogeneous and at the same time approximately electrically neutral; such conditions are encountered in high-power lasers (Sect. 14.4.2). In such cases, it is not expedient to employ the general system of equations (2.22, 20, 43) for finding the spatial distributions of plasma density and field $\boldsymbol{E} = -\nabla\varphi$, because Poisson's equation (2.43) then involves a small difference $n_+ - n_e$ of relatively large quantities. Two slightly different quantities n_e and n_+ must be replaced with a single one, $n \approx n_e \approx n_+$; this operation reduces the system of three quations to a system of two, (2.46) and (2.40).[12] If we assume μ_e = const, (2.40) reduces to $\mathrm{div}\,(n\boldsymbol{E}) = 0$. If the discharge is burning in a gas flux moving at a velocity $\boldsymbol{u}$, the convective term $n\boldsymbol{u}$ must be added to the plasma flux under the div sign in (2.46).

The elimination of $n_+ - n_e$ from the system is in accord with the actual causal relationship of the phenomena. When the space charge is small, the field in a nonhomogeneous conducting medium is determined by current distribution and

[12] The current distribution is quasistationary even in nonstationary plasma processes, owing to high relaxation rate of the bulk charge (Sect. 9.2.2). Consequently, the replacement of (2.39) with (2.40) is justified.

boundary conditions, not by local charges. Quite the opposite: the space charge adjusts itself to field inhomogeneities and can be determined, after the field has been found, using electrostatics equation (2.33).

An nonhomogeneous plasma column is also formed between plane electrodes at moderate $pL \lesssim 10\,\mathrm{Torr\,cm}$, because there is not enough length for the formation of a homogeneous one (considered above). A sharp inbalance in creation and loss of electrons in the plasma may be observed. In the case shown in Fig. 8.9, the largely insignificant ionization in the column is nevertheless much stronger than the recombination and diffusion losses. One-dimensional calculations ignoring losses and diffusion [8.10] (and then confirming they are small) led to the same conclusion. The plasma region in narrow (in the sense of pL values) gaps is not an analogue of the positive column.

When nonlocal effects are taken into account in a one-dimensional model of a narrow gap, similar to [8.10], the longitudinal structure of the discharge remains qualitatively the same [8.20]. This is not surprising, in view of what we said in Sect. 8.5.4.

8.7 Heating of the Gas and Its Effect on the Current-Voltage Characteristic

The entire energy gained by electrons from the field under stationary conditions is transferred to the gas as a result of collisions with atoms and molecules. The energy released in $1\,\mathrm{cm}^3$ per second is $jE = \sigma E^2$. The Joule heat is partly consumed to excite molecular vibrations; if these vibrations rapidly relax (this is mostly the case, nitrogen constituting an important exception), one can speak of a unifying gas temperature T even in molecular gases.

8.7.1 Transport of Joule Heat by Heat Conduction

In a tube without gas inflow, the heat is transported to the walls, which can be assumed to have the room temperature T_0. The gas temperature smoothly falls off from the axis to the walls (Fig. 8.16). The density of heat flux into the wall is $J_R = -\lambda(\partial T/\partial r)_{r=R}$, where λ is the thermal conductivity. If we operate with temperature averaged over cross section, the loss of energy from $1\,\mathrm{cm}^3$ of the gas each second $2\pi R J_R/\pi R^2$, is equal, up to a numerical coefficient, to $\lambda(T-T_0)/R^2$. It can be written in the form $Nc_{pl}(T-T_0)\nu_T\,\mathrm{erg/(cm^3 s)}$, where c_{pl} is the specific heat at constant pressure per molecule and ν_T is the inverse to the time of heat removal from the volume. The *heat removal frequency* is $\nu_T = \chi/\Lambda_T^2$, where $\chi = \lambda/Nc_{pl}$ is the thermometric conductivity and $\Lambda_T \approx R/\sqrt{8} \approx R/2.8$. It is similar to diffusion frequency $\nu_{\mathrm{dif}} = D/\Lambda^2$ (heat conduction and diffusion are described by the same equations). A slight difference between Λ_T and $\Lambda = R/2.4$ follows from the averaging procedure.

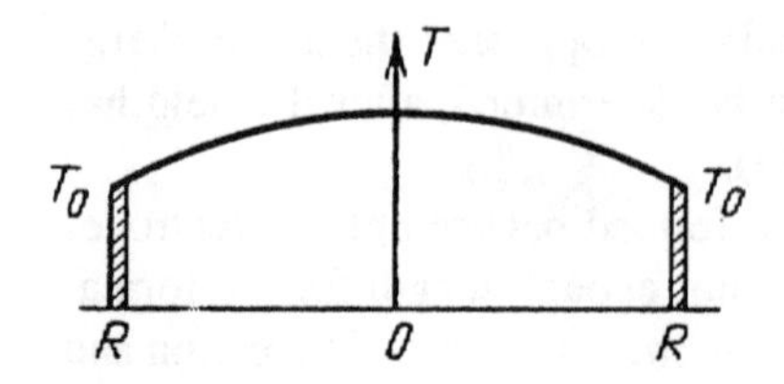

Fig. 8.16. Variation of gas temperature in a tube heated to a moderately high temperature

8.7.2 Convective Heat Transport

Another mechanism is possible for transporting heat out of a discharge: pumping of gas through the discharge volume, used in modern high-power laser systems (Sect. 8.1.2). This mechanism is *convective cooling*. Actually, a given mass of gas undergoes no cooling at all. Quite the opposite, a macroscopic portion of the gas is heated when it moves through the discharge; its temperature is gradually increasing in time and along the path. What is meant is the removal of heat from the discharge volume. If we again talk in terms of temperature T averaged over the flow length L_1, the rate of heat removal from the discharge volume can be written in the familiar form: $Nc_{p1}(T - T_0)\nu_F$. Now T_0 is the temperature of the gas entering the discharge, and $\nu_F \approx 2u/L_1$, where u is the flow velocity. The factor "2" takes into account that, on average, the heat is transported "half the distance". In a longitudinal discharge (Fig. 8.18), L_1 coincides with the interelectrode spacing L.

8.7.3 Energy Balance in the Gas

For greater clarity, let us start with the nonstationary balance equation. We take into account that the energy transfer from electrons to molecules is much faster than the outward transport from the gas. Hence, we consider the balance of electron energy to be stationary even if the gas temperature varies with time. The simplified equation for gas temperature is

$$Nc_{p1}\frac{dT}{dt} = jE - Nc_{p1}(T - T_0)\nu_{T,F} \ . \tag{8.23}$$

Under stationary conditions, the mean temperature is found from the equality

$$Nc_{p1}(T - T_0)\nu_{T,F} = jE \equiv w \ . \tag{8.24}$$

8.7.4 Dropping $V - i$ Characteristic

Experiments show that the $V - i$ curve of a diffusion-controlled discharge is not a horizontal but a slightly declining curve: as the current increases, the voltage slowly decreases. This effect results from gas heating. The current density is greater at the axis than at the walls, because the electron density there is greater (the field being constant over the cross section). The energy release and gas temperature at the axis are also higher than at the walls. However, pressure levels off in space (velocities are usually strongly subsonic even in discharges with flowing gas). Hence, the density of the gas is lower in regions of higher temperature. The ionization frequency being actually a function of E/N, not of E/p, a lower field is required to sustain ionization in the main part of current cross section; the voltage is also reduced. The law of fall-off can be evaluated using (8.24), $p = NkT = \text{const}$, and an approximate condition $E/N \sim ET \approx$ const that follows from (8.18). We find

$$j/j_0 = (E_0/E)^{3/2}(E_0/E - 1)\,. \tag{8.25}$$

Here E_0 is the field required to maintain a very weak current $j \to 0$ when the gas heats up negligibly and its temperature does not deviate from the room value T_0;

$$j_0 = N_0 c_{p1} T_0 \nu_{T,F}/E_0\,, \quad w_0 = j_0 E_0\,. \tag{8.26}$$

These are characteristic scales of the current density and the rate of Joule heat release. With this energy release, the gas becomes heated to a temperature T twice as high as that of the walls.

8.7.5 Stable and Unstable States

When the $V - i$ characteristic is a dropping one, the load line often intersects it not at one but at two points (Fig. 8.17). One of the states, namely, the upper one, is unstable; hence, it cannot be realized. Indeed, if current fluctuates upward, a lower voltage is required to sustain it than the one that is prescribed by the load line for the new current value. The result is a disbalance between ionization and removal of electrons; ionization begins to rise, the discharge resistance decreases, so that the current starts to grow until the state reaches the lower intersection point. If the current fluctuation is negative, $\delta i < 0$, electrons begin to disappear until the discharge burns out. The lower state is stable. If $\delta i > 0$, $\delta n_e > 0$, voltage drops below the necessary value, and enhanced decay brings ionization back to the original state.

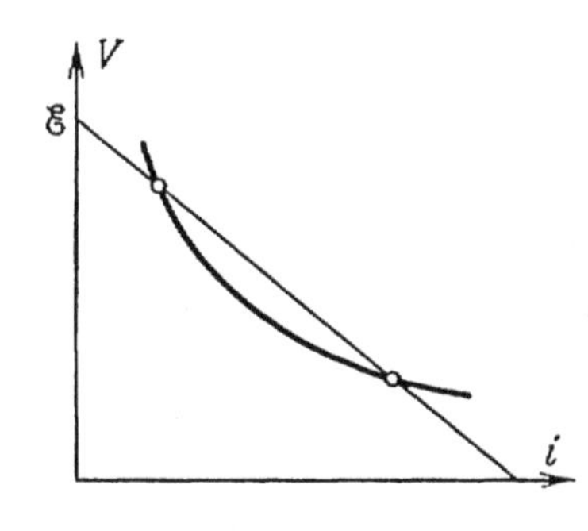

Fig. 8.17. $V - i$ characteristic, including gas heating effects. The loading curve is also shown

8.7.6 Gas Temperature and the Scales of Electric Parameters in Diffuse Glow Discharge

One of the most typical properties of this discharge is a sharp difference between the electron and gas temperatures. We have already discussed why the electron temperature is necessarily high: electrons are to ionize atoms. The same factors determine the order of magnitude of E/p in the column: $E/p \approx 0.2$–$20\,\mathrm{V/(cm\cdot Torr)}$. The gas temperature T is determined by the balance of energy release and heat removal, (8.24). The relative increase over the wall temperature (room temperature) is

$$(T - T_0)/T_0 = w/w_0 = jE/j_0E_0 \ .$$

The scales of energy release w_0 and current density j_0 are given by (8.26). They correspond to doubling the temperature. The thermal conductivity of most gases at $T \approx 300$–$600\,\mathrm{K}$ is $\lambda \approx (2\text{–}5) \cdot 10^{-4}\,\mathrm{W/(cm\cdot K)}$. In monatomic gases, $c_{pl} = (5/2)k$, and in two-atom gases, $(7/2)k$. With intermediate values $\lambda = 3 \cdot 10^{-4}$, $c_{pl} = 3k$, the thermometric conductivity is $\chi = 220/p\,\mathrm{cm^2/s}$ [p in Torr]. In tubes (or channels) of $R = 1\,\mathrm{cm}$, the heat transport to the walls is characterized by a frequency $\nu_T \approx 1.3 \cdot 10^3/p\,\mathrm{s^{-1}}$. For a typical value $E/p = 3\,\mathrm{V/(cm\cdot Torr)}$, the energy release and current density scales are $w_0 = 0.5\,\mathrm{W/cm^3}$ and $j_0 = 170/p\,\mathrm{mA/cm^2}$. According to (2.6, 7), the electron density scale for $\nu_m = 3 \cdot 10^9 p\,\mathrm{s^{-1}}$ is $n_e^0 \approx 6 \cdot 10^{11}/p\,\mathrm{cm^{-3}}$.

Experience shows that discharge rarely preserves a diffuse form if the gas in it is heated appreciably, say, to twice the original temperature. *Contraction* transforms the column into a *filament* with sharply increased current density and gas temperature; this stage is intermediate to the transition of a glow discharge into an arc at still greater current. The scales given above characterize the upper bounds on the realization of the weakly ionized cold plasma of the diffuse glow discharge. The higher the pressure, the lower this upper bound in current and electron density, the more intensive the heating of the gas at a given current. Therefore, low pressure is favourable for sustaining nonequilibrium weakly ionized plasma, and high pressure (of the order of atmospheric) is favourable for equilibrium plasma; this is indeed supported by discharge experience.

In convective heat transport, the scales depend on pressure differently: ν_F, j_0, and n_e^0 are independent of pressure, and $w_0 \propto p$. For example, for $L_1 = 10\,\mathrm{cm}$, $u = 50\,\mathrm{m/s}$, $\nu_F \approx 10^3\,\mathrm{s^{-1}}$, we find $w_0 \approx 0.4p\,\mathrm{W/cm^3}$. Taking a more realistic value, $E/p \approx 10$, we have $j_0 \approx 40\,\mathrm{mA/cm^2}$ and $n_e^0 \approx 1.5 \cdot 10^{11}\,\mathrm{cm^{-3}}$.

8.7.7 Positive Column in Nitrogen

For a number of reasons, among which we find applications to gas lasers and plasma chemistry, nitrogen attracts the attention of researchers in electronic, vibrational, and ion-molecular processes in weakly ionized non-equilibrium discharge plasmas. Experiments and efforts toward their theortical interpretation [8.21–26] indicate both the diversity and the complexity of ionization mechanisms in the positive column plasma. These mechanisms not always satisfy the simplest scheme of electron impact ionization of atoms and molecules in the ground state, as the general description of discharge columns often assumes. Any attempt to interpret the observed values of E/N in the column immediately reveal the inadequacy of this mechanism.

Table 8.4 gives an idea of the state of discharge on the axis of the column in extra-pure nitrogen within a long tube of 1.6 cm radius [8.21]. The last column of the table gives the effective ionization rate constant, $k_{i,eff}$ found from the relation $q_{loss} = k_{i,eff} N n_e$ which is equivalent to (8.20). Even if we take into account high vibrational temperature $T_v \approx 5000\,K$, formulas (5.40, 41) give k_i several orders of magnitude lower. Stepwise ionization does not give anything significantly greater. In order to explain the observed ionization rates by electron impacts, we need $E/N \approx (7\text{–}8) \times 10^{-16}\,Vcm^2$ ($E/p_{20} \approx 25$ V/cm Torr). In fact, even lower values of E/N were observed at pressures of tens of Torr, down to $(1.5\text{–}2) \times 10^{-16}\,Vcm^2$ [8.22, 23].

According to current understanding, associative ionization of the type $N_2 + N_2 \rightarrow N_4^+ + e$ takes place in such weak fields, involving all molecules in the upper vibrational ($v \geq 32$) and/or upper electronic states. The latter are produced not ordinarily, through electron impact, but in collisions of two vibrationally excited molecules with $v \geq 16$. The population at the upper vibrational levels rapidly increases owing to the intense exchange of vibrational quanta in molecular collisions.

8.8 Electronegative Gas Plasma

8.8.1 Attachment-Controlled Discharge

In some cases the main mechanism of removing electrons is via attachment that is not accompanied by detachment, so that attachment acts in a straightforward manner. This situation takes place in short-pulse discharges, and at an early stage ($t < 10^{-5}\text{–}10^{-3}$ s) of longer discharges while a sufficient number of molecules active with respect to detachment has not yet built up. Pressure must not be too low ($p \gtrsim 10$ Torr), in order to avoid diffusional losses. For recombination to be dominated by attachment, the current and density of electrons must not be too high, $n_e < 10^{12}\text{–}10^{13}\,cm^{-3}$.

Most often, attachment proceeds via the dissociative mechanism, with energy consumption (Sect. 4.4). The reaction $e + CO_2 \rightarrow CO + O^-$, which is the main one in laser mixtures of $CO_2 + N_2 + Ne$, requires 3.85 eV.[13] For this reason, the frequency ν_a and the coefficient of attachment $a = \nu_a/v_d$ increase quite considerably with E/p, but not as sharply as the ionization frequency and ionization coefficient ν_i and α : the ionization potential is several times greater (Figs. 8.18, 19) [8.27–29]. The molecules ionized in the laser mixture are CO_2, having the lowest

---

[13] O^- ions then join CO_2 molecules and form stable CO_3^- complexes.

Table 8.4. The state of nitrogen plasma in the positive column in a tube at $p = 3.9$ Torr [8.21] [a]

i [mA]	j [mA/cm^2]	E [V/cm]	T [K]	N [10^{16}cm^{-3}]	E/N [10^{-16}V/cm^2]	E/p_{20} [V/(cm Torr)]	n_e [10^{10}cm^{-3}]	N_A [10^{12}cm^{-3}]	N_{35} [10^{13}cm^{-3}]	ε_v [eV]	$k_{i,eff}$ [10^{14}cm^{-3}]
10	1.3	41	420	9.0	4.5	15	0.36	1.2	1.6	0.23	2
75	9.5	23	650	5.8	4.0	13	2.9	1.2	3.6	0.60	5

[a] p_{20} corresponds to N and $t = 20°$ C ;
N_A is the density of metastables $A^3\sum_u^+$;
N_{35} is the calculated density at the vibrational level with $v = 35$;
ε_v is the mean vibrational energy of molecules

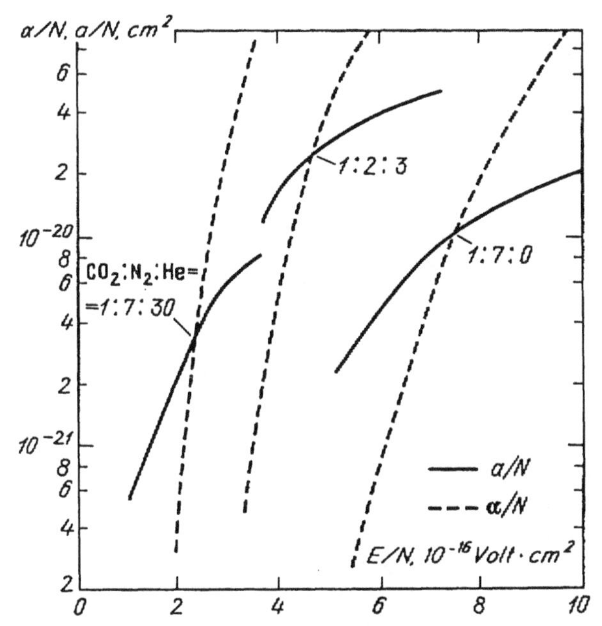

Fig. 8.18. Coefficients of ionization and attachment for several ratios of laser mixtures (CO_2 :N_2 :He); calculated on the basis of the kinetic equation [8.27, 28]

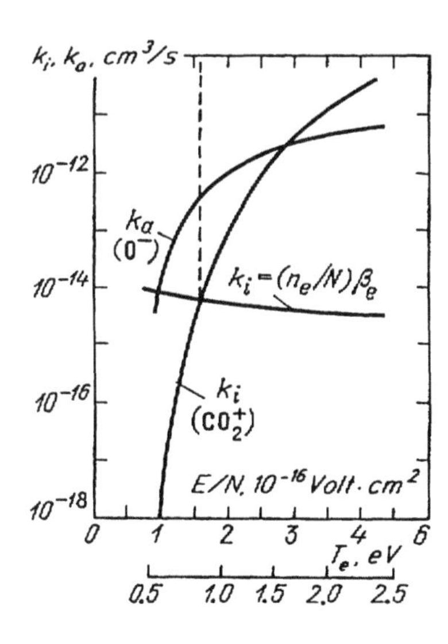

Fig. 8.19. Ionization and attachment rate constant and T_e calculated using the kinetic equation for the mixture CO_2 :N_2 :He= 1 : 7 : 12 [8.29]

ionization potential, $I_{CO_2} = 13.8$ eV. The steady state corresponds to the equality of production and removal rates for electrons:

$$\nu_i(E/p) = \nu_a(E/p) \, , \quad \alpha(E/p) = a(E/p) \, . \tag{8.27}$$

The point of intersection of the ionization and attachment curves determines the value of E/p necessary for sustaining the *attachment-controlled discharge*. Similar results for another important electronegative gas, namely air, were given in Sect. 7.2.5. The values of E/p and the corresponding ionization and attachment frequencies are summarized in Table 8.5. These data are of special interest: they indicate upper bounds, since detachment reduces losses and facilitates plasma sustainment. Thus for the 1 : 7 : 12 mixture the curves k_i and k_a in Fig. 8.19 intersect at $E/N = 2.8 \cdot 10^{-16}$ V·cm², $T_e = 1.6$ eV, which corresponds to an attachment-controlled discharge. If detachment is intensive, that is, if the discharge is recombination-controlled, then $k_i = (n_e/N)\beta_e$. If $n_e/N = 10^{-7}$, we obtain $E/N = 1.65 \cdot 10^{-16}$, $T_e = 0.9$ (for this point, $\beta_e = 6 \cdot 10^{-8}$ cm³/s).

For Table 8.5 also gives experimental data [8.27] that agree quite satisfactorily with calculations. Measurements were conducted in pulsed discharge between plane copper electrodes 29 cm² in area, separated by gaps from 1.2 to 4.2 cm, at pressures from 100 to 1200 Torr. The discharge voltage at such high pressure is high, up to 10 kV. The cathode fall is 200 V, so that almost the entire potential drop is across the positive column. The quasistationary voltage was built up at the electrodes in less than 0.1 μs from the moment of ignition, and was sustained for 10 μs. In a given mixture, the voltage and E/p were independent of current when the latter varied over several orders of magnitude: $j/p \sim 10^{-1}$–10^2 mA/(cm²Torr). (The $V - i$ curve was strictly horizontal.) Water vapour added at a concentration of a fraction of 1 % resulted in an increase of the voltage due to enhanced attachment.

8.8.2 Charge Kinetics Affected by Detachment

If the pressure is not too low ($p > 10$ Torr), which is typical of laser discharge, diffusional charge loss is of secondary importance. The bulk processes involving negative ions, which on the whole determine the density of electrons and the

Table 8.5. Field and frequencies of ionization in attachment-controlled discharges

$CO_2 : N_2 : He$ [a]	E/p, V/(cm·Torr) calculated	ν_i/p, 10^4 s^{-1}Torr^{-1}, calculated	E/p, V/(cm·Torr), experimental
1 : 7 : 30	7.9	0.052	9.1
1 : 1 : 8	8.6	0.35	9.5
1 : 7 : 12	9.2	8.2	
1 : 2 : 3	16.5	1.1	17
1 : 2 : 1			23.5
1 : 7 : 0	25		28
1 : 0 : 9			6
nitrogen [b]			22.3
nitrogen + 1 % H_2O			27
air	41	30	

Calculation results were taken from [8.28], with two exceptions: the 1 : 7 : 12 mixture [8.29], and air (Sect. 7.2.5). Experimental data were taken from [8.27].

[a] – the ratio of the numbers of molecules

[b] – technical-grade nitrogen (not too pure: contaminated with O_2, etc.)

conductivity of the mixture, are described by a system of kinetics equations for charge densities n_e, n_+, n_-:

$$dn_e/dt = k_i N n_e - k_a N n_e + k_d N n_- - \beta_e n_e n_+ ,$$
$$dn_-/dt = k_a N n_e - k_d N n_- - \beta_- n_- n_+ , \tag{8.28}$$
$$dn_+/dt = k_i N n_e - (\beta_e n_e + \beta_- n_-) n_+ .$$

Here $k_i N = \nu_i$, $k_a N = \nu_a$, k_d is the detachment rate constant for any molecule (we assume that the concentration of active molecules has stabilized), and β_e and β_- are the electron-ion and ion-ion recombination coefficients, respectively. Only two of the differential equations are independent; the third one is equivalent to the electroneutrality condition $n_e + n_- = n_+$. Nonstationary processes, say plasma decay, can also be analyzed using these equations.

8.8.3 Effective Recombination Coefficient

If detachment processes are fast, thus greatly compensating for attachment, system (8.28) can be approximately reduced to a single equation of kinetics for electron density [8.30]. Assume that negative ions decay much faster than they recombine: $k_d N \gg \beta_- n_+$. The result is an approximate *dynamic equilibrium* of attachment and detachment: $k_a n_e \approx k_d n_-$. The ratio of the numbers of electrons and negative ions stays constant, $n_-/n_e \approx k_a/k_d \approx \eta$, even though the numbers n_e and n_- may slowly change. Under such conditions, $n_+ \approx n_e(1+\eta)$, and the third equation of (8.28) gives

$$\frac{dn_e}{dt} = \frac{k_i N n_e}{1+\eta} - (\beta_e + \beta_- \eta) n_e^2 . \tag{8.29}$$

The first term on the right can be interpreted to indicate that less than one electron is produced in an ionization act. The rate of recombination of electrons appears to increase because an electron merging with a molecule disappears in the course of ion-ion recombination. If we consider plasma decay without the first production term, the quantity

$$\beta_{\text{eff}} = \beta_e + \beta_- \eta , \quad \eta = k_a/k_d \tag{8.30}$$

plays the role of the *effective recombination coefficient.* But if the steady state is considered, the equation for the balance of electron numbers can be written in the form

$$k_i N n_e = (1+\eta)(\beta_e + \beta_- \eta) n_e^2 = \beta'_{\text{eff}} n_e^2 . \tag{8.31}$$

This equation looks as if ionization proceeds in a normal way at the true frequency $\nu_i = k_i N$: negative ions are completely absent, but electrons recombine with a still higher effective coefficient

$$\beta'_{\text{eff}} = (1+\eta)(\beta_e + \beta_- \eta) , \quad \eta = k_a/k_d . \tag{8.32}$$

8.8.4 $V - i$ Characteristic, Charge Composition of Plasma, and Detachment Rate

We can only hypothesize on which particles in the discharge actively destroy negative ions (Sect. 4.4.3). At present the data on detachment rate is obtained indirectly, by comparing the experimentally obtained $V - i$ curve of a stationary discharge in an electronegative gas with calculations made under specific assumptions on k_d. An example of such analysis [8.31] which gave also the charge composition of the plasma is shown in Fig. 8.20. Triangles and circles illustrate experimentally measured current-voltage characteristics [8.32].

The discharge (longitudinal) was maintained in a plane channel of $5 \times 15\,\mathrm{cm}^2$ cross section, 46 cm long in the field direction (Fig. 8.18). In order to remove heat from the discharge volume, the gas was pumped in the same direction at a velocity 100 m/s. The flow does not greatly affect the discharge but prevents overheating of the gas. According to experimental $V - i$ curve, the value of E/p depends on current and pressure only slightly, being equal on the average to $E/p \approx 7\,\mathrm{V/(cm \cdot Torr)}$ ($E/N \approx 2.1 \cdot 10^{-16}\,\mathrm{V \cdot cm^2}$).

The same figure shows the results of $V - i$ curve calculations on the basis of stationary equalities (8.28). The rate constants k_i and k_a are taken from Fig. 8.19, assuming $\beta_e = \beta_- = 10^{-7}\,\mathrm{cm^3/s}$ and taking into account such details as the potential drop at the electrode and gas heating, and choosing a sequence of values of k_d from 0 to ∞. Obviously, experimental data are in contradiction with both extreme assumptions: no detachment ($k_d = 0$), and detachment completely compensating for attachment ($k_d = \infty$). The best fit is found for the value $k_d = 0.9 \cdot 10^{-14}\,\mathrm{cm^3/s}$. If the detachment cross section is assumed to be of the order of the collision cross section of molecules, the resulting number of active molecules is about 10^{-4} of the total number, which seems to be reasonable. The experimental value $E/p \approx 7$ corresponds to rate constants $k_i = 1.5 \cdot 10^{-14}\,\mathrm{cm^3/s}$, $k_a = 6 \cdot 10^{-14}\,\mathrm{cm^3/s}$. If $p = 20\,\mathrm{Torr}$, $N = 6.7 \cdot 10^{17}\,\mathrm{cm^{-3}}$, the frequencies are

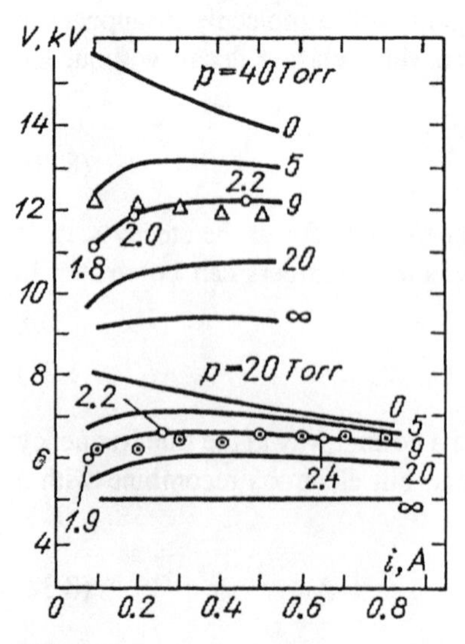

Fig. 8.20. Experimental [8.32] and calculated [8.31] $V - i$ curves of discharges in a 1 : 7 : 12 mixture. *Numbers* to the *right* of the *curves* mark the values of k_d in units of $10^{-15}\,\mathrm{cm^3/s}$ ($k_d = 0$ denotes pure attachment, $k_d = \infty$ indicates no attachment or no negative ions). Experimental data (*dots* and *triangles*) are best fitted by the curves calculated with $k_d = 9 \times 10^{-15}\,\mathrm{cm^3/s}$. *Numbers* along these *curves* are the values of E/N in units of $10^{-16}\,\mathrm{V\,cm^2}$

$\nu_i = 10^4\,s^{-1}$, $\nu_a = 4 \cdot 10^4\,s^{-1}$. The plasma contains a considerable number of negative ions: $\eta = n_-/n_e = k_a/k_d \approx 5.3$. The effective coefficient of electron recombination $\beta'_{eff} \approx 40\beta_e \approx 4 \cdot 10^{-6}\,cm^3/s$ is very high, 40 times the actual value (owing to attachment). Detachment compensates for about 80 % of attachment.

The $V - i$ characteristic of a stationary discharge in an air-filled tube is a clear illustration of the effect of detachment (Fig. 8.14); this was discussed in Sect. 8.6.3.

8.9 Discharge in Fast Gas Flow

Gas is pumped through a discharge volume in order to remove the Joule heat and prevent overheating of the gas itself and of the walls of the discharge chamber. Experiments reveal different degrees of the response of stationary discharges to gas flow but invariably confirm that as the flow velocity increases, the discharge voltage can only increase to a greater or lesser extent. We conclude that charge losses must increase and be compensated for by enhanced ionization. Several mechanisms for the effect of flow can be proposed.

8.9.1 Turbulence Transport of Charges to the Walls

The gas in a sufficiently fast flow becomes turbulent. *Small-scale turbulence* is often deliberately introduced, especially in high-power lasers, because this improves discharge stability and increases the upper limit of energy input. The effect of turbulent stirring of plasma volumes is similar to ambipolar diffusion. It transports more strongly ionized volumes from central regions to the walls while carrying less ionized volumes from the periphery toward the axis. This mechanism may prove to be efficient in a discharge that is diffusion controlled when flow is absent (that is, discharge at reduced pressure, in narrow tubes, and in narrow channels). The effect can be described by adding the turbulence diffusion coefficient to the ambipolar diffusion coefficient D_a.

Gas dynamics offers several empirical formulas for the former coefficient (the effect of weak ionization on turbulence is unlikely to be appreciable). Thus, if u [cm/s] is the mean flow velocity and a [cm] is the tube radius or half the height of a plane channel, then

$$D_T \approx 0.009 au\,cm^2/s\,. \tag{8.33}$$

For instance, in the experiment described in Sect. 8.8.4, $a = 1.5\,cm$, $u = 10^4\,cm/s$, $D_T = 225\,cm^2/s$. This coefficient is several times greater than $D_a \approx 75\,cm^2/s$ at $p = 40\,Torr$; nevertheless, the frequency of total "diffusional" losses, $\nu_{dT} = (D_a + D_T)/(2a/\pi)^2 \approx 120\,s^{-1}$, is less than that of bulk losses, $10^4\,s^{-1}$, so that turbulence diffusion plays a minor role under these specific conditions.

8.9.2 Convective Charge Transport

The loss of charge from the current channel due to the entrainment by the gas flow may be significant if the discharge length L_1 along the gas flow is small. This situation occurs in transverse discharges when the length of at least one electrode is much shorter than the interelectrode gap. The "convective transport frequency" is $\nu_F \approx 2u/L$. Presumably, this mechanism was dominant in the experiments [8.33] with air whose results plotted in Fig. 8.21. The cathode was a narrow strip $0.4 \times 40\,\text{mm}^2$, transverse to the flow, the anode was a large plate, and the gap width was 3 cm. If the positive column is characterized by $L \sim 1$ cm, then $u \sim 100$ m/s entails $\nu_F \sim 10^4$, which is comparable to the bulk loss frequency. The values of E/p, evaluated using the data of Fig. 8.21, increase from 15 to 25 in the range of u from 50 to 180 m/s. If a transverse discharge is extended along the flow, with geometry close to that in Fig. 8.1a, the discharge voltage is almost independent of the flow velocity. Indeed, the arrival of charges from the left in some middle section of the channel and their removal to the right do not disturb the overall charge balance.

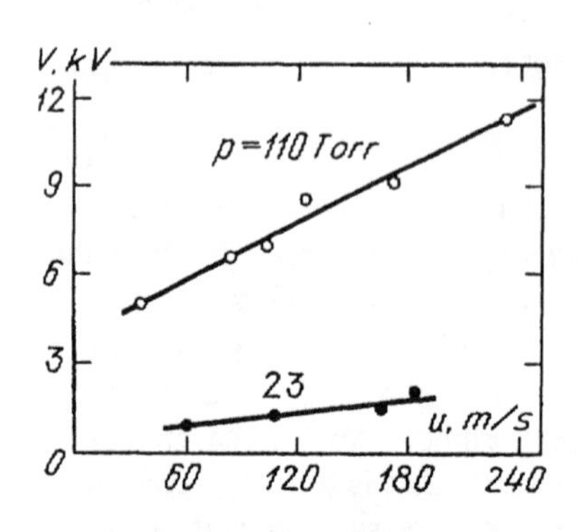

Fig. 8.21. "Burning" voltage of transverse discharge in air as a function of flow velocity. *(1)* p = 23 Torr, *(2)* p = 110 Torr [8.33]

8.9.3 The Effect of Flow on the Accumulation of Molecules Active with Respect to Detachment

This is similar to the time effect discussed in Sect. 8.8.1. If a macroscopic gas element moving through the discharge zone spends little time there, it cannot accumulate the number of active molecules that corresponds to the established conditions. Hence, the faster the motion of the gas through the discharge zone, the weaker the compensation by detachment of the electron attachment, which is equivalent to enhanced attachment losses. If the characteristic time of buildup of a stationary concentration of active molecules is of order 10^{-5}–10^{-3} s, the flow velocity must affect the discharge in electronegative gases if the time of flight of a gas particle through the discharge zone, L/u, is also of this order of magnitude. Presumably, this effect could accompany the convective transport in the experiment of Fig. 8.21.

No appreciable effect of flow velocity on discharge sustainment voltage has been observed in long, longitudinal self-sustaining discharge systems (Fig. 8.22).

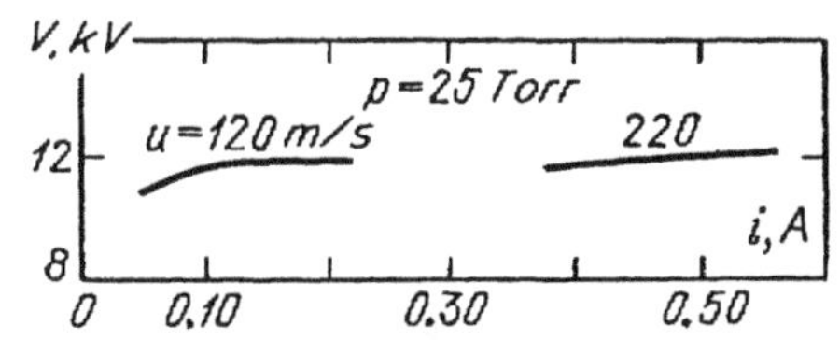

Fig. 8.22. $V - i$ characteristic of longitudinal discharge with gas flow. Laser mixture $CO_2:N_2:He = 1:2.5:15$, $p = 25$ Torr; *(1)* $u = 120$ m/s *(2)* 220 m/s [8.34]

In these experiments [8.34], the discharge was burning in a plane channel (Fig. 8.1b) 5.5 cm in height, 76 cm wide and with its length oriented along the flow (i.e. with an electrode gap spacing) of $L = 65$ cm. Even for $u = 220$ m/s. the time of flight is very large: $\tau_F = L/u = 3 \cdot 10^{-3}$ s. The concentration of active molecules has evidently stabilized. Under these conditions, the diffusion and turbulence loss frequency is not high: $D_{eff} = D_a + D_T \approx 650\,\text{cm}^2/\text{s}$, $\nu_{dT} \approx 220\,\text{s}^{-1}$. If the discharge length is smaller, the voltage is observed to increase as the flow velocity increases (in a self-sustained discharge). The dependence of V on u is always stronger in non-self-sustaining discharges (Sect. 14.4.5, Fig. 14.9). It should be mentioned that the effects of flow velocity on the characteristics of longitudinal discharges are not yet known sufficiently well. Discharges in transverse flow are analyzed in the review [8.35].

8.10 Anode Layer

8.10.1 Production of Ions

There are no positive ions at the metal anode surface because they are not emitted by the anode and are repelled by it. The anode is separated from the electroneutral plasma of the positive column by a layer of negative space charge in which the magnitude of the field decreases towards the column.[14] The electron current density j_e changes in the layer by a negligible amount $(\mu_+/\mu_e)j$, while the ionic current density in the plasma increases from zero to $j_{+c} = (\mu_+/\mu_e)j$. The ionic current j_{+c} flowing into the positive column is formed as a result of charge production in the anode layer, in response to a very small number of ionization acts of one electron, μ_+/μ_e. Indeed,

$$\frac{dj_+}{dx} = \alpha j_e \approx \alpha j\ ; \quad j_{+c} \approx j \int \alpha\, dx\ ; \quad \int \alpha\, dx = \mu_+/\mu_e\ .$$

This number is three orders of magnitude smaller than the number of generations of electrons produced in the cathode layer. Consequently, the anode fall V_A is much lower than the cathode fall.

[14] The sign of anode drop may be reversed at very low pressure.

8.10.2 Potential Drop and Current Density

According to early measurements in tubes [8.3, 36] at $p \lesssim 1$–10 Torr, $V_A \approx 10$–20 V and the anode fall is in each gas comparable with the ionization potential. The anode current density is of order 10^{-4}–10^{-3} A/cm^2; fields $E/p \approx 200$–600 V/(cm·Torr) were measured in nitrogen near the anode, which corresponds to the layer thickness $d_A \sim 0.05/p$ cm (of the order of the electron free path length l). Recent studies demonstrated that at medium pressures, the anode fall increases as the pressure is increased, the increase being greater in electronegative gases (Fig. 8.23). According to [8.37], the current density at the anode is independent of the value of current: $j/p^2 \approx 4.3 \cdot 10^{-4}$ A/(cm·Torr)^2 in nitrogen at $p \approx 5$–30 Torr, and $j/p^2 \approx 2.7 \cdot 10^{-4}$ in air at $p \approx 8$–60 Torr. (The figures refer to discharges between plane steel electrode disks 1.6 cm in diameter.)

The measured values of j/p^2 are quite close to the normal values at the cathode (Table 8.3). It appears that if the current column diameter is not less than the interelectrode distance, the normal cathode layer imposes the value of j on the entire column, including the anode. As follows from the calculations [8.8], even the radial distribution $j(r)$ originating at the cathode is repeated in such cases at the anode (Fig. 8.6). The anode fall calculated for nitrogen (Fig. 8.6) gives a good fit to the measured values (Fig. 8.23). The anode spot is small at low current. Numerical experiment [8.8] indicates that the spot radius r_A is determined by the spreading due to the transverse (radial) diffusion of electrons, and is related to the anode layer thickness ($pd_A \approx 0.25$ Torr·cm) by a formula typical of diffusion: $r_A \approx (D_e d_A/v_d)^{1/2} \approx d_A(T_e/V_A)^{1/2}$.[15]

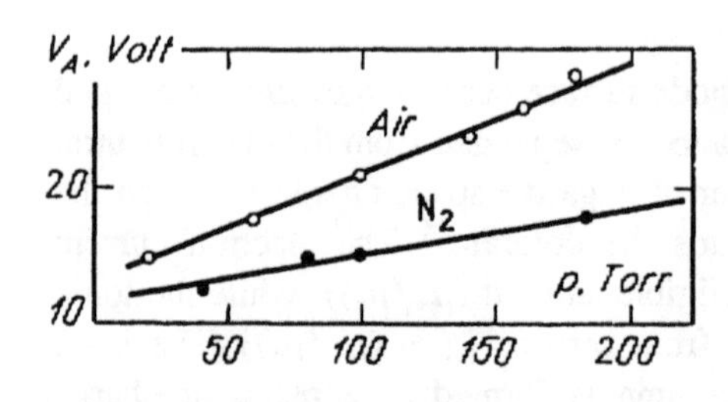

Fig. 8.23. Anode fall for discharges in nitrogen and in air as a function of pressure [8.37]

8.10.3 Do the Ions Produced in the Anode Layer Pierce the Positive Column?

Quite frequently, they do not. The probability that an ion perishes is μ_e/μ_+ times greater than that for an electron, since an ion drifts through the column longer by just this factor. In the example given in Sect. 8.6.2 [see (8.19)], the probability is 0.2 per 1 cm of column length, that is, a 5 cm long column does not let ions through. Here we ignore all possible processes of resonance charge transfer in which one ion is instantaneously replaced with another, indistinguishable from

[15] The results of [8.8] do not support the theoretical model suggested in [8.37].

it – the ion cannot really be said then to have perished in the usual sense of the word.

If a longitudinal discharge is maintained in a flow of gas, the flow is usually directed from the cathode to the anode, in order to immediately blow out of the discharge the perturbations generated in the anode region. The velocities that are employed in actual systems, $u \approx 100$–$200\,\mathrm{m/s}$, are comparable with the ion drift velocities v_{+d}. For example, in a self-sustained discharge in the mixture $CO_2 : N_2 : He = 1 : 7 : 12$ at $E/p \approx 7\,\mathrm{V/(cm \cdot Torr)}$, the velocity of the main ions CO_2^+ is $v_{+d} \approx 200\,\mathrm{m/s}$. Fast flow decelerates ions, and blows them out completely if $u > v_{+d}$. In contrast to their natural behavior, ions are entrained then against the field, towards the anode. This effect cannot, however, stop the current in the discharge, because it is carried mainly by electrons [8.9].

9. Glow Discharge Instabilities and Their Consequences

9.1 Causes and Consequences of Instabilities

The homogeneous state of the glow discharge positive column is quite often unstable, especially if the discharge is maintained in large volume, or at high pressure, when current is high and Joule heat is intensively released. Random perturbations grow catastrophically and the plasma switches to a different, spatially inhomogeneous state. The inhomogeneous modes caused by *instabilities* have been known for a long time: *striations*, that is, when the positive column is formed of bright and dark layers alternating along the current direction, and *contraction* of the plasma into a bright *current filament*. In fact, these effects have recently attracted special attention owing to the difficulties that they cause for the development of high-power gas lasers: instabilities destructively affect the laser generation. The suppression of the tendency to the *filamentation* of discharge constitutes the central, and the most difficult, problem in the development of high-power electrical discharge lasers (Chap. 14). High stability can be efficiently achieved only if the nature of discharge processes and the mechanisms of instabilities are well understood. Recent efforts in this field have essentially expanded our knowledge concerning such a classical object as glow discharge.[1]

9.1.1 Phenomenological Attribute of Stability (or Instability)

Quite often, plasma instability is apparent under visual observation. Differences in luminosity are caused, first of all, by unequal electron density. Consequently, the factors that cause inhomogeneity are related to the processes that control the density of electrons, their production, removal, and spatial transfer. The symbolic equation of kinetics of electrons is

$$\frac{dn_e}{dt} = Z_+ - Z_- \,. \tag{9.1}$$

In the steady state, the production and removal rates are equal: $Z_+ = Z_-$. The rates depend on n_e and other parameters: electron temperature, field, density of negative ions and excited atoms, and so on; the dependence may be given by

[1] Inevitably, we can dwell only briefly on the most general, key aspects of this extensive topic. Further discussion may be found in the reviews [9.1–5], where a detailed bibliography can be found.

differential equations, such as (8.28). Therefore, only a qualitative picture of the functional relationship between Z_+, Z_- and n_e can be outlined in the general case. The point of intersection of symbolic functions $Z_+(n_e)$ and $Z_-(n_e)$ (see Fig. 9.1) corresponds to a steady-state electron density $n_e^{(0)}$ which is ultimately deermined by external conditions: e.m.f. of power supply unit, geometry, and the resulting magnitude of the discharge current.

The stability of the steady state can be deduced on the basis of the mutual arrangement of curves in its neighbourhood. The state is stable if the removal curve Z_- passes above the production curve when $n_e > n_e^{(0)}$ and below it when $n_e < n_e^{(0)}$ (see Fig. 9.1a) because the system returns to equilibrium in response to an accidental deviation. Otherwise the state is unstable (shown in Fig. 9.1b): if n_e fluctuates upward, production becomes greater than removal and the number of electrons grows still further. The important factor is the mutual arrangement of the Z_+ and Z_- curves, not their absolute positions.

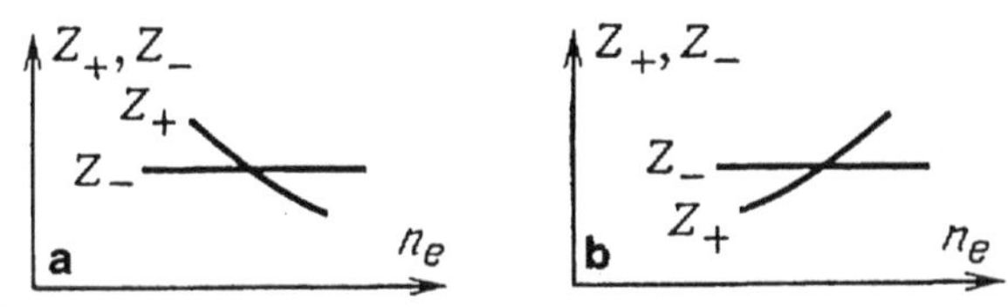

Fig. 9.1. Qualitative of electron production and removal rates in the neighborhood of (a) stable and (b) unstable states

9.1.2 Stabilizing and Destabilizing Factors

The above arguments allow the construction of a qualitative classification of the effects of various factors on stability. Let free electrons be created in the ionization of atoms from the ground state by electron impact, the electron temperature being insensitive to n_e; in this case, $Z_+ = \nu_i(T_e)n_e \sim n_e$. The bulk recombination $Z_- = \beta n_e^2$ helps stabilize the discharge, since it imparts to the removal rate a stronger, quadratic dependence on n_e. The discharge is stabilized by the external resistance: if n_e and the current increase, the voltage, electron temperature, and hence, ionization frequency ν_i all decrease. Owing to this chain of relations, the function $Z_+(n_e)$ increases with n_e less steeply than in proportion to n_e. For this reason, a homogeneous discharge controlled by ambipolar diffusion (so that $Z_- = \nu_d n_e \sim n_e$) is stable unless destabilized by factors to be discussed below. The same is true of discharges controlled by recombination in the bulk. Diffusion and heat conduction help level off the inhomogeneities in particle densities and temperature and are thus among the stabilizing factors.

Gas heating is a destabilizing factor. As the pressure levels off quickly in a gas, a local increase in gas temperature is accompanied with a drop in density (thermal expansion). This effect does not directly influence field strength, but the ratio E/N and electron temperature that depends on E/N both increase. The result is enhanced ionization and locally increased conductivity, current density, and Joule heat release. The gas is therefore heated even more. This is the so-called *thermal instability*; it is the most dangerous and wide-spread instability in gas lasers.

Stepwise ionization and accumulation of metastable atoms and molecules also destabilize the discharge. As n_e increases, more excited particles are created and ionization from the ground state is supplemented with ionization of excited particles which is an easier process (owing to a smaller electron binding energy). Therefore, the ionization rate Z_+ grows with n_e steeper than $\nu_i(T_e)n_e$. Later we will introduce some other mechanisms of discharge destabilization. In fact, a state is stable or unstable depending on which of the stabilizing and destabilizing factors comes out the winner.

9.1.3 Longitudinal and Transverse Instabilities and Effects They Produce

The growth of instability (i.e., a catastrophic increase of an initially small perturbation) is a nonsteady process. It must lead to something. Two possibilities can be suggested – and are indeed realized. (1) The system comes to a new, more stable state. (2) Steady state is never achieved; however, a nonsteady process cannot evolve eternally in one direction, so that a periodically changing state is established.[2]

Chains of causual links between various processes involved in the evolution of perturbations and their final results depend on the orientation of inhomogeneities with respect to a chosen direction in the discharge space, namely, the electric field and current vector. If the discharge parameters, say, electron density, vary along the field (Fig. 9.2), case (2) is realized: striations are formed, with regions of increased and reduced n_e and other discharge parameters alternating in space and time. Sometimes *domains* are formed, that is, individual regions with modified parameters (e.g., strong field domains) that are periodically generated and propagate along the current direction. The growth of perturbations perpendicular to the field (Fig. 9.2) results in discharge contraction and in formation of filaments with sharply increased current density. The discharge current mostly flows in these filaments, and increases. In tubes, symmetry results in a single axial filament; in plane channels, there may be several filaments (which often "dance").

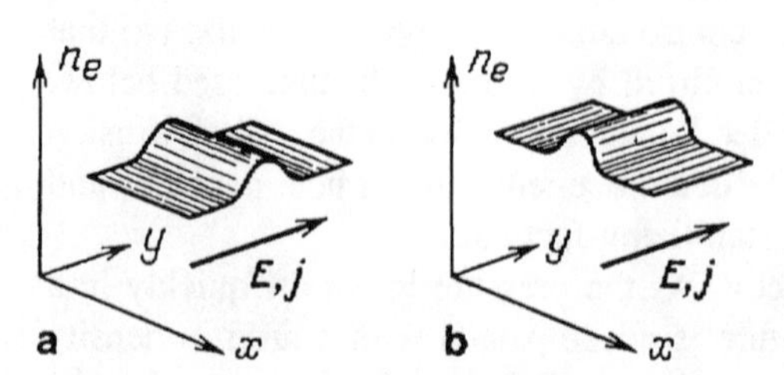

Fig. 9.2. Longitudinal (a) and transverse (b) perturbations of electron density

[2] In principle, irregular (stochastic) periodicity is also possible.

9.1.4 Principles of Stability Analysis

As a rule, it is extremely difficult to monitor the behavior of a discharge from a small perturbation to the final result of its evolution. Even the formulation of an adequate physical system of equations is often a tough proposition, to say nothing of the mathematical difficulties of solution. Typically, one is satisfied with a much less ambitious task: to find a region of the values of parameters of the steady state that correspond to stable and unstable states, and to calculate the initial rate of growth of small perturbations in the *linear approximation*. Sometimes it is possible to analyze independently the final, steady or periodic, inhomogeneous state. In fact, the solution of the former problem supplies considerable information, because, in general, the initial rate of evolution gives an idea of the time scale of transition to the final state via a nonlinear stage.

Let a homogeneous steady state of a discharge system correspond with the parameters $n_e^{(0)}$, $T_e^{(0)}$, etc. For stability analysis, the equations describing the behaviour of n_e, T_e, ... are linearized by substituting expressions of type $n_e = n_e^{(0)} + \delta n_e, \dots$ for parameters and assuming the perturbations δn_e to be small. The solution of the linear system is sought in the form of a plane wave, $\delta n_e = (\delta n_e)_a \exp[i(\omega t - \boldsymbol{K} \cdot \boldsymbol{r})]$, ... , where the wave vector $\boldsymbol{K}$ is assumed to be oriented along or normally to the field (oblique waves can be decomposed as vectors). Physically, this allows the following interpretation. Let us be interested in the response of the discharge to a perturbation of spatial size Λ and a given orientation (Fig. 9.2). By virtue of the linearity of the equations, a perturbation can always be written as a Fourier integral in plane waves, and each component can be analyzed independently. Clearly, the wave vectors $\boldsymbol{K}$ are then directed along the coordinates along which change n_e, T_e, ... ; the wavelengths best represented in the integral are of order Λ, or $K \sim \Lambda^{-1}$.

The substitution of wave solutions into linearized equations yields a system of m homogeneous algebraic equations for m amplitudes $(\delta n_e)_a$, $(\delta T_e)_a$, ... Setting the determinant of the system to zero yields the dispersion equation that relates ω to $\boldsymbol{K}$. If at least one of its m complex roots is such that $\Omega = \mathrm{Re}\, i\omega > 0$, perturbations grow exponentially, as $\exp(\Omega t)$. If there are several such roots, the instability growth rate is the greatest among Ω. The characteristic time of instability development is $\tau = \Omega_{\max}^{-1}$. Usually, long-wave perturbations grow at the fastest rate: they are less prone to levelling off by diffusion and heat conduction. The upper bound on wavelengths is imposed by the smallest dimensions of the discharge volume (tube diameter, channel height).

9.2 Quasisteady Parameters

9.2.1 Fast and Slow Processes

The state of the plasma, especially in an electronegative molecular gas, is described by a considerable number of parameters: n_e, n_-, n_+, T_e, T, vibrational temperature T_V (which may differ from the translational gas temperature T as a result of delayed vibrational relaxation), N, E, and the density N^* of excited molecules. The parameters obey nonstationary equations which are in general interrelated [one example is a reduced system (8.28)]. An analysis of the required determinant of order m for $m \sim 10$ would be unthinkable. The problem is alleviated by the fact that some plasma processes are fast and some are slow. When considering instability due to a specific process, developing over a time τ, we can approximately treat faster processes as relaxing instantaneously, that is, as quasisteady, and treat slower processes as frozen, that is, as processes not developing when the specified perturbations grow.

9.2.2 Time Scales of Various Processes

These are illustrted in Table 9.1. The spreading of space charge $\varrho = e(n_+ - n_e - n_-)$ in a medium of constant conductivity σ is defined by the continuity, electrostatics, and Ohm's equations:

$$\frac{\partial \varrho}{\partial t} + \operatorname{div} \boldsymbol{j} = \frac{\partial \varrho}{\partial t} + \sigma \operatorname{div} \boldsymbol{E} = \frac{\partial \varrho}{\partial t} + 4\pi\sigma\varrho = 0 \,,$$
$$\varrho(t) = \varrho(0) \exp(-t/\tau_\sigma) \,, \quad \tau_\sigma = 1/4\pi\sigma \,. \tag{9.2}$$

The time τ_σ of space-charge decay is called the *plasma relaxation* or *Maxwell time*. The pumping time of molecular vibrations is similar to the gas heating time, provided the vibrational heat capacity c_V and the density of molecules N_M are introduced. The vibrational relaxation is characterized by the time of energy transfer between vibrational and translational degrees of freedom, τ_{VT}. Pressure levels off in space at the velocity of sound c_s. The electron thermal conductivity in weakly ionized gas is approximately equal to $\lambda_e \approx (5/2)kn_eD_e$; the corresponding thermometric conductivity is $\chi_e \approx D_e$.

9.2.3 Quasisteady Behavior of Current

The main instabilities of discharge plasma are connected with the kinetics of production and removal of electrons, changes in gas temperature, and excitation of electronic and vibrational levels of atoms and molecules. Table 9.1 shows that these processes are much slower than the space-charge decay. The latter process can therefore be regarded as fast. Assume that n_e and σ are somehow redistributed in space in the course of the evolution of the instability with a characteristic time $\tau \gg \tau_\sigma$. Generalizing (9.2) to the case of variable σ, we have

$$\partial\varrho/\partial t + 4\pi\sigma\varrho + \boldsymbol{E} \cdot \nabla\sigma = 0 \,. \tag{9.3}$$

The space charge ϱ that is absent in the homogeneous stationary state is itself a perturbation. As all other perturbations, it changes at the rate of violation of homogeneity in the plasma. Hence, the first term in (9.3) is of order ϱ/τ, much less than the second term, ϱ/τ_σ. The last two terms, both of large magnitude, add up to $\operatorname{div}\boldsymbol{j}$; they compensate each other up to a relatively small quantity ϱ/τ.

Table 9.1. Time scales of various discharge processes under conditions typical of discharge in cw CO_2 lasers ($CO_2 + N_2 + He$); $p \sim 10$–100 Torr, $N \sim 10^{18}\,10^{-3}$, $T \sim 300$–500 K, $E/p \sim 10$ V/(cm·Torr), $T_e \approx 1$ eV, $T_V \approx 2000$–5000 K, $n_e \sim 10^{10}\,\mathrm{cm}^{-3}$, $\Lambda \sim 1$ cm [a)]

Process	Characteristic time	Duration, s
1. Space-charge relaxation	$\tau_\sigma = 1/4\pi\sigma$	10^{-10}–10^{-9}
2. Collisional energy transfer		
(1) electron tempeature relaxation	$\tau_u = 1/\nu_m\delta_1$	10^{-9}–10^{-8}
(2) gas heating	$\tau_T = Nc_{p1}T/\sigma E^2$	10^{-3}–10^{-2}
(3) pumping of molecular vibrations	$\tau_V = N_M c_V T_V/\sigma E^2$	10^{-3}–10^{-2}
(4) vibrational relaxation	τ_{VT}	10^{-4}–10^{-2}
3. Collision kinetics		
(1) ionization	$\tau_i = (k_i N)^{-1}$	10^{-5}–10^{-4}
(2) attachment	$\tau_a = (k_a N)^{-1}$	10^{-6}–10^{-5}
(3) detachment	$\tau_d = (k_d N)^{-1}$	10^{-6}–10^{-5}
(4) electron excitation	$\tau^* = (k^* N)^{-1}$	10^{-6}–10^{-4}
(5) electron-ion recombination	$\tau_{rec}^e = (\beta_e n_+)^{-1}$	10^{-4}–10^{-3}
(6) ion-ion recombination	$\tau_{rec}^- = (\beta_- n_+)^{-1}$	10^{-4}–10^{-3}
4. Transport processes		
(1) pressure levelling (sound)	$\tau_s = \Lambda/c_s$	10^{-5}–10^{-4}
(2) heat conduction of the gas	$\tau_\chi = \Lambda^2/\chi$	10^{-2}
(3) ambipolar diffusion	$\tau_{da} = \Lambda^2/D_a$	10^{-2}
(4) electron heat conduction	$\tau_\chi^e = \Lambda^2/\chi_e$	10^{-5}

[a)] Taken from [9.1]. More realistic values are given in some places

Hence, the derivative $\partial\varrho/\partial t$ in (9.3) can be neglected, and we can set $\operatorname{div} \boldsymbol{j} = 0$ at each moment of time, just as in truly steady-statse conditions.

In one-dimensional cases, current density is always spatially homogeneous and depends only on time. The electric circuit is not closed in the direction perpendicular to $\boldsymbol{E}$; hence, the transverse current cannot flow even if the conductivity in this direction undergoes a change (as in the case of filamentation). The transverse current is zero because the drift and diffusion components cancel out (Sect. 2.6). The latter component is not included in (9.2, 3).

In the case of longitudinal inhomogeneities (striations), $\boldsymbol{j}$ remains constant along the field. If we impose on longitudinal perturbations the plane wave form, then $\boldsymbol{K}\delta\boldsymbol{j} = 0$. If $\boldsymbol{K} \parallel \boldsymbol{E}$, $K \neq 0$, then $\delta\boldsymbol{j} \equiv 0$, that is, the current is constant both in space and in time.

9.2.4 Quasisteady Behavior of Electron Temperature

The attainment of mean electron energy due to collisions is one of the fastest processes, especially in molecular gases. Table 9.1 shows that only the space-charge relaxation is a faster process. As a result, the electron temperature remains quasi-steady during the development of most instabilities. It tracks the slower variations of other quantities. In the simplified energy balance equation for the electron, (2.12) [taking into account (2.17)], we have $d\varepsilon/dt \sim T_e/\tau$. If $\tau \gg \tau_u$, then large terms (of order T_e/τ_u) in the right-hand side cancel out with high accuracy. Therefore, T_e is related to E/N by an expression implied by (2.12), provided the electron energy is quasisteady. Formula (2.15) is again derived, although here we do not assume ν_m and δ to be independent of T_e:

$$E/N = \left(3mkT_e\tilde{\nu}_m\tilde{\nu}_u/2e^2\right)^{1/2} , \quad \tilde{\nu}_{m,u}(T_e) = \nu_{m,u}/N . \tag{9.4}$$

To avoid confusion, in this chapter we denote the fraction of energy lost by an electron by δ_1.

9.2.5 Violation of the Condition of Quasisteady T_e

This occurrence is possible if the field varies appreciably over one electron temperature relaxation length Λ_u, (2.18). In order to find when this may happen, we compare the orders of magnitude of various terms in the general equation for the electron energy (2.52), where the spatial inhomogeneity was taken into account, in contrast to (2.12). Let us compare the convective term in dT_e/dt that contains $\boldsymbol{v}_d\nabla T$ (the electron pressure term has the same order of magnitude), the heat conduction transfer rate in an electron gas, and the rate of energy transfer to molecules. As $kT_e \approx eE\Lambda_u$, the ratios of these terms are $1 : \Lambda_u/\Lambda : \Lambda/\Lambda_u$. Very small-scale inhomogeneities, with $\Lambda \ll \Lambda_u$, level off via heat conduction quite rapidly. If $\Lambda \gg \Lambda_u$, local energy transfer to molecules is dominant, and T_e is quasistationary. This occurs in molecular laser discharges (at $p = 30$ Torr, the electron path length is $l \sim 10^{-3}$ cm, $\delta_1 \sim 10^{-2}$, $\Lambda_u \sim 10^{-2}$ cm, and $\Lambda \sim 1$ cm). If, however, $\Lambda \sim \Lambda_u$ and all three terms are of the same order of magnitude, no local quasisteady relation exists between T_e and E. The electron temperature varies along the direction of E with a delay, owing to the relatively slow relaxation. This situation is typical for striations in inert gases when $l \sim 10^{-2}$ cm, $\delta_1 \sim 10^{-4}$–10^{-5}, $\Lambda_u \sim 1$ cm, and Λ is on the order of the tube diameter (~ 1 cm).

9.3 Field and Electron Temperature Perturbations in the Case of Quasisteady-State T_e

9.3.1 Existence of the Potential of the Field

A nonsteady process may generate variable magnetic fields connected with changes in the electric field and possible changes in conduction current. It is not difficult to estimate, using Maxwell's equations (3.13, 14) and the data of Table 9.1, that the induced *vortex* electric field constitutes a negligible fraction of perturbations δE. The perturbed electric field has a *potential* and satisfies to a very good approximation the stationary equation curl $\boldsymbol{E} = 0$. As follows from this equation, only the component $\boldsymbol{E}$ along the direction of change of all parameters can vary in space in one-dimensional cases. This is not surprising.

Field perturbations are caused by space charges and the polarization field is oriented along the direction of displacement of charges. For plane waves, we have $[\boldsymbol{K} \times \delta \boldsymbol{E}] = 0$, whence $\delta \boldsymbol{E} \parallel \boldsymbol{K}$ for $K \neq 0$.

9.3.2 Transverse Inhomogeneities

If $\boldsymbol{K} \perp \boldsymbol{E}$, then $\delta \boldsymbol{E} \perp \boldsymbol{E}$. A perturbation of the field magnitude,

$$\delta|E| = \sqrt{\boldsymbol{E}^2 + (\delta \boldsymbol{E})^2} - E \approx (\delta \boldsymbol{E})^2/2E ,$$

is of second order of smallness, and can be safely ignored. All parameters are invariable along the drift direction and nothing impedes the establishment of T_e. The electron temperature is quasistationary. The field magnitude being practically constant, T_e may change only as a result of changes in gas density N. This also takes place when the gas is heated. Both electron temperature and ionization rate increase in response to thermal expansion.

According to (8.4), a perturbation in T_e is related to the perturbation in density by the formula

$$\delta T_e/T_e = -\xi \delta N/N \ , \quad \xi = 2/(1 + \hat{\nu}_m + \hat{\nu}_u) \ , \quad \hat{\nu} \equiv d \ln \nu / d \ln T_e \ . \qquad (9.5)$$

The numerical coefficient ξ is always positive; its value is of order unity. If $l(T_e) = \text{const}$ and $\delta_1(T_e) = \text{const}$, then $\nu_{m,u} \propto \sqrt{T_e}$, $\xi = 1$.

9.3.3 Relation Between Longitudinal Perturbations of T_e, E, and n_e

In the case of longitudinal inhomogeneities, the field is perturbed so that $\delta \boldsymbol{E} \parallel \boldsymbol{E}$, but the current remains unaffected. Therefore, $\delta E/E = -\delta\sigma/\sigma$.[3] As a rule, longitudinal instabilities develop via nonthermal mechanisms which act much faster than any processes changing temperature or density in the gas. Hence, N is a frozen parameter in this case. As follows from (9.4),

$$\delta T_e/T_e = \xi \delta E/E \qquad (9.6)$$

and the temperature T_e is higher where the field is stronger. By varying $\sigma = e^2 n_e / m\nu_m$ at $N = \text{const}$,

$$\delta\sigma/\sigma = \delta n_e/n_e - \hat{\nu}_m \delta T_e/T_e = -\delta E/E \ ,$$

and eliminating $\delta E/E$ via (9.6), we obtain a relation between electron temperature and density [9.1]:

$$\delta T_e/T_e = -(2/\hat{\nu}'_u)\delta n_e/n_e \ , \quad \hat{\nu}'_u \equiv 1 + \hat{\nu}_u - \hat{\nu}_m \ . \qquad (9.7)$$

The factor $(2/\hat{\nu}'_u) = 2/(1 + \hat{\delta}_1)$ is definitely positive and is of order unity. Therefore, the field and electron temperature are reduced at points of increased

[3] We neglect the electron diffusion current j_{dif}, because if $\Lambda \gg \Lambda_u$ and T_e is quasisteady, j_{dif} is small in comparison with the drift current. Indeed, (2.41, 24, 19) imply that $j_{dif}/\sigma E \sim \Lambda_u/\Lambda \ll 1$.

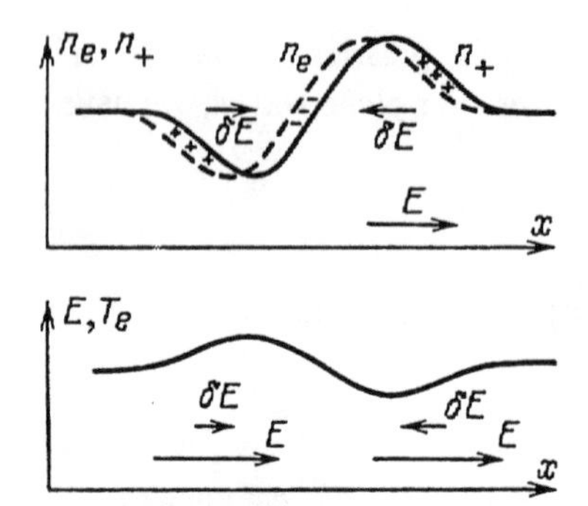

Fig. 9.3. Reduction of E and T_e where n_e increases

electron density, and vice versa. The physical mechanism producing this effect is illustrated in Fig. 9.3. It is assumed that the size of inhomogeneity, Λ, is much greater than the Debye radius d (at $T_e \approx 1$ eV and $n_e \approx 10^{10}\,\mathrm{cm}^{-3}$, $d \approx 10^{-2}$ cm) so that the densities n_e and n_+ are perturbed "in synchrony". The shift of n_e (dashed curve) with respect to n_+ (solid curve) is caused by drifting of electrons. The polarization field δE adds to the external field E in the region of the dip in n_e. The field δE is subtracted from E in the region of the crest.

9.4 Thermal Instability

This instability constitutes a universal mechanism which perturbes homogeneous discharges at elevated pressures and sufficiently high currents in atomic and molecular gases, both in tubes and in plane channels. It results in discharge contraction, and in the formation of current filaments in which the degree of ionization and the gas temperature are much higher than in ordinary glow discharges. The instability develops from transverse inhomogeneities. The field E remains homogeneous along the current direction; its magnitude changes only because the voltage at the electrodes decreases (as a result of current increase and increased voltage drop across the external resistor).

Thermal instability was first studied before the high-power laser problem arose; it was analyzed in discharges controlled by diffusion of charges to the tube walls [9.6].

The instability mechanism clarified in Sect. 9.1.2 is reflected in the following closed chain of causal links that can be started at any one step:

$$\delta n_e \uparrow \rightarrow \delta(jE) \uparrow \rightarrow \delta T \uparrow \rightarrow \delta N \downarrow \rightarrow \delta(E/N) \uparrow \rightarrow \delta T_e \uparrow \rightarrow \delta n_e \uparrow \ . \qquad (9.8)$$

Upward (downward) directed arrows symbolize an increase (decrease) in a quantity. The rate of growth of perturbations is limited by the slowest step, namely, the heating of the gas. The electron density, electron temperature, and also the relation of gas temperature to density at constant pressure, $p = NkT =$ const, can be considered quasisteady, corresponding to the instantaneous value of E/N.

9.4.1 Threshold and Growth Rate

We will determine the critical (threshold) conditions for the appearance of instability and evaluate its growth rate. The crucial parameter is the gas temperature T, which obeys the energy balance equation. The balance equation (8.23) is written for space-averaged quantities but this is not an obstacle. Small-scale spatial inhomogeneitiess are not very "dangerous" because heat conduction dissipates them rapidly. Perturbations of wavelength Λ of the order of the volume size are especially dangerous. Heat loss from the entire volume is nevertheless taken into account in (8.23) (this is a stabilizing factor).

An equation for perturbations δT is most easily constructed by varying (8.23) against the background of steady state (8.24). The mass conservation law (constancy of the rate of flow) implies that $N\nu_F \propto Nu$ = const in the case of convective transport. Also $\nu_T \propto \chi \propto 1/N$ in the case of heat conduction. The product $N\nu_{T,\mathrm{F}}$ is not perturbed in either case. In the term $jE = \sigma E^2$, we will vary only the factor that varies most steeply, n_e. The collision frequency is assumed to be approximately constant, because, according to (9.5), the variations of factors in the varied product $\nu_m = N\tilde{\nu}_m(T_e)$ have opposite signs and partially cancel each other out. Assuming after varying that $\delta T \sim \exp(\Omega t)$, we find

$$\Omega = \nu_T^0 \frac{\delta \ln n_e}{\delta \ln T} - \nu_{T,\mathrm{F}}\,, \quad \nu_T^0 = \frac{\sigma E^2}{N c_{p1} T} = \frac{\gamma - 1}{\gamma}\frac{\sigma E^2}{p}\,, \tag{9.9}$$

where $\nu_T^0 = \tau_T^{-1}$ is the inverse of the characteristic time of doubling the gas temperature at a nonperturbed value of T, and γ is the adiabatic exponent. In order to relate n_e to T, we describe the removal of electrons in terms of an effective recombination coefficient (8.32). The electron density n_e is quasisteady, because the production and removal of electrons are processes that are fast in comparison with gas heating:

$$n_e = \nu_i\left[T_e(E/N)\right]/\beta'_{\mathrm{eff}}\,, \quad N = p/kT\,. \tag{9.10}$$

Let us vary these equalities assuming $\beta'_{\mathrm{eff}}(T_e)$ = const; this is quite legitimate. In view of (9.5), we have

$$\frac{\delta \ln n_e}{\delta \ln T} = -\hat{\nu}_i \frac{\delta \ln T_e}{\delta \ln N} = \xi\hat{\nu}_i\,, \quad \hat{\nu}_i = \frac{d \ln \nu_i}{d \ln T_e}\,. \tag{9.11}$$

The electron density is much more sensitive to heating than T_e because $\hat{\nu}_i$, in contrast to ξ, is a large number. Its physical meaning is quite clear in the case of the Maxwell distribution of electrons, when

$$\nu_i \propto \exp(-I/kT_e)\,, \quad \hat{\nu}_i = I/kT_e \approx 5 - 10\,. \tag{9.12}$$

Substituting (9.11) into (9.9) and assuming for the sake of brevity $\xi = 1$, we obtain a very simple and clear formula:

$$\Omega = \nu_T^0 \hat{\nu}_i - \nu_{T,\mathrm{F}}\,, \quad \hat{\nu}_i = d \ln \nu_i / d \ln(E/N)\,. \tag{9.13}$$

The above definition of $\hat{\nu}_i$ is valid if we assume $T_e \propto E/N$. This is convenient when $\nu_i(E/N)$ is approximated by Townsend's function.

According to (9.13), an instability develops ($\Omega > 0$) only when the current and heat output exceed a certain threshold:

$$jE > Nc_{p1}T\nu_{T,F}/\hat{\nu}_i \approx (0.1 - 0.2)Nc_{p1}T\nu_{T,F} . \tag{9.14}$$

According to (8.24), this threshold corresponds to heating the gas by 10 to 20 %. As a result of a sharp dependence of ionization frequency on $T_e \propto E/N \propto T$, this very increase in temperature produces a strong enhancement of the ionization rate (by a factor of ē), thereby "launching" the system. Heat removal stabilizes the process: if increased, it raises the instability threshold. In the case of the heat conduction mechanism of cooling, as in tubes, we have $\nu_T \equiv \tau_\chi^{-1} \propto p^{-1}$, and the threshold $(jE)_{cr}$ is independent of pressure. Since $E/p \approx$ const, the threshold current is $j_{cr} \propto p^{-1}$. Experiments confirm this result: a homogeneous discharge is hard to obtain at high pressures, and at smaller currents the column contracts to the axis of the tube. In convective heat transfer, $(jE)_{cr} \propto p$ and the threshold current $j_{cr}(p) \approx$ const. The threshold is higher, the higher the flowrate of the pumped gas. When the energy release rate is appreciably greater than the threshold, the instability growth rate $\Omega \approx \nu_T^0 \hat{\nu}_i \sim 10^3$–$10^4\,s^{-1}$ (Table 9.1), that is, the thermal instability develops over 10^{-4}–10^{-3} s. The same conclusion follows from an analysis of the nonlinear stage of filament evolution [9.3].

9.4.2 Molecular Gas with Retarded Vibrational Relaxation

One example is nitrogen. In such cases, thermal expansion is delayed. Joule heat is first accumulated in vibrational degrees of freedom of molecules; as perturbations grow (provided the appropriate threshold has been exceeded), this heat transforms at an accelerating rate into the translational energy of the gas, because the vibrational relaxation accelerates with increasing temperature. The ionization rate also intensifies. What happens resembles a thermal explosion in which the chemical reaction goes faster and faster as temperature grows, so that the growth of temperature is constantly accelerated [9.4].

9.4.3 Stabilizing Effect of Ionization Produced by External Sources

The current and energy input into a discharge may be substantially increased without producing thermal instability by using a non-self-sustaining discharge in which the gas is ionized by an external source (especially by a fast electron beam) – see Chap. 14. The stabilizing effect stems from a sharp weakening of the inverse dependence of the rate of "self-sufficient" ionization on gas density.[4] Let an external source produce S pairs of ions per cm^{3}s in the volume. Now the balance equation for the number of electrons (9.10) takes the form

$$S + \nu_i\left[T_e(E/N)\right] n_e = \beta'_{eff} n_e^2 , \quad N = p/kT . \tag{9.15}$$

[4] A linear theory of stability of a non-self-sustaining discharge was first formulated in [9.7].

Let us vary these equalities, setting for brevity $T_e \propto E/N$ and $\xi = 1$. Let $S \sim N$, as is typical for ionization by an electron beam. We obtain

$$\frac{\delta \ln n_e}{\delta \ln T} \approx \frac{\hat{\nu}_i - S/\nu_i n_e}{1 + 2S/\nu_i n_e} < \hat{\nu}_i \, .$$

Now, the electron density is less sensitive to heating; in the limit $S \gg \nu_i n_e$ when $n_e \approx \sqrt{S/\beta'_{\text{eff}}}$, it even decreases in response to heating as $N^{1/2} \propto T^{-1/2}$. In the expression for the growth rate

$$\Omega = \nu_T^0 \frac{\hat{\nu}_i - S/\nu_i n_e}{1 + 2S/\nu_i n_e} - \nu_{T,\text{F}} = \frac{\nu_T^0 \hat{\nu}_i}{1 + 2S/\nu_i n_e} - \frac{S \nu_T^0}{2S + \nu_i n_e} - \nu_{T,\text{F}} \, , \tag{9.16}$$

the first, destabilizing term has decreased in comparison with (9.13). In addition to the heat removal term, $\nu_{T,\text{F}}$, another stabilizing term (the second in the last expression) appears in the formula. If $S \gg \nu_i n_e$, that is, if the discharge is strongly non-self-sustaining, the threshold values of current and energy input (i.e., ν_T^0) are greater for the same heat removal $\nu_{T,\text{F}}$ than in the self-sustaining discharge. Actually, instability may arise when the rate of external ionization becomes almost equal to that of ordinary ionization, that is, when the non-self-sustaining discharge becomes nearly self-sustaining. If $S \ll \nu_i n_e$, (9.16) transforms into (9.13).

9.4.4 Enhanced Stability of Discharge in Oscillating Fields

Other conditions being equal, a *high-frequency capacitance discharge* (Chap. 13) is more stable with respect to thermal instability than a constant current discharge; this is an experimental fact. As a result of the steep dependence of ν_i on E, ionization in oscillating fields occurs at the moment of field maxima. When the field is weaker, the plasma mostly decays. The balance of charge production and removal in a steady-state discharge is maintained only when averaged over a period. The electron density then slightly oscillates around the steady-state mean value $\bar{n}_e$. Thermal instability can develop only during ionization. At these moments, the field is higher than the constant field that would produce a constant density n_e equal to the mean density $\bar{n}_e$ in the oscillating field. Indeed, the stronger ionizing field must compensate for insufficient ionization at the moments of weak field. In fact, the slope of the function $\nu_i(E)$ and the value of $\hat{\nu}_i$ are smaller in stronger fields [see (9.12)]. According to (9.14), the threshold energy release (called *maximum energy input*) is correspondingly higher.

The energy threshold for instabilities is found to be even higher in the constant current non-self-sustaining discharge in which the gas is ionized by *short repeated high-voltage pulses* and the energy is supplied by the dc current. The stabilizing factor is the same, but the effect is better pronounced because the amplitudes of pulses separated with relatively long pauses must be even higher in order to maintain the same mean electron density. By using this principle of discharge sustainment, it is possible to raise the energy input by an order of magnitude without disturbing its homogeneity, and thus develop a high-power laser (Sect. 14.4.5). The above arguments on stability improvement are supported by calculations [9.8].

9.5 Attachment Instability

9.5.1 Stability of Attachment-Controlled Discharges

Attachment as such does not lead to instability. The attachment frequency ν_a increases with E/N not as steeply as ionization frequency, and $\hat{\nu}_a < \hat{\nu}_i$. In the case of longitudinal inhomogeneities, E/N and T_e depend on n_e inversely (Sect. 9.3.3), so that the stable situation of Fig. 9.1a holds for pure attachment. Indeed, neglecting spatial inhomogeneity and varying the electron kinetics equation $\partial n_e/\partial t = (\nu_i - \nu_a)n_e$ about the point of steady state $\nu_i = \nu_a$, and assuming $\delta n_e \propto \exp(\Omega t)$, we find

$$\Omega = \left(\frac{\partial \nu_i}{\partial T_e} - \frac{\partial \nu_a}{\partial T_e}\right) n_e \frac{\delta T_e}{\delta n_e} = \frac{\delta \ln T_e}{\delta \ln n_e}(\nu_i \hat{\nu}_i - \nu_a \hat{\nu}_a) \ , \quad \nu_i = \nu_a \ . \tag{9.17}$$

For instance, the ν_i and ν_a curves of the mixture $CO_2 : N_2 : He = 1 : 7 : 12$ intersect at $T_e = 1.6\,\mathrm{eV}$ (Fig. 8.19). The ionization potential $I_{CO_2} = 13.3\,\mathrm{eV}$. The frequency of dissociative attachment of an electron to a CO_2 molecule is $\nu_a \sim \exp(-U/kT_e)$, where $U = 3.85\,\mathrm{eV}$ (Sect. 4.4.1). At the intersection point, $\hat{\nu}_i = 8.3$, $\hat{\nu}_a = 2.4$. In accordance with (9.7, 17), $\Omega < 0$. Perturbations damp out quite rapidly: $|\Omega| \sim 10\nu_a$.

Let a transverse perturbation appear. Heating and thermal expansion are considerably slower than the ionization-attachment kinetics so that the gas density N can be considered frozen. As a result, $\delta \ln T_e/\delta \ln n_e = 0$ (Sect. 9.3.2), and $\Omega = 0$ in (9.17). This uncertain equilibrium is stabilized by external resistance (Sect. 9.1.2). The stability of attachment-controlled discharges is confirmed by the experiment [9.9] described in Sect. 8.8.1.

9.5.2 Mechanism of Instability

The attachment instability can be appreciable only if detachment compensates for attachment to a considerable degree. The electron temperature of the steady state is then substantially lower than that at which $\nu_i = \nu_a$ because ionization must compensate for only a small difference between attachment and detachment. The longitudinal perturbations may grow because the detachment rate is independent of T_e. Let the electron density increase slightly at some point in the discharge. The electron temperature at this point decreases. The ionization rate obviously decreases, but this factor may affect the balance of the numbers of electrons less than the weakening of attachment. The latter factor leaves the (unchanged) detachment uncompensated and electrons keep entering the plasma from decaying negative ions. The number of electrons increases, and E and T_e decrease still further. This effect is represented by a chain of causal links:

$$\delta n_e \uparrow \rightarrow \delta T_e \downarrow \rightarrow \nu_a \downarrow \rightarrow (Z_+ - Z_-) \uparrow \rightarrow \delta n_e \uparrow \ . \tag{9.18}$$

9.5.3 Growth Rate

The unsteady process of interest is described by a system of two kinetics equations from (8.28), for n_e and n_-. The third equation is equivalent to the charge neutrality condition $n_+ = n_e + n_-$. Their joint analysis [9.1, 2] leads to very complicated expressions. Let us take an essentially simpler case where positive ions recombine much more slowly than attachment and detachment proceed and than the instability develops [9.10]. In this case the parameter n_+ can be regarded as frozen, $\delta n_- = -\delta n_e$, so that only one kinetics equation, that for n_e, survives from (8.28). By varying it under the assumption of spatial homogeneity, as in Sect. 9.5.1, we find the instability growth rate:

$$\Omega = \frac{\delta \ln T_e}{\delta \ln n_e}(\nu_i \hat{\nu}_i - \nu_a \hat{\nu}_a) - (\nu_a + \nu_d) \ , \quad \frac{\delta \ln T_e}{\delta \ln n_e} < 0 \ , \tag{9.19}$$

where $\nu_d = k_d N$ is the negative ion detachment frequency. Taking into account (9.7) and an approximate equality of attachment and detachment velocities, $\nu_a n_e \approx \nu_d n_-$, we can rearrange Ω in the form

$$\Omega \approx \nu_a \left[\frac{2\hat{\nu}_a}{\hat{\nu}'_u} \left(1 - \frac{dk_i/dT_e}{dk_a/dT_e} \right) - \frac{n_+}{n_-} \right] \ . \tag{9.20}$$

The form of the first term in (9.19) coincides with that of (9.17) but now $\nu_i \ll \nu_a$ so that it can be positive. As follows from (9.20), this is possible only if $dk_i/dT_e < dk_a/dT_e$. In the mixture $CO_2 : N_2 : He = 1 : 7 : 12$, this inequality holds if $T_e < 1.05\,\mathrm{eV}$ [9.2, 11] (Fig. 8.19). If T_e is considerably lower than this value, the first term in (9.19, 20) is a sufficiently large positive quantity and may become greater than the second term which is proportional to n_+/n_-. It is necessary, however, that the ratio n_+/n_- (it determines the stabilizing effect) be not too high.

Calculations [9.2, 11] have determined the boundaries of the instability region for the mixture given above: $n_e < 10^{10}\,\mathrm{cm}^3$, $10^{-1} < n_-/n_e < 10$, $T_e < 1\,\mathrm{eV}$. Recombination, which always plays a stabilizing role, is too strong for the instability to occur if $n_e > 10^{10}$ (Sect. 9.1.2). If $n_-/n_e < 10^{-1}$, that is $n_+/n_- > 10$, then the reservoir of bound negative charges whose release by detachment could result in destabilization is too small. If $n_-/n_e > 10$, the number of positive ions at the given electron density is too high, recombination is strong, and a high ionization rate is required to compensate for it, at $T_e > 1\,\mathrm{eV}$. The destabilizing effect is thereby eliminated. According to (9.20), the time scale of the evolution of instability, if it occurs, is of the order of the attachment time $\tau_a = \nu_a^{-1}$.

9.5.4 Domains

The attachment instability typically results in domain formation. If a fluctuation $\delta n_e > 0$ or $\delta T_e < 0$ occurs in an unstable state with a relatively high number of negative ions, these ions immediately start to decay. Numerous electrons appear, the conductivity sharply increases, and the field at this spot decreases. The result is a *weak field domain.* If a fluctuation $\delta n_e < 0$ or $\delta T_e > 0$ occurs in an unstable state with relatively low n_-, attachment increases just as irresistibly, the conductivity drops, and the field increases. A *strong field domain* is thus formed. Domains move towards the anode with a phase velocity which may be several orders of magnitude lower than the electron drift velocity. Effects of spatial inhomogeneity must be taken into account. The manifestations of these effects will be discussed in Sect. 9.7, when considering the motion of striations. As a rule, domains are periodically initiated, travel, and then disappear. They are not visually detectable, but instruments do record current oscillations. No domains are observed in pure Ar, He, and N_2, where no attachment occurs.

9.6 Some Other Frequently Encountered Destabilizing Mechanisms

9.6.1 Stepwise Ionization

We will demonstrate this effect with the diffusion-controlled discharge. We again drop for the sake of simplification and brevity, the terms that take into account the spatial inhomogeneity of plasma, and write the kinetics equations for the densities of electrons and excited atoms:

$$dn_e/dt = k_i N n_e + k_i^* N^* n_e - \nu_{da} n_e \ , \quad k_i = k_i(T_e) \ , \tag{9.21}$$

$$dn^*/dt = k^* N n_e - k_2 N^* n_e - \nu_d^* N^* \ , \quad k^* = k^*(T_e) \ . \tag{9.22}$$

Here $k_i = \nu_i/N$, $k^* = \nu^*/N$ are the rates of ionization and excitation of atoms from the ground state, k_i^* and k_2 are the rates of ionization and deexcitation of excited metastable atoms, and ν_d^* is the frequency of losses of excited atoms not caused by electron impact (losses due to diffusion toward the walls and deactivation by collisions with atoms). Equation (9.22) takes into account that $k_i^* \ll k_2$ because an electron needs sufficient energy for ionization but does not need it for deactivation. If the instability develops more slowly than the excited atoms decay (in fact, this is a typical situation), the density N^*, being quasisteady, tracks the growth of n_e:

$$N^* \approx N^*[n_e(t)] \approx k^* N n_e/(k_2 n_e + \nu_d^*) \ . \tag{9.23}$$

Let us take the variations of (9.21, 23) at the steady-state point, neglecting the dependence of k_i^* and k_2 on T_e in comparison with that of k_i and k^*. Assuming $n_e \propto \exp(\Omega t)$, we find

$$\Omega = \frac{\nu_d^* k_i^* N^{*^2}}{\nu^* n_e} + \frac{\delta \ln T_e}{\delta \ln n_e} \left(\hat{\nu}_i \nu_i + \hat{\nu}^* k_i^* N^* \right) . \tag{9.24}$$

The first positive term $\Omega^{(+)}$, reflecting the ionization of the accumulating excited atoms, is quadratic in N^*; it results in the destabilizing effect of the stepwise ionization. Considered as a function of n_e, it reaches maxima at the values

$$(n_e)_{opt} = \nu_d^* / k_2 \ , \quad (N^*)_{opt} = k^* N / 2k_2 = N_{eq}^* / 2 \tag{9.25}$$

that are optimal for instability to occur. Here N_{eq}^* is the equilibrium Boltzmann density of excited atoms, corresponding to T_e. It satisfies the detailed balance principle $k^* N n_e = k_2 N_{eq}^* n_e$. The quantity

$$\Omega_{max}^{(+)} = k_i^* k^* N / 4k_2 = k_i^* (N^*)_{opt} / 2 = k_i^* N_{eq}^* / 4 \tag{9.26}$$

gives the upper bound to the instability growth rate. Instability would evolve at this rate under the most favorable conditions, if no stabilizing factors were at work. Far from the boundaries of the stability region, however, Ω is of the order of $\Omega_{max}^{(+)}$ even if such factors are present.

In order to evaluate the orders of magnitude of the quantities of interest, consider an example, Let $p = 3$ Torr, $N = 10^{17}\,\mathrm{cm}^{-3}$. In a tube of radius $R = 1$ cm, the frequency of ambipolar diffusion of charges towards the walls (this diffusion controls the discharge in inert gases) is $\nu_{da} \approx 5 \cdot 10^3\,\mathrm{s}^{-1}$. Therefore, the ionization frequency is also close to this value: $\nu_i \approx \nu_{da}$. The excitation frequency $\nu^* \approx 10^5\,\mathrm{s}^{-1}$ is greater than ν_i because $E^* \approx 3I/4$; $k_i^* \sim 10^{-9}\,\mathrm{cm}^3/\mathrm{s} \gg k_i \approx 5 \cdot 10^{-14}\,\mathrm{cm}^3/\mathrm{s}$ because the binding energy of excited atoms $I - E^* \approx I/4$; $k_2 \approx 10^{-8}\,\mathrm{cm}^3/\mathrm{s}$, $N_{eq}^* \approx 10^{13}\,\mathrm{cm}^{-3}$, $\nu_d^* \approx 3 \cdot 10^3\,\mathrm{s}^{-1}$. Hence, $(n_e)_{opt} \approx 3 \cdot 10^{11}\,\mathrm{cm}^{-3}$, $(N^*)_{opt} \approx 5 \cdot 10^{12}\,\mathrm{cm}^{-3}$, $\Omega_{max}^{(+)} \approx 3 \cdot 10^3\,\mathrm{s}^{-1}$. The growth rate is of the order of $\nu_{da} \approx \nu_i$.

As for the second ionization term in (9.24), it is negative for longitudinal perturbations, as in (9.17, 19), and its effect is stabilizing. This effect has to be overcome for instabilities to occur; hence, the region of possible occurrence of instability is correspondingly limited. Additional analysis of the stabilizing effect of ionization is given in Sect. 9.7.2. The effect disappears in the case of transverse perturbations, when T_e is related to n_e only through thermal expansion.

9.6.2 "Maxwellization" of Electrons

Electrons lose large amounts of energy when colliding with atoms and exciting (or ionizing) them. The result is a considerable depletion of the spectrum in the range of energies sufficient for ionization, and a reduction of the ionization frequency per electron. At greater concentrations, electrons begin interacting with one another, exchanging large portions of energy. The spectrum tends to a maxwellian form in which the fraction of fast electrons is larger than when electrons collide only with atoms and gain no energy from them. The ionization frequency increases, the electron density grows, and the spectrum becomes even more maxwellian. If n_e is sufficiently high, at higher currents, this instability may lead to contraction and striation formation, just as stepwise ionization does.

9.7 Striations

9.7.1 Observations

The positive column of discharge in tubes is striated (Fig. 9.4) much more often than is detectable by the naked eye. As a rule, striations are travelling. In inert gases, they move at velocities $v_{ph} \sim 100\,\mathrm{m/s}$ (at $p \sim 10^{-1}$–$10\,\mathrm{Torr}$) from the anode to the cathode. The intensity of light emission oscillates at frequencies of about 1 kHz, so that the eye cannot resolve the oscillations and the column appears homogeneous. Striations may not be moving, revealing the alternation of bright and dark layers along the tube, which is how they were discovered. Striations appear fixed when the discharge incorporates a local, permanent source of strong perturbations, for example, a probe at high negative potential or a sharp change in tube cross section. Sometimes the region adjacent to the cathode

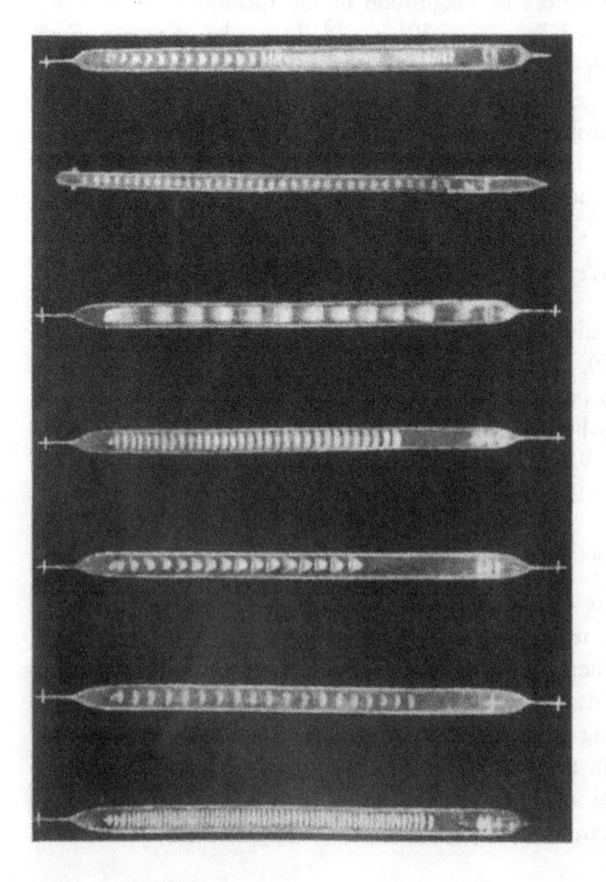

Fig. 9.4. General appearance of striated discharge

plays the role of such a perturbation. Fixed striations build up away from the perturbation point towards the anode and are gradually damped out. Sometimes only the first few striations fade out at the cathode, while the remaining ones are preserved. Typically, the length of one striation, that is, the distance between the corresponding points of neighbouring striations, equals several tube radii. This is true both for fixed and for running striations.

Striations survive in a limited range of current values, gas species, pressure, and tube radius. The amplitude of luminosity oscillations, striation wavelength, and the velocity of striation propagation depend on the same parameters. The amplitude of luminosity oscillations, that to some extent reflects the amplitude of electron density oscillations, is not high at the boundary of the existence region. The oscillations are very nearly sinusoidal. The realtionship between small- and large-amplitude striations is almost the same as between acoustic and strong nonlinear waves in gas dynamics. Beyond the region of existence of striations, the positive column is stable and homogeneous, although external perturbation agents can generate striations in a small neighbourhood beyond the natural region of existence. Extensive experimental data on striations can be found in the reviews [9.12–15].

9.7.2 Conditions of Occurrence

Striations are manifestations of ionization oscillations and waves. This means that periodic changes in electron density are caused not by redistribution of a fixed number of electrons (as in plasma waves), but by alternation of regions of predominant production and removal of electrons. Instability mechanisms leading to striation originate with ionization processes. A striated discharge may be caused by stepwise ionization, the maxwellization of the electron distribution function (Sect. 9.6), and by any agent causing enhancement in longitudinal inhomogeneities. Stepwise ionization is effective only at moderate currents, when the metastable atoms produced by electron impact decay as a result of diffusion to the walls. At greater currents (at $i \gtrsim 100\,\mathrm{mA}$ in inert gases), n_e becomes so high that metastables are destroyed by electron impact. As follows from (9.23), their density then stops growing with increasing n_e; hence, instability will definitely develop in response to the nonlinear dependence of the gas ionization rate on n_e. In this process, stepwise ionization is replaced by maxwellization or some other mechanism.

Running striations constitute a wave process in which the decisive role is played by spatial plasma inhomogeneities (namely, longitudinal). An important length scale inherent in plasmas is the relaxation length of electron temperature (energy spectrum) in the field: $\Lambda_u = v_d \tau_u \approx l/\sqrt{\delta}$. It is this scale, in addition to another geometric scale, namely, the tube radius R, that largely dictates the occurrence of spatial inhomogeneities and their scales. Actually, this situation arises because the steep increase of ionization rate with T_e in the case of longitudinal perturbations is a stabilizing factor [see (9.24) and the remark at the end of Sect. 9.6.1]. Indeed, since j is constant, the field drops at points where

n_e increases (Sect. 9.3.3). If T_e decreases together with E, the ionization rate is reduced and the positive increment of n_e is damped out.

The electron temperature reaches the value dictated by the given field over a relaxation length Λ_u. This stabilizing factor (the quasisteady behaviour of T_e in response to varying E) is at its most effective in the case of long waves or small wave numbers K, $K\Lambda_u \ll 1$. According to (9.7), $\delta \ln T_e/\delta \ln n_e \approx -2$ and the stabilizing factor in (9.24) ($\approx 2\hat{\nu}_i\nu_i$) is substantially greater than $\Omega^{(+)}_{\max}$ given by (9.26). Destabilization caused by stepwise ionization cannot overcome the stabilization due to ionization. Stabilization due to ionizations weakens at short wavelengths, $K\Lambda_u > 1$, because T_e does not have enough time to decrease to the value corresponding to the reduced field. If the wavelength is very short ($K\Lambda_u \gg 1$), T_e is completely insensitive to oscillations in E, and ionization does not lead to stabilization. This mechanism bounds the striation wavelength from above by the electron energy relaxation length. Striations always have $K\Lambda_u$ greater than one. However, wavelengths are bounded from below as well. Waves which are too short in comparison with the tube radius R cannot survive either, because perturbations in n_e are actively destroyed by longitudinal ambipolar diffusion. As a result, striations can exist only in a certain interval of wavelengths. The most favourable conditions for striations in inert gases are $K\Lambda_u \approx 5$–10, $KR \sim 1$.

In molecular gases, where the fraction δ_1 of energy transferred to a molecule by a colliding electron is quite high, the relaxation length Λ_u is very short. The dimensions of those disturbances in n_e, which are not stabilized by ionization, are too small. Hence these disturbances may easily be destroyed by diffusion, and the formation of striations is obstructed. The situation with monatomic inert gases is different since $\delta_1 = 2m/M$ is a very small quantity and the length Λ_u is large, especially at low pressures in the range 10^{-2}–10 Torr. The case of Λ_u exceeding R is favourable for the occurrence of large-scale spatial inhomogeneities. Indeed, striated discharges are typical for inert gases, although narrow striations are also observed in hydrogen. Note that the study of striations started at the beginnning of this century, but the insight into the true nature of this widespread phenomenon was not gained until the 1960s.

9.7.3 Theory of Low-Amplitude Ionization Waves

The theory is based on linearizing the equations describing the behaviour of the electron gas in a plasma. In the hydrodynamics approximation, these equations are (2.44) for n_e, (2.52) and (2.53) for T_e, and (2.54) for E. These equations must contain a destabilizing factor, and if this factor is the stepwise ionization, the right-hand side of (9.21) [with N^* according to (9.23)] can be used as the charge source in (2.44). The dispersion relation for $\omega(K)$ is found as explained in Sect. 9.1.4, with the waves regarded as directed along $\boldsymbol{E}$. This calculation was consistently carried to the final result in [9.13, 16]. The calculation [9.17], involving maxwellization as the destabilization mechanism at elevated currents, gave not only a qualitative but also a good quantitative fit to the measured dependence of striation wavelength on frequency and to the boundaries of striation existence in argon.

The system of equations (2.44, 52, 54) are of second order in t, so that a quadratic equation is obtained for ω. If, however, the wave solution is substituted for δT_e into (2.52), one finds that the temporal and convective terms are in the ratio $3\omega/5Kv_e \approx 3v_{ph}/5v_d$. Here $v_{ph} = \omega/K$ is by definition the phase velocity of the wave. The experimental data and the solution that follows demonstrate that the velocity of motion of striations is much lower than the drift velocity of electrons, so that there is every reason to drop the term with $\partial T_e/\partial t$. The dispersion relation then becomes of first order. As follows from (2.44), in the case of a stepwise ionization mechanism, it differs from (9.24) only in the term due to the ambipolar diffusion in the plasma. This diffusion tends to level off the disturbances and exerts a stabilizing influence, so that the contribution to $\mathrm{i}\omega$ is

$$\mathrm{i}\omega = \Omega^{(+)} + (\delta \ln T_e/\delta \ln n_e)\left(\hat{\nu}_i\nu_i + \hat{\nu}^* k_i^* N^*\right) - D_a K^2 . \tag{9.27}$$

The destabilizing term $\Omega^{(+)}$ has already been discussed in Sect. 9.6.1.

Let us introduce several simplifications, in order to avoid a cumbersome relation between δT_e and δn_e in (9.27). Let ν_m, μ_e, $\nu_u = \delta_1 \nu_m$ be independent of T_e (in a monatomic gas, $\delta_1 = 2m/M = \text{const}$). We neglect the ambipolar diffusion flux in (2.54) because $D_a \ll D_e$. Then $n_e v_e = -j/e = \text{const}$ and we need not vary the product $n_e v_e$ (Sect. 9.2.3). Let us ignore the loss of energy spent on ionization in comparison with energy transfer to atoms. This is true if $\nu_i \ll \nu_u$, that is, if pressure is not too low, because $\nu_i \approx \nu_{da} \propto p^{-1}$ and $\nu_u \propto p$. Taking variations of (2.54, 52) (with $\partial T_e/\partial t = 0$) around the point of uniform steady state and taking account of $kT_e \approx eE\Lambda_u$ via (2.19), we find the relation between the complex perturbation amplitudes:

$$\delta E/E = (-1 + \mathrm{i}K\Lambda_u)\delta n_e/n_e , \tag{9.28}$$

$$\left[(5/2)K^2\Lambda_u^2 + 1 + \mathrm{i}(5/2)K\Lambda_u\right] \delta T_e/T_e = \delta E/E - \delta n_e/n_e , \tag{9.29}$$

$$\frac{\delta \ln T_e}{\delta \ln n_e} = \frac{-2 + \mathrm{i}K\Lambda_u}{(5/2)K^2\Lambda_u^2 + 1 + \mathrm{i}(5/2) \cdot K\Lambda_u} . \tag{9.30}$$

These formulae are significant and support the qualitative conclusions of Sect. 9.7.2. The relations become practically quasisteady in the limit $K\Lambda_u \ll 1$. They transform to those given by (9.6, 7) ($\hat{\nu}_m = 0$, $\hat{\nu}'_u = 1$, $\xi = 2$). The stabilization due to ionization in (9.27), where $\delta \ln T_e/\delta \ln n_e \approx -2$, is very strong. Long waves are damped out. In the general case, the waves of δn_e, δE, and δT_e are shifted in phase with respect to each other. We will presently see that these phase shifts ensure the motion of the waves. For the sake of simplification, consider the limiting case $K\Lambda_u \gg 1$ (the only one that can be realized experimentally in monatomic gases), and separate the real and imaginary parts of (9.30):

$$\frac{\delta \ln T_e}{\delta \ln n_e} = -\frac{2}{5K^2\Lambda_u^2} + \mathrm{i}\frac{2}{5K\Lambda_u} . \tag{9.31}$$

The real negative term which is proportional to the stabilizing factor becomes very small, so that undamped waves become possible. The imaginary part determines the frequency, proportional to wavelength: $\omega \propto K^{-1}$. The stabilization due to ionization, proportional to K^{-2}, becomes weaker, the shorter the wavelength. However, excessively short waves are stabilized quite well by diffusion, whose effect is proportional to K^2 [see (9.27)]. Therefore, the overall stabilizing factor denoted here by $\Omega^{(-)}(K)$ has a minimum at a certain K_{opt}. If $\Omega^{(+)} > \Omega^{(-)}_{\text{min}}$, the waves are allowed and the equation $\Omega^{(+)} = \Omega^{(-)}(K)$ limits the interval of possible wavelengths around K_{opt} on both sides.[5] The formulae derived from (9.27, 31) give the following wave number K_{opt} that is optimal for driving the oscillations:

$$K_{\text{opt}}\Lambda_D \approx (\Lambda_D/\Lambda_u)^{1/2}(2\hat{\nu}_i/5)^{1/4} ,$$

where $\Lambda_D = R/2.4$ is the diffusion length for a tube of radius R. In monatomic gases at $p \sim$ 1–10 Torr and R = 1 cm, $\Lambda_u \sim l/\sqrt{\delta_1} \approx$ 1–10 cm, $\Lambda_D \approx 0.3$ cm, and $\hat{\nu}_i \approx 10$. The optimal striation wavelength comes to several tube diameters, in agreement with experiments; at the same time, $K_{\text{opt}}\Lambda_u \approx$ 5–10, which justifies the approximation $K\Lambda_u \gg 1$.

9.7.4 Why Do Striations Move?

Let us clarify the physical reason for the motion of striations, which typically run from the anode to the cathode. Density gradients in real, relatively short waves are considerable. The relative separation of charges that exists, despite a high degree of charge neutrality, is determined not by drift but by diffusion (Fig. 9.5). On the slopes where n_e decreases towards the cathode (as x increases), the arising polarization field δE is added to the constant nonperturbed field E. On the slopes where n_e increases as x increases, δE is subtracted from E. For this reason, the δE wave is shifted by one quarter of a wavelength to the cathode with respect to the δn_e wave [Fig. 9.5, formula (9.28)]. The energy release in the electron gas, jE, has the same distribution as E, because j = const. This means that the maxima of jE coincide with those of E and are shifted by a quarter of a wavelength with respect to the maxima of n_e.

Among the various components of the electron energy balance (Sect. 9.2.5), heat conduction plays the dominant role for short waves with $K\Lambda_u \gg 1$. Electronic heat from an enhanced source $\delta(jE) = j\delta E$ spreads in both directions and the T_e profile very nearly repeats those of the heat and E source. The δT_e wave is almost in phase with the δE wave [this is clear from (9.28–31) also]. The ionization rate $\nu_i(T_e)n_e$ is also almost in phase with it because it is much more

[5] The conclusion that the instability growth rate Ω is positive in some narrow interval of short wavelengths with $K\Lambda_u \gg 1$ was made in a well-known review [9.12] for a similar problem, disregarding stepwise ionization or any other destabilizating mechanism. It was repeated in the review [9.18], citing [9.12]. This conclusion was a result of neglecting electronic heat conduction which causes damping of short-wave perturbations.

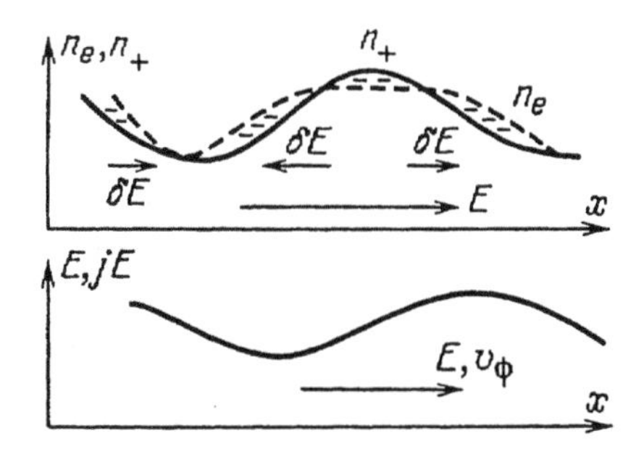

Fig. 9.5. Electron density and energy deposition distributions in moving striations: the case of short waves, when electron diffusion dominates the drift

sensitive to T_e than to n_e. Hence, ionization is most intensive at the piont of instantaneous equilibrium of n_e on the x axis, where $\delta n_e \approx 0$ and E is maximal. A quarter of a period later δn_e increases to the maximum value, that is, the crest of n_e moving in the direction of $\boldsymbol{E}$ arrives at this point.

9.7.5 Evaluation of the Velocity and Frequency of Striations

This is easier done not in a formal way, by extending the calculations of Sect. 9.7.3, but directly, on the basis of the arguments given above. The essence of the phenomenon is then better understood. Very short δn_e and δE waves are shifted by a quarter of a period. The relation between these parameters in complex representation is purely imaginary. According to (9.28),

$$\delta E/E \approx \mathrm{i} K \Lambda_u \delta n_e / n_e \,. \tag{9.32}$$

From the balance of the additional heat release and heat spreading due to heat conduction in (2.50, 51), we find $j\delta E \approx K^2 \lambda_e \delta T_e$. Substituting $j = e n_e \mu_e E$, $\lambda_e = (5/2) k n_e D_e$, and $T_e = E \Lambda_u$, we obtain

$$\frac{\delta T_e}{T_e} \approx \frac{2}{5K^2\Lambda_u^2}\frac{\delta E}{E} \approx \mathrm{i}\frac{2}{5K\Lambda_u}\frac{\delta n_e}{n_e} \,. \tag{9.33}$$

Owing to the temperature leveling via heat conduction, the relative change in T_e is much smaller than that of the field. It is also smaller than the change in plasma density, although here the difference is less pronounced.

The additional ionization rate due to enhanced heating is

$$\delta(\nu_i n_e) \approx n_e (d\nu_i/dT_e)\delta T_e = n_e \hat{\nu}_i \nu_i (\delta T_e/T_e) \,.$$

In a quarter of a period, that is, in a time $t \approx (K v_{ph})^{-1}$ where $v_{ph} = \omega/K$ is the phase velocity of the wave, n_e grows by a quantity δn_e of the order of the amplitude: $\delta(\nu_i n_e) t \approx \delta n_e$. Hence,

$$v_{ph} \approx \hat{\nu}_i \nu_i K^{-1} |\delta \ln T_e / \delta \ln n_e| \,.$$

After substituting here (9.33), we obtain the velocity and frequency of striations:

$$v_{ph} \approx \tfrac{2}{5}\hat{\nu}_i \nu_i / K^2 \Lambda_u \,, \quad \omega \approx \tfrac{2}{5}\hat{\nu}_i \nu_i / K \Lambda_u \,. \tag{9.34}$$

The frequency of oscillations of n_e and of other quantities in striations is proportional to the wavelength, and the velocity is proportional to the wavelength squared. The ionization frequency serves as frequency scale because striations are typically characterized by $K\Lambda_u \approx 10$ and $\hat{\nu}_i \approx I/kT_e \approx 10$. If the pressure in inert gases is not high, electrons are mainly removed to the walls by ambipolar diffusion, so that $\nu_i \approx \nu_{da} = D_a/\Lambda_D^2 \sim D_a/R^2$. Hence, ω is of the order of the diffusion frequency. In more rigorous solutions [9.13, 17], the qualitative core of the expressions for $\omega(K)$ and v_{ph} is the same as in (9.34). For argon in a tube of $R = 1.5$ cm, at $p = 0.5$ Torr and $i = 3.6$ mA, experiments and calculations [9.17] show that as the frequency $\omega/2\pi$ increases from 1.4 to 2 kHz, the striation wavelength $2\pi/K$ increases from 6 to 9 cm, and the velocity v_{ph}, from 80 to 180 m/s. Estimates yield correct orders of magnitude.

It is worthy of note that the group velocity of striations $v_{gr} = \partial\omega/\partial K < 0$, that is, v_{gr} points against the phase velocity and (in the framework of the chosen approximation) has equal magnitude. Therefore, any marking feature, such as a brighter region due to a pulsed local disturbance, runs towards the anode, in contrast to the striations themselves. Low-amplitude waves obeying linear theory typically survive under conditions that correspond to moderate penetration into the instability region. The stabilizing factors are still strong there and slow down by buildup of amplitude. Well-developed nonlinear waves of high amplitude can also arise far from the boundaries.

9.7.6 High-Amplitude Striations

Now we can conclude that striations are *ionization waves* moving through the gas, that is, alternating regions of enhanced and weakened ionization originating with the waves of enhanced and weakened field. In general, the processes in high-amplitude striations are the same as in weak-field waves but assume a greater role. High-field regions grow into abrupt potential jumps, and the field in more distant (on the anode side) regions may not just vanish, but may even become negative.

Ionization in high-field regions becomes so intensified that the electron density increases by an order of magnitude. This very bright, relatively narrow zone is known as the striation *head*. Ionization stops beyond it and electrons vanish owing to diffusion to the walls. Their density gradually decreases by the same order of magnitude until the next striation head. This dark zone is much longer than the bright one because electrons disappear at a slower rate than they are produced. As n_e falls off, the field is restored [$j(x) \propto n_e E = \text{const}$] and the process repeats itself. Qualitatively, this picture is similar to the stationary structure around cathode region of a static discharge (Sect. 8.5.3 and footnote 9 on p. 193). The cathode layer with negative glow corresponds to the striation head, and the Faraday dark space, to the extended region up to the next head. Instead of such a head, the positive column with moderate field and emission intensity is formed (provided the column is not striated).

9.7.7 Experiment and Its Interpretation

Figure 9.6 shows the results of probe measurement of the instantaneous distributions of n_e, T_e, and potential φ along the x axis of a strongly striated discharge tube [9.19]. The distribution of E, obtained by differentiating the $\varphi(x)$ curve, is also plotted. Measurements were carried out along the axis of a long tube. Striations move from the anode to the cathode at a velocity $v_{ph} = 60\,\text{m/s}$. The above distributions move as a whole from left to right at this velocity. The length of one striation is $d = 5.5\,\text{cm}$. The distribution of light intensity approximately follows that of n_e. Estimation of the space charge using double differentiation of the $\varphi(x)$ curve shows that the degree of charge neutrality is high even at points of high gradients: $|n_e - n_+| \lesssim 10^{-4} n_e$. On the average, $\langle E/p \rangle = 0.22\,\text{V/(cm·Torr)}$ along the striation and the column as a whole.

The maximum of E and of the ionization rate is localized at the point of steepest drop in n_e, on the right-hand slope of the crest. For this reason, the peak of $(n_e)_{max}$ and the entire striations moves to the right, to the cathode. The electron diffusion current is very high on the right-hand slope of n_e: $eD_e|dn_e/dx| \approx 530\,\text{mA/cm}^2$ ($D_e \approx 2.5 \cdot 10^6\,\text{cm/s}$); it is directed against the actual current i. With the Bessel radial profile $n_e(r)$ taken into account, the current density on the tube axis is $j = 2.3i/\pi R^2 = 77\,\text{mA/cm}^2$. The diffusion current is exceeded by the drift current $e\mu_e n_e E = 530+77 \approx 610\,\text{mA/cm}^2$ by just the amount necessary to supply the required density of j. Hence, for $\mu_e = 1.7 \cdot 10^5\,\text{cm}^2/(\text{s·V})$ we find $(E/p)_{max} \approx 8\,\text{V/(cm·Torr)}$, which is 36 times the average $\langle E/p \rangle$. When the

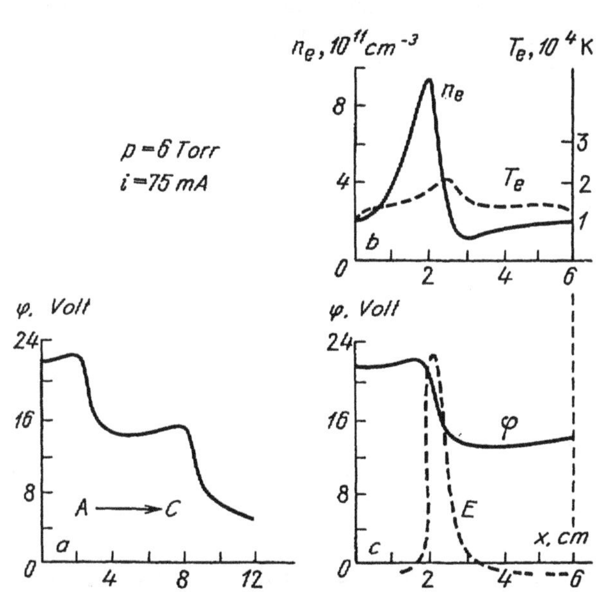

Fig. 9.6. High-amplitude striations in argon in a tube of radius $R = 0.85\,\text{cm}$ at $p = 6\,\text{Torr}$ and $i = 75\,\text{mA}$. (a) Potential distribution along the axis; $A \to C$ is the direction from anode to cathode, (b) distribution of n_e and T_e within one period, (c) potential distribution on the same scale as in (b), superposed on the coordinate axis with the (b) curve. Dashed curve is the field obtained by differentiating the potential [9.19]

mechanism of removal of electrons is their diffusion to the walls, their density beyond the striation head (where production is impeded) must fall off obeying the formula $v_{ph} dn_e/d|x| = -\nu_{da} n_e$. Hence,

$$d \approx (v_{ph}/\nu_{da}) \ln(n_{e,max}/n_{e,min}) \; . \tag{9.35}$$

In the experiment, $\mu_+ \approx 250\,\text{cm}^2/(\text{s}\cdot\text{V})$, $T_e \approx 1.3\,\text{eV}$, $D_a = \mu_+ T_e = 320\,\text{cm}^2/\text{s}$, $\nu_{da} = 2.4^2 D_a/R^2 = 2.6 \cdot 10^3\,\text{s}^{-1}$. The measured values of v_{ph}, d, and density drop $n_{e,max}/n_{e,min} \approx 10$ are in perfect agreement with estimate (9.35); hence the picture obtained experimentally is deciphered quantitatively. It must be emphasized that the electron distribution function in strong-field striations with abrupt field jumps, and also in weak-field striations with $K\Lambda_u \gg 1$, is of a nonlocal type, that is, it is determined not just by the local instantaneous field E, but also depends on the behaviour of potential in the space-time neighbourhood (as in the cathode layer, Sect. 8.4.10). This effect can in itself lead to a striated discharge and may affect the structure of strong-field striations [9.20].

9.7.8 Why Is the Striated State Favored?

A layered state must lead to some gains, since the positive column, which has the unequivocal possibility of being homogeneous (this state would ensure that all the conservation laws hold), nevertheless "prefers" a layered state. The gain is that the voltage V and power output iV in a striated column are lower, at the same current, than in a homogeneous column. The production and removal of electrons in high-amplitude striations are separated both in space and in time and are balanced only when averaged over the entire length of a striation, or over a period t_1 of oscillations in a given cross section of the tube. The slope of the function $\nu_i(E)$ is so steep that if the field is strong but acts for only a short time, a lower value of the integral $\int E\,dt$ is required over a period t_1 in order to produce a certain number of electrons to compensate for their removal. The voltage drop ΔV over the length d of a striation is

$$v_{ph} \int_0^{t_1} E\,dt = \int_0^{d} E\,dx = \Delta V \; ,$$

which is less than the drop over the same length of a homogeneous column with constant field.

The "usefulness" of a striated discharge resembles the usefulness of concentrating the entire function of discharge sustainment in the cathode layer (Sect. 8.4.1). The following clarification can be given: It is easier for an electron to gain the energy necessary for ionization if it crosses the required potential difference over a short path in a strong field. It will then take part in a smaller number of collisions obstructing its acceleration. From this standpoint, it is more profitable to distribute the potential over a set of abrupt steps than spread it over the same length at a constant gradient. The phenomenon of striations satisfies the principle of minimal power (Sect. 8.4.8). In fact, it was given an independent, more profound interpretation based on analyzing the processes themselves. Striations are formed because under certain conditions the homogeneous state becomes unstable.

9.8 Contraction of the Positive Column

A systematic study of the transition of a diffuse glow discharge to a contracted state has been carried out only for cylindrical tube geometry. In working with plane channels, measurements are typically limited to recording the critical current at which the homogeneous "burning" of the discharge discontinues. Conditions for studying discharge in plane channels are very unfavourable: the filament geometry is not always reproducible in successive experiments, the spot where a filament will form is not known in advance, and the filaments "dance". The ideal symmetry of tubes imposes rigid constraints on the process. It can be hypothesized, however, that the main properties of filaments are the same regardless of discharge geometry.

9.8.1 Experimental Results

Let us look at the results of one study [9.21] where fairly complete information on the phenomenon was obtained. The discharge was sustained in neon at p = 75–100 Torr in a tube of radius R = 2.8 cm. The field strength was measured by probes, the radial distribution of electron density was found from the bremsstrahlung produced by the scattering of electrons by neutral atoms, which is proportional to $n_e N$. The radial gas density profile $N(r)$, related to the gas temperature distribution $T(r)$ via constant pressure $p = NkT$, was taken into account. This is important because the gas was heated quite strongly. To find $T(r)$, the heat conduction equation for the true distribution of heat sources $j(r)E$ was solved. The temperature of the tube controlled, its wall being kept at T_0 = 300 K. The electron temperature on the axis was found from the measured field and from $N(0)$ using (4.13), which describes the electron energy balance. The momentum transfer cross section for neon depends quite weakly on energy and equals approximately $\sigma_m \approx 2.3 \cdot 10^{-16}\,\text{cm}^2$; the gas is monatomic, $\delta_1 = 2m/M$.

Figure 9.7 shows the $V-i$ characteristic of the positive column. The left-hand part (up to the jump) corresponds to the diffuse discharge that fills the entire tube at the degree of homogeneity typical of a diffusion-controlled discharge for which $n_e(r) \sim J_0(2.4r/R)$ (Sect. 8.6.2). The $V-i$ curve is of "dropping" type; this is characteristic of diffusion-controlled discharges at relatively high current, which strongly heats the gas (Sect. 8.7.4). The field and voltage decrease in jumps at the critical current which corresponds to the sudden transition of a homogeneous column to a contracted mode. A brightly luminous filament appears at the axis, while the rest of the tube volume gets darker. The transition occurs at a current density, averaged over the tube cross section, of about 5.3 mA/cm^2. The critical current density at the axis is 2.3 times greater, 12 mA/cm^2, in accord with the

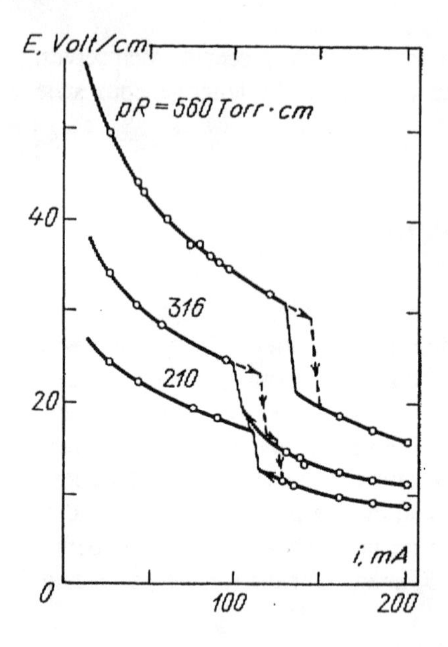

Fig. 9.7. $V - i$ characteristic of discharge in a tube containing neon in the region of transition from diffuse to contracted form. Tube radius $R = 2.8$ cm; *(1)* $pR = 210$ Torr·cm; *(2)* $pR = 316$; *(3)* $pR = 560$. The solid curve in the region of the jump was recorded as current was decreased, and the dashed curve, as it was raised [9.21]

Bessel $j(r)$ profile. The critical electron density at the axis $n_e(0) \approx 10^{11}$ cm^{-3}. A hysteresis is observed: The transition to the contracted discharge during current increase takes place at a somewhat greater current than the transition to diffuse current during current reduction. This is an indication of the double-valuedness of states in the region of the jump transition and of their possible metastability.[6]

Figure 9.8 shows radial distributions of n_e, and Table 9.2 lists discharge parameters at the axis, all at $p = 113$ Torr. Note how the current channel is sharply contracted, judging by its conductivity or n_e, and how the degree of

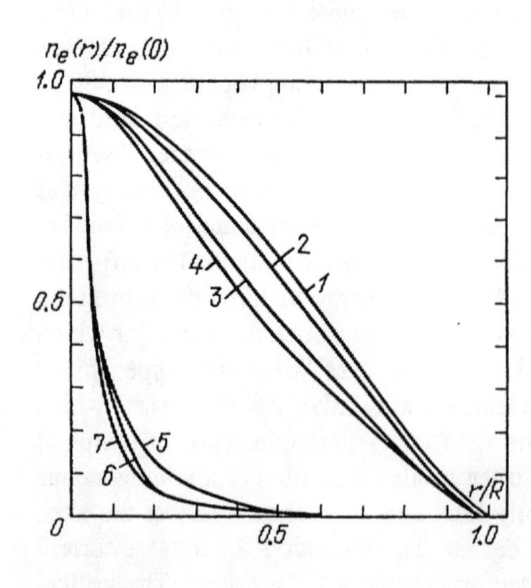

Fig. 9.8. Profiles of n_e measured under the conditions of Fig. 9.7. *(1)* $i/R = 4.8$ mA/cm; *(2)* 15.4; *(3)* 26.8; *(4)* 37.5; *(5)* 42.9; *(6)* 57.2; *(7)* 71.5 mA/cm. The transition from diffuse to contracted state occurred in the region between *(4)* and *(5)*

[6] We may recall the liquid-vapor phase transition (superheated liquid, supercooled vapor).

Table 9.2. Discharge parameters in a $R = 2.8$ cm tube at $p = 113$ Torr; the transition occurs at a current of 105–115 mA [9.21]

i, mA	$T_e(0)$, eV	$T(0)$, K	$n_e(0)$, 10^{11} cm^{-3}
13.5	3.0	440	0.12
43	3.3	650	0.39
75	3.6	840	0.93
96	3.7	930	1.2
120	3.0	1200	54
160	2.6	1300	72
200	2.5	1400	93

ionization jumps up by nearly two orders of magnitude. The effective filament radius [at the $n_e = n_e(0)/2$ level] is less than the tube radius by a factor of 20. The gas temperature at the axis is increased in the transition, and its fall-off from the axis is steeper in the contracted discharge than in the diffuse mode. The half-maximum point of the temperature drop between the axis and the walls lies at the point (0.7–0.8) R for a diffuse discharge and at the point $0.5R$ for the contracted discharge. This is natural: in the latter case, heat sources are concentrated at the axis. Upon contraction, the electron temperature slightly decreases because the field decreases. A very similar picture was also observed in argon.

The behaviour is somewhat different when a molecular gas is added to the inert host (Fig. 9.9) [9.22]. As usual, the $V - i$ curve first falls owing to gas heating, but then starts to rise. This is apparently related to the transition from diffusional to recombination losses as n_e increases. The $V - i$ characteristic of a bulk recombination-controlled discharge is typically rising (Sect. 8.6.3). The diffusion mode corresponds to very low values of $E/p \approx 0.12$ V/(cm·Torr). It appears that this is a result of (stepwise) ionization of metastable nitrogen atoms, since the production of electrons is thereby facilitated. After the jump,

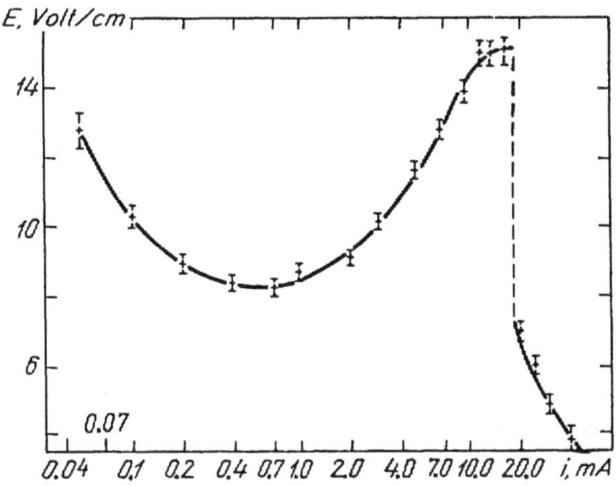

Fig. 9.9. $V - i$ characteristic of discharge in a tube of $R = 5$ cm in Xe + 0.12 % Ne, $p = 120$ Torr [9.22]

in contracted mode, a current filament of 0.5 mm radius is formed at the axis. An admixture of nitrogen somehow stabilizes the discharge in pure xenon: the transition occurs at much lower current, 1 mA instead of 19 mA, in the xenon-nitrogen mixture.

9.8.2 What Is Necessary for Contraction Onset

For the majority of electrons to be concentrated in a narrow channel in the vicinity of the axis (or at the spot where the filament was formed in a plane channel), at least two conditions must be met. *First,* electrons must be produced mostly where their density is high, i.e., there must be some *nonlinearly increasing* dependence of production rate on n_e, steeper than the usual one, $\nu_i n_e$ with $\nu_i(E) \approx$ const. The ionization frequency ν_i must fall appreciably from the axis to the periphery. If ν_i is independent of radius r, as we find for diffuse discharges, electron sources are uniformly distributed in volume in proportion to n_e, and electrons themselves fill the cross section more or less uniformly. Let us emphasize that only the dependence on n_e can produce the fall-off of ν_i away from the axis. Contraction leaves the longitudinal field homogeneous over the cross section, because curl $\boldsymbol{E} = 0$.

Second, the removal of electrons must be of *bulk* type, and it must be sufficiently fast for electrons liberated in a filament not to diffuse far laterally. Otherwise the diffusion (ambipolar) would make the electrons fill up the volume, even if the sources were concentrated on the axis. Electrons must disappear close to the place of production. As a result, contraction occurs only at sufficiently high currents and high n_e, when bulk recombination has a higher rate than recombination at the walls. The filament radius is determined by the greater of two lengths: the radius of the region where electron production sources are concentrated, and the distance

$$R_{\max} \approx \sqrt{D_a \tau_{rec}} \approx \sqrt{D_a / \beta n_e}\ , \quad \text{or} \quad R_{\max} \approx \sqrt{D_a \tau_a}$$

to which an electron can travel by diffusion during the recombination or attachment lifetime. The condition for contraction to be possible is that $R_{\max} \ll R$. This gives the estimate $n_e > 10^{11}\,\mathrm{cm}^{-3}$, which is supported by experimental data. The fact that contraction occurs in pure inert gases points to the formation of molecular ions such as Ne_2^+ and Ar_2^+, otherwise recombination would be too slow.

9.8.3 Mechanisms of Nonlinear Production

These are the same mechanisms that result in diffuse discharge instabilities: thermal instability, stepwise ionization, Maxwellization. All three ensure that the ionization frequency, that is, the ionization rate per electron, increases with n_e. This behaviour was explained in Sect. 9.6 for the case of stepwise ionization and Maxwellization. For the thermal mechanism, ν_i as a function of n_e is given by

$$\nu_i \propto \exp(-I/kT_e) \propto \exp\left[-\text{const}/(T_0 + \text{const}\, n_e\right]\ ,$$

where T_0 is the gas temperature in the case of very weak current (room temperature). This formula is implied by the relations: $T_e \propto E/N \propto T$ at $p \propto NT$ = const, and $T - T_0 \propto \sigma E^2 \propto n_e$. All three mechanisms manifest themselves at densities n_e greater than about $10^{11}\,\mathrm{cm}^{-3}$; it is fairly difficult to choose the main one among them. They may even act simultaneously, although experiments identify one predominant mechanism in some specific cases. Thus the thermal mechanism was not included in the interpretation of the experiments in neon (and argon) described in Sect. 9.8.1, although the overheating of the gas was tremendous, by a factor of 3.

The hysteresis observed in the transition from the diffuse to the contracted phase and back (Fig. 9.7) is a consequence of the two-valuedness of states in the transition zone. There are two ways to realize a given current i : it can be distributed at a low density over the entire cross section, or it may be concentrated at the axis in a small fraction of the cross section. In both cases, the heat transfer is balanced out by the heat released by the current. The transition from one state to the other is delayed because it starts only at a rather considerable penetration into the two-valued region, where one of the states is unstable. This is the case in which the instability growth rate is sufficiently large and the transition is triggered by a small fluctuation (see footnote 6 on p. 240).

9.8.4 Contraction of Discharge in Gas Flow

A discharge in plane channels with an intensive through-put of laser mixture clearly manifests the dependence of the maximum energy input, at which filamentation sets in, on the stream velocity u. The higher u, the shorter the residence time of a particle in the discharge zone, the shorter the time available for gas heating, and for the evolution of the instability, the more stable the discharge, and the higher the current can be increased before its homogeneous distribution breaks down. A clear-cut correlation of the homogeneity breakdown and gas temperature at the plane channel outlet is observed in a longitudinal discharge in plane channels (Fig. 8.1b). Contraction was observed each time the temperature rose by approximately 100° C. (These are the arguments in favour of the thermal nature of contraction under these conditions [9.23].)

A clear picture of filamentation under such conditions is shown in a unique photograph in Fig. 14.6. The methods of suppressing contraction developed for high-power laser design are described in Sect. 14.3, 14.4.

9.8.5 Filament and Arc

Although the contraction of glow discharges is often referred to as "arcing", the filament plasma is different from the typical equilibrium plasma of arc discharges with $T_e \approx T \approx$ 6000–10000 K. The characteristics of a filament lie between those of the strongly nonequilibrium plasma of diffuse glow discharges and the equilibrium plasma of arc discharges. The temperatures T_e and T in a filament differ appreciably: $T_e \approx (1-3)\cdot 10^4$ K, $T \approx (2-3)\cdot 10^3$ K; greater differences exist, however, in a glow discharge. The electron density $n_e \sim 10^{13}$–$10^{14}\,\mathrm{cm}^{-3}$, and the current density is greater than in the glow discharge but lower than in the arc. The field strength is lower than in the glow discharge but higher than in arcs. One feature in common with an arc discharge is the coexistence of the current region and currentless surrounding region, the longitudinal field being the same in both regions. In this sense, the thin column of the arc is also "contracted".

10. Arc Discharge

10.1 Definition and Characteristic Features of Arc Discharge

The discharge known as "the arc" is, as a rule, self-sustaining, with a relatively low cathode potential fall (of the order of the ionization or excitation potential of atoms, that is, about 10 eV). This characteristic distinguishes the arc discharge from the glow discharge, in which the cathode fall is hundreds of volts. The small cathode fall results from cathode emission mechanisms that differ from those in the glow discharge. These mechanisms are capable of supplying a greater electron current from the cathode, nearly equal to the total discharge current. This factor eliminates the need for considerable amplification of the electron current, which is the function fulfilled by the high cathode fall in glow discharges. Arc cathodes emit electrons as a result of *thermionic, field electron, and thermionic field emission.* More complex, combined processes of electron production at the cathode may also exist.

The arc discharge is characterized by large currents, $i \sim 1\text{–}10^5$ A, much greater than the typical currents of glow discharges, $i \sim 10^{-4}\text{–}10^{-1}$ A. The cathode current density is also greater than in glow discharges. It may be $10^2\text{–}10^4$ A/cm^2 for some modes of arc discharge and $10^4\text{–}10^7$ A/cm^2 for other modes. Note for comparison that the normal current density of 155 A/cm^2 on a copper cathode in air at a pressure $p = 1$ atm, (this is high for a glow discharge) corresponds to the lower limit of the arc range. As a rule, arcs burn at low voltage: not exceeding 20–30 V for short arcs, and as low as several volts in some cases. In many cases, but not always, $V - i$ characteristics of arcs are of the falling type.

Arc cathodes receive large amounts of energy from the current and reach high temperature, either over the entire cathode area or just locally, usually for short time intervals. They are *eroded* and suffer vaporization. The emission spectrum of the cathode region of a glow discharge coincides with the spectrum of the gas in which the discharge burns, but arc spectra show the lines of vapor of the electrode material. In fact, *vacuum arcs* burn in the vapor of the vaporized metal. As for the state of the plasma in the positive column, that is, the region between the layers adjacent to the electrodes, arc plasma may be in thermal *equilibrium* but quite often it may be *nonequilibrium*. (Thermal equilibrium of plasma is to be discussed in Sect. 10.11.) This characteristic depends on gas pressure. We can say that equilibrium plasma is found only in arcs, but nonequilibrium plasma is characteristic both of glow and arc discharges, the latter burning at low pressure.

10.2 Arc Types

Almost all forms of dc discharge, except the glow discharge, can be subsumed under the definition of the arc, as the discharge with low cathode fall. Consequently, quite a few discharge modes are classified as arcs. The classification may be based on the characteristics of cathode processes, plasma state in the positive column, or the medium (gas or vapor of cathode materials) in which the current is sustained.

10.2.1 Arc with Hot Thermionic Cathode

The cathode of such arcs gets heated as a whole to a temperature about 3000 K, or even higher, so that the high current of the arc is simply the result of intense thermionic emission. The current flows through a comparatively large area on the cathode and its density is $j_C \sim 10^2$–10^4 A/cm^2. The arc is anchored at a fixed spot on the cathode surface and the current spot is stationary. Only refractory materials, which vaporize with great difficulty, can withstand this high temperature: carbon (graphite, coal, carbon black), which does not melt at all at normal pressure (its boiling temperature $T_{boil} \approx 4000$ K), tungsten, which is especially widespread in practical devices ($T_{melt} \approx 3700$ K, $T_{boil} \approx 5900$ K), molybdenum, zirconium, tantalum, etc. Arcs with hot tungsten cathodes are employed in apparatus (often high-pressure units) requiring long service life (low erosion) of electrodes such as plasmatrons, welding machines, and certain types of arc furnaces.

10.2.2 Arcs with External Cathode Heating

This is a particular case of hot thermionic cathode, with the cathode heated not by the discharge current, but by an external source. The discharge is therefore *non-self-sustaining*. The required temperature is lowered by employing *activated* cathodes, as in electron tubes. An arc with external heating differs from a vacuum diode in that the gap is filled with a conducting gaseous medium. As the current increases and heats the cathode more intensively than the supply of external energy, the discharge may switch to the self-sustaining mode. Arcs of this type are employed in some low-pressure devices, and in thermionic converters of thermal energy into electric energy.

10.2.3 Arcs with Cold Cathode and Cathode Spots

In such arcs, the current flows through one or more spots that appear and disappear, and move rapidly and randomly on the cathode surface. The current density within a spot is very high, $j_C \sim 10^4$–10^7 A/cm^2. The metal at the point of short-time localization of a spot heats up greatly and is eroded and vaporized, but the cathode as a whole (and the spot's neighborhood) remain relatively cold. Spots always form on low-melting-point metal cathodes (copper, iron, silver, liquid mercury, etc.), which would not withstand the temperature required to work in the hot thermionic cathode mode (for Cu, $T_{melt} \approx 2570\,K$). At weak currents and low pressures, however, spots appear on refractory metals (W, Mo, etc.), as well. The main mechanism of electron emission of *cathode spots* seems to be thermionic field emission.

10.2.4 Vacuum Arc

This is an arc with cathode spots, initiated between electrodes in vacuum but burning in the metal vapor produced by intense erosion and vaporization of electrodes and immediately filling the discharge gap. Vacuum arcs are used in vacuum circuit breakers for high-current electrical equipment an important area of application of arc discharges.

10.2.5 High-Pressure Arc

By high pressure we mean pressures above $p \sim 0.1$–$0.5\,atm$, at which the plasma of the positive column is typically in equilibrium. Especially frequent among arcs of this type are arcs at atmospheric pressure, including those in open air. The column of the atmospheric pressure arc is a typical and widespread example of a *dense low-temperature equilibrium plasma* sustained by electric field. The temperature is usually $T \approx 6000$–$12{,}000\,K$, although higher temperatures, up to $50{,}000\,K$, are achieved under special conditions (Sect. 10.2.8).

10.2.6 Very-High-Pressure Arc:

$p \gtrsim 10\,atm$. This mode, belonging to the group of high-pressure arcs, deserves special attention. A plasma of such high density emits so intensively that up to 80–90 % of the released Joule heat is converted into radiation in the arc column. The yield of radiation is much lower at, say, atmospheric pressure. This property found an important application: the development of (super) high-pressure lamps. In these lamps, the arc burns in xenon or mercury vapors, which have optimal radiative characteristic and high efficiency of electroc-to-light power conversion.

10.2.7 Low-Pressure Arcs

These arcs burn at $p \sim 10^{-3}$–$10^{0}\,Torr$, so that the plasma in the positive column is strongly nonequilibrium, not differing in principle from the glow discharge plasma both with respect to "temperature gap" ($T_e \gg T$) and to the degree of ionization (which is much lower than the equilibrium value). The ionization is, however, higher than in glow discharge, because arcs burn at much higher current.

10.2.8 Special Modes

This group includes highly unusual modes, such as the so-called *Gerdien* arc found in 1922. This arc burns in a water vortex that squeezes the current channel away from a narrow diaphragm through which the channel passes. As a result of water evaporation, the arc is practically burning in water vapor. The highest temperature ever observed in an arc discharge, 50,000 K, was obtained at the channel axis of an improved Gerdien arc at a current of 1.5 kA and diaphragm diameter 2.5 mm.

10.3 Arc Initiation

10.3.1 Methods of Arc Initiation

The easiest way to start an arc is to connect the electrodes to a suitable power supply capable of providing sufficiently high current, bring the electrodes into contact, and then separate them. The electrodes may become red-hot at the point of contact, partly vaporize, and produce emission, so that at the moment of separation the arc strikes in the vapor, which is usually ionized more easily than the gas. The vapor is subsequently replaced by the gas (if it is there).

To initiate a high-current arc, an auxiliary anode can be employed. It is inserted between the main electrodes so as to touch the cathode, and is rapidly removed after applying the initial voltage.

When the cloud of ionized vapor, formed at the moment of short-circuit, reaches the main anode, the arc starts burning.

It is also possible to apply a high voltage sufficient for gas breakdown through the arc gap to stationary electrodes. In this case, the power supply and the external circuit must allow the burning of the arc discharge in accordance with the general $V - i$ curve (Fig. 8.4) and the loading curve. Normally, the usual mains voltage of 220 V is adequate for starting low-pressure arcs and mercury lamps in this manner. In the latter, a high pressure builds up gradually as the mercury evaporates; the first stage is a glow discharge that grows into an arc as the cathode heats up.

10.3.2 Transition from Glow to Arc Discharge

The transition is described by the segment FG of the $V - i$ curve in Fig. 8.4. The transition is caused by cathode heating as the current gradually increases and raises the current density of the abnormal glow discharge. With refractory (thermionic) cathodes, the transition is more or less smooth. If the cathode is made of a low-melting-point metal that behaves in the arc as a cold material, the glow discharge is suddenly transformed into an arc and cathode spots appear instantaneously. This occurs at lower currents ($i \sim 0.1$–1 A) than for thermionic cathodes ($i \sim 10$ A).

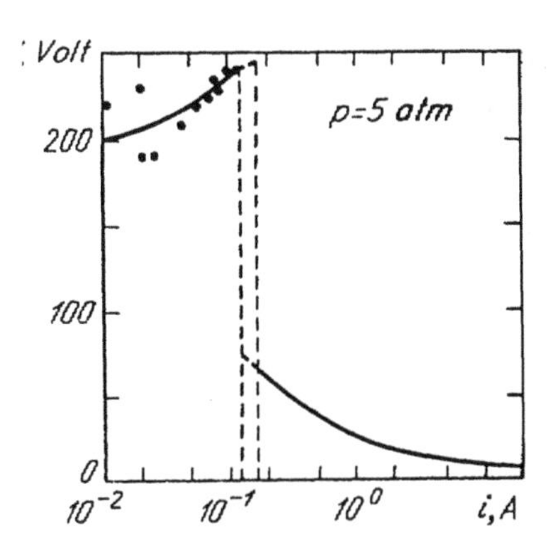

Fig. 10.1. $V - i$ characteristic of a xenon lamp, p = 5 atm., in the region of transition from glow to arc discharge [10.1]

Figure 10.1 shows the $V - i$ characteristic of the discharge in xenon lamp at $p = 5$ atm, at the transition stage [10.1]. At this high pressure, the glow discharge was stabilized by a high additional external resistance. At $i \approx 10^{-2}$–10^{-1} A, an abnormal glow discharge burns. At $i \approx 0.1$–0.2 A, it becomes unstable and is converted to an arc. The voltage at the electrodes then falls abruptly, while the current increases. In the unstable transition region, the same discharge current ($i \approx 0.2$ A) requires 250 V if the sustaining mechanism is secondary emission and multiplication of electrons in the cathode fall, but only 70 V if arc emission mechanisms are at work. The latter mechanism is more "economical", so that a spontaneous jump to arc discharge occurs; this is a clear illustration of instability manifestation (in this case, of the cathode process; see Chap. 9).

10.3.3 Short-Term Interruption of Current

Arcs with hot or cold cathodes behave quite differently when the current is interrupted. In the former case, the discharge can be restruck after a relatively long interval (up to 1 s for carbon electrodes) without the need to bring the electrodes together. In the latter case, even a very short interruption produces an irreversible effect. An arc with copper electrodes does not recover after a break of 10^{-3} s, and a mercury cathode arc, after 10^{-8} s.

10.4 Carbon Arc in Free Air

This is a classical example of the arc discharge. This mode, the first to be discovered (Sect. 1.4), is known as the voltaic arc. A moderate-current carbon arc is started by separating the initially contacting carbon electrodes, and a high-current arc is started using an auxiliary electrode (Sect. 10.3.1). This discharge was given the name "arc" because if the electrodes are arranged horizontally, the heated current channel bends upward since its central part floats owing to the Archimedean (buoyancy) force. The carbon arc at atmospheric pressure belongs to the class of high-pressure hot-cathode arcs. Figure 10.2a is a photograph of a vertically arranged arc which is axisymmetric.

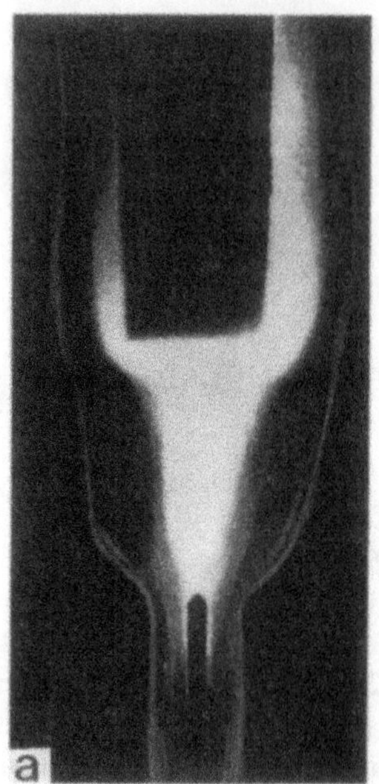

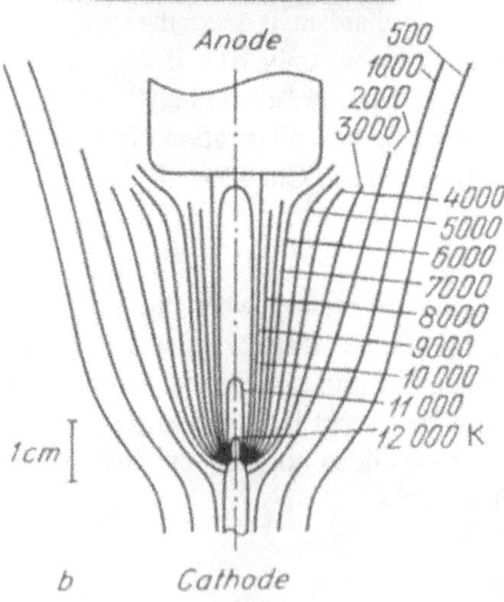

Fig. 10.2. Carbon arc in air at a current of 200 A: (a) a Toepler photograph, (b) measured temperature field [10.1]

$V - i$ characteristics of a carbon arc are shown in Fig. 10.3. The cathode fall was approximately 10 V, the anode fall about 11 V, that is, 21 V together, the rest falling across the positive column. With the separation between electrodes $L > 0.5$–1 cm, the arc burning voltage increases linearly with L, indicating a constant longitudinal potential gradient in the column. For instance, for $i = 7$ A, the field strength in a long homogeneous column is $E \approx 22$ V/cm. The field decreases with increasing current. At a certain current, the arc voltage drops to a lower level, the $V - i$ curve becomes almost horizontal, and a characteristic

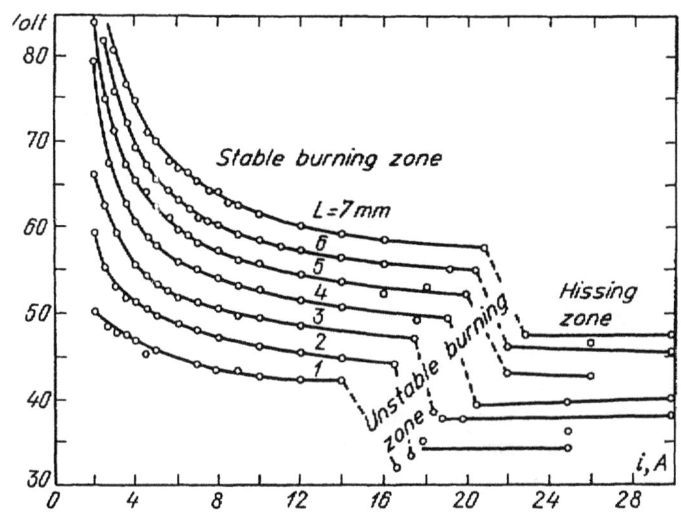

Fig. 10.3. $V-i$ characteristic of carbon arc in air. Values of L indicate the distance between electrodes [10.1]

hissing noise is heard. The sound is caused by intense vaporization of the anode in rapdily moving anode spots where the current density is high.

The plasma column in atmospheric-pressure air is in quasi-equilibrium. Figure 10.2b shows the recorded temperature distribution in the arc shown in Fig. 10.2a; $L = 4.6\,\text{cm}$, $i = 200\,\text{A}$. In the axial region, $T \approx 10{,}000\,\text{K}$, and at the maximum near the cathode, $T \approx 12{,}000\,\text{K}$. The radius of the high-temperature (hence, ionized and mostly electrically conducting) channel is about 0.5 cm; the channel slightly diverges toward the anode. The cathode temperature $T_\text{C} \approx 3500\,\text{K}$ and that of the anode, $T_\text{A} \approx 4200\,\text{K}$ (prior to "hissing"). The arc is anchored to the tip of the tapered cathode. The current density at small currents, $i \sim 1$–$10\,\text{A}$, is $j_\text{C} \approx 470\,\text{A/cm}^2$ at the cathode and $j_\text{A} \approx 65\,\text{A/cm}^2$ at the anode. As the current increases, j_C increases to $5 \times 10^3\,\text{A/cm}^2$at $i \approx 400\,\text{A}$, but further growth of current does not increase the cathode current density: the area of the cathode spot increases accordingly.

10.5 Hot Cathode Arc: Processes near the Cathode

10.5.1 Functions of the Cathode Layer

In principle, the cathode layer's functions are the same is in a glow discharge. In the absence of external heating, the cathode layer creates conditions for the *self-sustainment* of the current (in this particular case, a strong current). But this function is fulfilled in a different manner. The number of pairs of charges produced in the cathode layer of a glow discharge is such that their reproduction is ensured by the secondary emission due to the ionic flux. In the case of thermionic emission, the ions produced in the cathode layer must supply the cathode with the energy required to maintain the proper temperature. This method of extracting electrons from the cathode by ions, not by knocking them out individually, but by heating the metal, is more efficient. Secondary emission produces only $\gamma \sim 10^{-3}$–10^{-1} electrons per ion, so that the farction of ionic current at the cathode in a glow discharge is slightly less than unity, $1/(1+\gamma)$, and that of electronic current is $\gamma/(1+\gamma) \sim 10^{-3}$–$10^{-1}$. Analysis shows that thermionic emission in a hot cathode arc provides $S \approx 0.7$–0.9 of the total current, and ions carry to the cathode $1 - S \approx 0.1$–0.3 of this current. Hence, $S/(1-S) \approx 2$–9 electrons are emitted by the cathode per ion. This high efficiency of ions can be achieved only with a high current that greatly heats the cathode.

The difference in the value of cathode fall V_C is a result of the large difference in the ratios of the electronic and ionic currents at the cathode in glow and arc discharges. Practically the entire current in the electrically neutral plasma that follows the cathode layer is carried by electrons. Several generations of electrons must be produced in the cathode layer of a glow discharge in order to raise the fraction of electronic current from $\gamma/(1+\gamma) \sim 10^{-3}$–$10^{-1}$ to unity, and this calls for a voltage of hundreds of volts, because 30 to 50 V are needed on average to produce a pair of ions in a weakly ionized gas. The fraction of electronic current in the cathode layer of an arc discharge must be raised by only $1 - S \approx 10$ to 30 %. Not even one additional generation is required here. As a result, the arc cathode fall is of the order of, or even lower than, the ionization potential. At the same time, the cathode layer is unavoidable. The ionic fraction of current in the electrically neutral plasma of the positive column is negligibly small, $\mu_+/(\mu_e + \mu_+) \sim 10^{-2}$. This weak ionic current could not heat up the cathode even together with the purely thermal energy flux. Indeed, the fraction of the current carried by the ions is enhanced in the layer from very nearly zero to 10–30 %.

The arc cathode layer serves its function mostly as the region where ions are copiously produced and where the ions acquire from the field the kinetic energy that they carry to the cathode and add to the energy from other sources, namely, the potential energy of ion neutralization and the heat flux from the plasma. This is not the only role of the layer. It also serves other purposes, all of them interrelated. The gas temperature at the cathode surface coincides with the metal temperature and is at most half that in the positive column (Fig. 10.4). Charges cannot be produced here at all if only *thermal* ionization is at work. Note that the production rate in the layer must be even higher than in the column because a strong ionic current must be generated. The effect is achieved in the cathode layer via a nonthermal mechanism of imparting energy to the emitted electrons. Furthermore, the strong electric field produced at the cathode surface reduces the work function through the Schottky effect and hence facilitates thermionic emission. Both the field and the cathode fall arise, in their turn, in response to an enhanced ion production rate in the layer and the accumulation of positive space charge owing to different removal rates of electrons and ions.

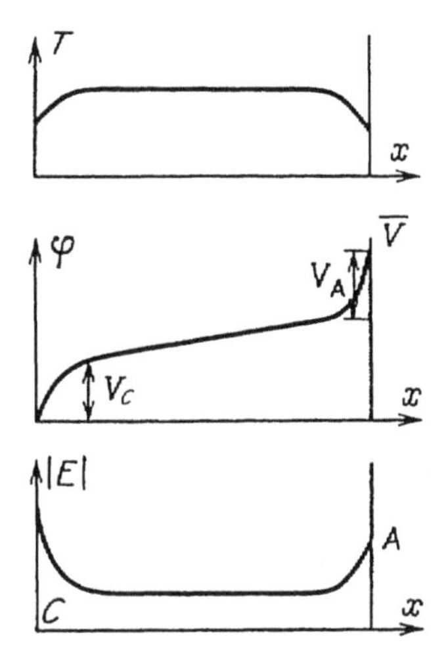

Fig. 10.4. Distributions of temperature, potential, and field in the arc from cathode to anode

10.5.2 Structure of the Cathode Layer

The main feature involved is the high current density at the cathode. A very high positive space charge is formed at the cathode. It entails a rather abrupt drop of field and potential in a very thin layer whose thickness is even less than one free path length of ions and electrons. This collisionless layer consumes a considerable fraction of the cathode potential fall, and its surface is separated from the positive column by a more extended intermediate layer of quasineutral plasma where the field is much weaker than at the cathode and where enhanced ionization of atoms by electron impact takes place. The electrons are accelerated in the collisionless layer to at least a considerable fraction of the ionization potential (Fig. 10.5). The remaining energy is gained in the region of collisions with other electrons. Here lies the main source of the production of ions that carry the current towards the cathode.

The electronic and ionic components of the total current remain unchanged, and equal to the cathode values, in the collisionless layer containing no charge sources. The fraction of the electronic current increases in the adjacent part of the cathode layer from S to $\mu_e/(\mu_e + \mu_+) \approx 1$, while the fraction of the ionic current falls from $1 - S$ to $\mu_+/(\mu_e + \mu_+) \ll 1$. Charge densities $n_e \approx n_+$ increase toward the positive column, as a result of intensive charge production. Electrons here undergo scattering collisions and some of them are returned to the cathode, forming a small *back current*. Although decelerated in the collisionless layer, these electrons partly reach the cathode. This effect resembles the electron flux from the plasma to a negatively charged probe. The magnitude of the back current is described by the same formula due to Langmuir (6.2).

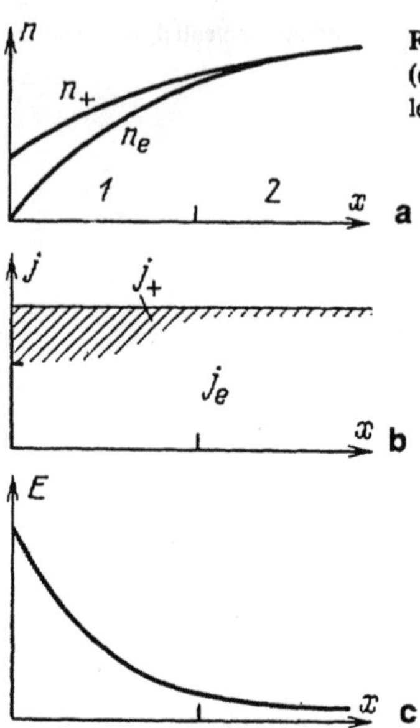

Fig. 10.5. Distributions of (a) charge density, (b) current, and (c) field in the cathode layer of an arc discharge. *(1)* collisionless layer; *(2)* quasineutral layer

10.5.3 Field at the Cathode

Consider the collisionless layer at the cathode. The x axis is directed from the cathode to the anode. The electronic and ionic current densities in the layer are constant and equal to

$$j_e = Sj = en_e v_e \ , \quad j_+ = (1 - S)j = en_+ v_+ \ . \tag{10.1}$$

Assume that the entire cathode potential fall V_C is concentrated across the collisionless layer. The field E_1 at the boundary with the plasma is much lower than the field E_C at the cathode boundary. The energies and velocities of charges are determined by the potential difference they fall through:

$$v_e = (2eV/m)^{1/2} \ , \quad v_+ = \left[2e(V_C - V)/M\right]^{1/2} \ . \tag{10.2}$$

We substitute (10.1, 2) into Poisson's equation

$$-\frac{d^2V}{dx^2} = 4\pi e(n_+ - n_e) = \frac{4\pi j}{\sqrt{2e}}\left[\frac{(1-S)\sqrt{M}}{\sqrt{V_C - V}} - \frac{S\sqrt{m}}{\sqrt{V}}\right] . \tag{10.3}$$

Remarking that $d^2V/dx^2 = (1/2)d(E^2)/dV$, integrating (10.3) once with the boundary condition $E = E_1 \approx 0$ at $V = V_C$, and determining the field at the cathode boundary where $V = 0$, we find that it has a value given by

$$E_C^2 = \frac{16\pi j}{\sqrt{2e}}\left[(1-S)\sqrt{M} - S\sqrt{m}\right]\sqrt{V_C} \ . \tag{10.4}$$

This formula was derived by *McKeown* in 1929.

The positive space charge due to ions helps enhance the field at the cathode. The negative charge of electrons partly compensates for this effect. However, the degree of compensation is very small for $S \approx 0.7$–0.9 which is realistic for any arc. The second term in (10.4), describing the effect of electrons, comes to at most several per cent of the first and thus can be safely dropped. In numerical form,

$$E_C = 5.7 \cdot 10^3 \, A^{1/4}(1 - S)^{1/2} V_C^{1/4} j^{1/2} \, , \quad h = 4V_C/3E_C \, , \qquad (10.5)$$

where A is the atomic mass of the ion; E_C [V/cm]; V_C [V]; j [A/cm^2]; and h is the thickness of the layer under discussion. We easily find h by integrating (10.3) twice without the electron term. The situation in the collisionless part of the cathode layer is similar to the process in the vacuum diode (Sect. 6.6.1) and (10.5) corresponds to the "three-halves power" law (6.15). The role of charge emitter in this "diode" is played not by the cathode but by the plasma adjacent to the layer; as in the case of a negative probe (Sect. 6.6.2), this plasma supplies the layer with positive ions.[1]

Equalities (10.4) or (10.5), which are in fact valid for the cathode spot as well, form a unified set of equations together with other relations describing processes at the cathode. However, for purposes of evaluation, unknown values can be taken from experiments. For example, with the parameters typical for hot-cathode arcs $j = 3 \times 10^3$ A/cm^2, $S = 0.8$, $V_C = 10$ V, $A = 28$ (nitrogen) we obtain $E_C = 5.7 \times 10^5$ V/cm, $h = 2.3 \cdot 10^{-5}$ cm. By Schottky's formula (4.13), this field reduces to work function by $e\Delta\varphi = 0.27$ eV. At $T = 3000$ K, the thermionic emission rate increases by a factor of $\exp(e\Delta\varphi/kT) \approx 3$.

10.5.4 Energy Balance at the Cathode and the Fraction of Ionic Current

The cathode temperature, which is the main factor determining the thermionic emission current, and the ratio of ionic to electronic components are found from the system of equations describing the energy balance at the cathode, the charge production in the cathode layer, (10.4), etc. Calculations of this type [10.2] are highly imperfect, because the processes are complex and numerous, and the data on important parameters are incomplete. The results obtained are mostly of qualitative value. A representative estimate of one of the most interesting quantities, S, can be obtained by constructing the energy balance in a simplified manner.

Each ion carries to the cathode the kinetic energy required in the cathode fall V_C. A part of it, β_1, is transferred to the cathode upon impact; β_1 is known as

[1] The behavior shown by (10.5) is typical of currents so strong that the space-charge layer is very thin and ions move there without collisions. If the current is weak and the layer is extended, with ions in drift, then $dE/dx = 4\pi j_+/\mu_+ E$. Hence, $E_C \approx (8\pi j_+ h/\mu_+)^{1/2}$, $V_C \approx 2E_C h/3$. Instead of (10.5), we obtain $E_C = (12\pi j_+ V_C/\mu_+)^{1/3}$, and the current in a "gas-filled" diode, being space-charge limited, is given, instead of (6.15), by $j_+ \approx (9\mu_+/32\pi)V^2/h^3$, as in the cathode layer of a glow discharge [see (8.14)].

the *accommodation coefficient*. When an ion is neutralized, the energy released is equal to the ionization potential I of the resultant atom, minus the work function φ which is spent on removing the neutralizing electron from the metal. A part of it, β_2, is also transferred to the cathode. The fractions $1-\beta_1$ and $1-\beta_2$ of energy remain with the outgoing atom. Not much is known about the accomodation coefficients β_1 of kinetic and β_2 of potential energy but certain similarities and indirect arguments make it possible to assume that both are of order unity. With each emitted electron, the cathode loses the energy $e\varphi$ (plus a small energy $2kT_C$ carried by the emitted electron).

If we assume that the energy supplied by ions to the cathode is completely spent on electron emission, and set $\beta_1 = \beta_2 = 1$, we find [10.1]

$$j_e\varphi = j_+(V_C + I - \varphi)\ , \quad S = \frac{j_e}{j_e + j_+} = \frac{V_C + I - \varphi}{V_C + I}\ . \tag{10.6}$$

For example, for $I = 14\,\mathrm{V}$, $V_C = 10\,\mathrm{V}$, and $\varphi = 4\,\mathrm{V}$, we have $S = j_e/j = 0.83$, $j_+/j = 0.17$. These are reasonable values that do not contradict other data. However, the actual energy balance has other components as well. The cathode receives from the plasma the heat conduction flux Q_T due to the temperature difference between plasma ($T \approx 6000$–$12{,}000\,\mathrm{K}$) and cathode ($T_C \approx 3000\,\mathrm{K}$). Another component is the radiant flux Q_{rad}. The heat entering the emitting surface of the cathode is transported into the metal by the heat conduction flux Q_H because the opposite part of the cathode body is colder (owing to cooling). Some part $Q_{rad,C}$ is radiated away by the cathode.

The dimension of Q is that of power [W]. Dividing Q by arc current i [A], we denote $q = Q/i$ [V]. The meaning of each respective q is the energy per electron charge transported by the electric current (numerically, it is energy in eV). The energy balance at the cathode, calculated per unit transported charge, is (for $\beta_1 = \beta_2 = 1$)

$$q_H = (1 - S)(V_C + I - \varphi) + q_T + q_{rad} - S\varphi - q_{rad,C}\ . \tag{10.7}$$

When all these factors are taken into account, the value of S remains of the same order of magnitude as the estimate in (10.6). Detailed calculations of the balance, involving other equations for quantities in (10.7), demonstrated that as the arc current increases, the temperature and current density at the cathode grow but the cathode fall, the field at the cathode, and the fraction of ionic current are all reduced [10.2].

Note that when the fraction of ionic current is calculated, one often evaluates energy balance not at the cathode itself, via (10.6), but in the cathode layer [10.2]. It is assumed that the entire energy gained by electrons within the cathode fall goes into ionization, that is, into the production of ions that subsequently reach the cathode. This gives

$$j_e V_C = j_+ I\ , \quad S = \frac{I}{I + V_C}\ , \quad \frac{j_+}{j} = \frac{V_C}{I + V_C}\ . \tag{10.8}$$

The same numerical values that were taken to evaluate S via (10.6) now yield $S = 0.58$, $j_+/j = 0.42$. Presumably, (10.8) gave an overestimate of the value of the ionic current because, in view of exciting collisions, the production of one pair of ions consumes more energy, on the average, than one ionization potential.

10.5.5 Results of Measurements

Figure 10.6 plots the cathode fall measured by different methods (tungsten cathode, helium at $p = 1$ atm) as a function of arc current i. The calorimeter method consists in measuring calorimetrically the heat flux (power) transported by heat conduction from the cathode surface into the metal, Q_H [W]. The cathode fall V_C is obtained from this measurement and from the known current i by employing theoretical considerations concerning the energy balance at the cathode. Figure 10.6 also shows the measured ratio $q_H = Q_H/i$ known as the *voltaic equivalent of heat flux* into the cathode. It is related to V_C by the energy balance formula (10.7). In another method, one measures the decrease in voltage across electrodes when they are brought closer together. It is assumed that immediately before the moment of shorting, after which the voltage drops abruptly to a fraction of one volt (this corresponds to the contact resistance), the positive column disappears completely and the voltage equals simply the sum of the cathode and anode falls. The latter fall is measured separately and is subtracted. All methods point to a reduction in V_C as the current increases, although numerical results sometimes quantitatively disagree. It seems that the result of probe measurements is the most reliable.

Figure 10.7 shows the radial temperature distributions at the end-face current surface of a tungsten rod-shaped cathode 0.6 cm in diameter and 2.5 cm long. The temperature increases to 4000 K at the center. Note how the high-temperature current area grows as the current increases. The surface temperature is mostly measured by measuring surface brightness. The current density at the tungsten cathode is 10^3–10^4 A/cm^2. These parameters are found from the known total current and measured temperature distribution, which indicates the area occupied by the cathode current. The calculation is checked with the thermionic emission current formula. The fraction of ionic current cannot be measured directly. It is

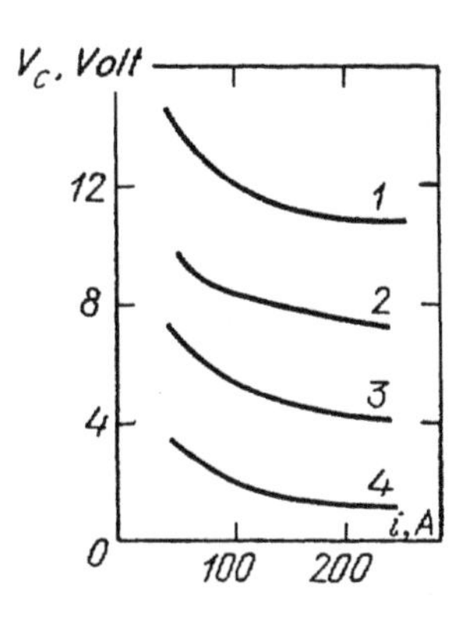

Fig. 10.6. Cathode fall of potential V_C as a function of arc current using tungsten cathode in helium atmosphere, $p = 1$ atm.. Measured by *(1)* probe, *(2)* calorimetric method, and *(3)* by bringing electrodes close together. Curve *(4)* is the voltaic equivalent of the thermal flux into the cathode, q_H [10.2]

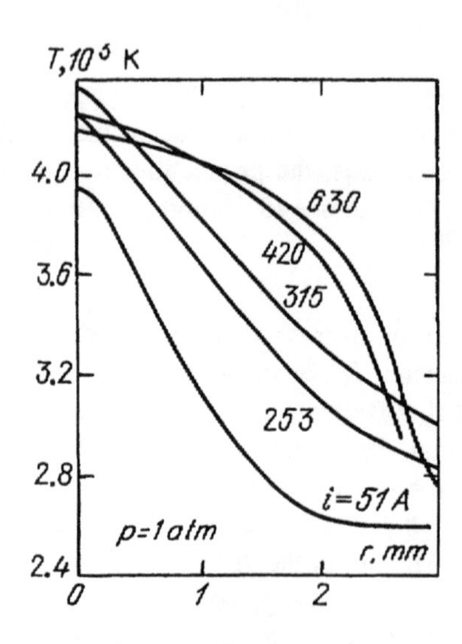

Fig. 10.7. Temperature distribution along the radius of the end-face surface of a tungsten rod cathode in argon at 1 atm. [10.2]

extracted by analyzing the energy balance and other measured quantities. The analysis gives $1 - S \approx 0.1$–0.3.

One of the problems that are very important for exploiting high-power electric arc devices is that of *erosion* of electrodes, especially cathodes. Even refractory metals undergo destruction and vaporization. The presence in the gas of small amounts of oxygen and water vapor, even in concentrations as low as 0.1 %, leads to oxidization and sharply enhances erosion. This is especially true for such widespread arc environment as air. The cathode resistance to erosion is characterized by the *specific erosion,* defined as the mass lost per coulomb of charge passed through the arc. The specific erosion depends on a number of conditions. For tungsten rod cathodes in inert gases at currents of hundreds of amperes and moderate-to-high pressures, it is of order 10^{-7} g/C. For example, this means that a cathode loses 0.1 g per hour at a current of 300 A.

10.5.6 Hollow Cathode

At low pressures, $p \lesssim 1$ Torr, cathode spots are formed even on tungsten (Sect. 10.6), with material in a spot being eroded much more strongly than in the hot cathode mode. Consequently, rod cathodes of refractory metals are used in practical devices only at high pressures. Designs based on the hollow cathode principle are used at low pressures. A *hollow cathode* in its simplest form is a segment of a tube to whose inner surface the arc is anchored, in similarity to a glow discharge (Sect. 8.5.4). Very high durability of refractory-metal cathodes has been achieved in arc devices with a gas blown through a hollow cathode, creating a specific gas environment inside it [10.2]. Such devices manifest record-low specific erosion of 10^{-9}–10^{-10} g/C.

10.6 Cathode Spots and Vacuum Arc

Highly localized current centers appear on the cathode when it is necessary to carry a considerable current, but the cathode cannot, for one reason or another, be heated as a whole to a high temperature. Such reasons may lie in low melting point of the metal (this is a typical situation), in a relatively low current that can cause emission from the cathode only when concentrated to a small area, or in low ambient pressure,. In the latter case, it is necessary to inject into the gap some number of atoms (vapor) of cathode material, otherwise there can be no ionic flux transporting energy to the cathode and the cathode emission cannot be sustained. However, a high concentration of energy is at the same time the condition for efficient vaporization of the metal. Indeed, if (roughly) $i < 1$–$10\,\text{A}$, $p < 1$ Torr, cathode spots are formed even on refractory metals, which behave as thermionic cathodes at higher currents and pressures (as described in Sect. 10.5). In contrast to anchoring to a spot, the latter effect is known as *diffuse anchoring* of the arc to the cathode. In the case of low-melting-point metal cathodes, spots are formed at any pressure and any current.

10.6.1 Basic Experimental Facts

a) Evolution of spots. At the earliest stage of the discharge, small, fast spots moving randomly and independently of one another are formed on the cathode. Their size $r \sim 10^{-4}$–$10^{-2}\,\text{cm}$, and velocities are from 10^3 to $10^4\,\text{cm/s}$. They cause negligible, presumably nonthermal, erosion of the surface. It has been suggested that erosion is caused by microscopic explosions of tiny protrusions on the surface, in response to current concentration in the metal itself. The arrangement of spots changes in a time of about $10^{-4}\,\text{s}$. Small spots merge into larger and less mobile spots moving at velocities of about 10–$10^2\,\text{cm/s}$. The erosion in such enlarged spots is considerably more intense and is of thermal nature because it is caused by heating and vaporization of macroscopic segments of the surface at the expense of the energy injected by the ionic current originating from the vapor plasma.

b) Threshold current and spot multiplication. The current through an individual spot cannot be too low. There is a minimum, threshold current for a spot, and hence for the arc as a whole: $i_{\text{min}} \sim 0.1$–$1\,\text{A}$. As the current decreases to i_{min}, it becomes concentrated in a single spot; if $i < i_{\text{min}}$, the arc burns out. It was empirically found that for many nonferromagnetic materials, $i_{\text{min}} \approx 2.5 \times 10^{-4} T_{\text{boil}} \sqrt{\lambda}$ [A], where λ [W/cmK] is the heat conduction coefficient; and T_{boil} is in degrees Kelvin [K]. The physical meaning of this relation is not yet clear; this is also the case for much in the behavior of cathode spots. On the average, the current through a single spot is from 1 A (on liquid mercury) to 300 A (on tungsten). As the discharge current increases, the number of spots grows. As a rule, the spots do not appear on fresh places, but multiply by "splitting". The spot lifetime with respect to splitting depends on current growth rate. If $di/dt \sim 10^5$–$10^7\,\text{A/s}$, it is on the order of $10^{-5}\,\text{s}$. Sometimes a spot may decay and vanish.

c) Current density. If the current through a spot $i \approx 10$ A and the spot radius $r \approx 10^{-3}$ cm, the current density is $j \approx i/\pi r^2 \approx 3\times10^6$ A/cm^2. Spot current densities reported by different authors lie in the range $j \sim 10^4$–10^7 A/cm^2, and even reach 10^8 A/cm^2 for copper. The data on j are extremely uncertain, imprecise, and contradictory. Typically, j is measured when only one spot survives, with known spot current i, but the difficulty lies in determining the spot size. When high-speed photography is used, the current spot size is identified with that of the luminous region, even though this is doubtful. The spot size may be found after switching the arc off from the erosion left on the cathode surface (autograph technique); here again, one cannot be sure that identification is unambiguous.

d) Vaporization. Cathode spots that receive the high-concentration energy flux transported by the ionic current become sources of intensitve vapor jets. Jet velocities are of the order 10^5–10^6 cm/s. The specific erosion within large spot clusters on copper reaches 10^{-4} g/C. We have already mentioned that erosion is much less intense in individual spots at the early stages of the discharge; for example, 5×10^{-7} g/C has been reported for copper.

e) Temperature and number densities of particles. The data on metal surface temperature within a spot are very contradictory. The temperature is found either by measuring brightness or by evaluating the pressure and density of vapor at the surface. The values reported for spots on mercury range from 700 to 2000 K, and on copper, from 2400 to 3700 K. The vapor density evaluated on the basis of saturation pressure is about 10^{17}–10^{19} cm^{-3} (calculations for Cu give T closer to 3700 K, and n_e closer to 10^{19} cm^{-3}). The concentration of neutrals, determined by measuring the attenuation of a probe electron beam in the vacuum arc at an iron cathode, was found to be 10^{16}–10^{17} cm^{-3}. Measurements of charged particle number density over cathode spots (by Stark broadening and by the ratio of intensities of the spectral lines of copper atoms and ions) gave $n_e \approx 5\times10^{17}$ cm^{-3}, which is typical of dense low-temperature plasma. Hence, the vapor density and the degree of ionization are quite high. Similar measurements lead to the value of electron temperature $T_e \approx 1$–2 eV.

f) Cathode fall. There are two ways to measure cathode fall: by probes, and by "bringing electrodes together" (Sect. 10.5.5), including the recording of the burning voltage of a short vacuum arc. It is assumed that the anode fall of a vacuum arc is small. Observations indicate that the anode fall is especially low for metals with low metastable levels. This is an argument in favor of stepwise ionization of atoms in the metal vapor. Table 10.1 lists some parameters of cathode spots of several metals. We have included either averaged values or their spread if the results of different authors are very divergent or a quantity is strongly dependent on measurement conditions. Among the multitude of data accumulated over decades of studies, the table selects, where possible, those measurement results that have been obtained since 1960.

Table 10.1. Characteristics of cathodes with cathode spots

Metal	Cu	Ag	Zn	Hg	Fe	W
Ionization potential [V]	7.68	7.54	9.36	10.39	7.83	7.98
Potential excited metastables [V]			4	4.7–4.9		
Cathode fall [V]	15–21	12–16	10–11	8–9.5; 19[a]	17–18	16–22
Boiling point [K]	3510	2436	1046	630	3045	5900
Thermal conductivity [W/cm K]	4.1	4.17	1.13	0.104	0.84	1.67
Threshold spot current [A]	1.6	1.2	0.3	0.07	1.5	1.6
Average current to a spot [A]	75–200	60–100	9	0.5–2	60–100	100–300
Current density [A/cm^2]	10^4–10^8		3×10^4	10^4–10^6	10^7	10^4–10^6
Specific erosion at a current of 100–200 A [g/C]	10^{-4}	1.3×10^{-4}				1.3×10^{-5}
Jet velocity [km/s]	15	9	3–5	1–4	9	13–30

[a] Two modes are observed: $V_C \approx 8$–9.5 V in one, and 19 V in the other

g) $V - i$ characteristic. $V - i$ characterisitcs of "metal" vacuum arcs are plotted in Fig. 10.8. In contrast to many other arcs, all these are rising characteristics. No explanation of this fact has been found. Details of experimental results (and attempts at interpreting them) can be found in [10.4–6], where bibliographies are also given.

h) Spot in a magnetic field. One curious phenomenon was noticed by *Stark* in 1903, who observed the behavior of cathode spots on the surface of liquid mercury cathodes: in a magnetic field tangential to the surface, the spot moves in the direction opposite to the magnetic force $[\boldsymbol{i} \times \boldsymbol{H}]$ acting on the current. No satisfactory explanation of this effect has been found, even at the qualitative level (despite numerous attempts).

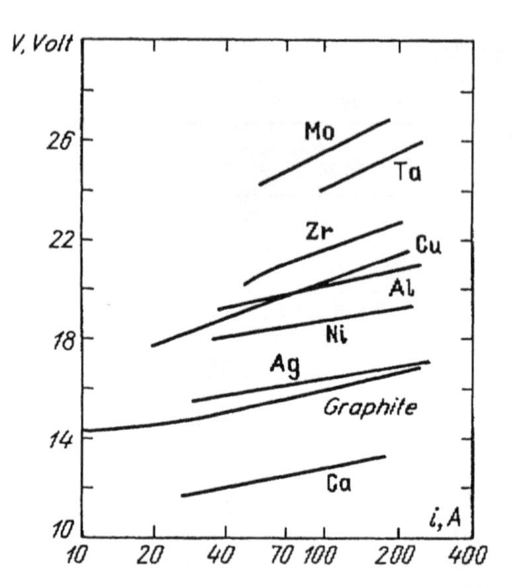

Fig. 10.8. $V - i$ curves of low-current vacuum arcs with various materials as cathodes. Electrode diameter is 1.27 cm and separation 0.5 cm [10.3]

10.6.2 The Status of the Theory of Cathode Spots

Very few phenomena in gas discharge physics have generated a comparable number of hypotheses, models, and theoretical frameworks to those devoted to cathode spots. None of them explains the totality of experimental facts. The theories offer descriptions (sometimes correct ones) of some parts of the picture but leave many a "Why?" unanswered. Why do spots split? Why do they "run"? Why do they shift in the "wrong" direction in a magnetic field? Why such tremendous current densities [about 10^8 A/cm^2, assuming these values are reliable]? And so forth. In 1968 one arc discharge researcher counted as many as 17 different – and even mutually exclusive – interpretations of cathode spot effects [10.4], and a host of new papers have been published since. This situation is definitely caused by the extreme complexity and entanglement of the picture in which solid state, surface, interfacial, plasma, electric, and thermal processes are closely intertwined.[2]

What happens in the spot area greatly resembles the picture found at a hot cathode. The spot receives the ionic current. Ions bring in the kinetic energy acquired in the cathode fall. Together with the potential energy released in neutralization, this kinetic energy goes to the heating up of the metal. This factor combines with the effect of strong field (Sect. 10.6.3) to produce emission and, on the other hand, vaporization of the metal. In a vacuum gap, vapor creates the medium that is ionized by emitted electrons accelerated in the cathode fall. This,

[2] The exceptional practical importance of cathode spot phenomena focused the attention of specialists and led to an impression that the cathode spot si a uniquely complex object. In fact, numerous complex and unexplained aspects of gas discharges are passed by simply because they seem to be of secondary importance. The situation may well change with the appearance of a practical stimulus, as happened with glow discharge instabilities (Chap. 9).

in turn, constitutes the source of ionic current. The process of evaporation from a localized heated center depends on the outflow of energy by heat conduction into adjacent metal layers, on other components of the energy balance, etc. Quite a few elements of this picture allow calculation in reasonable agreement with experiments, although theories included, until recently, a number of experimentally determined parameters.

The hydrodynamic model of a cathode spot, in which the complex of processes is reconstructed in a self-consistent form and no empirical parameters are used, has been fully developed [10.7]. It also describes cathode spots on film cathodes. Such cathodes are often used in real systems. It is not infrequent that the cathode spot which is seemingly localized on the massive cathode actually burns on the oxide coating or a coating of other containments. Inhibited heat transfer from the film to the massive cathode facilitates local heating. As the film under the spot burns out, the spot tends to jump to another area on the film, rather than to continue burning on the surface of the massive cathode. This is one of the causes of spot displacement, in addition to the jumping of the discharge from one microscopic protrusion to another after thermal destruction of the former.

10.6.3 Emission Mechanism

This is definitely a point of crucial importance. Among familiar emission mechanisms, only thermionic field emission can supply current densities $j \sim 10^6\,\mathrm{A/cm^2}$ without assuming the existence of superhigh temperatures and fields at the surface.[3] Table 4.8 shows that for $T = 3000\,\mathrm{K}$ and work function $\varphi = 4\,\mathrm{eV}$, the emission current density reaches the required order of value $10^6\,\mathrm{A/cm^2}$, even without assuming that the field near microscopic protrusions on the surface is enhanced. The survival of such tips in a well-developed spot appears to be doubtful because they would be the first to melt, splash out, and vaporize.

The spot current density j, the ionic current fraction $1 - S$, the field E at the surface, and the metal temperature T are related by a system of equations describing the emission, $j_e = Sj = F(T, E)$, the field according to (10.5), $E = C\sqrt{j_+} = C\sqrt{j(1-S)}$, and the energy balance. *Lee* [10.8] carried out in 1959 a series of computer calculations using the first two equations and fixing T, work function φ, and the constant C roughly corresponding to copper ($A = 63$, $V_C = 10\,\mathrm{V}$). These computations, which determined the possible sets of spot parameters, indicate that results not in dire contradiction with experimental data can be obtained on the basis of a thermionic field emission model together with equation (10.5) for the field. Several such sets are listed in Table 10.2. It must be remarked that the fraction of ionic current which is then required at low j and E

[3] Even before quantum mechanics had been developed, Langmuir hypothesized in 1923 that fields due to space charge may pull electrons from the cathode spot of the arc. When the quantum-mechanical field emission of electrons became known (1928; Sect. 4.6.3), McKeown further developed Langmuir's hypothesis and derived (10.4) for the field at the cathode.

Table 10.2. Parameters of arc cathode spots, consistent with current densities j_e and j_+ and with the field at the cathode E, for given T and φ [10.8]

T [K]	φ [V]	E [V/cm]	j [A/cm^2]	j_+/j
3000	4	3.3×10^7	4×10^6	0.30
3000	4	2.8×10^7	2×10^6	0.45
3000	4	0.8×10^7	1×10^5	0.83
3500	4	2.6×10^7	4×10^6	0.23
3500	4	2.3×10^7	2×10^6	0.30
3500	4	0.8×10^7	1×10^5	0.45
3000	3.5	2.5×10^7	4×10^6	0.20
3000	3.5	2.1×10^7	2×10^6	0.25
3000	3.5	0.6×10^7	1×10^5	0.45

appears to be unrealistically high. On the other hand, the cathode fall produces a strong field at the highest values of j and E (when the fraction j_+/j is realistic) only if it is concentrated in a layer of thickness $h \approx V_C/E \approx 10/3.3 \times 10^7 \approx 3 \times 10^{-7}$ cm, that is a mere of ten atomic diameters! In order to obtain 10^8 A/cm^2, we need $E \approx 1.5 \times 10^8$ V/cm, $h \approx 6 \times 10^{-8}$ cm, which is hardly imaginable (being of the order of the thickness of surface layer). On some other models and theories, see [10.4, 6]; however, the one outlined above is favored now as it seems capable of explaining $j \sim 10^6$ A/cm^2.

As shown in [10.7], this model covers copper-type materials with moderate volatility. In non-volatile materials (tungsten) or easily volatile materials (mercury), the parameters of the cathode spot in a vacuum arc cannot be explained in these terms. At 6000 K, which is required for the evaporation of tungsten (without which there would be no ions), the thermionic emission current density would be much higher than the observed values. No field is then generated at the cathode to pull electrons away. The negative space charge at the cathode cancels the strong field and limits the spot current as it does in vacuum diodes (Sect. 6.6). In contrast, thermionic emission and thermionic field emission are negligible in mercury where even the critical temperature is as low as 1753 K. Unrealistically high fields are necessary for electron field emission to be solely responsible for producing this current. As conjectured, in [10.9], no conventional cathode layer exists in a mercury arc and an *electric double layer* of thickness less than one free path length is formed close to the surface. According to this model, energy deposited in the plasma by ions accelerated in the double layer produces conditions for thermal ionization and formation of a *plasma cathode*.

10.6.4 Explosive Emission

This is a suitable place to describe this phenomenon discovered by *Mesyats* in 1966 [10.10], because it has common features with the processes in the cathode spot of vacuum arcs. Experiments demonstrated that in a pulsed breakdown of a vacuum gap with a very thin tip a sharp current rise is accompanied, after some delay from the moment of applying the voltage, with an explosion of the tip point and an ejection of a cathode plume, that is, a blob of plasma. Similar processes take place on massive cathodes as well, but explosions occur on microscopic protrusions on the surface. These effects are studied by high-speed oscilloscopes, and by photoelectron and electron optic techniques with time resolution of 10^{-10}–10^{-9} s. On a tungsten tip with tip radius 0.2 μm and with field at the tip 1.2×10^8 V/cm, the current delay and explosion "induction" time are 1 ns. The time lag is the longer, the weaker the field and (in particular) the lower the current.

As a result of repeated application of voltage pulses about 10 ns long and with 1 ns rise time, the tip is eroded and smoothed but new microscopic tips are formed on it, with the tip field enhanced by an order of magnitude. Now the delay of breakdown and of current by 1 ns takes place at the macroscopic field of 10^7 V/cm at the tip. If the cathode is massive, the field at microscopic tips is enhanced by a hundredfold and more, so that the same effect is obtained at a macroscopic field of 10^6 V/cm. The plasma blob expands in all directions at a velocity of about 2×10^6 cm/s, consuming about $(2\text{–}3) \times 10^{-3}$ g/s. The current density from a tip surrounded with plasma reaches 10^8 to 10^9 A/cm^2, the current being 10^2 to 10^3 A from an area of about 10^{-6} cm^2.

The metal explodes because a large amount of Joule heat is released at the end of the (micro) tip by the current of field emission which transforms to thermionic field emission as the tip heats up. The fact that these currents are indeed high has been established experimentally. Thus it was possible to obtain, without tip destruction (i.e., only at the expense of field emission), currents up to 4×10^9 A/cm^2 with voltage pulses 5 ns long, on a tip of 0.1 μm radius and 10° angle at the open of the cone. When electrons are ejected from the surface, they are replaced by new ones from inside the metal, so that a current this enormous density passes through the metal at the emitting surface. It is this current that heats the tip to the explosive vaporization of microscopic protrusions; the delay is the time necessary to accumulate the required energy.

It is interesting to identify the mechanism of emission at that stage after explosion during which current is usually sharply enhanced. Analysis shows that the number of electrons transported in the current pulse is grater than the number of atoms in the plasma blob by a factor of 10–10^3. Therefore, the ionization of the vaporized material is not crucial. Presumably, the processes are similar to what occurs in the cathode layer of the arc: the plasma plume is polarized, positive space charge is formed at the metal surface, the field is further enhanced, and even more intense thermionic field emission develops, as in the cathode spot of vacuum arc. The effect of explosive emission is exploited in devices for generating high-power nanosecond electric pulses and electron beams [10.10]. Equipment has already been developed in which a power of the order of 10^{13} W was achieved.

10.7 Anode Region

Processes at the anode are also complex and very diverse. As on the cathode, the arc may be anchored to the anode in two ways.

a) Diffuse anchoring. In this mode, the current is spread over a relatively large area of the anode, at a density $j \sim 10^2\,\text{A/cm}^2$. Material erosion is negligible in this mode because energy flux densities at the surface are not very high.

b) Anode spots. A spot is formed when the anode is small and the growing current is forced to occupy its edges, "awkward" areas, lateral patches, etc. At a certain current the discharge is destabilized and contracts at the anode surface. Current density in the spot reaches $j \sim 10^4$–$10^5\,\text{A/cm}^2$. As a rule, the number of spots increases with increasing arc current and pressure. Sometimes many spots are formed, arranged in symmetric regular patterns. The spots may move, also following regular trajectories such as concentric circles; anode spots are very bright and eject vapor jets.

10.7.1 Potential Fall

a) Space-charge layer. The anode fall is composed of two parts. One part reflects the formation of a negative space-charge layer at the very anode surface because the anode repels positive ions towards the cathode. This component of the anode fall arises to compensate for the absence of ionic current in the neighborhood of the anode, via an increased degree of ionization of gas (vapor) atoms or a corresponding slight increase in electronic current. Its magnitude is of the order of ionization or excitation potential for vapor or gas atoms. This value of anode fall may prove sufficient in the case of diffuse anchoring. The anode fall may then be very small or even negative, 1 to 3 V. This situation is observed when the anode surface is large, a dense highly ionized plasma is adjacent to it, and there is no need to supply additional energy for sustaining the electron current to the anode.

b) Geometric fall. If the anode surface area is small, less than the cross section of the positive column, or if the arc is strongly contracted on the anode, the current channel must undergo compression from the column to the anode. The current is carried almost entirely by electrons. The total current $i \approx n_e \mu_e E S_{ch}$, where S_{ch} is the channel cross section, can be conserved in the course of channel contraction if the electron density or the drift velocity (i.e., field), or both these quantities simultaneously, increase. Additional field, i.e., additional voltage, is needed to produce additional electrons and thus compensate for recombination enhanced by increased n_e, and to enhance drift. This is how the second component of potential fall appears in the anode region. Its origin is rather geometric, not connected with the formation of space charge at the electrode; strictly speaking, it should not have been added to the anode fall although the two components are sometimes added together in the literature (in fact, they cannot always be separated experimentally, which is the case in glow discharges).

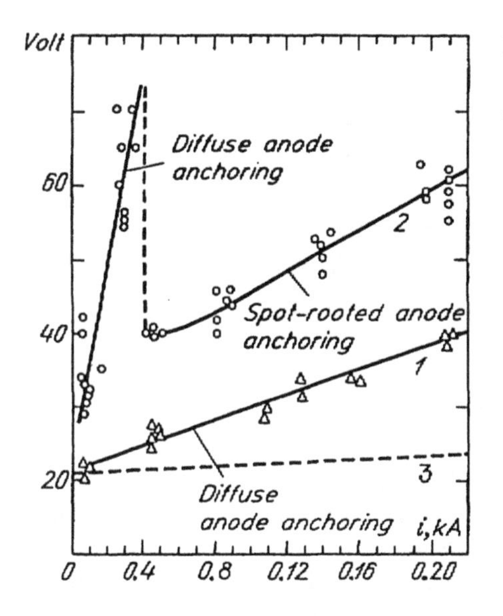

Fig. 10.9. $V - i$ characteristics of vacuum arcs with large- and small-area anode (*lower* and *upper* curves, respectively). Dashed line shows the cathode fall [10.11].

c) Effect of anchoring on $V - i$ characteristic. Figure 10.9 plots $V - i$ characteristics of vacuum arcs with large and small cathodes. A large anode results in duffise anchoring, and the anode fall is very small at low currents. The arc voltage is concentrated almost entirely in the cathode layer. In fact, subtraction of this fall (dashed line) from the total voltage gives the anode fall. As the current increases, the anode fall gradually rises. If the anode is small, the anode component of voltage grows sharply as the current increases, but then homogeneity collapses and the anode spot is formed. The transition is accompanied with an abrupt drop in voltage as a result of reduction in anode fall. Spot anchoring is more economical (in the sense of lowering the required voltage) than diffuse anchoring.

10.7.2 Carbon Arc Anode

The anchoring of carbon arc in air is diffuse up to a certain current limit ($V - i$ curve in Fig. 10.3), $j \approx 40\,\text{A/cm}^2$; the total anode fall is 36 V, of which 16 V is the "geometric" component and20 V falls in the region of negative space charge. The anode spot appears at $i \approx$ 15–20 A, with $j \approx 5 \times 10^4\,\text{A/cm}^2$; the arc begins to hiss (Sect. 10.3). The anode fall is then reduced by about 10 V. The spot moves at a velocity 300 m/s. As current increases further, a jet of hot vapor is ejected from the spot. The spot temperature is 4200 K. If the carbon anode contains admixture of salts or oxides of cerium or other rare earth elements, a deep very bright crater is formed at currents about 100 A, ejecting a plume of flame. The crater surface brightness corresponds to that of the Sun, emitting up to $6\,\text{kW/cm}^2$. Up to 70 % of the power from not too long arcs is emitted by the anode. This characteristic is employed in illumiination engineering (in searchlights). The anode fall at metal anodes in air reaches 1.5 to 2 times the ionization potential of the vapor atoms.

10.7.3 Energy Balance

Energy is released in that area of the anode where the current flows. It is composed of kinetic energy that electrons acquire in the anode fall and of the binding potential energy released when electrons neutralize the positive charge of metal ions. This component is equal to the work function. All in all, it comes to about 10 eV per electron of the current. Hence, the energy flux density in a spot at $j \sim 10^4$ A/cm^2 is $10^4 \times 10 = 10^5$ W/cm^2. The anode spot temperature in vacuum metal arcs is $T \approx 2700$–3300 K.

10.8 Low-Pressure Arc with Externally Heated Cathode

Discharges of this type are used in gas-filled rectifier diodes, in thyratrons etc. As a rule, oxide cathodes are employed. Gas-filled diodes are filled with argon or a mixture of Xe and Kr at a pressure of about 1 Torr (rectifier diodes with mercury vapor were also produced some time ago). The arc voltage is 10 to 20 V, the current is $\sim$ 1 A. A considerable part of the voltage is the cathode fall. The anode fall is 2–3 V. The potential across the positive column (which is not really a "column" because the discharge vessel is wide) is small.

10.8.1 The Purpose of Gas Filling

Let us compare the arc using an externally heated cathode, sustained between plane electrodes, with a similar device completely devoid of gas. This is the *vacuum diode* discussed in Sect. 6.6.1 for which the three-halves power law (6.15) is valid. Assume that the emission current of the cathode is $j_{em} = 0.1$ A/cm^2. Owing to the limiting effect of space charge, the current j of the same order of magnitude could be obtained with an interelectrode gap of 1 cm if a voltage $V = 1200$ V were applied to the electrodes.

If the diode is gas-filled, the gas breaks down and turns into plasma. In inert gases and with large transverse dimensions of the vessel, charge loss of the plasma is small and $E/p \approx 0.3$ V/cm Torr is sufficient for sustaining the electrically neutral positive column. If $p = 1$ Torr and $L = 3$ cm, the voltage across the plasma decreases by only 1 V. The anode potential, minus this and also minus a small anode fall, is carried quite well by the conducting plasma to the cathode. Most of the voltage is concentrated across the cathode layer, or rather, across its collisionless part at the metal surface, which is equal to, or shorter than, the free path length of charged particles. If we apply the same vacuum diode formula (6.15) to the "evacuated" collisionless layer $h = 0.05$ cm thick, we find that a mere 18 V are needed to carry the current $j = 0.1$ A/cm^2, instead of 1200 V.

The effect of gas filling consists in the compensation of the negative space charge of electrons in most of the diode space by ions that are copiously produced in the process of ionization of the gas by electrons. As a result, the passage of electrons emitted from the cathode on their way to the anode is facilitated.

10.8.2 Cathode Layer as a Vacuum Gap with Bipolar Current Limited by Space Charge

The function of cathode fall in a non-self-sustaining arc is very different from that in ordinary arcs. Here the cathode need not be heated up by ionic current, thereby sustaining the cathode emission. This function is served by the heating power source. What is left to the cathode layer is the acceleration of thermal electrons so that they sustain the necessary degree of ionization in the plasma. The losses in plasma being rather low, a relatively low ionization rate is sufficient. As a result, the cathode fall is of the order of the ionization or excitation potential (if ionization is stepwise), or even lower (Sect. 10.8.3; this is the so-called *low-voltage* arc). As for the thickness of the collisionless layer in the cathode fall, it grows to a value corresponding to the current dictated by the entire circuit. This current must be transported through the layer containing the space charge.

The process in the collisionless layer of a non-self-sustaining arc differs from those both in an ordinary diode and in the collisionless layer of self-sustaining arcs. In an ordinary diode, the cathode is hot but no ionic current flows. The space charge is negative; the field is practically zero at the cathode but sufficiently large at the anode (Sect. 6.6.1). In the layer of a self-sustaining arc, the ionic current from the plasma (where the field is practically absent) is weaker than the current from the cathode but not weak enough for the space charge to turn negative. The charge is positive so that the field at the cathode is very intense. This situation corresponds to the function of the layer: indeed, it is necessary to ensure strong emission and pull out all the emitted electrons. This fact is reflected in McKeown's equation (10.5).

In a *non-self-sustaining arc,* just as in a self-sustaining one, there is no field on the side of the well-conducting plasma. The plasma sends ionic current in the direction of the cathode. However, this current must not be too strong, because the cathode does not require any additional stimulation from the discharge (as long as the discharge current j is less than the emission current j_{em} produced by heating). Quite the opposite, the emission current must be limited by space charge. As a result, the space charge at the cathode is negative and the field is nearly zero, as it is in a diode with a externally heated cathode. The field thus vanishes at both ends of the layer. Hence, the magnitude of the field must reach a maximum somewhere in the middle and the space charge must reverse sign. This behavior is shown in Fig. 10.10.

The field distribution in the cathode layer is described by (10.1–3). After a single integration under the condition $E = 0$ at the plasma boundary of the layer, they yield relation (10.4) for the field at the cathode. In the case under consideration, the field is weak. Assuming $E_C = 0$ in (10.4), we find the ionic-to-electronic current ratio necessary for this:

$$j_+/j_e = (1 - S)/S = \sqrt{m/M} \approx 3 \times 10^{-3} - 10^{-2} \,. \qquad (10.9)$$

If the ionic current were to increase substantially in comparison with this estimate, a positive space charge would arise and a field would appear at the

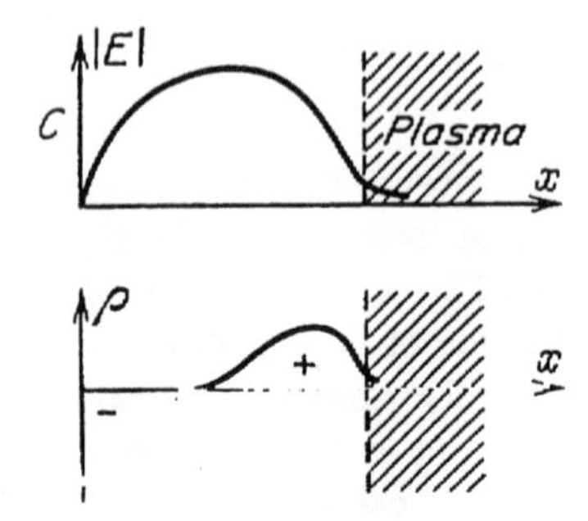

Fig. 10.10. Field (*top*) and space-charge (*bottom*) distributions in the cathode layer of an arc with auxiliary cathode heating

cathode. This indeed occurs when the discharge current exceeds the emission current and a transition to the self-sustaining arc begins. Experiments show the ionic current in a non-self-sustaining arc to be greater by a factor of 2 to 3 than that predicted by (10.9); nevertheless, its contribution to the total current is merely about one per cent.[4] At the same time, the presence of ions affects the magnitude of the space-charge-limited current to a much greater degree. Slowly moving ions considerably neutralize the negative space charge of electrons and thereby facilitate the motion of electrons from the cathode to the layer-plasma boundary.

Let us perform the appropriate calculation. We integrate (10.3) once, as we did in Sect. 10.5.3, apply the boundary condition at the cathode $E(0) = 0$, and make use of (10.9). We find

$$\frac{dV}{dx} = \left\{ 16\pi j \sqrt{\frac{m}{2e}} \left[\sqrt{V_C - V} + \sqrt{V} - \sqrt{V_C} \right] \right\}^{1/2} . \tag{10.10}$$

The integration of (10.10) reduces to a quadrature. We assign the result to the anode boundary of the layer $x = h$, where $V = V_C$, and solve the obtained relation for current:

$$j = k \frac{1}{9\pi} \sqrt{\frac{2e}{m}} \frac{V_C^{3/2}}{h^2} , \quad k = \frac{9}{16} \int_0^1 \frac{dz}{(\sqrt{1-z} + \sqrt{z} - 1)^{1/2}} = 1.86 . \tag{10.11}$$

The current, which here is *bipolar,* in contrast to the *unipolar* current in the diode (carried by charges of the same sign), is greater by a factor of 1.86 than implied by the Child-Langmuir law (6.15). This is a manifestation of the neutralizing effect of ions.

[4] The result (10.9) is illustrative in another respect as well. It gives an idea of the lower limit of the number of ionizations that an electron must perform when entering plasma from the cathode fall layer. Were it not for the ion flux to walls and other losses, each electron would have to perform only $\sqrt{m/M} \sim 10^{-2}$ acts of ion production.

10.8.3 Experiment

The voltage and layer thickness of the cathode fall were measured in experiments with arc discharges in argon at $p = 0.05$ Torr. the oxide cathode had an active surface of $2.9\,\mathrm{cm}^2$. The voltage remained constant, $V \approx 13\,\mathrm{V}$, in the current inteval from $i = 1\,\mathrm{A}$ to $i = 4\,\mathrm{A}$, and the thickness of the cathode fall layer decreased in this interval from $h = 0.031$ to $0.012\,\mathrm{cm}$. Formula (10.11) (for $j = 0.35$–$1.4\,\mathrm{A/cm^2}$) gives in this current range $h = 0.025 - 0.012\,\mathrm{cm}$, in good agreement with experimental data [10.6]. Figure 10.11 shows $V - i$ curves of an arc in argon at a higher pressure. The current is passed between an oxidized filament cathode and a plane nickel anode. The $V - i$ curve is again horizontal in the interval $i \approx 0.5$–$2.5\,\mathrm{A}$, but the voltage is 7 to 8 V lower than the argon excitation potential (low-voltage arc). It is typical for non-self-sustaining arcs that the voltage decreases as the pressure increases in a certain interval. The constancy of V points to a decrease in the layer thickness h as the current increases.

All experiments show that if the arc current is not too small, the arc voltage remains unchanged as long as the discharge current does not exceed the emissin current supplied by cathode heating. A rise of the $V - i$ curve begins when $i > i_{em}$. This rise in V is caused by the need to produce additional emission from the cathode. Therefore, the horizontal $V - i$ segment that is typically employed in gas-filled diodes stretches the further on the current axis, the stronger the cathode heating.

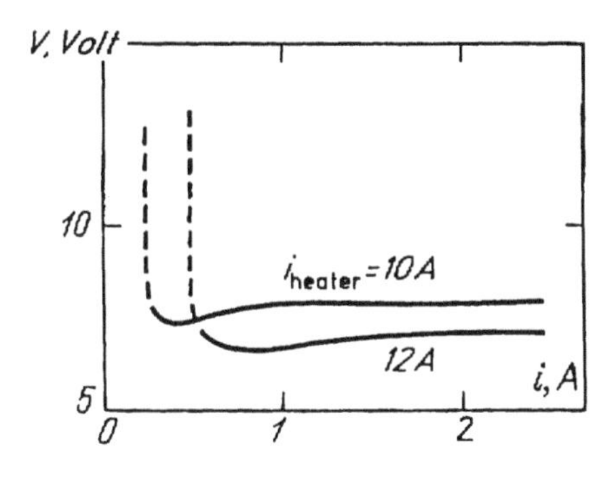

Fig. 10.11. $V - i$ characteristic of an arc with auxiliary cathode heating in a spherical vessel of 5 cm radius in argon at a presure of several Torr. Electrode spacing was 1 cm. The *upper* curve represents lower heating current and lower cathode temperature [10.12]

10.9 Positive Column of High-Pressure Arc (Experimental Data)

10.9.1 Stabilization

Considerable power is released in the arc column. For example, in air at $p = 1$ atm. and $i = 10\,\mathrm{A}$, the field is $E = 20\,\mathrm{V/cm}$. Hence, the power released per cm of arc length is $W = Ei = 200\,\mathrm{W/cm}$. For the process to become steady state, this energy must be transported away from the discharge volume. The mechanism responsible for transporting the Joule heat from the current channel is heat conduction (and radiation, at very high pressure). The subsequent fate of the heat flux depends on the arc design. If the arc is enclosed in a gas-filled tube (this is typical for research systems), the heat goes into walls that have to be cooled to prevent their destruction. Quite often (in experiments and applications) the arc column is cooled by a stream of cold gas (sometimes even liquid). A vortex flow is especially efficient because it drives the hot gas safely away from the walls. If an arc burns in a free atmosphere, its heat dissipates in the ambient atmosphere via convective flows. If the arc is short, the energy mostly goes into the electrodes which may be specially cooled. In all these approaches to realizing a steady-state burning, one speaks of an arc *stabilized* by walls, or fluid flow, or electrodes.

10.9.2 Degree of Equilibrium of Plasma

This depends on the gas, the pressure, and the current. At $p \lesssim 0.1$ atm. and $i \sim 1$ A, the plasma is invariably not in equilibrium. The plasma of air, other molecular gases, and metal vapor at $p \gtrsim 1$ atm. are in equilibrium at practically any current. The equilibration is caused by intensive energy exchange between electrons and molecules through excitation of vibrations and rotations, and by large elastic scattering cross sections of electrons in metal vapor. In Hg vapor, equilibrium sets in at $p \gtrsim 0.1$ atm. The temperature separation is greater in inert gases because the scattering cross sections of electrons by atoms are relatively small. The separation is eliminated only at high currents since the electron-ion interaction becomes strong and the degree of ionization is high. Thus in argon at $p = 1$ atm., the electron and gas temperatures are found to coincide ($T_e \approx T \approx 8000$ K) only when $i > 10$ A, so that $n_e > 3 \times 10^{15}\,\text{cm}^{-3}$. If the currents are smaller, T is approximately one half of T_e (Fig. 10.12). The equilibrium in He is even more difficult to achieve: a considerable temperature

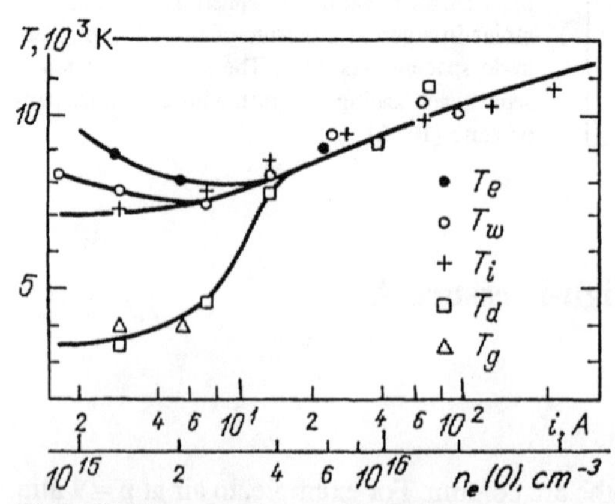

Fig. 10.12. Temperature separation in the positive arc column in argon, or Ar with an admixture of H_2 at $p = 1$ atm. as a function of current density or electron density. T_e is the electron temperature, T_w corresponds to the population of the upper levels, the ion temperature T_i is related to n_e by the Saha formula, T_g is the gas temperature, and T_d corresponds to the population of the lower levels [10.6]

gap is observed at $p = 1$ atm., from small currents up to $i \sim 100$ A ($T_e \approx 9000$–10,000 K, $T \approx 4000$–5000 K). The separation vanishes only at $i \approx 200$ A, when $n_e \approx 5 \times 10^{16}\,\mathrm{cm}^{-3}$ ($T_e \approx T \approx 10,000$ K).

10.9.3 Radial Temperature and Density Distributions of Electrons

Both of these quantities reach a maximum on the column axis and fall off towards the walls. But since equilibrium ionization is a very steep function of temperature, $n_e \propto \exp(-I/2kT)$, the electron density falls off much more quickly away from the axis than the temperature does (T falls more or less uniformly from $T_{max} \sim 10,000$ K on the axis to $T_w \sim 1000$ K at the walls if they are well cooled). The intensity of light emission behaves as n_e. As a result, the conducting, very bright arc column is a relatively thin channel on the axis. This geometry is illustrated in Figs. 10.13, 14. Note that the steepness of the falloff of $n_e(r)$ is masked by the logarithmic scale chosen for n_e in the figure.

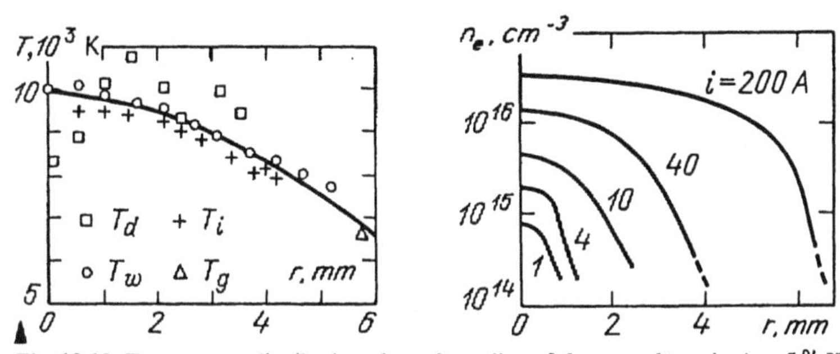

Fig. 10.13. Temperature distribution along the radius of the arc column in Ar + 5 % H_2 at $p = 1$ atm., $i = 50$ A, under nearly equilibrium conditions when all temperatures become equal. Symbols as for Fig. 10.12 [10.6]

Fig. 10.14. Radial distribution of n_e in the arc column in Ar + 5 % H_2 [10.6]

10.9.4 $V - i$ Characteristics

The field in the column is found by changing the interelectrode gap L at a constant current. If $L > 0.5$–1 cm, the voltage is a linear function of L so that $E = dV/dL \approx$ const. A long column is therefore longitudinally homogeneous. If the pressure is increased at a fixed current, the field is enhanced (Fig. 10.15). The enhancement is caused by increased radiative losses and, possibly, by a certain increase in heat transfer from plasma to the walls, dictating enhanced power per unit length, $W = Ei$. At equal currents, a stronger field is required to sustain the plasma in a tube than in an arc burning in free atmosphere, because the transfer is more intense and more power is needed (Fig. 10.16). The field is higher in hydrogen than in other gases owing to its greater thermal conductivity and to more intensive heat transfer from the column. To supplement Figs. 10.13, 14, Fig. 10.17 gives the $V - i$ characteristic of the same mixture.

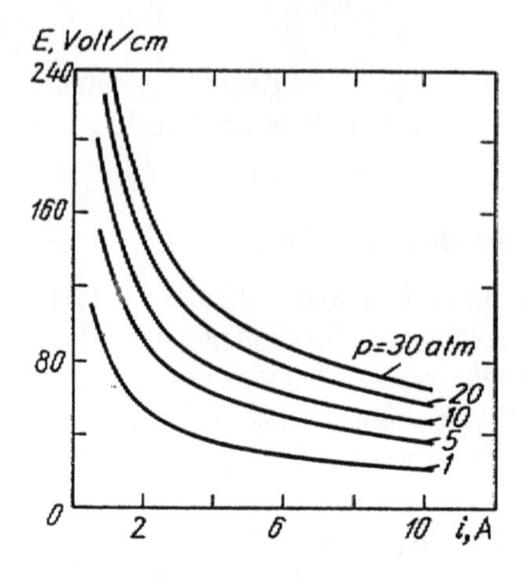

Fig. 10.15. $V - i$ characteristics of positive arc columns in air at various values of pressure [10.6]

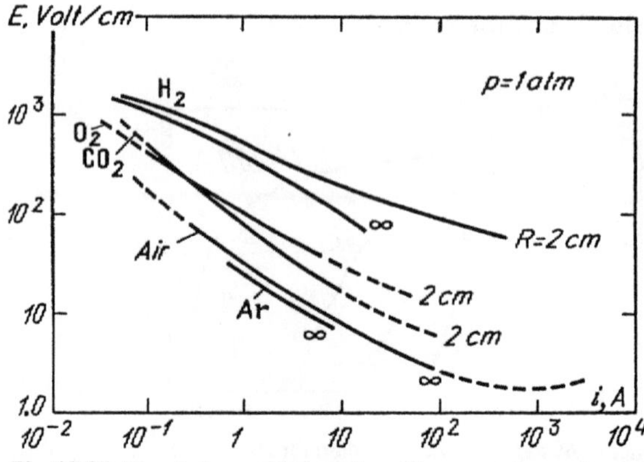

Fig. 10.16. $V - i$ characteristics of positive arc columns in various gases at atmospheric pressure in tubes of radius $R = 2$ cm, and in free gas ("$R = \infty$") [10.12]

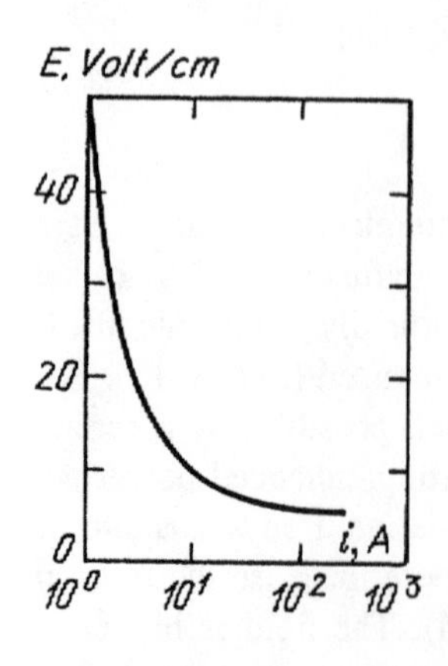

Fig. 10.17. $V - i$ curve of the positive column in Ar + 5 % H_2 [10.6]

10.9.5 Radiation of the Column

The emissivitiy and luminous flux of the column depend on the gas, its pressure, and current. To a certain approximation, they are proportional to n_e^2. In nitrogen and air at p = 1 atm., radiative losses make up from one to several percent of the power input. In argon at 1 atm., the losses become appreciable (greater than 15%) at power $W > 150$ W/cm. Among a number of gases, mercury vapor has exceptionally high radiant properties. An experimental analysis of the power balance in a mercury arc in a tube of 4.1 cm internal diameter and 50 cm length at $p = 0.88$ atm., $i = 6$ A, $E = 5.8$ V/cm, has demonstrated that out of the power $W = 35$ W/cm, 10 W/cm are transported by heat conduction into walls, 18 W/cm are radiated away, and the remaining 7 W/cm appear to be transported out of the plasma in the resonance lines of Hg at $\lambda = 1850$ and 2537 Åand absorbed in the quartz walls [10.6]. In fact, the 10 W/cm transported by heat conduction from the tube filled with mercury vapor at high pressure was found to be independent of tube diameter and vapor pressure.

The radiation from the arc column is especially intense at very high pressure $p \gtrsim 10$ atm. (of course, at sufficiently high power levels), particularly in Hg, Xe, and Kr. This feature is exploited in mercury and xenon lamps. Here are empirical formulas for the radiative power of the arc column in a number of gases [10.6].

$$\begin{aligned}
&\text{Hg: } p \gtrsim 1\,\text{atm.}, && W_{\text{rad}}[\text{W/cm}] = 0.72\{W[\text{W/cm}] - 10\} \\
&\text{Xe: } p = 12\,\text{atm.}, && W_{\text{rad}} = 0.88(W - 24)\,, \; W > 35 \\
&\text{Kr: } p = 12\,\text{atm.}, && W_{\text{rad}} = 0.72(W - 42)\,, \; W > 70 \\
&\text{Ar: } p = 1\,\text{atm.}, && W_{\text{rad}} = 0.52(W - 95)\,, \; W > 150\,.
\end{aligned}$$

10.10 Plasma Temperature and $V - i$ Characteristic of High-Pressure Arc Columns

Dense equilibrium low-temperature plasmas attract the attention of physicists and engineers probably even more than *weakly ionized inequilibrium plasma* does. They are in laboratories and exploited in experimental and industrial equipment. nowadays it is possible to generate such plasma in fields of any frequency range (Chap. 11) but the *arc plasma generation* remains the simplest, easiest, and most widespread technique. The prime characteristic of an equlibrium plasma is its *temperature*, and the problem is to understand what factors determine it, how it is related to the electrical parameters of the discharge (electric *current* and *power*), and how the $V - i$ curve of the column reflects these relations.

10.10.1 Thermal Ionization

In some respects, a description of the state of an equilibrium plasma, and of its energy balance that determines the field necessary to sustain it in a steady-state manner, is simpler than in the case of the nonequilibrium plasma of glow discharge. There is no need to go into details of complex mechanisms and kinetics of production and removal of charges. The electric conductivity σ of the plasma is uniquely determined by its temperature and pressure (Fig. 10.18). As a rule, the pressure is known simply as an experimental parameter.

The very process of ionization differs from that which takes place in weakly ionized nonequilibrium plasmas. Molecules are ionized in the latter case by electrons that have obtained the required energy directly from the field. In thermal ionization, the effect of the field is as if "depersonalized". The field pumps energy into the electron gas as a whole. Electrons are thermalized through collisions with one another and through subsequent maxwellization of their distribution. The gas is ionized by those electrons that acquired sufficient energy not from the field but in the exchange with other particles. *Thermal ionization* proceeds quite independently of the way by which energy flows into plasma.

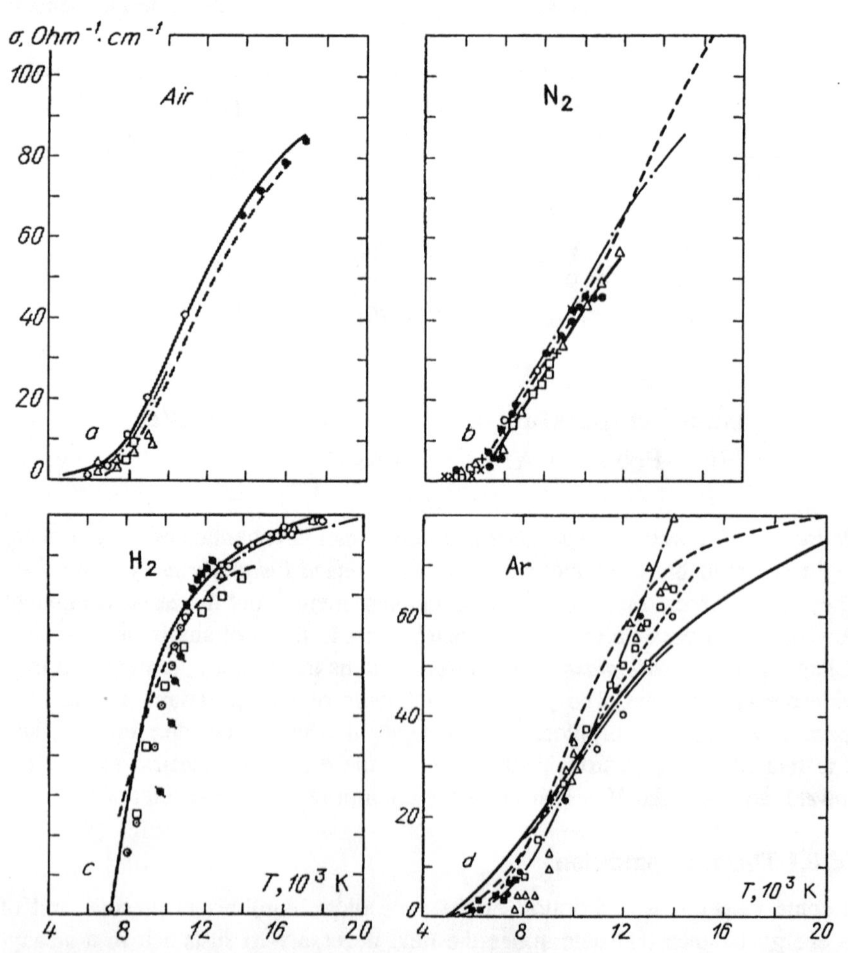

Fig. 10.18. Electrical conductivity of equilibrium plasmas in air, N_2, H_2, and Ar, at p = 1 atm.. Symbols are measurements of various authors [10.13]; solid curves: calculations of [10.14]

10.10.2 Equations of Equilibrium Plasma Columns

Consider a long cylindrical plasma column in a longitudinal field E. Let the arc burn steadily in a nonmoving gas enclosed in an externally cooled tube of radius R. Such conditions are quite frequent in discharge apparatus; actually, the model gives a good idea of the state in the conducting channel, even if the arc burns in free atmosphere or in a gas flow, because the temperature on the discharge axis is not very sensitive to external conditions. We will be interested here in moderately high pressures, say, atmospheric pressure, and moderately high currents, so that the plasma temperature does not exceed 11,000–12,000 K. In such cases the radiative losses are usually much lower than the conductive transport of heat from the column and can be neglected.

The electric field in a column which is homogeneous over its length is constant in cross section because curl $\boldsymbol{E} = 0$. The radial distributions of conductivity σ, current density $j = \sigma E$, and Joule heat sources $w = jE = \sigma E^2$ [W/cm^3] are determined only by temperature distribution, via $\sigma(T)$. The energy balance in the plasma is described by the equation

$$-\frac{1}{r}\frac{d}{dr}rJ + \sigma(T)E^2 = 0\,, \quad J = -\lambda\frac{dT}{dr}\,, \tag{10.12}$$

where λ is the thermal conductivity (Figs. 10.19, 20).

The boundary conditions to this equation are: at $r = R$, $T = T_w$, where T_w is the wall temperature; for reasons of symmetry, $dT/dr = 0$ at $r = 0$. The

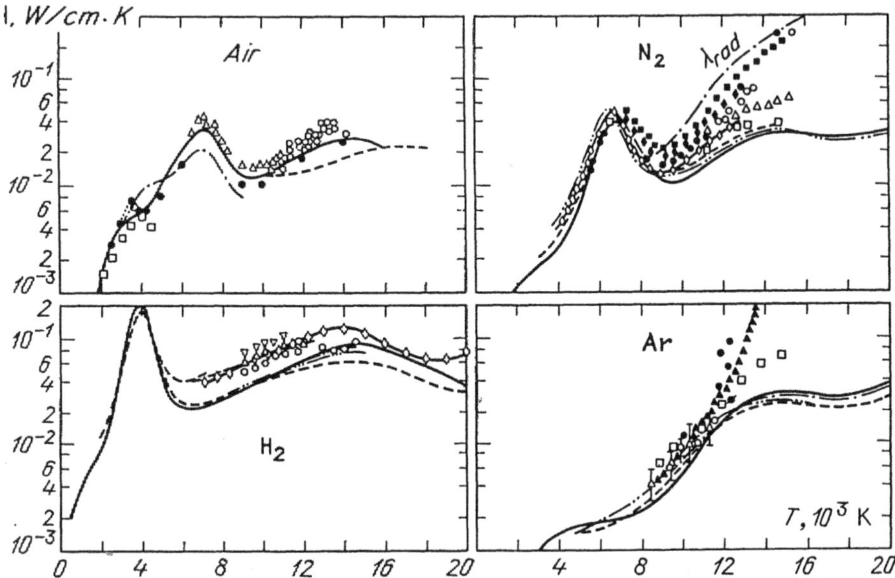

Fig. 10.19. Heat conduction of equilibrium plasmas in air, N_2, H_2, and Ar at p = 1 atm.. Symbols are measurements of various authors [10.13]; solid curves: calculations of [10.14]; λ_{rad} is the radiative thermal conductivity

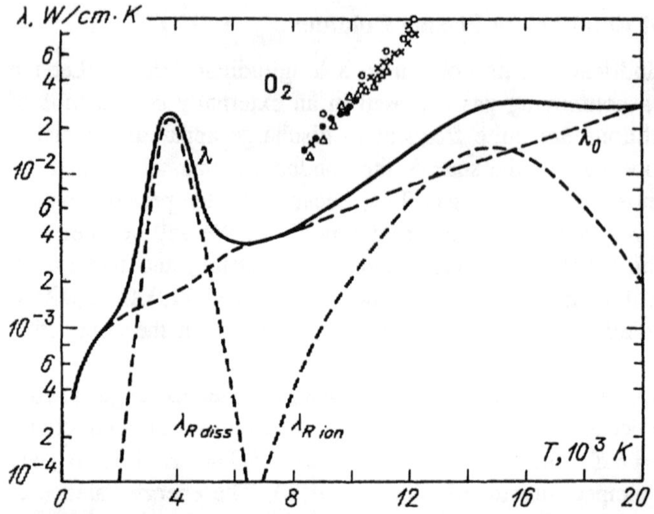

Fig. 10.20. Components of the overall thermal conductivity λ of oxygen at $p = 1$ atm.. λ_0 corresponds to ordinary transport; $\lambda_{R,\text{dis}}$ and $\lambda_{R,\text{ion}}$ are reactive components connected with the transport of dissociation and ionization energy (e.g., atoms arrive at a cold spot from a hot one and recombine, so that the potential energy of dissociation acquired in the hot spot is deposited at the cold one). Curves are from calculations of [10.14]; dots represent measurements of a number of authors [10.13]

temperature of a conducting plasma is much higher than T_w, so that we can safely assume $T_w = 0$. The discharge current is

$$i = E \int_0^R \sigma 2\pi r \, dr \, . \tag{10.13}$$

The current is controlled experimentally and thus is a prescribed parameter. The field is found by solving the problem as formulated, the problem being well defined if the material characteristics $\sigma(T)$ and $\lambda(T)$ are known. This gives us the $E(i)$ curve of the column. If we introduce the heat flux potential $\Theta = \int_0^T \lambda \, dT$ (Fig. 10.21) it becomes sufficient to work with a single material function, $\sigma(\Theta)$, instead of two. Equation (10.12) is known as the Elenbaas-Heller equation (obtained in 1934).

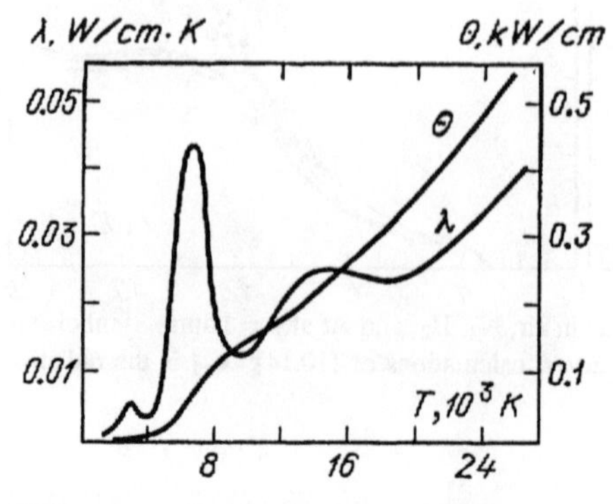

Fig. 10.21. Thermal conductivity λ and heat flux potential Θ in air at 1 atm

10.10.3 The Channel Model and the Principle of Minimum Power

The nonlinear nature of the real function $\sigma(\Theta)$ does not allow the general analytic solution of (10.12). Various formal methods based on linearizing $\sigma(\Theta)$ and on splitting the integration domain into subdomains, and various numerical methods have been developed since the 1930s when the arc column problem was first formulated [10.13]. In order to understand better the nature of the relationships, we turn to the *arc channel model* suggested by *Steenbeck* in 1932, which gives an essentially correct picture.

The conductivity is vanishingly small at not very high temperatures. At $T \approx$ 4000–6000 K it becomes appreciable and grows steeply as T increases. The heat flux makes the temperature fall off towards the walls more or less uniformly. The current, however, is vanishingly small everywhere except at the tube axis, where the temperature is sufficiently high (Fig. 10.22). The channel model is fairly obvious, in view of this picture. Let us introduce the effective radius r_0 of the conducting channel and assume approximately that $\sigma = 0$ outside the channel, at $r > r_0$, so that no current flows there. Conductivity inside the channel at $0 < r < r_0$ is high and approaches the value $\sigma_{\max} \equiv \sigma(T_{\max})$ that corresponds to the temperature at the axis of $T_{\max} \equiv T(0)$. The channel model reduces to replacing the actual distribution $\sigma(r)$ by a stepwise curve (dashed lines in Fig. 10.22).

In this approximation, the approximate expression for the current, (10.13), transforms to

$$i = \sigma_{\max} E \pi r_0^2 \,, \tag{10.14}$$

and equation (10.12) is readily integrated in the zero-current zone $r_0 < r < R$. We approximate the boundary conditions by $T_0 \equiv T(r_0) \approx T(0) \equiv T_{\max}$ at the channel boundary and by $T_w = 0$ at the wall and find

$$\Theta_{\max} = \frac{W}{2\pi} \ln \frac{R}{r_0} \,, \quad W = \frac{i^2}{\pi r_0^2 \sigma_{\max}} \,, \quad \Theta_{\max} = \int_0^{T_{\max}} \lambda \, dT \,, \tag{10.15}$$

where $W = iE$ is the power released per cm of the column.

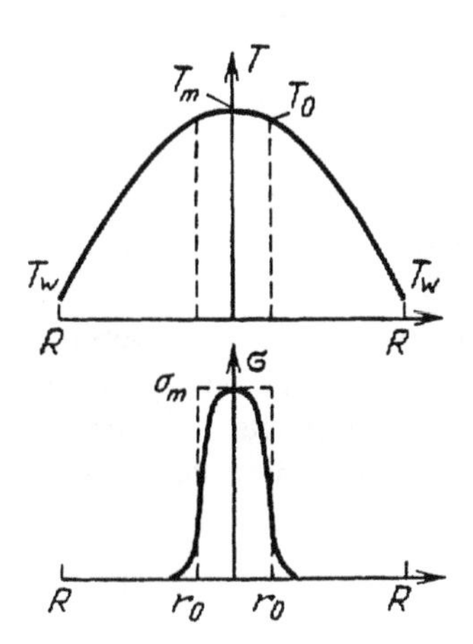

Fig. 10.22. Distribution of temperature T and conductivity σ along the arc column radius. Dashed curve replaces the $\sigma(r)$ profile with the "step" of the channel model

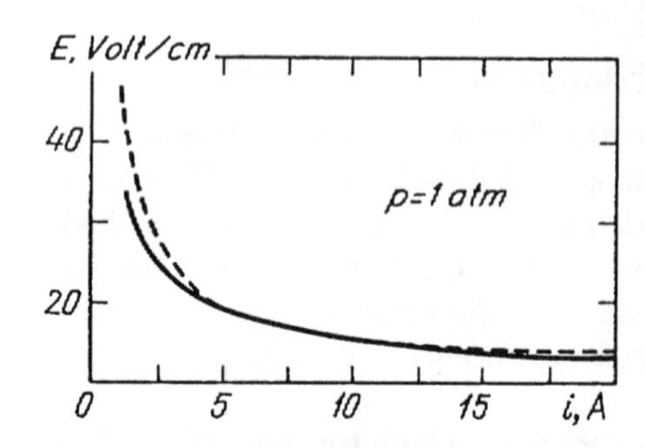

Fig. 10.23. $V-i$ characteristic of arc column in nitrogen in a tube of radius $R = 1.5$ cm. Solid curve: experiment; dashed curve: calculation employing the minimum power principle [10.1]

The equations (10.14, 15) contain three unknown variables T_{max}, r_0, E; the current i and tube radius R being externally prescribed parameters. To add a relation that is lacking, Steenbeck suggested the use of the *principle of minimum power*. For given i and R, the temperature distribution in the tube (and hence the plasma temperature T_{max} and channel radius r_0) must reach such values that the power W and field $E = W/i$ be minimum. Arc calculations based on the minimum power principle fit experimental data quite well (Fig. 10.23); this was one of the reasons why the principle enjoyed a certain popularity (see also Sect. 8.4.8).

However, the validity of this principle never ceased to be an open question for researchers; it is still being scrutinized and discussed today. A search for justification led to links with nonequilibrium thermodynamics. As for the *arc column* equations, the principle was proved to hold. Nevertheless, the possibility of its application must be verified by special analysis in each specific case. Thus an unwise application of the minimum power principle to *inductively coupled rf* and *microwave* discharge modes in the framework of models that were quite similar to the channel model led to erroneous results, the mistakes being masked by the seemingly satisfactory agreement with comon sense and with experimental data (at least in certain parameter ranges) [10.15].

10.10.4 Energy Balance in the Current Channel

We shall not be pursuing any further the matter of the minimum power principle to see what agents determine the plasma temperature. We do not need supplementary principles. Indeed, the entire information is contained in (10.12, 13). Having started the approximate solution with modeling the function $\sigma(r)$ by a rectangle, we have to complete it in a consistent manner. Integration of (10.12) in the currentless zone yielded (10.15). Now we look at the current channel. The power W released in the channel is transported through its boundary by the heat flux J_0, $W = 2\pi r_0 J_0$. Despite the assumption made in deriving (10.14, 15), the heat flux from the channel is determined by an actual, even if small, temperature difference across the conducting channel, $\Delta T = T_{max} - T_0$. To an order of magnitude, $J_0 \sim \lambda_{max}\Delta T/r_0$, where $\lambda_{max} \equiv \lambda(T_{max})$. If this estimate is improved by integrating (10.12) in the interval $0 < r < r_0$, again assuming the uniform distribution of heat sources σE^2, we obtain

$$4\pi\Delta\Theta = W \approx 4\pi\lambda_{max}\Delta T\ , \quad \Delta\Theta = \Theta_{max} - \Theta_0\ . \tag{10.16}$$

10.10.5 Quantitative Definition of the Notion of "Channel" and Closing of the System of Channel Equations

The new relation (10.16) introduces a new unknown: ΔT or T_0. However, we have not yet given the definition of the "*channel*". A speculative introduction of its characteristics, namely radius r_0 and temperature $T_{\max}$, is not yet a basis for calculating tnem. We have first to establish quantitatively how to differentiate between media regarded as conducting and nonconducting. We cannot know the radial distribution of current in the column until the problem is solved; hence, it is natural to draw the conditional boundary at that point on the current density distribution $j(r) \propto \sigma(r)$ where it decreases by a predetermined factor in comparison with the maximum value on the channel axis. If the steep slope of the function $\sigma(T)$ is taken into account, this convention corresponds to a not too large temperature drop ΔT.

For example, assume the arc current to be small, the temperature and degree of ionization of plasma to be low, and electron-atom collisions to affect the resistance more than electron-ion collisions. Then $\sigma \sim n_e$, and if ionization is thermodynamically in equilibrium,

$$\sigma(T) = C \exp(-I/2kT) , \quad C(T) \approx \text{const} . \tag{10.17}$$

To be specific, we assume that $\sigma_0 \equiv \sigma(T_0)$ is less than $\sigma_{\max}$ by a factor of $\bar{e}$; recalling that $I/2kT \gg 1$, we find

$$\Delta T = T_{\max} - T_0 \approx (2kT_{\max}/I)T_{\max} . \tag{10.18}$$

For conductivity, the main role at greater currents, when ionization exceeds 1 %, is played by electron-ion collisions; the form of $\sigma(T)$ is then changed. Nevertheless, it can be approximated by the same formula (10.17) in a certain temperature range if an effective ionization potential I_{eff} is found by approximating the $\sigma(T)$ curve. Thus for air, nitrogen, and argon at p = 1 atm., the same interpolation formula can describe σ with accuracy sufficient for evaluations in a temperature range typical for arc discharge:

$$\begin{aligned} &\sigma \approx 0.83 \times 10^2 \exp\left(-36{,}000/T[\text{K}]\right) \text{ Ohm}^{-1}\text{cm}^{-1} \\ &T \approx 8000 - 14{,}000\,\text{K} , \quad I_{\text{eff}} \approx 6.2\,\text{eV} . \end{aligned} \tag{10.19}$$

The condition that σ change substantially in a relatively narrow temperature range $\Delta T \ll T$, for example, by a factor of two, can be written in a general differential form, without specifying the form of the function $\sigma(T)$:

$$\frac{1}{\sigma}\left(\frac{d\sigma}{dT}\right)_{T=T_{\max}} \times \Delta T \approx 1 , \quad \frac{\Delta T}{T_{\max}} \approx \left(\frac{d \ln \sigma}{d \ln T}\right)^{-1}_{T=T_{\max}} . \tag{10.20}$$

With $\sigma(T)$ given by (10.17), (10.20) turns into (10.18). Having eliminated ΔT from (10.16, 20), we arrive at the equality

$$4\pi\lambda_{\max}\sigma_{\max} = W(d\sigma/dT)_{T=T_{\max}} \tag{10.21}$$

that closes the system of equations of the channel arc model and defines the maximum temperature T_{max}. It is remarkable that the formal application of the principle of minimum power, $(dW/dr_0)_{i=\text{const}} = 0$, to functional relations (10.14, 15) gives exactly (10.21). Since the validity of the principle for the arc discharge had been proved (even if in a very complex manner), this result supports the validity of simple, illustrative arguments in Sects. 10.10.4, 5.[5]

Nevertheless, these arguments contain an element of arbitrariness implied by the selection of a specific ratio σ_{max}/σ_0 equal to $\bar{e}$ in (10.18) or to unity in the right-hand side of (10.20). No such uncertainty is present in the approximate integral relation

$$\int_0^{T_{max}} \sigma\lambda\, dT = W\sigma_{max}/4\pi\ , \tag{10.22}$$

which is more general than the differential formula (10.21) and corresponds, by virtue of its derivation (Sect. 11.2), to the best replacement of the current distributed in the tube cross section by the current concentrated in a channel.

10.10.6 Plasma Temperature

Any one of the approximate relations (10.18) together with (10.16), or (10.21), or (10.22) (in the order of improving accuracy) determine the maximum plasma temperature T_{max} in the arc column as a function of the power input per unit length, W. If the power is fixed, T_{max} is independent of the tube radius and, by virtue of (10.22), depends only to a small extent on the heat conduction characteristics of the gas in the electrically nonconducting peripheral region. In the crude approximation (10.21) and (10.16) with (10.18), T_{max} is completely independent of this factor.

The final temperature T_{max} is such that the temperature drop in the energy release region, closely tied to T_{max} via (10.20) and via the conductivity $\sigma(T)$, ensures the balanced heat transport to the zone *outside the current channel*. The subsequent fate of energy hardly affects the maximum temperature (if the power is fixed). We can conclude that temperature is *equally insensitive to the method of cooling the arc* and to the organization of the discharge environment (free atmosphere or a flow of cold gas). Only the power input into the plasma, which depends (quite steeply) on cooling intensity, exerts a direct effect on T_{max}. The

---

[5] The reader must beware of the error encountered in some theoretical papers in which $\sigma(T)$ is approximated by a stepwise function $\sigma = 0$ for $T < T_0$, $\sigma = \text{const}$ for $T > T_0$ and the step coordinate T_0 is rigidly fixed. Indeed, one is tempted to conclude from Fig. 10.18 that $T_0 = 9000$–$10{,}000$ K. In fact, the temperature of the conductivity onset T_0 increases or decreases with the maximum plasma temperature T_{max} when the current and electric power are varied. While it is admissible to ignore some fraction of current beyond the main current channel, it is a grave error if a definite absolute amount of current is left beyond this channel. An absurd situation may result if, say, $T_0 = 9000$ K and $T_{max} = 9010$ K: almost the entire current flows outside the channel; or, at the other extreme, $T_0 = 9000$ K but $T_{max} = 15{,}000$ K: a predominant part of the channel carries a negligible fraction of total current and thus does not constitute a channel at all.

greater the amount of heat that can be transported out of the cold gas, the greater the power that can be introduced into it without violating the steady state.

As a result of the steep dependence of conductivity on temperature, high temperatures can be achieved only by a disproportionate increase in power. This is clear from general formulas (10.21, 22) but becomes even more convincing if an explicit dependence $T_{max}(W)$ is found by prescribing the function $\sigma(T)$ (with the true ionization potential at low temperatures and the effective one at higher temperatures):

$$T_{max} = \sqrt{(I/8\pi\lambda_{max}k)W} \,. \tag{10.23}$$

On the average, the temperature increases more slowly than $W^{1/2}$ because heat conduction generally increases with increasing T. Let us make an evaluation for a carbon arc in atmospheric air at $i = 200\,\text{A}$, with measurements showing that $T_{max} \approx 10{,}000$–$12{,}000\,\text{K}$ (Fig. 10.2). According to the $V - i$ curve of Fig. 10.16, in these conditions the field is likely to be $E \approx 2.5\,\text{V/cm}$, whence $W = 500\,\text{W/cm}$. Assuming $\lambda_{max} = 1.5 \times 10^{-2}\,\text{W/cm K}$ (Fig. 10.19) and $I_{eff} = 6.2\,\text{eV}$ [from (10.19)], we find from (10.23) that $T_{max} = 9800\,\text{K}$, in reasonable agreement with experimental data.

When the current is relatively low ($i = 10\,\text{A}$) in air at $p = 1$ atm., the $V - i$ curve of Fig. 10.15 gives $E = 20\,\text{V/cm}$, $W = 200\,\text{W/cm}$. Formula (10.23) gives an estimate $T_{max} \approx 7000\,\text{K}$ [$\lambda_{max} \approx 2 \times 10^{-2}\,\text{W/cm K}$; $I_{eff} \approx 10\,\text{eV}$ is now higher because the degree of ionization is low: $x = 2 \times 10^{-4}$].

10.10.7 Column Characteristics

The problem is solved using the formulas given above. For a given W, we find T_{max} from (10.23) or the more general (10.21, 22), and then find r_0 from the first relation of (10.15). The second formula of (10.15) now gives i. The field is $E = W/i$. The similarity law holds: at a fixed power W the channel radius r_0 is proportional to the tube radius R, and the entire distribution $T(r)$ varies in a similar manner as R is varied. The calculation illustrated in Fig. 10.23 uses (10.21).

The parameter that is varied and measured directly in experiments is not the power, but the current; hence, the relations governing the column behavior will become clear if all its characteristics are plotted as functions of the current. Thus we can choose $\sigma(T)$ in the form (10.17). To simplify the formulas, we also take $\lambda = \text{const}$, $\Theta = \lambda T$; this does not introduce substantial distortions.[6] Making use of (10.23, 15, 17), we find

$$\sigma_{max} = \left(IC/8\pi^2\lambda k T_{max}^2\right)^{1/2} (i/R) \,, \tag{10.24}$$

$$T_{max} = \frac{I/2k}{(1/2)\ln(8\pi^2\lambda C k T_{max}^2/I) - \ln(i/R)} \,. \tag{10.25}$$

[6] Actually, anomalies due to the nonmonotonic behavior of $\lambda(T)$ may arise in molecular gases at low temperatures $T \approx 6000$–$8000\,\text{K}$ (Figs. 10.19, 20).

As the current increases, the conductivity grows almost proportionally, so that plasma passes this current. However, σ is a steep function of temperature so that T grows much more slowly. The power is related to temperature by the energy balance in the channel and thus also grows slowly with i. Correspondingly, the field $E = W/i$ decreases:

$$W = \frac{8\pi\lambda k T_{max}^2}{I} \approx \frac{\text{const}}{[\text{const} - \ln(i/R)]^2}, \tag{10.26}$$

$$E = \frac{8\pi\lambda k T_{max}}{I}\frac{1}{i} \approx \frac{\text{const}}{i[\text{const} - \ln(i/R)]^2}. \tag{10.27}$$

Formula (10.27) gives the $V - i$ characteristic of the column, i.e., the voltage decreases with increasing current. The channel radius is

$$r_0 = R(\sigma_{max}/C)^{1/2} = R\left(I/8\pi^2\lambda k T_{max}^2 C\right)^{1/4}(i/R)^{1/2}. \tag{10.28}$$

Roughly speaking, $r_0^2 \propto i$, so that a current increase results in a greater area, not in a greater current density $j \sim i/r_0^2 \propto T_{max}$, which increases as slowly as temperature does. All the quantities T_{max}, σ_{max}, W, E are functions of the ratio i/R. If the tube has a smaller radius, the same power is developed at a proportionally lower current.

A new item appears in the plasma energy balance at temperatures above 11,000–12,000 K: radiative losses. To compensate for these losses in the steady state, an additional power and stronger field are needed, and the $V - i$ curve develops a positive slope.

10.10.8 How to Achieve Extreme High Temperatures

If this aim is pursued, it is necessary to ensure intensive heat removal in order to be able to inject high power into the plasma. It is difficult to achieve this effect only by increasing the current: it would have to be too high. Heat removal can be intensified by reducing the tube radius (thus increasing the temperature gradient), or by blowing a fast flow of fluid [gas or, better still, water as in the Gerdien arc (see Sect. 10.2.8)]. Very high temperatures of several tens of thousands of degrees are easier to achieve in short pulsed discharges, because a very high current pulse can be produced and the energy accumulates in the plasma before heat removal starts (nonsteady process).

In order to generate and study a high-temperature steady state arc column, the discharge is passed through stacked, cooled and insulated copper washers with hole diameters of several millimeters. Copper washers alternate with dielectric ones having the same hole diameter. The washers form a long well-cooled small-diameter tube. An axial temperature up to 15,000 K has been obtained, in this way, in nitrogen at $p = 1$ atm.

10.11 The Gap Between Electron and Gas Temperatures in "Equilibrium" Plasma

The energy exchange in electron-atom collisions in weakly ionized nonequilibrium plasmas with $T_e \gg T$ is "unilateral": energy is transferred only from the electrons to the heavy particles. The exchange in equilibrium plasmas is bilateral. In the case of ideal equality, $T_e = T$, the electrons gain from heavy particles in some collisions exactly the same amount of energy that the former pass on to the latter in other collisions. The equality $T_e = T$ is violated in the presence of an electric field. Actually, only electrons gain energy from the field to any appreciable extent. They transfer it to heavy particles that later participate in transporting the energy away from the gas (into the walls, etc.). This form at relay transport functions owing to positive temperature differences between electrons and the gas, $T_e - T$, and between the gas and the walls, $T - T_w$. In Sect. 10.10 we neglected the difference between T_e and T and worked in terms of the common temperature T. Let us try to check the validity of this assumption.

10.11.1 Energy Balance Equation of Electrons Interacting with the Field and the Heated Gas

This is a generalization of (2.12):

$$\frac{3}{2}k\frac{dT_e}{dt} = \left[\frac{e^2E^2}{m\nu_m^2} - \frac{3}{2}\delta k(T_e - T)\right]\nu_m \, . \tag{10.29}$$

As before, δ in (10.29) must be interpreted as the fraction of energy that an electron transfers, on the average, to a colliding heavy particle if the latter has negligible energy. If $T_e \gg T$, the equation transforms into (2.12). If the field is zero, stable equilibrium, $T_e = T$, is implied by this equation.

10.11.2 Criterion of Equilibrium

Applying (10.29) to steady-state conditions [electron temperature builds up quite rapidly (Table 9.1) and is practically always quasistationary] we obtain the relation between the temperature gap and the field:

$$\frac{T_e - T}{T_e} = \frac{2e^2E^2}{3\delta kT_e m\nu_m^2} = 2 \times 10^2 A \left\{\frac{E[\mathrm{V/cm}]l}{T_e[\mathrm{eV}]}\right\}^2 \, . \tag{10.30}$$

In this last transformation, we changed from collision frequency to electron path length $l = v/\nu_m$ and set $mv^2/2 = 3kT_e/2$. In addition, we assumed that $\delta = 2m/M$, where M is the atomic mass and A is the atomic weight. If the gas were molecular before heating, the presence of small number of remaining molecules slightly intensifies the exchange (δ is increased) and reduces the temperature gap. Formula (10.30) can be used to evaluate the extent of nonequilibrium caused by the difference between the electron and ion temperature.

The field necessary to sustain nearly equilibrium plasma is determined by the energy balance of the ionized gas as a whole (Sect. 10.10). Having found E and $T \approx T_e$ for a given current either by calculations or experimentally, we find from (10.30) the actual temperature gap that ensures a steady transfer of Joule heat from electrons to the gas. If the relative gap exceeds the admissible limit of, say, 50 %, the assumption of equilibrium plasma has to be dropped from calculations and one has to consider the system of equations for the energy of electrons, (10.30), and of the gas, (10.12) – but with the true conductivity, which now depends not on T, but on T_e.

The ionization equilibrium in the plasma may also be disturbed. In that case the kinetics of electron density must also be analyzed, as in Chap. 8. This is indeed the situation with processes in the contracted current filaments that are formed in glow discharges (Sect. 9.8).

The plasma of high-pressure arcs may be very highly ionized. The Coulomb scattering of electrons by ions plays an essential, and often dominant, role. In the general case, with (2.8) taken into account, we have

$$l^{-1} = N\sigma_m + 1.3 \times 10^{-13} n_e (T_e[\mathrm{eV}])^{-2}\ \mathrm{cm}^{-1}\,, \qquad (10.31)$$

where the first term refers to neutral atoms. The Coulomb logarithm is assumed to equal $\ln \Lambda = 4.5$, which corresponds to $T_e \approx 10^4$ K, $n_e \sim 10^{16}$–$10^{17}\ \mathrm{cm}^{-3}$. Collisions with ions are dominant if the degree of ionization exceeds the level

$$n_e/N > 0.77 \times 10^{13} \sigma_m (T_e[\mathrm{eV}])^2 \qquad (10.32)$$

of order 10^{-2}. For instance, in $Ar + 5\,\%H_2$ at $p = 1$ atm. we find that at a current $i = 50$ A, $T \approx T_e \approx 10^4$ K and $n_e \approx 2 \times 10^{16}\ \mathrm{cm}^{-3}$ (Figs. 10.13, 14); $N \approx 7.2 \times 10^{17}\ \mathrm{cm}^{-3}$, $n_e/N \approx 0.028$. According to Fig. 10.17, $E \approx 6$ V/cm. Collisions are coulombic, $l \approx 3 \times 10^{-4}$ cm, and the temperature gap is $(T_e - T)/T_e = 3.5\,\%$, so that in this sense the plasma is fairly well in equilibrium.

10.11.3 Why a Lower Field Is Required, Other Conditions Being Equal, to Sustain Equilibrium Plasma Rather than Nonequilibrium

As an example, the field in the positive column of an atmospheric-pressure nitrogen arc burning in a cooled tube of $R = 1.5$ cm in diameter is $E = 10$ V/cm at a current $i = 10$ A (Fig. 10.23). In this arc, $T \approx 8000$ K and the gas number density $N \approx 10^{18}\ \mathrm{cm}^{-3}$; $n_e \approx 2 \times 10^{15}\ \mathrm{cm}^{-3}$, $n_e/N \approx 2 \times 10^{-3}$, and collisions of electrons with neutral atoms are predominant. For the same densities of the gas, ($p \approx 30$ Torr) and the same radius in a glow discharge column in nitrogen, $E/p \approx 3.5$ V/cm Torr (Fig. 8.14) and the field $E \approx 100$ V/cm is stronger by an order of magnitude.

Formula (10.30) also leads to this picture if it is applied (in order to equalize conditions) to the case of negligible collisions with ions. The relative temperature gap is proportional to $(E/N)^2$. In the example given above, the arc temperature gap is less by two orders of magnitude than in a glow discharge, because the electron temperature varies over a rather narrow interval. It cannot be lower than about 1/10 of the ionization potential I, otherwise there would be no free electrons.

The physical reason for the described difference in temperature gaps lies in the difference between the mechanisms by which electrons acquire the energy I necessary to ionize an atom.In a nonequilibrium weakly ionized plasma, electrons acquire energy directly from the field: $e^2E^2/m\nu_m^2 \sim (E/N)^2$ in each collision. In equilibrium plasma, electrons obtain energy from other particles, including heavy ones. Each electron of the ensemble gains from the field in a collision an essentially smaller portion of energy since $(E/N)^2$ is less by two orders of magnitude. Electrons feed energy to atoms, and then all particles "pool" their energy to concentrate it on colilsions into the energy of some electrons (in the "tail" of the Maxwellian distribution) which produce ionization. In short, the field in nonequilibrium plasma must bring an electron's energy up to I, and in equilibrium plasma, only to $kT_e \ll I$.

10.11.4 When Is Plasma in Equilibrium?

Each electron thus gains much less energy from the field in an equilibrium plasma than in nonequilibrium one, but the gas gets heated up to a much higher temperature, so that a greater energy release is required, $w = jE$. So much heat is released because electrons are numerous and the current (not the field) is strong. This is quite clear in the energy balance equation of the gas (10.12), if we use (10.30) to rewrite $\sigma E^2 = n_e e^2 E^2/m\nu_m^2$ in terms of the rate of energy transfer from electrons to heavy particles:

$$\sigma E^2 = \frac{3}{2}\delta(T_e - T)\nu_m n_e = -\frac{1}{r}\frac{d}{dr}r\frac{d\Theta}{dr} \sim \frac{\Theta}{R^2} \,. \tag{10.33}$$

A small gap corresponding to $T \approx T_e$ is achieved at high n_e (at high current). The same effect is achieved by increasing the collision frequency ν_m, thereby intensifying the exchange of energy. This is one of the reasons why plasma is more often in equilibrium at high pressures. The second reason, which sometimes is even more important, is the slowing down of the diffusion losses of electrons; this factor facilitates the increase in the degree of ionization to the equilibrium level corresponding to T_e.

11. Sustainment and Production of Equilibrium Plasma by Fields in Various Frequency Ranges

11.1 Introduction. Energy Balance in Plasma

Equilibrium plasma is formed in steady-state (and in sufficiently long-pulsed) *high-pressure* discharges. Fields in all four frequency ranges are currently employed to generate this plasma: dc, rf, microwave, and optical. The words "sustainment of plasma" will mean a process in which the energy of the field is continuously released in a certain mass of gas, so that the plasma state is preserved. The *generation (production)* of plasma is defined here as a process of continuous production of plasma, whereby fresh masses of cold gas are continuously turned into plasma. Every process of plasma sustainment by electric field can be used to generate it, by blowing cold gas through the volume where the discharge is sustained in order to obtain a continuous plasma jet. Such generators of dense low-temperature plasma – *plasmatrons* – are widely used in physics research and in industrial applications.

This chapter outlines the processes of sustainment and generation of equilibrium plasma in various fields; we will emphasize the common features and see how the main physical problem is solved: the determination of *plasma temperature* depending on the characteristics of the applied external field.

11.1.1 Energy Balance Equation

In the general case, the gas temperature obeys gas dynamics equations that must also take into account non-hydrodynamic mechanisms of energy transfer: heat conduction and radiation. However, quite a few practically important processes are such that even if the plasma is moving, its velocity is subsonic and the pressure is constant in time and space. The gas dynamics energy equation is then transformed into an equation for temperature involving specific heat at constant pressure c_p:

$$\varrho c_p dT/dt = -\operatorname{div} \boldsymbol{J} + \sigma\langle E^2\rangle - \Phi\,, \quad \boldsymbol{J} = -\lambda\nabla T\,. \tag{11.1}$$

Here $\varrho = NM$ is the mass density of the gas, related to temperature by the condition of constant pressure, $p = (N+n_e)kT$, and λ is the thermal conductivity. The derivative d/dt refers to a finite mass of the gas. If this mass moves at a velocity $\boldsymbol{u}$, then

$$dT/dt = \partial T/\partial t + (\boldsymbol{u}\cdot\nabla)T\,, \tag{11.2}$$

where $\partial/\partial t$ refers to a fixed point of space. We consider, in addition to constant fields, only sinusoidal fields. In this case, σ in the expression $\langle \boldsymbol{j} \boldsymbol{E} \rangle = \sigma \langle \boldsymbol{E}^2 \rangle$ for energy released in $1\,\text{cm}^3$ per second stands for high-frequency conductivity (Sect. 3.4), and angle brackets $\langle\,\rangle$ denote the averaging of a quantity over one oscillation period that is assumed to be sufficiently short. The quantity Φ in (11.1) describes radiative losses. As a rule, we neglect them because their role is not significant at atmospheric pressure and at $T < 11{,}000$–$12{,}000\,\text{K}$.

11.1.2 Conservation Law for Total Energy Flux in Steady-State Static Discharges

The energy balance equation can be analyzed and solved for $\boldsymbol{E}$ regarded as a constant parameter only in the case of a constant and homogeneous field, as in the arc column (Sect. 10.10). In the general case, the electric field distribution is to be determined together with the temperature distribution. The field satisfies Maxwell's equations (Sect. 3.3) which contain electrodynamic material characteristics σ and ε which are functions of temperature. The temperature and electric field are thus described by a system of interrelated equations.

Typically, the plasma temperature in the zone where the field energy is released is independent of whether the gas is moving or not. To determine it and its dependence on the applied field we can analyze *static modes*, with the discharge burning in *stationary* gas [11.1]. We will consider steady-state static discharges neglecting radiative losses. In this case (11.1) transforms to

$$-\text{div}\,\boldsymbol{J} + \sigma\langle \boldsymbol{E}^2 \rangle = 0\,, \quad \boldsymbol{J} = -\lambda \nabla T \tag{11.3}$$

[cf. (10.12)].

Let us turn to the equation of energy balance of the electromagnetic field, (3.21), which follows from Maxwell's equations. In stationary conditions,

$$\text{div}\,\langle \boldsymbol{S} \rangle = -\sigma\langle \boldsymbol{E}^2 \rangle\,, \quad \boldsymbol{S} = (c/4\pi)[\boldsymbol{E}\cdot\boldsymbol{H}]\,, \tag{11.4}$$

where $\boldsymbol{S}$ is the flux density vector of electromagnetic energy. Equation (11.4) indicates that the flux is attenuated because the energy supplied by the electromagnetic field is dissipated in the medium. Combining (11.3) and (11.4), we obtain

$$\text{div}\,\left(\boldsymbol{J} + \langle \boldsymbol{S} \rangle\right) = 0\,, \tag{11.5}$$

which indicates that the total flux of thermal and electromagnetic energy has no sources. The amount of electromagnetic energy entering a volume and dissipated within it is exactly equal to the amount of heat leaving the volume.

11.1.3 Fluxes Integral

In the one-dimensional case, (11.5) is integrable, yielding the first integral of the system of the equations of field and plasma energy, known as the "*fluxes integral*":

$$r^n \left(J + \langle S \rangle \right) = \text{const} . \tag{11.6}$$

Here $n = 0$ for flat geometry, $n = 1$ for cylindrical, and $n = 2$ for spherical. The integration constant is found from the boundary conditions.

11.2 Arc Column in a Constant Field

To demonstrate a greater generality of relations that prescribe plasma temperature in any equilibrium discharge, we return to the problem, discussed in Sect. 10.10, of determining the temperature in the arc column. We approach it from a new standpoint, using the fluxes integral (11.6). By virtue of symmetry conditions, radial fluxes (both thermal and electromagnetic) vanish at the column axis, that is,

$$J_r + S_r = 0 \, , \quad -\lambda dT/dr = -S_r = cE_z H_\varphi / 4\pi \, , \tag{11.7}$$

where the flux S_r is expressed through (11.4) in terms of the electric field E_z directed along the axis and the magnetic field H_φ; if current is linear, H_φ has azimuthal orientation. Electromagnetic energy flows into the current channel from the outside across the lateral surface, is dissipated in the plasma, and is carried back to the outside by heat conduction flow. The "field" interpretation of energy transformations in terms of the electromagnetic concepts is equivalent to the concept of the release of Joule heat by the current.

We express E_z in (11.7) in terms of H_φ using Maxwell's equations (3.13):

$$E_z = \frac{c}{4\pi\sigma} \frac{1}{r} \frac{d}{dr} r H_\varphi \, , \quad S_r = -\frac{c^2}{16\pi^2 \sigma} \frac{H_\varphi}{r} \frac{d}{dr} r H_\varphi \, . \tag{11.8}$$

Now we multiply (11.7) by σ so as to have all temperature-dependent quantities on one side of the equation. Integrating in r from the axis to tube walls ($r = R$), we find the relation between plasma temperature T_m on the axis and the radial distribution of magnetic field:

$$\int_{T_w}^{T_m} \sigma(T) \lambda(T) dT = \frac{c^2}{16\pi^2} \int_0^R \frac{H_\varphi}{r} \frac{d}{dr} (r H_\varphi) dr \, , \tag{11.9}$$

where $T_w \approx 0$ is the temperature of the inner wall of the tube surrounding the column.

The exact integral relation (11.9) can be approximated by using the *channel model* for calculating the right-hand side. Then we are able to express the known function of the sought temperature T_m in the left-hand side in terms of the arc current i to which the magnetic field is proportional. According to (11.8), H_φ outside the channel is $H_\varphi = H_1(r_0/r)$, where $H_1 = 2i/cr_0$ is the field at the boundary of the conductor $r = r_0$; inside the channel, $H_\varphi = H_1(r/r_0)$. Substituting these expressions into (11.9) and integrating, we arrive at (10.22),

$$\int_{T_w \approx 0}^{T_m} \sigma(T)\lambda(T)dT = \frac{i^2}{4\pi^2 r_0^2} = \frac{W\sigma_m}{4\pi}, \tag{11.10}$$

mentioned in Sect. 10.10.5. It closes the set of equations of the channel model of arc discharge without any assumptions about the form of $\sigma(T)$ and about the numerical relation between conductivities at the points $r = 0$ and $r = r_0$; besides, it makes the simplified channel model an excellent approximation to the true situation. We will see that relations of type (11.10) also give the temperature of plasma in other types of discharge.

11.3 Inductively Coupled Radio-Frequency Discharge

11.3.1 Introductory Remarks

Radio-frequency inductively coupled discharges are becoming increasingly more widespread as a method of plasma generation in laboratories and industry, although the arc generation method is still the most common method of producing dense low-temperature plasmas. The principle of inductively coupled plasma generation was outlined in Sect. 7.6. It has been used to develop *electrodeless plasmatrons,* which have important advantages in comparison with arc plasmatrons. The plasma of electrodeless systems is pure, while in arc plasmatrons the contamination of plasma by products of electrode erosion is inevitable. This characteristic is decisive for the progress of *plasma* production of high-purity compounds and high-purity granulated refractory materials, and so forth. The service life of electrodeless plasmatrons is virtually unlimited while high-power arc plasmatrons suffer from rapid erosion and failure of electrodes. Obviously, however, the operation of a high-power rf discharge is much more complicated than a dc discharge. The required power sources (high-power generators of Megahertz range) are also more complex, more expensive, and troublesome.

The magnitude and spatial distribution of plasma temperature in inductively coupled discharges have two aspects of interest. As in the case of dc arc discharges, they are important for clarifying the physical relations governing the sustainment of plasma by the field and for practical applications. Furthermore, a new problem arises, not encountered in dc discharges: the discharge (as a load) must be coupled to the rf generator, otherwise the generator cannot function efficiently. Such electrical parameters of the plasma load as its ohmic resistance, self-inductance, and mutual inductance that characterize the magnetic coupling to the inductor (all of them affect the functioning of the electric system as a whole) are directly linked with the magnitude and distributions of temperature and currents in the discharge.

The schematic circuit of an inductively coupled discharge was shown in Fig. 7.18. *Inductor* designs vary, using from one or two to many turns in the coil. The current flowing through the inductor after the power source has been switched on is often not associated with electric fields high enough to initiate breakdown in atmospheric-pressure gas: they are sufficient to sustain an already burning discharge, but not to start it. For example, an auxiliary electrode may be introduced for a short time into the discharge tube. This electrode may be heated up by the Foucault current induced by the rf field, and it may vaporize. Heated and therefore rarefied metal vapor or gas thereby undergoes breakdown. After the discharge in the main gas has been fired, the auxiliary electrode is removed.

11.3.2 Equations of Discharge in a Long Solenoid

An analysis of the idealized one-dimensional process gives a fairly detailed idea of the energy and electrodynamic characteristics of an inductively coupled discharge, as it did in the case of the long arc column. Let a dielectric tube of radius R be inserted into a long solenoid coil (Fig. 11.1). The plasma is sustained at the expense of the Joule heat of the circular current induced by the oscillating magnetic field of the rf current in the coil. The steady state is maintained via the heat conduction transport of energy to the cooled tube walls. The radial temperature distribution in the gas within the tube, schematically shown in Fig. 11.1, is described by (11.3):

$$-\frac{1}{r}\frac{d}{dr}rJ_r + \sigma\langle E_\varphi^2\rangle = 0\,, \quad J_r = -\lambda\frac{dT}{dr}\,. \tag{11.11}$$

In form, it coincides with (10.12) for an arc column, but now the electric field is azimuthal and the magnetic field is directed along the axis. In the MHz frequency range and at atmospheric pressure, high-frequency conductivity (3.23) does not differ from the dc plasma conductivity, the polarization and displacement currents are small in comparison with the conduction currents, and the complex dielectric permittivity is purely imaginary (Sect. 3.5.1). In cylindrical geometry, Maxwell's equations (3.13, 14) without displacement current and with E and $H \propto \exp(-i\omega t)$ take the form

$$\frac{dH_z}{dr} = \frac{4\pi}{c}\sigma E_\varphi\,, \quad \frac{1}{r}\frac{d}{dr}rE_\varphi = \frac{i\omega}{c}H_z\,. \tag{11.12}$$

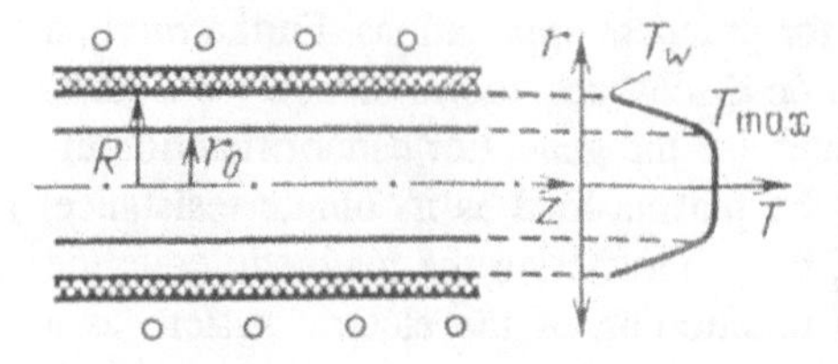

Fig. 11.1. Induction discharge in a tube of radius R placed inside a long solenoid; r_0 is the discharge radius. The radial temperature distribution is given on the *right*

The boundary conditions to the system (11.11, 12) are as follows. By virtue of symmetry, at $r = 0$ we have $J_r = 0$, $E_\varphi = 0$. At $r = R$, we have $T = T_w \approx 0$. The magnetic field in a nonconducting cold medium at the tube wall is the same as inside an empty solenoid [11.2],

$$H_z(R) \equiv H_0 = (4\pi/c)(I_0 n) , \tag{11.13}$$

where I_0 is the current in the coil and n is the number of turns per unit length. The complex amplitudes I_0 and H_0 can be assumed real. The phase shifts of the oscillating fields H_z and E_φ are counted off the phase of the field H_0 in the neighborhood of the coil.

11.3.3 Inductively Coupled Heating of Materials

This method is widely used in industry for hardening of metal parts, drying, melting, etc. For instance, a metal rod is inserted into an inductor, such as a solenoid coil. The metal is heated by induction currents. The field does not penetrate deep into the conductor because of the skin effect (Sect. 3.5.6), so that Joule heat is released only in the surface layer. Owing to the high thermal conductivity, however, the metal soon heats up as a whole. If the geometry is cylindrical (a long rod of radius r_0 and conductivity σ in a long solenoid coil), the field is described by the same equations (11.12).Since $H_z(r) = \text{const}$ in the nonconducting gap between the coil and the rod boundary, condition (11.13) is transferred directly to the surface of the conductor where $H_z(r_0) = H_0$. The equations are solved in Bessel functions of complex argument, which are tabulated for the cases of interest.

11.3.4 Model of Metallic Cylinder

This model states that the plasma conductor is similar to a metal conductor; the only difference is that now the rod "radius" r_0 and its conductivity σ_m [which is assumed to be constant, in the first approximation, and corresponding to the maximum plasma temperature T_m at the axis], are not known in advance. This model is a literal analogue of the channel model of arc discharge, being based on the same steep dependence of σ on T whereby the conductivity of the gas drops sharply where the plasma temperature falls off appreciably towards the tube walls. If the plasma temperature is sufficiently high, the skin effect forces the heat of the induction currents to be deposited in an annular layer, as in a metal. A plateau on the temperature distribution is formed in the middle of the ring as a result of conductive heating of the medium (Fig. 11.2).

The metallic cylinder model makes it possible to separate, at least partly, the solutions of the electrodynamic and thermal problems. In the former problem, σ_m and r_0 are treated as parameters. We will not write the general solution for a cylindrical conductor and only consider the case of strong skin effect when the field penetrates into the conductor to a small depth and the layer geometry is practically planar. The condition of validity of this approximation is $\delta \ll r_0$,

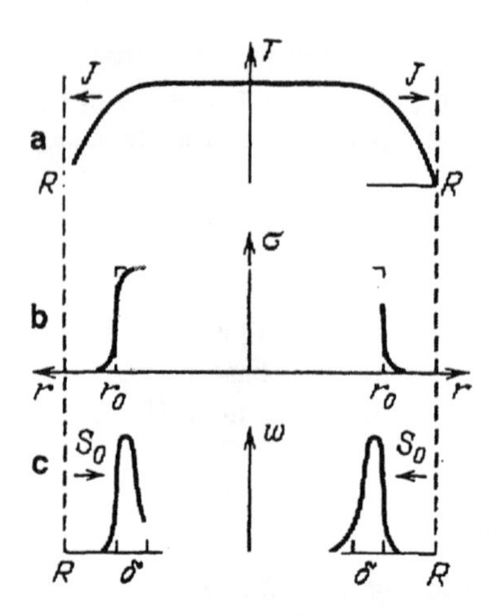

Fig. 11.2. Radial distributions of (a) temperature, (b) conductivity, and (c) Joule heat release in induction discharge. Dashed curves correspond to step function $\sigma(r)$ in the metallic cylinder model. *Arrows:* J, heat flux; S_0, electromagnetic energy flux; δ, skin layer thickness

where $\delta = c/\sqrt{2\pi\sigma\omega}$ is the skin-layer thickness (3.29); it is usually satisfied in inductively coupled discharges at frequencies in the range of the industrial value $f = 13.6\,\text{MHz}$. Measuring the x coordinate from the surface into the layer (against the radius r) and directing the axis y tangentially to the surface, we rewrite (11.12) in the form

$$\frac{dH_z}{dx} = \frac{4\pi}{c}\sigma E_y\,, \quad \frac{dE_y}{dx} = -\frac{\mathrm{i}\omega}{c}H_z\,. \tag{11.14}$$

Now $H = H_0$ at $x = 0$ and E_y, $H_z \to 0$ as $x \to \infty$. Equations (11.14) describe a plane monochromatic electromagnetic wave in a medium with purely imaginary complex dielectric permittivity. Let us write out their solution in a complex form, as in Sect. 3.5, and then convert to the real form:

$$\begin{aligned} H_z &= H_0 \exp\left[-\mathrm{i}(\omega t - x/\delta) - x/\delta\right] \to H_0 \mathrm{e}^{-x/\delta}\cos(\omega t - x/\delta)\,, \\ E_y &= H_0(\omega/4\pi\sigma)^{1/2} \exp\left[-\mathrm{i}(\omega t - x/\delta + \pi/4) - x/\delta\right] \to \\ &\to H_0(\omega/4\pi\sigma)^{1/2}\mathrm{e}^{-x/\delta}\cos(\omega t - x/\delta + \pi/4)\,. \end{aligned} \tag{11.15}$$

The amplitudes of H_a and E_a fall off exponentially into the plasma, and $E_a \ll H_a$ at the phase shift of $\pi/4$. For instance, in air at $p = 1\,\text{atm.}$ and $T = 10{,}000\,\text{K}$, $\sigma = 25\,\text{Ohm}^{-1}\text{cm}^{-1}$; at $f = 13.6\,\text{MHz}$, $\delta = 0.27\,\text{cm}$, $4\pi\sigma/\omega = 3.3 \times 10^6$ and $E_a = 5.5 \times 10^{-4} H_a$. The electromagnetic energy flux is directed into the conductor, normal to its surface, and equals

$$\langle S \rangle = S_0 \mathrm{e}^{-2x/\delta}\,, \quad S_0 = (c/16\pi)(\omega/2\pi\sigma)^{1/2} H_0^2\,. \tag{11.16}$$

The energy coming from the inductor into the conductor is

$$S_0 = 9.94 \times 10^{-2} \frac{(I_0 n[\text{A/cm}])^2 (f[\text{MHz}])^{1/2}}{(\sigma[\text{Ohm}^{-1}\text{cm}^{-1}])^{1/2}}\,\frac{\text{W}}{\text{cm}^2}\,. \tag{11.17}$$

The power released per unit length of a cylindrical conductor is $W = 2\pi r_0 S_0$. The sources of Joule heat are concentrated in the surface layer of effective thickness $\delta/2$.

11.3.5 Plasma Conductor Radius

The plasma conductor radius is readily linked to the plasma temperature T_m and power W, if we consider heat transfer across the nonconducting gap between the conductor and the tube. Integrating (11.11) with $\sigma = 0$, we find

$$\Theta_m - \Theta_w = (W/2\pi)\ln(R/r_0) \approx \Delta r S_0 . \tag{11.18}$$

As we have not specified the expression for W, (11.18) is similar to a similar equation (10.15) for arcs. The last transformation in (11.18) takes into account that in static induction discharges the cylindrical annulus in which the heat is deposited is nearly contiguous to the tube, so that the tube-plasma gap is narrow: $\Delta r = R - r_0 \ll R$. In this respect, an induction discharge subjected to skin effect behaves differently from a dc arc in which the current flows within a thin axial channel.

11.3.6 Plasma Temperature

The *metallic cylinder* model does not allow the calculation of the main factor, i.e., the *conductivity* of the plasma conductor or the *plasma temperature*. The reasons for this are clear now that we have discussed the channel model of arc discharge in Sect. 10.10. Temperature is determined by the energy balance in the energy deposition zone; hence, in order to analyze the transport of energy from this zone, one has to take into consideration the actual temperature drop in the model cylinder and draw a qualitative distinction between *conducting* and *nonconducting* media.

According to (11.11), the zone where the rf field does not penetrate and there are no heat sources is characterized by $T(r) = \text{const} = T_m$ (Fig. 11.2). The entire temperature difference cylinder, $\Delta T = T_m - T_0$, falls across the surface layer of thickness $\delta/2$, where the energy is released. The heat flux leaving the conductor is approximately $J_0 \approx 2\lambda_m \Delta T/\delta$ ($\lambda_m \equiv \lambda(T_m)$); it coincides with the electromagnetic energy flux S_0 from the inductor. This gives us the energy balance equation for the conduction plasma itself, by analogy to (10.16) for an arc discharge:

$$2\lambda_m \Delta T/\delta \approx S_0 , \quad 4\pi\lambda_m \Delta T \approx W(\delta/r_0) . \tag{11.19}$$

The function $\sigma(T)$ being the same as for the constant field in the arc, the condition specifying the temperature drop at which the conducting medium can be assumed to transform into "nonconducting" one [(10.18) or (10.20)] remains valid. Combining one of these relations with (11.19) we obtain an equation yielding the plasma temperature of the inductively coupled discharge. Thus, using (10.17) with effective ionization potential I (Sect. 10.10.5) and formulas (11.16) for S_0, (11.13) for H_0, and (3.29) for δ, we express the plasma conductivity of a steady discharge in terms of the parameter controlled in experiments, namely, the inductor current (ampere-turns):

$$\lambda_m \left(2kT_m^2/I\right) \sigma_m = (I_0 n/2)^2 \,. \tag{11.20}$$

The function $\sigma(T)$ being steep, T_m and λ_m vary in a narrow range, so that, approximately, $\sigma_m \propto (I_0 n)^2$. As the inductor current increases, the temperature increases slowly,

$$T_m = \frac{I/2k}{\ln(4\lambda_m kT_m^2 C/I) - \ln(I_0 n)} = \frac{\text{const}}{\text{const} - \ln(I_0 n)} \,, \tag{11.21}$$

in complete analogy with (10.25) for dc arc discharges. As in the latter case, a high temperature is hard to achieve: high ampere-turns are necessary, and also high power:

$$W = 2\pi r_0 S_0 \approx 2\pi R S_0 \propto H_0^2 \sigma_m^{-1/2} \propto I_0 n \propto \sigma_m^{1/2} \,. \tag{11.22}$$

the plasma temperature is independent of the field frequency (provided the skin layer is thin).

11.3.7 Exact Formula for Temperature

It is striking that, in addition to (11.6), for an inductively coupled discharge there is a second integral of the system of plasma energy equations and Maxwell's equations. This is so because equations without displacement current, (11.12), make it possible to recast the electromagnetic energy flux in differential form,

$$\langle S_r \rangle = -\frac{c^2}{32\pi^2 \sigma} \left\langle \frac{dH_z^2}{dz} \right\rangle = -\frac{c^2}{64\pi^2 \sigma} \frac{dH_a^2}{dr} \,, \tag{11.23}$$

where H_a is the real amplitude of magnetic field. We now substitute (11.23) into the fluxes integral (11.6) [where the constant is zero, as in (11.7) for the arc discharge], multiply the entire equality by σ, and integrate over r from 0 to R. Taking into account (11.13) and recalling that if the skin layer is thin, $H_a \approx 0$ at $r = 0$, we obtain the expression

$$\int_{T_w \approx 0}^{T_m} \sigma(T)\lambda(T)\, dT = \frac{c^2 H_0^2}{64\pi^2} = \left(\frac{I_0 n}{2}\right)^2 \tag{11.24}$$

which determines the plasma temperature as a function of inductor current; it was first derived in [11.3]. In contrast to (11.10), it is exact. In the case of arc discharges the right-hand side of (11.9) cannot be integrated, because (11.8) implies that S_r cannot be presented in purely differential form; however, common features of (11.24) and (11.10) are obvious. If $\sigma(T)$ is given in the form of (10.17) and $I_{\text{eff}} \gg kT_m$, then (11.24) can be reduced to equality (11.20), which we obtained above in a simplified way.[1]

[1] Footnote see opposite page

11.3.8 Examples of Calculations and Measurements

Let us use the formulas to make some estimates. Let the discharge burn in air at p = 1 atm., f = 13.6 MHz and plasma temperature T_m = 10,000 K; λ_m = 1.4×10^{-2} W/cm K. We have already mentioned in Sect. 11.3.4 that under these conditions σ_m = 25 Ohm^{-1} cm^{-1} and δ = 0.27 cm. The Joule heat is deposited in the ring of effective thickness $\delta/2$ = 0.14 cm. According to (11.19, 10.18), plasma is sustained in this plasma cylinder if the inflow of energy from the surface is S_0 = 250 W/cm^2. According to (11.16, 13), this flux is obtained if H_0 = 75 Oe and I_0n = 60 A/cm. The maximum amplitude of electric field at the external plasma boundary is found from (11.15) to be $E_a \approx$ 12 V/cm. The maximum density of the circular current in plasma is $j_a \approx \sigma_m E \approx$ 300 A/cm^2. The total circular current in plasma per unit column length is $j_a\delta \approx$ 80 A/cm. It is comparable to the inductor current per unit length, I_0n = 60 A/cm. Therefore, the mutual-induction effect of the plasma current on the operation of the rf generator circuit is fairly strong.

The heat flux potential in plasma is $\Theta_m \approx$ 0.14 kW/cm. Let the discharge burn in a tube of radius R = 3 cm. As follows from (11.18), the plasma-tube gap of Δr = 0.56 cm is relatively narrow. The power input per unit length of plasma column, given by (11.22), is $W \approx$ 3.8 kW/cm. These calculations indicate that for heating the plasma not to 10,000, but to 12,000 K the power and ampere-turns must be doubled. In fact, this high temperature was calculated with radiative losses not taken into account, so that an even higher power is required. The temperature in inductively coupled rf discharges does not normally exceed 10,000–11,000 K.

Radial distributions of temperature and electron density measured in experiments are plotted in Fig. 11.3. In xenon, the temperature is lower than in argon, since the ionization potential is lower. A small temperature drop in the central part of the discharge, about 500 K, is caused by radiative losses. The Joule heat is deposited only in the peripheral annular layer, while the radiative losses occur in the central part as well because plasma transparency turns them into bulk losses.

11.3.9 Threshold for the Existence of Equilibrium Plasma

If the current through the inductor starts to decrease from the values at which the skin layer thickness $\delta \ll r_0 \approx R$, the plasma temperature and conductivity decrease and δ increases [see (11.21, 22)]. When δ reaches a value of the order of r_0 or R, the skin effect becomes insignificant and the formulae used above are invalidated. In the opposite extreme, $\delta \gg R$, the magnetic field inside the

[1] Integrals containing the Boltzmann function are calculated approximately by a method suggested by Frank-Kamenetsky: $1/T$ in the exponent of the exponential is expanded in the neighborhood of the upper limit T_m,

$$1/T \approx 1/T_m + (T_m - T)/T_m^2 ,$$

and all slowly varying factors with $T = T_m$ are factored out of the integral.

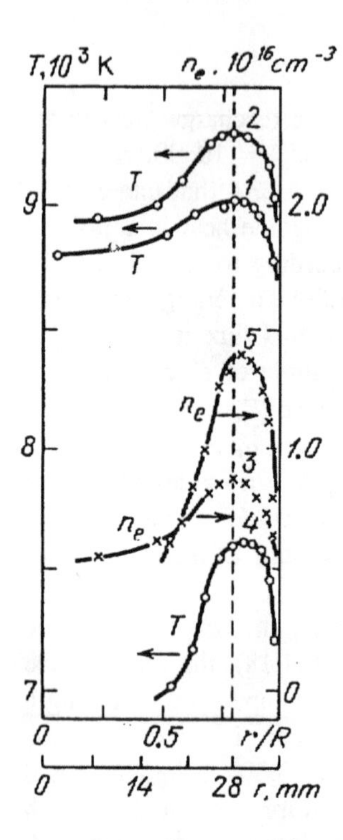

Fig. 11.3. Measured distributions of temperature (circles) and electron densities (crosses) in an induction discharge in a tube of $R = 3.5$ cm at frequency $f = 11.5$ MHz and $p = 1$ atm. [11.4]. Curve *(1)* argon, power input into plasma 4.7 kW; *(2)*, *(3)* Ar, 7.2 kW; *(4)*, *(5)* Xe, 6 kW

solenoid is homogeneous and equal to H_0, as it is in the absence of plasma. According to (11.12), the electric field is $E(r) = i\omega H_0 r/2c$. The power released per unit length of the plasma cylinder of radius r_0 is

$$W = \int_0^R \sigma\langle E^2\rangle 2\pi r\, dr \approx \frac{\pi\sigma_m\omega^2 H_0^2 r_0^4}{16c^2} = \frac{\pi^3\sigma_m\omega^2 r_0 (I_0 n)^2}{c^4}\,. \tag{11.25}$$

Now the temperature falls from T_m to T_0 (the effective plasma boundary) over the entire radius r_0, so that we find, in contrast to (11.19) and in similarity with (10.16), that

$$W \approx 4\pi r_0 \lambda_m \Delta T/r_0 \approx 4\pi\lambda_m \Delta T \approx 8\pi\lambda_m k T_m^2/I\,. \tag{11.26}$$

The plasma temperature cannot be allowd to drop too much since otherwise conductivity vanishes and energy is not released any more. Therefore, the power given by (11.26) is now more or less stable even if the conductivity decreases. As follows from (11.18, 24), the plasma radius decreases with decreasing temperature. We conclude that the inductor current or $I_0 n$, as implied by (11.25), depends inversely on plasma conductivity or temperature and field frequency. Over a wide range of conductivity values that covers the cases of both $\delta \ll R$ and $\delta \gg R$, $I_0 n$ and W as functions of σ_m and ω must have the forms shown in Figs. 11.4, 5 (if $\delta \ll R$, then $W \propto I_0 n \propto \sigma_m^{1/2}$).

The current $I_0 n$ as a function of σ_m or T_m passes through a minimum, which corresponds to the point of merger of the curves for the two extreme cases

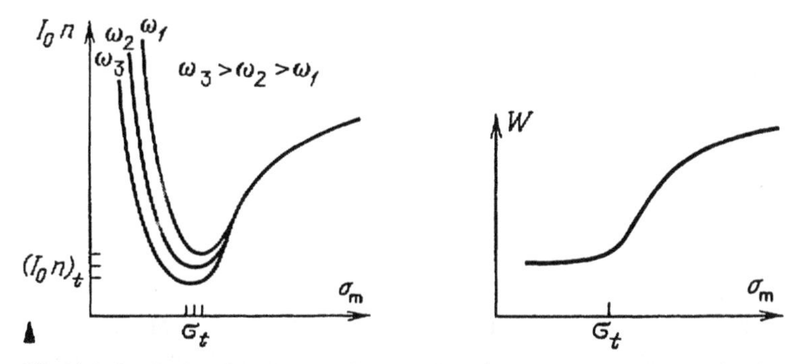

Fig. 11.4. Qualitative dependences of the number of ampere-turns per cm on the plasma conductivity. If there are these unequalities in the drawing, they may be omitted from subscription

Fig. 11.5. Qualitative dependence of power input into plasma on conductivity

discussed above, that is, to the condition $\delta \approx R$. There is a minimum (*threshold*) value of inductor current at which equilibrium plasma is still sustainable in inductively coupled discharge. It can be evaluated by extrapolating (11.20) up to the *threshold* temperature $T_m = T_t$ at which $\delta = R$:

$$(I_0 n)_t \equiv (I_0 n)_{min} \approx \left(\frac{4}{\pi} \frac{\lambda_t k T_t^2 c^2}{I \omega R^2} \right)^{1/2} . \qquad (11.27)$$

For example, in air at $p = 1$ atm. and for $R = 3$ cm, $f = 13.6$ MHz, we have $(I_0 n)_t \approx 10$ A/cm. The threshold temperature $T_t \approx 7000$–8000 K.

11.3.10 Stable and Unstable States

Figure 11.4 also demonstrates another property typical of equilibrium discharges of various types. At a given inductor current $I_0 > (I_0)_t$, other conditions being equal, *two* steady states of equilibrium discharge exist. One of them corresponds to high conductivity and considerable skin effect; the other corresponds to low conductivity and no skin effect. Only the former state is realizable experimentally. The states on the left-hand branch at $T < T_t$ *are unstable*. For instance, let the temperature fluctuate upwards. A lower inductor current than the actual one would be needed for sustaining the new state. Plasma starts to heat up until a state at $T > T_t$ is reached on the right-hand branch of the curve in Fig. 11.4. Similar arguments readily confirm that it is *stable*.

11.4 Discharge in Microwave Fields

11.4.1 Discharges in Waveguides

Long-lasting discharges were obtained at high pressures in the early fifties when high-power (kilowatt) continuously operating generators became available. A number of methods of feeding microwave field energy to plasmas are possible.

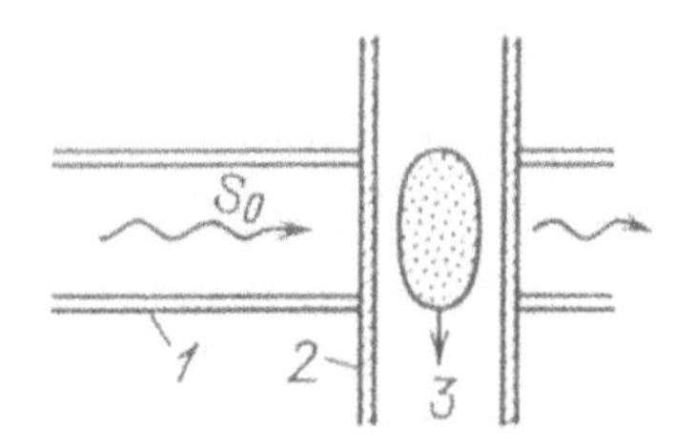

Fig. 11.6. Microwave discharge in a waveguide: *(1)* waveguide, *(2)* dielectric tube, *(3)* discharge plasma

A typical arrangement used is shown in Fig. 11.6. A dielectric tube, transparent to microwave radiation, passes through a rectangular waveguide. The plasma is maintained in the intersection region at the expense of dissipation of microwave energy. The released heat is transported away either conductively through cooled tube walls or, more frequently, by a gas blown through the tube. The latter version is a microwave plasmatron (Sect. 11.6).

Typically, the H_{01} mode of electromagnetic wave travelling in the waveguide is employed. In this mode, the electric field vector is parallel to the narrow waveguide walls (Fig. 11.7a). The field intensity in this direction is constant and it varies sinusoidally along the wider walls (Fig. 11.7b). The discharge takes place in the middle of the waveguide cross section, where the electric field has its highest value. The plasma column is elongated along the electric vector $\boldsymbol{E}$. The waveguide size is usually related to the frequency used. For $f = 2.5\,\mathrm{GHz}$ (wavelength in vacuum $\lambda_0 = 12\,\mathrm{cm}$), the wider wall width is 7.2 cm, the narrow wall is 3.4 cm wide. The dielectric tube is about 2 cm in diameter. The resulting plasma column is then about 1 cm in diameter.

The electric field of the incident wave induces an alternating current in the plasma column. The conduction current is not closed, or rather, it is closed by the displacement current. A rapidly varying current is in itself a source of electromagnetic radiation. The result is the scattering (and reflection) of the incident wave by the plasma conductor placed in a waveguide. The scattered wave interferes with the incident one; the resulting field sustains the plasma. A part of

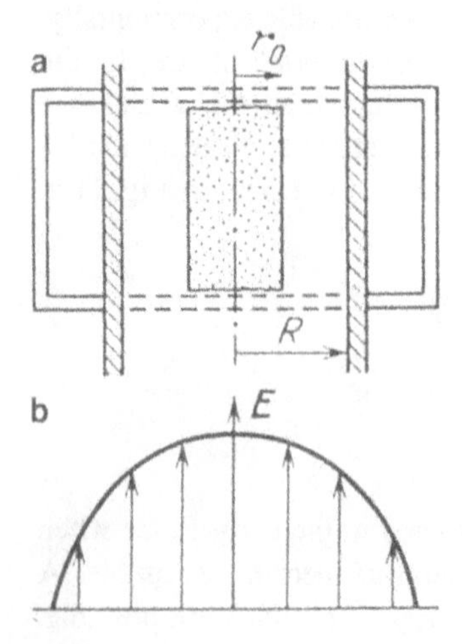

Fig. 11.7. Discharge in a waveguide maintained by an H_{01} mode: **(a)** cross section of the waveguide by the diametric plane of the tube (the plasma is shown *shaded*); **(b)** distribution of electric field along the wider wall

the incident wave is transmitted through the plasma. The power P_0 fed by the generator into the incident wave is thus divided between the reflected and the transmitted waves and is partly dissipated in the plasma. A thin well-conducting rod may dissipate half the incident power; one quarter is transmitted and one quarter is reflected.

The efficiency of the discharge device is greatly increased if the tube is followed by a reflector which returns the transmitted wave, making it pass through the plasma again. As a result, a standing wave is formed in the waveguide in the region of discharge. The reflector–discharge distance is adjusted so that the plasma column lies at the antinode of the electric field. In this way it was found possible to inject up to 80–90 % of the energy produced by the generator into the plasma [11.5].

This system, injecting 1 to 2 kW into a plasma in air at atmospheric pressure produces $T \approx 4000$ K, and up to 5000 K in nitrogen. A temperature gap arises in argon: ($T_e \approx 6500$–7000 K, $T \approx 4500$ K) because energy exchange between electrons and heavy particles in a monatomic gas is slow, especially if the temperature and electron density are not too high (cf. Sects. 10.9, 11). The temperature at a higher frequency ($\lambda_0 = 3$ cm) and in a waveguide of correspondingly smaller cross section is somewhat higher (6000 K in nitrogen). The temperature in microwave discharges is always lower than in rf discharges: the reflectivity of plasma sharply increases with increasing electron density (Sect. 11.4.4).

11.4.2 Discharge in Resonators

Major developments in this field were due to *P.L. Kapitsa* and coworkers, also in the early 1950s [11.6]. This work lead to the development of a continuously operating 175 kW generator operating at a frequency 1.6 GHz ($\lambda_0 = 19$ cm). Standing waves E_{01} were excited in a cylindrical resonator (Fig. 11.8). The electric field at the cylinder axis was directed along the axis and varied sinusoidally along the axis, with the maximum at the center of the cylinder, and the field diminishing radially. The discharge is started on the axis, in the region of maximum field. The plasma region is elongated parallel to the electric field; it forms a filament at high discharge power. The length of this filament reaches half the wavelength (about 10 cm), at a diameter of 1 cm. At high powers of plasma filament twists and floats upwards, pushed by the Archimedes force. It is stabilized by forcing the gas in the resonator to follow a helical trajectory, which prevents floating and makes the filament stable. It was thus possible to inject up to 20 kW of power

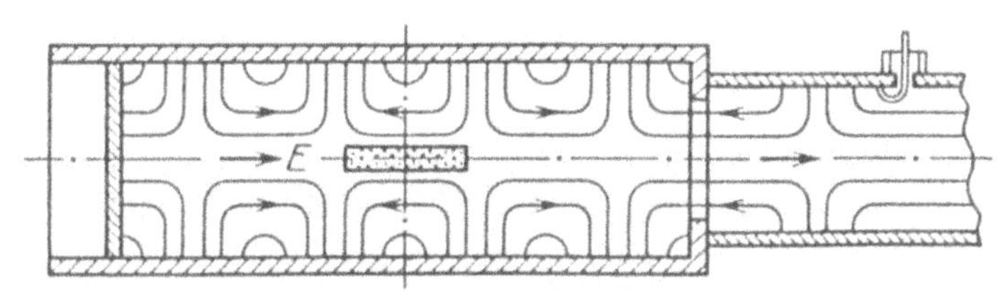

Fig. 11.8. Discharge in a resonator [11.6]. The lines of force of the electric field are shown for an E_{01} mode; the plasma filament is shown *shaded*

into a stabilized discharge in hydrogen, deuterium, and helium at pressures of one to several atmospheres, with the plasma temperature not normally exceeding 8000 K.

11.4.3 Discharges Sustained by a Plane Electromagnetic Wave

The process occurring in waveguides involves numerous details caused by complicated geometry and scattering of the wave by the plasma. These details mostly affect the quantitative rather than qualitative characteristics. To clarify the main features of equilibrium microwave discharges, consider a simple one-dimensional model. Assume that a plane electromagnetic wave passes through a plane dielectric wall transparent to microwaves and is incident on a plasma. The heat deposited in the plasma is transported conductively to the externally cooled wall, so that the steady state is maintained (Fig. 11.9).

At first glance, this model may seem extremely unrealistic; actually, it is not. In fact, we concentrate our attention on that segment of the tube and of the adjacent discharge surface which face the incident wave (Fig. 11.6). If the wave does not penetrate deeply into the plasma and the plasma temperature is quite high due to sufficiently high power (this is indeed realized), the curvature of the surface is unimportant. At any rate, this idealization preserves all the significant qualitative features, and the numerical values of the main parameters that are obtained give a good idea of the true values.

In plane geometry, the temperature deep inside the plasma tends to a constant value T_m as $x \to \infty$ and the wave damps out, that is, the constant in (11.6) is zero and

$$J + \langle S \rangle = 0\,, \quad J = -\lambda\, dT/dx\,. \tag{11.28}$$

In the microwave range, effects of *wave* nature (reflection, interference) become important, and the field is described by the wave equation of type (3.31) but with complex dielectric permittivity. Equations for the complex amplitudes E_y, H_z of the monochromatic field are

$$\frac{d^2 E_y}{dx^2} + \left(\varepsilon + \mathrm{i}\frac{4\pi\sigma}{\omega}\right)\frac{\omega^2}{c^2}E_y = 0\,, \tag{11.29}$$

where σ and ε are given by (3.23,24). The temperature and degree of ionization in microwave discharges are never high, and electron-ion collisions play an insignificant role in (10.32). In this case,

$$\sigma \propto (1-\varepsilon) \propto n_e \propto \exp(-I/2kT)\,, \tag{11.30}$$

where (in the case of a one-component gas) I is the true (not effective) ionization potential. The boundary conditions to system (11.3,29) or (11.28,29) with S defined by (11.4) state that deep inside the plasma ($x \to \infty$), $E = 0$ and that at the wall ($x = -x_0$), $T = T_w \approx 0$, and that the energy flux S_0 in the incident wave is given. The solution must determine the plasma temperature $T_m \equiv T(\infty)$

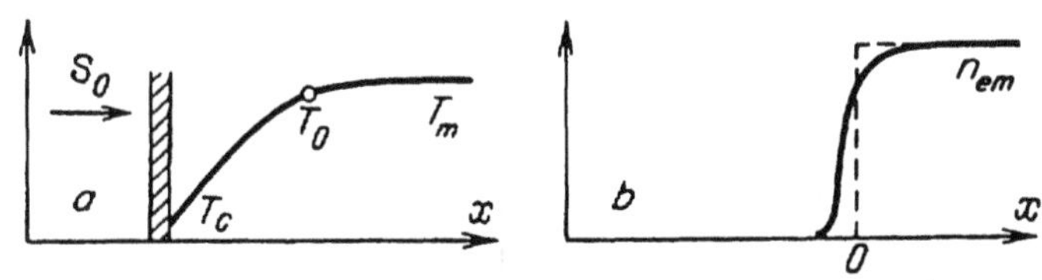

Fig. 11.9. Characteristics of a discharge sustained by electromagnetic waves: (a) Temperature distribution, transparent dielectric wall is *shaded*. (b) Corresponding electron density distribution. *Dashed line* is he approximation of $n_e(x)$ by a step

and that part of the electromagnetic energy flux S_1 which is spent on plasma maintenance [this is equivalent to finding the *reflection coefficient* of the plasma (for power of electromagnetic wave), $\varrho = (S_0 - S_1)/S_0$].

11.4.4 Approximate Solution

Following the arc channel model and the metallic cylinder model for inductively coupled discharges, we introduce the effective plasma boundary, which we place at $x = 0$ where $T = T_0$, and set $\sigma = 0$, $\varepsilon = 1$ for $T \leq T_0$ and $x \leq 0$ but $\sigma = \sigma_m$ and $\varepsilon = \varepsilon_m$ for $T_0 \leq T \leq T_m$ and $x \geq 0$; this corresponds to replacing the $n_e(x)$ distribution with a step (Fig. 11.9). The reflection coefficient for a wave that is in normal incidence from the "vacuum" onto a sharp homogeneous medium is [11.2]

$$\varrho = \frac{(n-1)^2 + \varkappa^2}{(n+1)^2 + \varkappa^2} . \tag{11.31}$$

The indices n and $\varkappa$ are expressed in terms of σ and ε by (3.34).

The part of the electromagnetic energy flux that penetrates the plasma is damped there, obeying (3.37):

$$d\langle S\rangle/dx = -\mu_\omega \langle S\rangle , \tag{11.32}$$

where the absorption coefficient μ_ω is given by (3.36) or (3.38). The energy of the wave is dissipated in the surface layer of thickness of order $l_\omega = \mu_\omega^{-1}$. Here the temperature drops from T_m to T_0. The energy balance in the layer is described by an approximate equality $S_1 = J_0 \approx \lambda_m \Delta T/l_\omega$, in complete analogy to the arc and rf discharges. The temperature T_m is found from the equation

$$S_0 [1 - \varrho(T_m)] = \lambda_m \left(2kT_m^2/I\right) \mu_\omega(T_m) , \tag{11.33}$$

which is similar to (11.20) for the rf discharge. The equation for heat transfer from plasma to walls across the "nonionized" gap, $\Theta_m - \Theta_w = S_1|x_0|$, yields the gap width x_0.

Tables 11.1, 2 illustrate the situation with the results of calculations of the electromagnetic characteristics of atmospheric-pressure air plasma and microwave radiation fluxes necessary to maintain various temperatures. Calculations were carried out for one of the frequencies used in experiments, f = 10 GHz

Table 11.1. Electrodynamic characteristics of air plasma at $p = 1$ atm. and frequency $f = 10\,\mathrm{GHz}$

T [10^3 K]	n_e [cm^{-3}]	ν_m [10^{10} s^{-1}]	σ [10^{10} s^{-1}]	ε	$\frac{4\pi\sigma}{\omega\lvert\varepsilon\rvert}$	n	$\varkappa$	l_ω [cm]	ϱ_0
3.5	6.6×10^{11}	7.5	0.13	0.78	0.33	0.89	0.14	1.7	0.0089
4.0	4.4×10^{12}	7.1	0.88	−0.53	3.3	0.81	1.1	0.22	0.28
4.5	1.6×10^{13}	6.6	3.3	−5.1	1.3	1.3	2.6	0.091	0.57
5.0	4.8×10^{13}	6.4	9.9	−18	1.1	2.1	4.7	0.050	0.71
5.5	9.3×10^{13}	6.0	19	−39	1.1	2.8	7.3	0.032	0.83
6.0	2.1×10^{14}	5.8	41	−88	1.0	4.3	11	0.022	0.88

Table 11.2. Microwave radiation fluxes necessary to sustain air plasma at $p = 1$ atm. and $f = 10\,\mathrm{GHz}$

T [10^3 K]	λ [10^{-2} W/cm K]	Θ [10^{-2} kW/cm]	S_1 [kW/cm^2]	ϱ	S_0 [kW/cm^2]	S_1^* [kW/cm^2][a]	S_0^* [kW/cm^2][a]
4.2	0.92	1.1	–	0.2	–	0.2	0.25
4.5	0.95	1.4	0.045	0.4	0.075	0.23	0.38
5.0	1.1	1.9	0.14	0.65	0.40	0.35	1.0
5.5	1.3	2.5	0.30	0.76	1.25	0.56	2.3
6.0	1.55	3.3	0.60	0.81	3.1	1.06	5.6

[a] Taking into account heat-conduction losses from the plasma column of radius $R = 0.3$ cm.

($\lambda_0 = 3\,\text{cm}$). The reflection coefficient (11.31) is denoted by ϱ_0, and a better value of ϱ (found as an approximation that takes into account the smearing of plasma boundaries, which reduces the reflection coefficient) was used [11.1].

As the temperature decreases, the absorption length of the electromagnetic waves, $l_\omega \propto n_e^{-1} \propto \exp(I/2kT)$, sharply increases. The plane wave model becomes meaningless when l_ω is comparable with the radius R of the discharge column. Progressively greater power is needed because of the plasma's "transparency" to the wave, $l_\omega > R$. This is similar to the situation with the rf discharge (Sects. 11.3.9, 10). The minimum *threshold* temperature T_t for microwave discharges is found from the condition $l_\omega(T_t) \approx R$. In air, $T_t \approx 4200\,\text{K}$ for $R = 0.3\,\text{cm}$ and it corresponds to *threshold* fluxes in the plasma of $S_{1t} \approx 0.2\,\text{kW/cm}^2$, and to the incident wave flux $S_{0t} \approx 0.25\,\text{kW/cm}^2$. These figures agree with experimental data. The reflection of the electromagnetic wave from the plasma, increasing as temperature increases, restricts the possibility of reaching high temperatures in microwave discharges. In air, the plasma temperature does not exceed 5000–6000 K.

11.4.5 Geometrical Optics Limit

In the general case of comparable real and imaginary parts of the complex dielectric permittivity, that is, $4\pi\sigma/\omega|\varepsilon| \sim 1$, the flux $\langle S \rangle$ cannot be given in purely differential form (11.23) or (11.32). This situation is typical for microwave discharges (Table 11.1). However, in the limit $4\pi\sigma/\omega\varepsilon \ll 1$ and $\varepsilon \approx 1$ (which becomes possible at high frequencies) the equation that is exactly valid for homogeneous media, (11.32), becomes meaningful for nonhomogeneous media as well because in this case electromagnetic waves are only weakly damped over a distance of the order of one wavelength, λ_0/n ($\varkappa \approx 2\pi\sigma/\omega \ll 1$, $n \approx \sqrt{\varepsilon} \approx 1$; Sect. 3.5.5). This case corresponds to the geometrical optics approximation. If $n \approx 1$, $\varkappa \ll 1$, then (11.31) implies that reflection is low: $\varrho \ll 1$. There then exists an exact second integral of Maxwell's equations and of the equations for energy; like (11.24), this integral makes it possible to express the plasma temperature in terms of the external field parameters. Substituting $\langle S \rangle$ given by (11.32) into (11.28), multiplying the result by $\sigma = c\mu_\omega/4\pi$, and integrating the result, we find the relation

$$\int_{T_w \approx 0}^{T_m} \sigma(T)\lambda(T)dT = \frac{c}{4\pi} S_0 \,, \tag{11.34}$$

which is akin to (16.10) for constant fields and to (11.24) for rf fields.

11.4.6 Quasistationary Field Limit (RF Discharge)

The rf discharges treated in Sect. 11.3, where it was possible to neglect the displacement current in Maxwell's equation, correspond to the opposite limiting case of $4\pi\sigma/\omega|\varepsilon| \gg 1$. As follows from the arguments of Sect. 3.5.6, in this case $n \approx \varkappa \approx \sqrt{2\pi\sigma/\omega} \gg 1$, and (11.31) implies that $\varrho \approx 1$. The incident wave is reflected from the plasma almost completely and the field penetrating into it is damped out over one skin layer thickness, $\delta = c/\sqrt{2\pi\sigma\omega}$. This does not mean, though, that only a small fraction of the energy supplied by the generator is injected into the plasma in rf discharges. First of all, it is meaningless to speak of inductively coupled rf discharges in *wave* terms, because the wavelength is much greater than the dimensions of the system. If, however, we insist on the "wave" standpoint, the effect can be interpreted to indicate that the "reflected" power is returned into the generator, which only has to replenish the small difference between the incident and reflected fluxes of electromagnetic energy. The situation with discharges in waveguides is different. For technical reasons, here we have to transport the wave reflected by the plasma away from the generator, so that reflection brings down the efficiency of utilizing the generator energy.

11.5 Continuous Optical Discharges

11.5.1 Specific Features of Optical Sustainment of Plasma

The discharge in the optical frequency range is a relatively novel phenomenon. Even the combination of these two terms – *optical discharge* – has been accepted only rather recently. Nevertheless, the term reflects the physical content of the process to the same extent that the long-familiar terms "microwave disharge" and "rf discharge" do. Dense equilibrium plasmas can be sustained in a steady state by optical radiation, just as by other constant and oscillating fields. Likewise, an optical plasma generator (*optical plasmatron*) can be obtained if cold gas is pumped through the discharge zone. The possibility of implementing these processes was given a theoretical foundation in [11.7], and a *continuously burning optical discharge* was realized in the laboratory in 1970 [11.8].

A fairly high optical power is required to sustain a plasma. If the objective is a long-term effect, as achieved in all other fields, only CO_2 laser radiation can be exploited at present, because this is the only practically available high-power cw laser. Fortunately, the wavelength of the IR radiation of CO_2 lasers is large, since the absorption coefficient of plasma for light falls off steeply with frequency. For instance, visible light is absorbed rather weakly in atmospheric-pressure plasmas (Sect. 11.5.3). Pumping the necessary power into a plasma in the visible light frequency range would require a power greater than that of CO_2 lasers by a factor of 10^2 to 10^3.

The optical method of supplying energy to plasma has a very distinctive feature. No structural elements are needed to carry the energy to plasma (electrodes, inductor, waveguide or resonator). Optical radiation can be transported by a light beam across empty space or through any gas; this is attractive and promissing for applications. In principle, an optical discharge can be initiated anywhere, far from any solid objects; the disharge can be made to travel along the beam, or it may be localized by *focusing* the radiation and thus stabilizing the discharge. The plasma can be moved through space by shifting the beam, for example, by moving the focus to which the discharge is "locked". As we have mentioned, an optical plasmatron can be designed by blowing a cold gas through the discharge zone stabilized by focusing the beam; this plasmatron produces a continuous plasma jet with a very high temperature. Even under standard conditions, the plasma temperature of optical disharge is substantially higher than in other discharges: 15,000–20,000 K. Finally, the high temperature of optical discharge can be used for developing stable light sources of very high brightness localized, in principle, in free space.

The possibility of generating a continuously burning optical disharge by non-laser light sources is extremely limited. The absorption coefficient in the optical frequency range being quite low, a very high degree of gas ionization (close to total single ionization of atoms) is required for appreciable energy deposition in a given volume. The plasma temperature must be correspondingly high, 15,000–20,000 K. But the source of energy sustaining the plasma must be at a temperature at least as high. Indeed, the second law of thermodynamics forbids free transfer of energy from a colder to a hotter body. It is thus impossible to sustain a temperature above the solar temperature of 6000 K by focusing rays from the sun with a mirror or lens, of even arbitrarily large diameter and concentrating arbitrarily high power. However, the absorption of light by plasma at this low temperature is so weak that even if a plasma were somehow initiated, it would immediately decay because of nonreplenished energy losses.

11.5.2 Experimental Arrangement

The beam of a CO_2 laser is focused by a lens (or a mirror) (Fig. 11.10). The lens is made of NaCl or KCl, since ordinary glass is opaque to the infrared line $\lambda = 10.6\,\mu$. To initiate the discharge, it is necessary to create the *seed plasma nucleus* (the problem of *initiation* has to be solved for all types of equilibrium discharge). This can be achieved by stimulating breakdown in the focal spot with an auxiliary system or with a tungsten wire introduced into the focal spot for a short time. The metal on the wire surface is slightly vaporized, the vapor is ionized and starts absorbing laser light. The wire is immediately removed and the discharge keeps burning in the gas. The plasma slightly shifts from the focal spot towards the laser up to the point where the intensity of the beam is just sufficient for to sustain the plasma. The size of the plasma region changes from 1 mm at a threshold power to 1 cm and more at higher power levels.

In order to be able to work with different gases and to vary their pressure, the beam is sent into a chamber through a salt window. If the laser power is

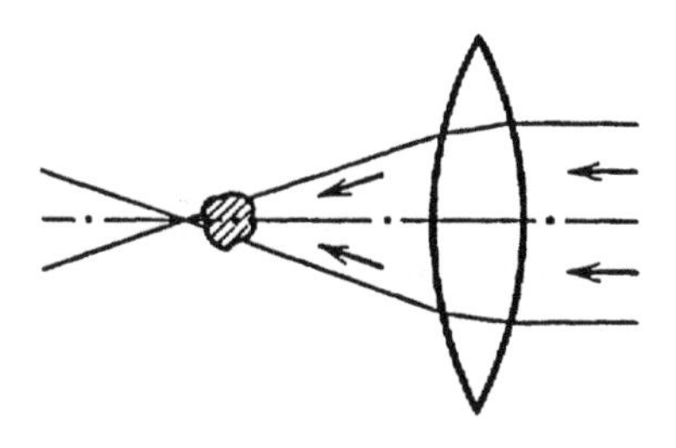

Fig. 11.10. Experiment for sustaining continuous optical discharge. Plasma (*shaded*) is shifted from the focal point towards the laser beam

high enough, a continuous optical discharge can be started in free air. Pioneer experiments [11.8] used a laser that was small by today's standards, 150 W. More powerful lasers were a rarity at the time (1970). The discharge was initiated in xenon at a pressure of several atmospheres, since theory predicted that the required laser power should be small. Now that researchers have considerably more powerful lasers, discharges are produced in different conditions, including in the open air. Actually, this last feat requires laser power that is high even by current standards, not less than 5 kW, and a high-quality beam (low-divergence beam) that allows good focusing.

11.5.3 Absorption of CO_2 Laser Radiation in Plasmas

The quanta of the CO_2 laser, $\hbar\omega = 0.117\,\mathrm{eV}$, are small in comparison with kT ($\hbar\omega/k = 1360\,\mathrm{K}$) and are thus absorbed in a strongly ionized optical-discharge plasma, by reverse bremsstrahlung mechanism, in electron-ion collisions. After doubly charged ions are taken into account, the formula for the coefficient of absorption of CO_2-laser quanta, with numerical coefficients, is [11.1]

$$\mu_{\omega(\mathrm{CO_2})} = \frac{2.82\times 10^{-29} n_e(n_+ + 4n_{++})}{\{T[\mathrm{K}]\}^{3/2}} \lg\left\{\frac{2.7\times 10^3 T[\mathrm{K}]}{n_e^{1/3}}\right\}\ \mathrm{cm}^{-1}\,. \quad (11.35)$$

An important point for what follows is the maximum that $\mu_\omega(T)$ has at $p = \mathrm{const}$ (Fig. 11.11). The maximum appears when the single ionization of atoms is almost complete. As the temperature increases further within a certain interval, the degree of ionization remains almost unchanged. Double ionization starts at considerably higher temperatures, while the gas density and hence $n_e \approx n_+$ at $p = \mathrm{const}$, keep decreasing as $1/T$. Therefore, prior to the onset of double ionization, $\mu_\omega \propto n_e^2 T^{-3/2} \propto T^{-7/2}$. A new strong increase of $\mu_\omega(T)$ occurs in the course of double ionization, but such high temperatures are not achieved in discharges.

The maximum value of the coefficient $\mu_{\omega,\max}$ increases with pressure somewhat more slowly than p^2. In air at $p = 1$ atm., $\mu_{\omega,\max} \approx 0.85\,\mathrm{cm}^{-1}$ and the minimum absorption length of the laser radiation is $l_{\omega,\min} \approx 1.2\,\mathrm{cm}$. For neodymium laser light, $\lambda = 1.06\,\mu$, $\mu_{\omega,\max} \approx 6\times 10^{-3}\,\mathrm{cm}$, and $l_{\omega,\min} \approx 170\,\mathrm{cm}$. These figures show clearly why short-wave radiation is not advantageous for sustaining a plasma: the transparency of the plasma is too great.

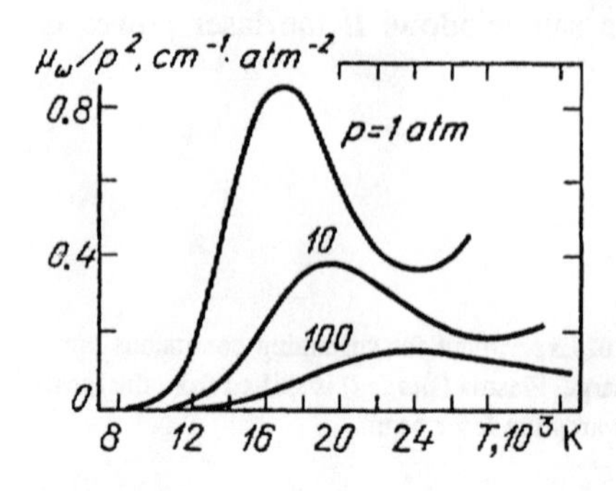

Fig. 11.11. Absorption coefficients for CO_2 laser radiation in air. The maximum absorption coefficients (from the *upper* to *lower* curve) are $\mu_{\omega,\max} = 0.85$, 38, 1600 cm^{-1}

11.5.4 Plasma Temperature and Threshold Power

Figure 11.10 shows the complex geometry of light beams supplying energy to the plasma, and of the discharge zone. As in the case of discharges in waveguides, the selection of a suitable one-dimensional model for analytical evaluations involves certain crude assumptions. Nevertheless, the main features of the process are in satisfactory agreement with the spherical model, especially if the beam is focused by short-focus optics and the beam power is not much greater than the threshold level (so that the plasma cannot shift away from the focal spot).

Consider a plasma sphere of radius r_0; high temperature is maintained in it by absorbing convergent spherically symmetric rays of total power P_0. As in all preceding cases, we assume that absorption is absent ($\mu_\omega = 0$) outside the sphere, where $T < T_0$, while inside the sphere the absorption coefficient μ_ω is constant and corresponds to the maximum temperature at the center, T_m. The possibility of approximating $\mu_\omega(T)$ by a step function is justified by the steep rise of this function to the maximum value. At the *threshold* power level, the dimensions of the plasma region are small and the plasma is transparent to the laser light. The power absorbed in the plasma is of the order of $P_1 \approx P_0\mu_\omega r_0$. It is transported out of the energy deposition zone by the heat conduction flux J_0.

In general, the temperature in optical discharges is high, and radiative losses may be considerable, although this is mostly true for pressures above 5 atm. and powers substantially greater than the threshold level. The size of the plasma region at the limit of plasma existence is so small that heat conduction losses, proportional to $4\pi r_0^2$, are greater than the radiative losses, which are proportional to the plasma volume $4\pi r_0^3/3$ (owing to the plasma transparency). Therefore,

$$P_1 = 4\pi r_0^2 J_0 \approx 4\pi r_0^2 \Delta\Theta / r_0 = 4\pi r_0 \Delta\Theta \ . \tag{11.36}$$

where $\Delta\Theta = \Theta_m - \Theta_0$ is the drop in the heat flux potential in the plasma. Beyond the absorbing sphere of radius r_0,

$$-4\pi r^2 d\Theta/dr = P_1 \ , \quad \Theta = P_1/4\pi r \ , \quad \Theta_0 = P_1/4\pi r_0 \ . \tag{11.37}$$

This expression takes into account that the gas at "infinity" is cold. As follows from (11.36) and the last equation of (11.37), $\Delta\Theta = \Theta_0$ or $\Theta_m = 2\Theta_0$. The substitution of the absorbed power $P_1 = P_0\mu_\omega r_0$ into the equation $P_1 = 2\pi r_0 \Theta_m$ gives the relation of the laser power to the temperature at the center of the plasma region:

$$P_0 = 2\pi\Theta(T_m)/\mu_\omega(T_m) \ . \tag{11.38}$$

As the function $\mu_\omega(T)$ has a maximum and $\Theta(T)$ is monotonic, the power P_0 passes through a minimum (Fig. 11.12). The minimum $P_{0,\min} \equiv P_t$ lies at a temperature T_t close to the point $\mu_{\omega,\max}$. This is the lowest (*threshold*) power of focused light beam that is sufficient for sustaining a steady-state plasma. The states on the left-hand branch of the curve $P_0(T_m)$, where $T < T_t$, are *unstable* and thus cannot be realized. Indeed, if the temperature increases, the actual power

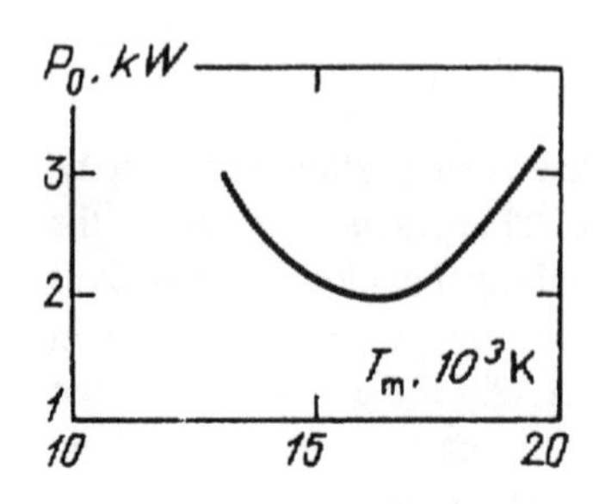

Fig. 11.12. Power of spherically convergent laser beam as a function of maximum plasma temperature of steadily sustained plasma. Air, $p = 1$ atm., CO_2 laser radiation

becomes greater than that required to maintain the new state. The plasma gets heated up until the state switches to the right-hand side. Similar arguments then readily show that these states are stable. This situation is typical for equilibrium discharges.

More detailed analysis is required to evaluate the plasma radius r_0. It cannot be found from (11.36–38). The point is that r_0 is connected with the dimensions of the light channel (in the spherical model, this is the focal spot radius ϱ_0). This radius was ignored in earlier arguments because we assumed that the rays coverge exactly at the center. The minimum plasma radius corresponding to threshold conditions equals $r_{0,\min} \equiv r_t \approx \sqrt{(4/3)\varrho_0 l_\omega}$ [11.1]. For $\varrho_0 \approx 10^{-2}$ cm, $l_\omega \approx l_{\omega\,\min} \approx 1$ cm, $r_t \approx 1$ mm.

11.5.5 Required Laser Power

The evaluation formula (11.38) predicts the laser power needed to sustain a continuously burning optical discharge under specific conditions. It corresponds to roughly $T_m = T_t$ and $\mu_{\omega,\max} = \mu_\omega(T_t)$. Thus the plasma temperature for the CO_2-laser radiation ($\mu_{\omega,\max} \approx 0.85\,\mathrm{cm}^{-1}$) in atmospheric-pressure air is $T_t \approx 18{,}000$ K and $\Theta_t \equiv \Theta(T_t) \approx 0.3$ kW/cm; the threshold is $P_t \approx 2.2$ kW. These figures are confirmed by experimental data. As the pressure increases, the threshold power decreases rather rapidly since $\mu_{\omega,\max}$ increases; this behavior is obsrved up to a certain limit at which radiative losses become important. The threshold is lower in gases with low ionization potential and low thermal conductivity (lower T_t and Θ_t). This is why xenon at $p \approx 3$–4 atm. was chosen for the initial experiments with a low-power laser: from 100 to 200 W would be sufficient then in Xe or Ar. About 300 kW would be needed to sustain a discharge in atmospheric-pressure air at the frequency of the neodymium-glass laser.

11.5.6 Why Unusually High Temperature Is Obtained in Optical Discharges

The temperature in arc and rf discharges in atmospheric-pressure air is about 10,000 K, in microwave discharges it is about 5000 K, and in optical discharges, about 18,000 K. In argon $T \approx 20{,}000$ K, twice that of arc and rf discharges. The reason lies in the transparency of the plasma to optical radiation, owing to the dependence $\mu_\omega \propto \omega^{-2}$.

At not too high frequencies, the field energy is efficiently dissipated even if ionization is not too high, that is, if temperature is moderately high. Moreover, the field does not penetrate a strongly ionized plasma (owing to skin effect or reflection), which diminishes dissipation and heating. Dissipation at optical frequencies is at its highest when the degree of ionization is very high. As the gas gets hotter, it remains transparent but absorbs more and more strongly, thereby facilitating further heating, until total single ionization is achieved. At very high pressures, however, it is very difficult to raise the temperature higher than 13,000–15,000 K, since radiative losses, going up with temperature, become considerable.

The unusually high temperature in an unenclosed gas, not shaded by walls, is a unique property of the continuous optical discharge which makes it possible to develop a light source of extremely high brightness: The optical discharge plasma emits blindingly bright white light, which cannot be viewed with the naked eye. Especialle high temperatures up to 25,000–30,000 K, are predicted for helium at $p \approx 1$–3 atm., since helium has an extremely high ionization potential [11.9]. This could be a high-power source of UV light. However, helium also manifests a high threshold, so that this source would not be inexpensive.

11.5.7 Measurement of Temperature and Thresholds

Figure 11.13 shows photographs of continuously burning optical discharge [11.10]. The temperature field is plotted in Fig. 11.14 [11.11]. The temperature was measured by recording continuum radiation in a narrow interval of wavelengths around $\lambda = 5125$ Å and the radiation in spectral lines of nitrogen atoms and ions. The center of the plasma region shifted 1.1 cm towards the beam. The temperature measured at the center at $p = 2$ atm. was 18,000 K in Ar, but 14,000 K in Xe (below the ionization potential). In H_2 at 6 atm., it was 21,000 K, and in N_2 at 2 atm. it was 22,000 K. The temperature always falls off in a monotonic manner from the center to the periphery of the plasma region. The region size is

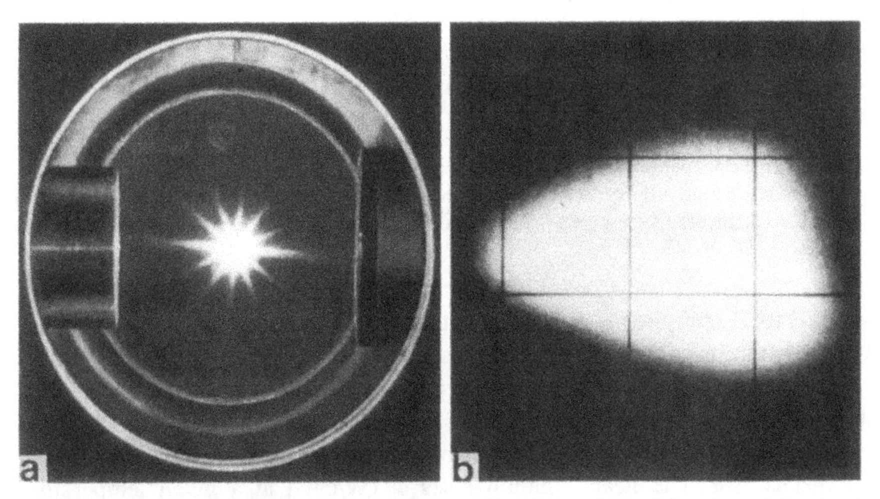

Fig. 11.13. Photograph of continuous optical discharge [11.10]. **(a)** General view through the chamber window (8 cm diameter); **(b)** enlarged image: 1 division: 1 mm, the beam travels from *right* to *left*

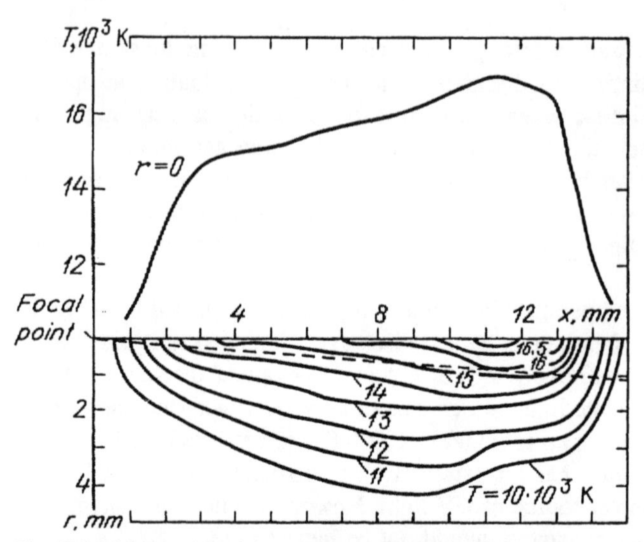

Fig. 11.14. Measured spatial temperature distribution in continuous optical discharge [11.11] in air at 1 atm.; CO_2 laser power is 6 kW and the beam travels from *right* to *left*. Effective boundary of the convergent light channel is shown by the *dashed line*. *Lower* part: isotherms, x is the optical axis, r is the radial distance from the axis. *Upper* part: distribution $T(x)$ along the beam axis

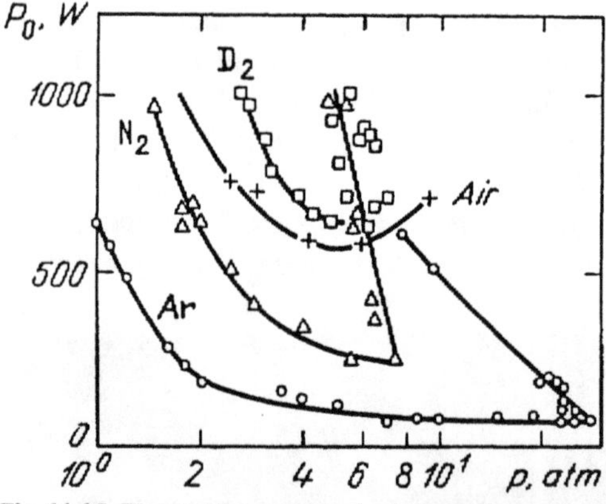

Fig. 11.15. Threshold power required to sustain continuous optical discharge in several gases (lower branches of the curves). Upper boundaries for the existence of discharge in N_2 and Ar are also shown (upper branches of the curves). The existence domain in P_0, p lies betwen the *upper* and *lower* curves [11.12]

usually 3 to 15 mm; the plasma is elongated parallel to the optical axis, following to some extent the shape of the light channel.

The threshold power as a function of pressure is plotted in Fig. 11.15 [11.12]. At $p \sim 1$ atm. the threshold power falls off steeply with p (lower curves), but the curves flatter out at $p \sim 5$ atm. This behavior is caused by a change in the energy loss mechanism. The heat radiation loses Φ [W/cm³] at a given temperature increase with pressure very nearly as the laser light absorption coefficient μ_ω, because both of these quantities are proportional to $n_e n_+ \approx n_e^2$. Under nearly threshold conditions, the energy release due to absorbed laser energy compensates for the radiative losses. Now $\mu_\omega S \approx \Phi$, where $S = P/\pi R^2$ is the intensity of laser radiation, and R is the radius of light channel in the region of plasma generation. If μ_ω and Φ depend on p in the same manner, the threshold power $S\pi R^2$ is a weak function of p. The thresholds in molecular gases are higher than in heavy inert gases: the transfer of the molecular dissociation energy contributing appreciably to the thermal conductivity (Figs. 10.19, 20).

When a laser beam is focused by a lens with a not very short focal length ($f \approx$ 8–10 cm or more), an upper limit to both P and p may be observed in some gases (Fig. 11.15). Steady-state discharges cannot be sustained at power levels P above the upper limit, and at pressures p above the point where the upper and lower curves converge (Fig. 11.15). This is a consequence of the attenuation of the laser beam when it penetrates the absorbing plasma. The beam intensity becomes insufficient for compensating radiative losses deep in the plasma [11.12]. If the beam is focused by a short-focus mirror with f = 1.5–2.5 cm, the upper limits on P and p are not observed (Fig. 11.16) [11.13]. In this case, the attenuation of the beam intensity due to light absorption upon penetration into the plasma is overcome by the intensity increase due to the large beam-convergence angle. As a result, the energy source supplying the making up for radiative loss is not depleted. Note that the laser beam in experiments [11.13] is sent vertically upward, which helps stabilize the discharge. The upward convective flux of heated gas caused by the Archimedes' force shifts the plasma toward the focal spot, into the zone of increased intensity. In experiments [11.12], the beam was horizontal so that the convective flow tended to displace the plasma out of

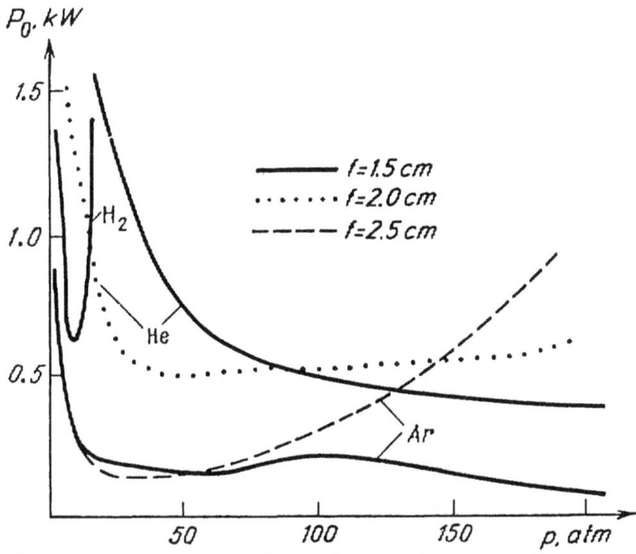

Fig. 11.16. Threshold power for sustaining continuous optical discharge at high pressures [11.13]. f is the focal distance of the focusing mirror

the light beam, thereby impeding the sustainment of discharge. Presumably, this factor may constitute a constraint on the domain of existence of the discharge.

11.5.8 Two-Dimensional Calculations and a One-Dimensional Model

The temperature field of the optical discharge is described by the balance equation for plasma energy (11.1), where it is advisable to replace $\sigma\langle E^2\rangle$ by $\mu_\omega S$. The geometry of the discharge is two-dimensional: the temperature $T(x,r)$ is a function of x along the optical axis and of radius r in the transverse direction (Figs. 11.10, 14). To simplify the problem, we assume that the laser intensity S has a constant distribution over the light beam cross section, and operates with power $P(x) = S\pi R^2$ and replace the absorption coefficient $\mu_\omega[T(x,r)]$ by its value $\mu_\omega[T(x,0)]$ on the axis. Then the energy release term is $\mu_\omega P/\pi R^2$, where P satisfies the equation $\partial P/\partial x = -\mu_\omega P$ – equivalent to (11.4, 32) for a variable cross section channel. In the heated zone, the shape of the light channel is essentially non-conical owing to the *refraction* of laser radiation in the plasma that it generates. Correspondingly, an equation is added to the system which describes the channel radius $R(x)$. An important factor is the *radiative heat transfer;* owing to this, the term Φ is not just the bulk radiation rate, but the difference between the emission and absorption of thermal radiation. To find Φ, the spectral *equation of radiative heat transfer* is used [11.14], and it is also included into the system of equations.

The solution of the above problem for the conditions corresponding to Fig. 11.14 [11.15] is in good agreement with the recorded temperature field and provides an explanation of one previously unexplained effect. The point is that usually a considerable part of the laser power passes through the plasma: 2.8 kW out of 6 kW in the present case. This would be amply sufficient for sustaining the discharge at the focal point, since only 2 kW is needed (this is the threshold for atmospheric air). Actually, two separated plasma regions are never observed, and the temperature at the point of the geometrical focus is considerably lower than at the center of the plasma region (Fig. 11.14). It was found that as a result of refraction, conically converging rays begin to diverge precisely at the point of $T \approx T_{max}$. Now everything is clear: the hottest plasma is at the narrowest part of the light channel, where the laser radiation reaches its maximum concentration. Obviously, this segment lies closer to the lens than the geometrical focal point. The smallest radius of the distorted light channel is greater by an order of magnitude than the radius of the focal spot in the absence of plasma.

The energy balance of a discharge sustained in atmospheric air is illustrative: out of 6 kW of laser power, the plasma consumes 3.2 kW. The absorption takes place in the central zone, where $T > 10{,}000$ K. Almost all of the absorbed energy is radiated away apart from the UV part of the radiation, representing 1.9 kW, which is absorbed in the region where $6000 < T < 10{,}000$ K. Out of the energy transported to "infinity", 2 kW is lost via heat conduction and 1.2 kW via radiative losses. The radiative heat transfer is thus quite intense in high-temperature gas.

Two-dimensional calculations are complicated and time-consuming. For this reason, a simplified one-dimensional model of a continuously burning optical discharge has proved to be quite useful [11.16]; it provides a speedier way of solving a number of problems and leads to a fairly satisfactory fit to experimental data (an example is given in Sect. 11.6.5). The one-dimensional model deals with the temperature distribution along the optical x axis. Correspondingly, the radial part of the heat flux divergence in (11.1) is replaced approximately by the term of radial heat-conduction losses $A\Theta/R^2$, where $\Theta(x)$ is the heat flux potential (Sect. 10.10.2), and $A \approx 1.5$–2 is a numerical coefficient whose value reflects the specific radial temperature profile. Radiative heat transfer is treated in this model using the radiative transfer equation. On the subjects of optical discharge see [11.1, 17, 18]; these references give the necessary bibliography.

11.6 Plasmatrons: Generators of Dense Low-Temperature Plasma

In Sect. 11.1 it was mentioned that any steady discharge can be used for developing a *plasmatron* by blowing cold gas through it. Plasmatrons of three types are currently employed in practice: *arc, rf,* and *microwave* systems (the *optical* plasmatron has not yet progressed left beyond the experimental stage).

11.6.1 Arc Plasmatrons

This type evolved long ago to the stage of industrial units, with powers ranging from hundreds of watts to thousands of kilowatts [11.19]. The design of high-power machines involves numerous engineering problems: stabilization, cooling, increasing the electrode service life, material selection, etc. The schematic diagram of a plasmatron is shown in Fig. 11.17a. The material to be processed is often used as the anode (Fig. 11.17b). Such systems are employed for cutting, welding, and arc-plasma melting of metals.

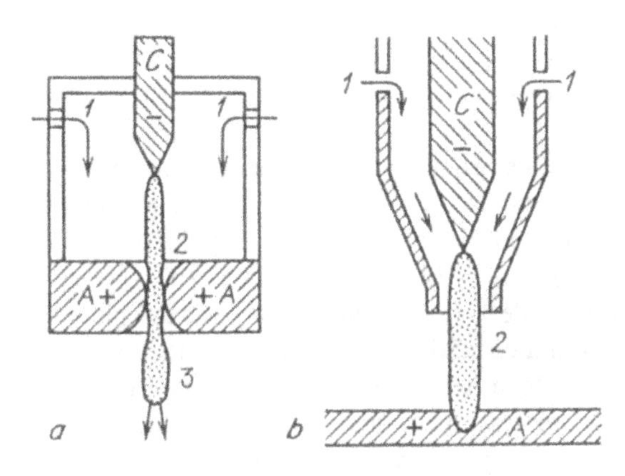

Fig. 11.17. Layout of an arc plasmatron: **(a)** Plasma jet *(3)* emerges from an orifice in the anode A; C is the cathode, *(1)* cold gas supply, *(2)* discharge. **(b)** The machined part (metal sheet to be cut) serves as the anode

11.6.2 RF Plasmatron

The basic scheme of an rf plasmatron is not very different from that of static discharge devices. The discharge burns in a tube placed inside an inductor coil. A cold gas is fed into the tube, and a plasma jet flows out of it (quite often, into the ambient atmosphere). The inductively *coupled plasma torch* in its modern form was designed by *Reed* in 1960 (Fig. 11.18). The most important design feature in it is the use of a *tangential gas inflow*. The point is that the plasma of a steady rf discharge extends nearly to the tube wall, so that the thermal load on the tube is very heavy. The situation is not greatly alleviated by simply pumping the gas axially. In the tangential inflow systems, the gas is pumped into the tube 20 to 30 cm upstream of the inductor tangentially to the cylindrical surface. The gas flow is helical. As a result of the centrifugal force, the pressure in the axial region is lowered and a vortex is formed. The longitudinal flow is weak here. The cold gas mostly flows along the tube within the peripheral cylindrical layer and squeezes the discharge away from the walls, thereby protecting them from the destructive effects of the high temperature. Gas flow rates are usually such that the average longitudinal velocity of the cold gas is about 1 m/s. The plasma jet is ejected into the atmosphere at a velocity of several tens of m/s. Inductively coupled plasmatrons whose power reaches tens and even hundreds of kilowatts are in operation today. Figure 11.19 shows a photograph of a discharge and plasma jet.

11.6.3 Microwave Plasmatrons

Typical designs of microwave plasmatrons are shown in Fig. 11.20. The one shown in Fig. 11.20a is not very different from the waveguide discharge scheme. A helical flow of gas is forced through the dielectric tube inserted through the waveguide. The discharge is squeezed away from the walls and the heat is transported out of the tube by the plasma jet. One advantage of microwave plasmatrons is their high efficiency: up to 90 %. This is achieved, among other things, by reflecting the wave transmitted through plasma back into it (Sect. 11.4.1). The plasma temperature here is not high, 4000–6000 K, less than in rf plasmatrons (9000 to 10,000 K) but this is sufficiently high for some applications. The plasma is as pure as in rf plasmatrons, in contrast to arc plasmas contaminated with products of electrode erosion. A power of several kilowatts is pumped into the plasma.

Figure 11.20b shows a different, quite promising arrangement aimed at producing high powers [11.5]. The discharge is sustained at the axis of a circular cross section waveguide of 5 cm radius and several tens of cm long. Microwaves of type E_{01} and wavelength in vacuum $\lambda_0 = 12.5$ cm are sent into the waveguide. The inner surface of the waveguide and the outer surface of the plasma column (its radius is about 1 cm) form a coaxial line for electromagnetic waves. The discharge is stabilized at the axis by a helical flow of gas pumped through the waveguide tube. The tube ends with a nozzle that lets out the plasma jet. The microwave power is almost completely absorbed by the plasma. The elec-

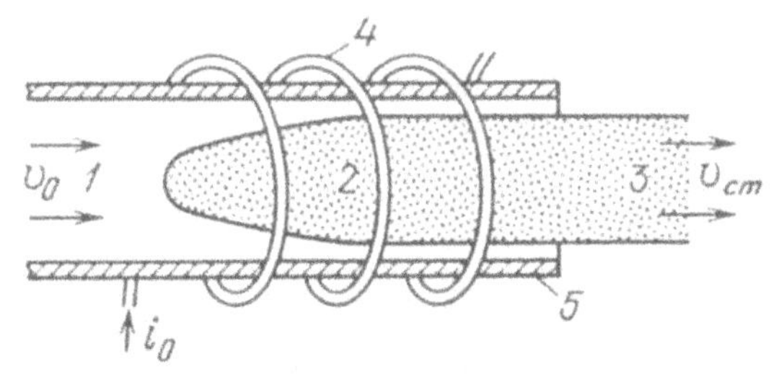

Fig. 11.18. Layout of an inductively coupled plasmatron. *(1)* cold gas inflow, *(2)* discharge, *(3)* plasma jet, *(4)* inductor coil, *(5)* dielectric tube

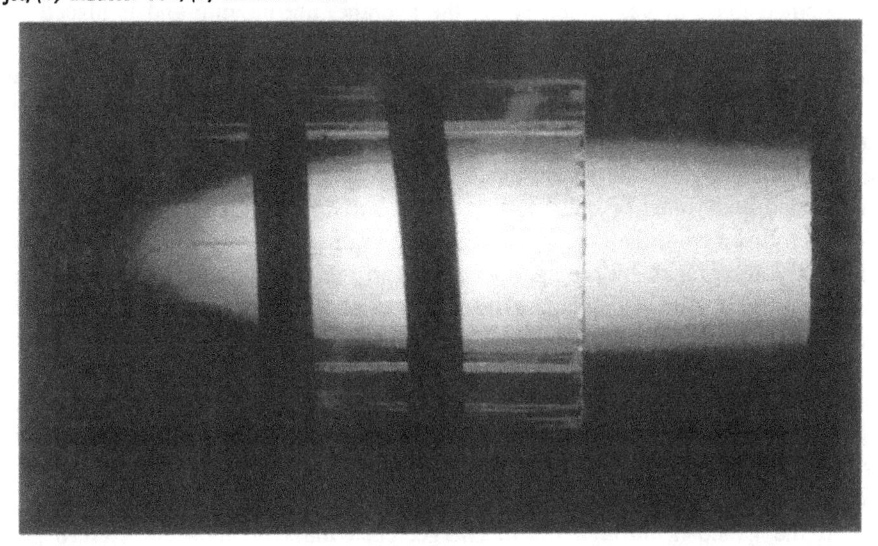

Fig. 11.19. Photograph of discharge and plasma jet. Tube diameter 6 cm; gas (air) flows from *left* to *right*; $p = 1$ atm.; flowrate $2 \times 10^3\ \mathrm{cm^3/s}$; power input into plasma 27 kW; temperature measured on the axis at the tube end 9800 K [11.20]

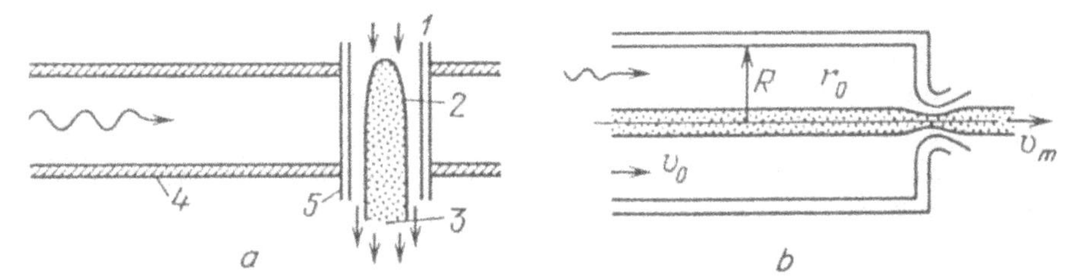

Fig. 11.20. Microwave plasmatrons. **(a)** Discharge *(2)* in dielectric tube *(5)* passing through a waveguide. *(1)* cold gas inflow, *(3)* plasma jet, *(4)* waveguide. **(b)** Coaxial design. Arrows show the gas flow and the electromagnetic wave. Discharge column and plasma jet are *shaded*

tromagnetic energy flux flows into the discharge column from the surface, along radii, just as in the inductively coupled rf discharge or in the resonator discharge; as microwaves travel along the tube, their energy is gradually dissipated. The plasma temperature is $T \approx 5000\,\mathrm{K}$.

11.6.4 Flow Around or Flow Through?

Two mutually exclusive idealized schemes can be used to begin analyzing the complex pattern of hydrodynamic flow in plasmatrons. We can assume that the discharge is burning not only in a region that is spatially defined, but also in a specified mass of gas which is, on the average, not moving and is placed in a gas flowing around it. This situation resembles the convective cooling of a solid heated from within. If the discharge ("heated body") is cylindrical, the flow around it can be illustrated with streamlines, as in Fig. 11.21. Heat conduction transfers the heat of the hotter volume to the contiguous layers of the gas flowing around it. Clearly, the temperature in the converging part of the stream just behind the heated body must be close to that of the body.

As follows from the analysis of all static discharges, the plasma temperature is mostly determined by the energy balance in the energy deposition zone. The temperature is hardly dependent on the fate of the energy once it has been transported beyond the discharge volume into the nonabsorbing, currentless zone. It can be expected, therefore, that the temperature in the discharge zone is not very sensitive to whether the heat goes through the walls cooled by a flow outside or carried away by the flow that is in direct contact with the fixed discharge volume. In other words, the discharge plasma temperature in the flow-around mode is approximately found by calculations for the appropriate static situation.

If the gs flows *through* the discharge, each mass of high-temperature gas leaving the plasmatron has resided in the discharge zone, and has had energy deposited directly in it. A qualitative representation of the flow is shown in Fig. 11.22 for the case of an inductively coupled plasmatron [11.1]. The heat is transported from the skin layer, where the energy is deposited, by the radial heat conduction flux. As a result of simultaneous longitudinal heat transport by the gas stream, the discharge surface [rather, an isotherm corresponding to the ionization temperature ("conduction onset point") $T_0 \approx 8000\,\mathrm{K}$] is tilted with respect to the axial flow. Cold-gas streamlines entering the high-temperature zone become tilted with respect to the axis and they undergo refraction in the layer where the temperature abruptly increases. This happens because of the cxpansion of the gas upon heating. The velocity component tangential of the isothermal surface

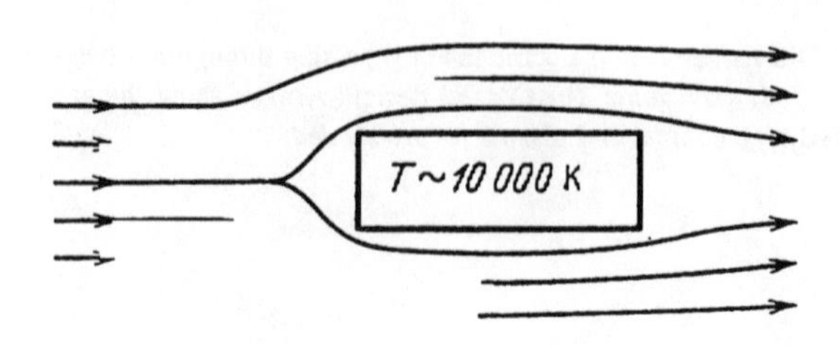

Fig. 11.21. Flow of cold gas around a static discharge burning in a specified mass of gas

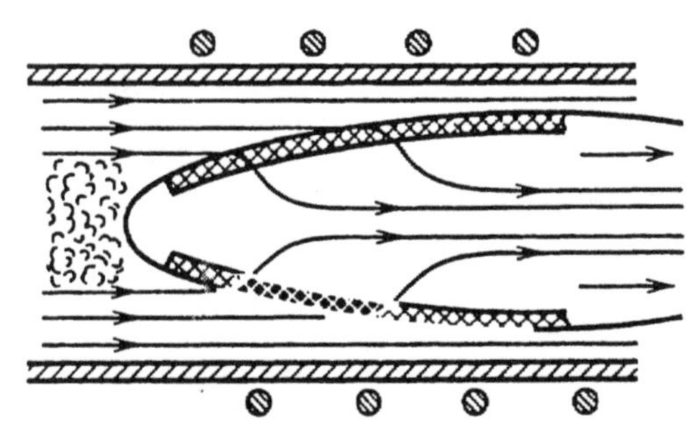

Fig. 11.22. Gas flow in an inductively coupled plasmatron if the gas is assumed to flow through the discharge region. The transitional region from the cold gas to the plasma and the layer of energy deposition are *shaded*. Gas streamlines are traced. Eddies are shown on the *left* in front of discharge

remains unchanged but the normal component greatly increases by virtue of the conservation of the mass flux density, $\varrho_0 v_n = \varrho_m u_m$. Here ϱ_0 is the density of the cold gas, v_n is the velocity at which the gas flows into the discharge zone along the normal, and ϱ_m and u_m are the density and normal component of the velocity of the hot gas emerging from the skin layer. The plasma temperature in the flow-through mode is also not very different from that in the static discharge [11.1, 18].

The actual flow pattern is intermediate between the diagrams of Figs. 11.21 and 11.22, and may sometimes be closer to one or other of the two version (Sect. 11.6.7).

11.6.5 Normal Velocity of Discharge Propagation

The crucial element in the interaction of gas flow and electrical discharge as shown in Fig. 11.22 is the process, in some segment of the shaded layer, where the field energy is dissipated and the cold gas is transformed into heated plasma. An ideal model for analyzing this process is that of a plane stationary system viewed in the reference frame fixed to a given element of the discharge front surface.

A cold gas flux $\varrho_0 v_n$ and a flux S of electromagnetic energy enter the discharge (Fig. 11.23). The dissipated energy is conductively transported counter to the gas stream, facilitating the heating of the cold gas to the "ionization" temperature T_0 at which it becomes capable of intensively absorbing the electromagnetic energy. In response to dissipation, the gas heats up from T_0 to T_m. The maximum plasma temperature T_m is not very different from that achieved in the corresponding static discharge; indeed, it is mostly determined by the energy balance in the energy deposition zone, where the temperature difference $\Delta T = T_m - T_0$ is small and the acceleration of the gas is insignificant. In a steady-state process, the gas flows into the discharge zone with a normal component of velocity v_n, related to T_m and S by the energy conservation law. If the entire field energy, released in this zone, is ultimately carried away by the flow, then $S = \varrho_0 v_n w(T_m)$, where w is the specific enthalpy of the gas (the process takes place at constant pressure).

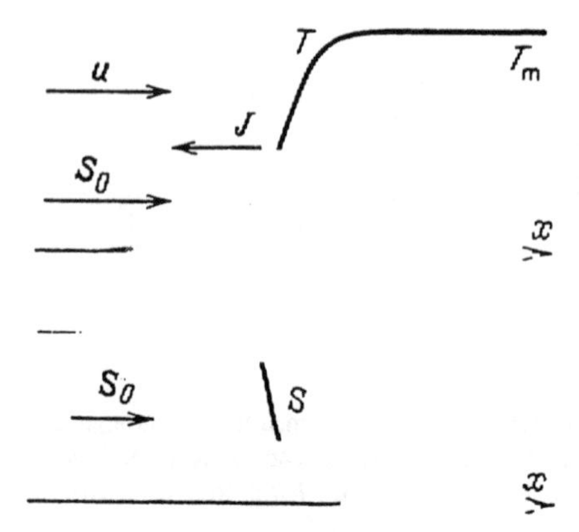

Fig. 11.23. Plane discharge mode in gas flow: temperature (T) and electromagnetic energy flux (S) distributions. *Arrows* indicate the directions of heat flux J, gas flow u, and electromagnetic energy flux S_0

The heat-conduction nature of v_n becomes evident it we resort to (11.19) which expresses the balance of the fluxes of the heat and electromagnetic energy in the dissipation zone of the latter, and to (10.18) which determines the temperature drop in this zone (its thickness is δ). We obtain

$$v_n = \frac{S}{\varrho_0 w(T_m)} = \frac{2\lambda_m \Delta T}{\delta \varrho_0 w_m} = \frac{4\lambda_m k T_m^2}{\delta \varrho_0 w_m I} \approx \frac{4\chi_m}{\delta} \frac{kT_m}{I} \frac{(c_p)_m T_m}{w_m} \frac{\varrho_m}{\varrho_0}$$

where $\chi_m = \lambda_m / \varrho_m (c_p)_m$ is the thermal diffusivity of the hot gas and $(c_p)_m$ is its specific heat at constant pressure; $w_m \equiv w(T_m)$. Normal velocities v_n are typically of the order of $v_n \sim 10 \div 100\,\text{cm/s}$. For details on discharge propagation, see [11.1, 18].

11.6.6 Optical Plasmatron

Let us consider a continuously burning optical discharge in gas streams (Fig. 11.24). The one-dimensional model outlined in Sect. 11.5.8 is quite suitable for clarifying a number of characteristics of this process. Let us approximate the convergent-divergent light channel by two adjacent truncated cones. Assume the flow to be straight; then the mass conservation law is simply $\varrho u = \varrho_0 v$, where ϱ and u are the density and velocity of the gas at a point x, and ϱ_0 and v are those in the cold incoming stream. Bearing this in mind and converting to spherical polar

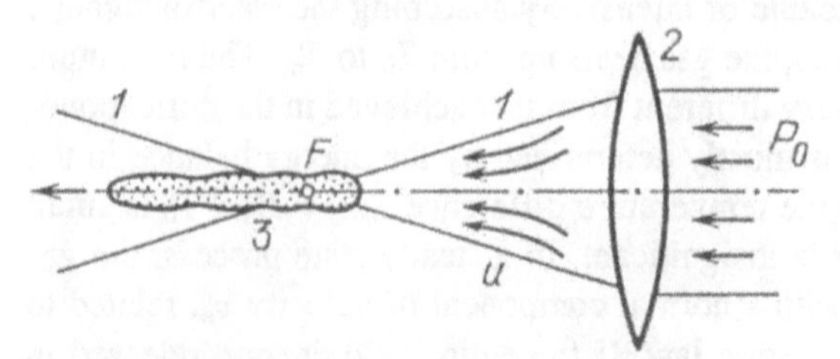

Fig. 11.24. Optical plasmatron: *(1)* light channel contour, *(2)* lens, *(3)* focal spot; *arrows* indicate the directions of laser beam P_0 and of gas flow u

coordinates, as they are more suitable for the cone geometry than cylindrical ones, we rewrite the steady-state equation for energy balance, (11.1, 2):

$$\varrho_0 v c_p \frac{dT}{dx} = \frac{1}{x^2}\frac{d}{dx}x^2\lambda\frac{dT}{dx} - \frac{A\Theta}{R^2} + \frac{P\mu_\omega(T)}{\pi R^2} - \Phi\,. \tag{11.39}$$

This equation is solved simultaneously with $dP/dx = -\mu_\omega P$ and the radiative heat transfer equations for Φ.

Figure 11.25 gives the result of calculations [11.16] for a case that has been experimentally analyzed [11.21]. The discharge plasma was photographed at for several values of v, the velocity of the cold gas stream. The luminous region is spindle-shaped. As the velocity increases, the flow "forces" the plasma into the focal spot while the hot luminous jet stretches beyond this point. At $v > 310$ cm/s, the plasma is blown out and the discharge disappears. As the flow is intensified, the maximum plasma temperature increases; this has also been confirmed by measurements.

Experiments have demonstrated that if the beam is focused by a long-focus lens with $f = 40$ cm, the discharge cannot be sustained at velocities lower than a

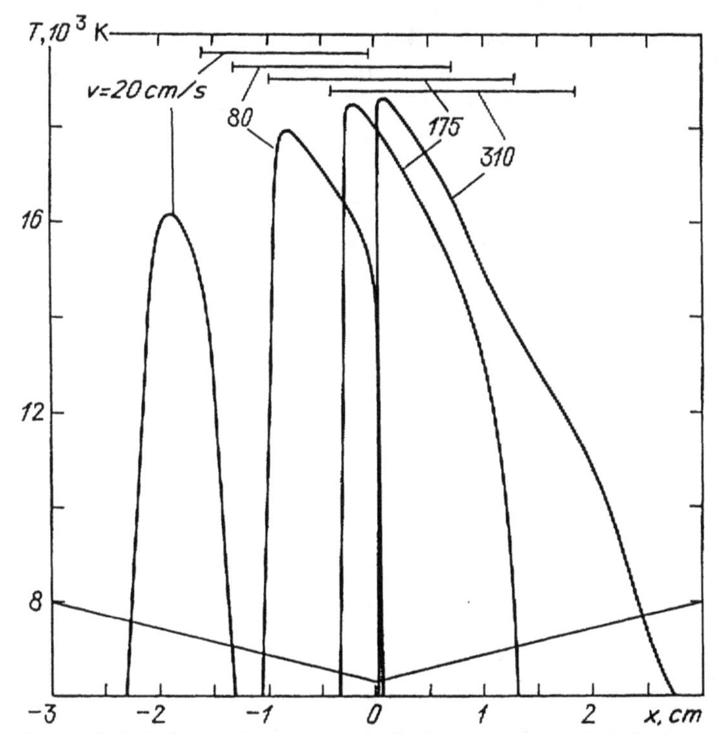

Fig. 11.25. Calculated axial temperature distributions in an optical plasmatron, based on (11.39) [11.16], and comparison with experimental data [11.21]. The beam and gas flow point from *left* to *right*, and the focus lies at $x = 0$. Argon, 1 atm. laser power $P_0 = 1.25$ kW, the focal length $f = 40$ cm, beam divergence 2 mrad. Heat-conduction loss coefficient $A = 2$. *Horizontal bars* show the length and position of the luminous region in the photographs given in [11.21]. Numbers at curves and bars indicate cold gas velocity

certain lower limit, $v \approx 5\,\mathrm{cm/s}$. If the lens has a short focus, say $f = 10\,\mathrm{cm}$, the discharge survives at any low gas velocity, and also in stationary gas. Calculations have also shown this effect, defining the domain in power P and velocity v in which the discharge is sustainable; the results fit the experimental data. We can conclude that the one-dimensional model of the phenomenon, as reflected in (11.39), is a good description (see also [11.17]).

The first two-dimensional calculations for continuous optical discharge in a gas stream, based on the system of gas-dynamics Navier-Stokes equations and energy balance equations (11.1,2), have now been performed. They take into account radiative heat exchange and the refraction of laser radiation in the plasma (Fig. 11.26). An interesting feature of the gas dynamics process was revealed by these calculations. The flow is found to be *unstable*. *Vortices* are generated downstream of the central part of the plasma; they are carried away by the flow and new ones are generated.

11.6.7 Fraction of the Flow Passing Through the Discharge

As we see in Fig. 11.26, the stream flows around the energy deposition zone, but nevertheless, some amount of gas flows through it. This was to be expected in view of the analysis presented in Sect. 11.6.4. Let us try to evaluate quantitatively the roles of the two elements in the flow pattern. Consider that part of the discharge front surface, facing the incoming flow, which is effectively the interface of the cold and hot regions. The gas density and temperature in the former are

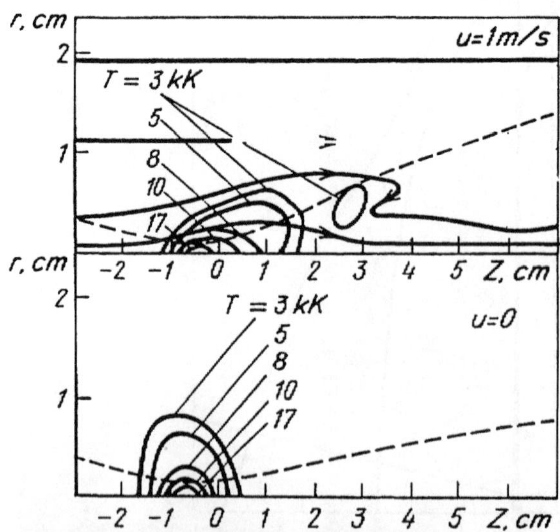

Fig. 11.26. Two-dimensional gas-dynamics calculation of gas flow in optical plasmatron [11.22]: Air, $p = 1$ atm., laser power $P_0 = 6\,\mathrm{kW}$, focal length $f = 15\,\mathrm{cm}$, velocity of uniform cold flow $u = 1\,\mathrm{m/s}$, z is the axial coordinate measured from the geometric focus point, and r is the radial coordinate. Lines of flow (with *arrows*), isotherms (marked by T [kK]), and refraction-distorted contour of the light channel (dashed curve) are traced. Below: isotherms in discharge with no flow is shown for comparison

ϱ_0, $T_{00} \approx 300$ K, and in the latter, ϱ_m, T_m; because the pressure is approximately constant $\varrho_0/\varrho_m = \mu_0 T_m/\mu_m T_{00} \sim 10^2$, where μ_0, μ_m are the molecular weights of the cold and hot (dissociated in the case of air and ionized) gases.

Any pressure drops Δp in a region of strongly subsonic flow are independent of the value of the pressure p and are determined by the dynamic pressure of the incoming flow: $\Delta p \sim \varrho_0 v^2$, where v is the velocity of unperturbed flow. The pressure at the critical point of a hot body is somewhat higher than the pressure at infinity. As a result, the velocity of the cold flow drops to a value v_n and the flow turns sideways, streaming around the hot region. The pressure decreases behind the energy deposition front, so that the gas penetrating the discharge is accelerated from v_n to u_m, as dictated by the law of mass conservation: $\varrho_0 v_n = \varrho_m u_m$. The velocity v_n at which the gas flows into the discharge front is the normal propagation velocity described in Sect. 11.6.5. The relations between the velocities v, v_n, and u_m are easily derived if we take into account that there is no scale for the dynamic pressure of hot gas except Δp or the dynamic pressure of the cold gas. This means that $\varrho_m u_m^2 \sim \varrho_0 v^2$. These two relations imply that $v/v_n \sim u_m/v \sim \sqrt{\varrho_0/\varrho_m} \sim 10$ while $u_m/v_n \sim \varrho_0/\varrho_m \sim 10^2$. In contrast to the idealized one-dimensional plane regime (Sect. 11.6.5), the cold gas impinges on the region of stationary discharge, not restricted transversely, not at the normal velocity v_n but at a velocity higher by a factor of $\sqrt{\varrho_0/\varrho_m} \sim 10$. From the total mass flux $\varrho_0 v \Sigma_0$ impinging on the section area Σ_0 of the hot region, a fraction on the order of $\varrho_0 v_n \Sigma_0$ penetrates the front surface. Therefore, the fraction of the incoming flow in the discharge is $v_n/v \sim \sqrt{\varrho_m/\varrho_0} \sim 10\,\%$; the remaining 90 % streams around the heated region as if around a solid body. This displays the effect of the two-dimensional nature of the flow. See details in [11.23].

12. Spark and Corona Discharges

12.1 General Concepts

12.1.1 Outward Features

The *spark discharge* (sparkover) occurs at voltages above the breakdown level at pressures above atmospheric (roughly) in gaps of about 1 cm and longer, that is, at $pd > 10^3$ Torr cm. The voltages required for the breakdown at such high values of pd are quite high, running to tens and hundreds of kV. The discharge is a rapid transient process, not a steady one, and is aptly described by the colloquial phrase "a spark jumps". *Lightning* is a discharge on a grandiose scale: it may be several kilometers long and is the electrical breakdown of the gap between a cloud and the ground, or between two clouds. A spark in laboratory conditions is but miniature lightning. When the breakdown voltage is reached, the interelectrode gap is pierced by a "fast as a lightning" light channel, sometimes zigzag, sometimes branching out, which immediately dies away. A spark is accompanied with a characteristic cracking sound, just as a lightning produces thunder. The sound is caused by the *shock wave.* It is generated by a sharp rise in pressure due to an intensive release of Joule heat in the spark channel when a high discharge current passes through it.

Many people will have seen – in photographs and in films – how a spark jumps across the gap between giant metal spheres supported by insulating columns (film directors are fond of this impressive performance). Such discharge devices are employed in high-voltage engineering (e.g., for voltage measurements). The discharge takes place at a certain minimum voltage that depends on the diameters of the spheres, the distance between them, and – slightly – on atmospheric conditions: pressure, temperature (rather, density), and humidity of the air. The system is calibrated and measurement parameters are standardized. If electrodes are separated by a not too thick dielectric plate of glass or cardboard, the spark may find a way through the plate, shooting a hole in it. Sparkovers occur both in the uniform field of plane gaps, being rooted at spurious points on the electrodes, and in strongly nonuniform fields: between the point and a plane, between a thin wire and a concentric cylinder, etc. In these last cases, the *corona* discharge precedes the sparkover if higher and higher voltages are considered.

A corona, that is, a weakly luminous discharge, appears in the neighborhood of a point or a thin wire, where the field is greatly enhanced. Ionization takes place only locally, and the gas there emits light. The electric current is closed by a flux of charges of a specific sign (depending on the polarity of the point) that are produced in the self-sustaining-discharge zone at the point and are dragged by a relatively weak field to the other electrode. No radiation appears from the outer region. A corona is typically observed at nearly atmospheric pressures, in air around high-voltage transmission line conductors, around lightning rods and the masts of ships ("Saint Elmo's fire"). To ignite a corona, a certain rather high voltage is required , which depends on the specific conditions. If the voltage is still higher, the remaining part of the gap breaks down and a spark jumps between the electrodes.

A large current, about 10^4–10^5 A, flows through the developed spark channel which is a good conductor. The voltage on the electrodes sharply decreases in response to voltage drop across the external resistance or as a result of the rapid discharge of the capacitor energizing the sparkover, so that the discharge burns out. If the voltage across the electrodes builds up again after the quenching of the discharge, the sparking is repeated. If the power supply is sufficiently powerful for sustaining a large current for a considerable time, the spark current produces a *cathode spot* and the spark transforms into an *arc* (Chap. 10). In fact, the state of the plasma in the channel of even a very transient spark discharge resembles the state in the arc column, so that it is sometimes legitimate to treat the final stage of the spark discharge as a pulsed arc.

The spark discharge is a multifacetted and complicated phenomenon. Its first stage is the process of the *streamer* or *leader* breakdown, which proceeds in a much more complex way than at low pressures where dark or glow discharges are initiated. After the conducting channel has been formed (this is also a multistage and "tricky" process), the discharge occurs. The capacitor is discharged the charge being transferred by the large current which flows across the gap that closes the circuit.

In one type of sparking, the sparks *"creep" along a dielectric* (glass or ebonite). They are produced when the end face of one electrode (e.g., a rod) contacts a dielectric plate and the other electrode is a metal plate on the other side of the dielectric. Discharge channels in the gas hug approach the dielectric, branching out from the rod and running around the plate to reach the metal on the opposite side. A branched trace may be seen on the plate, caused by the deformation of the material due to elevated temperature and pressure in the spark channel. The resulting pattern is known as *Lichtenberg figures*. the type of the pattern depends on the polarity of the rod, and its size is determined by voltage; the latter fact has been used for measurements and for studying atmospheric storm discharges.

12.1.2 Inapplicability of Townsend's Breakdown Model for High Pressure, Long Gaps, and Considerable Overvoltages

The breakdown mechanism based on the *multiplication of avalanches* via *secondary cathode emission* is predominant at low pressure, approximately at $pd < 200$ Torr cm. The corresponding theory, the principles of which were formulated by Townsend early in this century, explains a great number of observations. It gives a consistent interpretation of Paschen's dependence of the breakdown voltage V_t on pd, with its characteristic minimum (Sect. 7.2), and even gives a satisfactory quantitative fit to experimental data. With additional arguments concerning the accumulation of positive space charge in the gap in the consequent distortion of the external field by the increasing current, it is possible to trace, at least qualitatively, the initation of glow discharge from the onset of breakdown until the formation of the cathode layer. This is done by considering the transition from the dark Townsend to glow discharge, not by analyzing a sequence of steady states corresponding to progressively greater final currents, but by following the dynamics of the transition and the temporal growth of the current.

As the experimental techniques, equipment, and methods of analyzing short-lived transient processes were perfected, however, fresh facts were constantly revealed that could not be reconciled with the Townsend scheme. The study of individual avalanches and series of multiplying avalanches in the Wilson cloud chamber (*Raether* and his coworkers [12.1]), recording of visual patterns by photomultipliers and image converters, recording of oscillograms of breakdown current increasing with time, and frame-by-frame filming of the process by high-speed cameras led to considerable progress in understanding the nature of the breakdown.

At high pd and considerable overvoltages, the breakdown in plane gaps develops much faster than is predicted by the multiplication of avalanches through cathode emission. The secondary electron emission due to ion impact can be ignored because the duration of discharge is simply insufficient for ions to cross the gap. Even the photoemission mechanism is not sufficiently fast, because under these conditions the current-conducting channel is formed in the time of flight of an electron from the cathode to anode, or even faster. There is not enough time for the repetition of avalanches through cathode emission. High-speed filming made it possible to observe an ionized luminous channel that closes the gap immediately after the first powerful avalanche has passed.

The independence of the breakdown voltage on the material of the cathode, established by very accurate measurements, is evidence agains the participation of cathode processes in the breakdown mechanims. Under some conditions, the irrelevance of the cathode vis-a-vis sparking is fairly obvious, for example, in the breakdown between a point anode and a distant plane cathode. At the threshold voltage, the field at the cathode and over a considerable part of the gap on the cathode side is too low for electron multiplication to occur. It does occur only far from the cathode, in the neighborhood of the positive point, where the field is greatly enhanced. Lightning is another example. Other discrepancies between Townsend's theory and experimental data were also found. Thus there are such combinations as methylal or ester vapor and copper cathode that have extremely low secondary emission coefficients at the cathode, $\gamma_{\mathrm{eff}} < 10^{-8}$, so that the breakdown cannot be explained by the avalanche multiplication mechanism. There were no doubt by the late thirties that Townsend's theory fails completely at high pd and high overvoltages, that is, exactly at the conditions required to produce sparking.

12.1.3 Streamer Theory

The fundamentals of the new theory of spark breakdown were developed by *Loeb, Meek* and *Raether* [12.1, 12.2, 12.3] about 1940. The theory is based on the concept of the growth of a thin ionized channel (*streamer*) between the electrodes; the streamer follows the positively charged *trail* left by the *primary* intensive avalanche. Electrons of numerous secondary avalanches are pulled into the trail by the field. These avalanches are initiated by new electrons created by *photons* close to this trail. Photons are emitted by atoms that the primary and secondary avalanches have excited. Numerous results were reported in the course of subsequent efforts, and a number of details were found that greatly modified some of the initial concepts and estimates; nevertheless, the philosophy of the theory, reflecting the observatios, withstood the test of time. The story was not without complications, as when the importance of the Townsend mechanism was neglected – on inadequate grounds – in response to the achievements of the new theory (a detailed analysis of the situation is given in the review [12.4]). The Townsend mechanism does dominate at moderate values of pd (definitely so if $pd < 200$ Torr cm) and small overvoltages. As for the precise boundary values of pd at which the breakdown mechanisms replace each other, the information reported in the literature is very contradictory and rather poorly argued. According to [12.1], the transition in atmospheric air happens at $d \approx 5$–6 cm, that is, the streamer mechanism works only if $pd > 4000$ Torr cm. It is very likely that certain intermediate forms exist in the boundary range of $pd \approx 200$–5000 Torr cm, where neither cathode multiplication nor streamers are realized and bulk reactions serve as the secondary process: excitation of atoms by resonance radiation with subsequent associative ionization [12.5].

12.1.4 Leader

The breakdown in air gaps of many meters and in lightning discharges occurs via a growth of a *leader* from one electrode to the other: a thin channel that is conducting, orders of magnitude higher than the streamer channel. The leader process is a larger-scale event: it includes streamers as its elements. We will turn to leaders after Sect. 12.8 but now consider only streamers, since the process does not often go beyond them in short discharge gaps.

12.1.5 How Do We Define "Breakdown"?

The definition of "breakdown" is not so difficult at low pressures (Sect. 7.2). As a rule, the breakdown initiates a self-stustaining discharge and there is no need to distinguish between the two events. If the pressure is high, and especially if the field is non-uniform (one or both electrodes are points, wires, etc.), a self-sustaining current does not necessarily lead to catastrophic consequences. For example, the initiation of a corona discharge (which is self-sustaining) is not yet a breakdown. The current is quite low, and nothing special happens to the voltage, even though energy leakage through the corona is an undesirable effect for high-voltage transmission lines. The breakdown which is a real danger in high-voltage equipment is *short-circuiting*, that is, the formation of a highly conductive *spark channel;* this channel passes such a large current that the voltage across the discharge gap falls abruptly.[1]

12.2 Individual Electron Avalanche

An individual *avalanche* is a primary and inescapable element of any breakdown mechanism. Consider an avalanche in a uniform external field E_0 between plane electrodes. Let it be initiated by a single electron that leaves the cathode at the time $t = 0$. The x axis is directed from a point on the cathode to the anode. The radial distance from the x axis is denoted by r.

12.2.1 Numbers and Diffusional Spatial Distributions of Charges

Taking into account the possible formation of negative ions, we find the total numbers of electrons and ions increasing as the avalanche moves forward:

$$dN_e/dx = (\alpha - a)N_e\ , \quad dN_+/dx = \alpha N_e\ , \quad dN_-/dx = aN_e\ , \tag{12.1}$$

$$N_e = \exp\left[(\alpha - a)x\right]\ , \quad N_+ = \frac{\alpha}{\alpha - a}(N_e - 1)\ , \quad N_- = \frac{a}{\alpha - a}(N_e - 1)\ , \tag{12.2}$$

where α and a are the ionization and attachment coefficients. All the new electrons fly to the anode in a group at the drift velocity $v_d = \mu_e E_0$. However, diffusion makes the electron cloud spread around the central point $x_0 = v_d t$, $r = 0$. The electron density $n_e(x, r, t)$ in the cloud obeys the general diffusion equation (2.22, 20) in which the drift motion and electron production must both be taken into account. The solution of the equation takes the form [12.5]

$$n_e = (4\pi D_e t)^{-3/2} \exp\left[-\frac{(x - v_d t)^2 + r^2}{4D_e t} + (\alpha - a)v_d t\right]\ . \tag{12.3}$$

The density n_e decreases with distance from the moving center following a Gaussian law. The radius of the sphere on which the density is exactly $\bar{e}$ times less than that at the centre, $n_e(x_0, 0, t)$, increases with time (during the progress of the avalanche) by the characteristic diffusion law:

$$r_D = \sqrt{4D_e t} = \sqrt{4\frac{D_e}{\mu_e}\frac{x_0}{E_0}} = \sqrt{\frac{8\bar{\varepsilon}x_0}{3eE_0}}\ , \tag{12.4}$$

where $\bar{\varepsilon}$ is the mean random energy of electrons given by (2.23).

---

[1] In fact, voltage also drops slightly when a steady-state dark Townsend discharge is initiated: the overvoltage is removed. When a glow discharge is initiated, the voltage drops from the breakdown value to the steady burning voltage; if the positive column is short, V falls practically to the minimum breakdown voltage, which is fairly close to the cathode fall (Chap. 8).

The ions remain practically fixed during the time of flight of the avalanche to the anode. Hence, they accumulate at each point. The positive ion density is

$$n_+(x, r, t) = \int_0^t \alpha v_d n_e(x, r, t')dt' .$$

To obtain n_-, we have to replace α with a. The function n_e in the integral is given by (12.3). In the absence of attachment in the limit $t \to \infty$ and for regions not too far from the axis, an approximate calculation of the integral gives [12.5]

$$n_+(x, r) = \frac{\alpha}{\pi[r_D(x)]^2} \exp\left\{\alpha x - \frac{r^2}{[r_D(x)]^2}\right\} , \qquad (12.5)$$

where $r_D(x)$ is defined by (12.4). This result has a clear physical meaning. The ion density in the trail of the avalanche increases exponentially with the distance x from the cathode, in accord with the $\exp(\alpha x)$ law of multiplication of charges. In each cross section at a given x, the density decreases from the axis by the same Gaussian law for diffusion that governs the density of electrons (which produce the ions) at the moment when the centroid of the electron cloud passes through this section.

12.2.2 Visible Outlines of the Avalanche

In addition to ionizing the gas, electrons excite molecules, which remain practically fixed, as the ions do. The spatial distribution of excited particles that can give an image of the avalanche when emitting light is similar to the distribution of ions. A glance at (12.5) may suggest that the visible shape of the avalanche reflects the characteristic radius $r_D(x)$ of the distribution, that is, the outline of the avalanche is parabolic, $r = r_D(x) \propto \sqrt{x}$, and transforms smoothly into a spherical profile in the region of the electron head of the avalanche. This conclusion is not correct. Whatever the method of experimentally fixing the boundaries of the avalanche image (Sect. 12.2.3), this boundary inevitably corresponds to a definite absolute density of active particles (emitting molecules and ions), not their relative density. In general, this level is dictated by the sensitivity of the recording equipment. The sensitivity of instruments being sufficiently high, the minimum recordable density is much lower than the number density on the axis far from the cathode, where $\alpha x \gg 1$. As a result, the low number density at the visible outline $r_m(x)$ of the avalanche corresponds to a small value of the exponent in (12.5), much lower than αx. The outlines thus correspond to an approximately vanishing exponent; this curve is not parabolic, $r_D \propto \sqrt{x}$, but wedge-shaped:

$$r_m \approx r_D\sqrt{\alpha x} = \sqrt{8\bar{\varepsilon}\alpha/eE_0}\,x .$$

The wedge becomes rounded in the region of the avalanche head (Fig. 12.1).

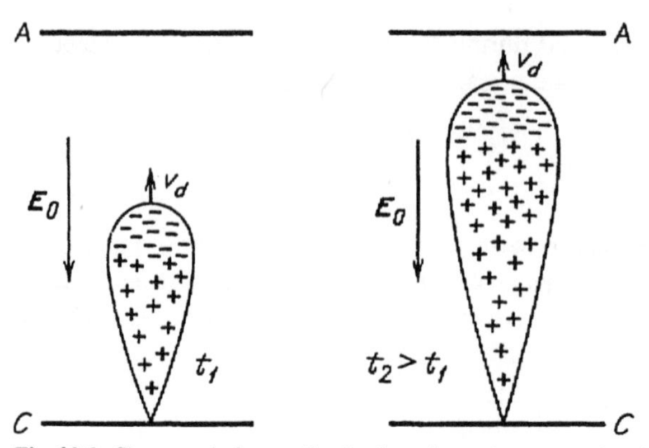

Fig. 12.1. Shape and charge distribution of an electron avalanche at two consecutive moments of time. *Arrows* indicate directions of external field E_0 and velocity of motion of the avalanche head, v_d

12.2.3 Elementary Current in a Circuit Containing the Discharge Gap

Let us calculate the current i in the circuit when only one electron moves in the gap between electrodes to which a voltage V is applied. We will then be able to find the current produced by the avalanche (Sect. 12.2.4). The external circuit is subjected to the electrostatic influence of the electron even when it is far from both electrodes. The metal of the electrodes is polarized by the field of the electron and positive *induced charges* appear on their surface (Fig. 12.2). The effect is the stronger, the smaller the charge–surface distance. Therefore, as the electron moves from the cathode to the anode, the induced positive charge on the anode surface increases and that on the cathode surface decreases (however, the sum of the induced charges on the two electrodes is e). Obviously, these changes are accompanied with a flow of charge in the connecting wires.

To find the induced charges, we make use of the *Ramo–Shockley theorem* of electrostatics published in 1938. A charge q_C on the surface of a conductor C, induced by a charge q, equals $q_C = -q(\varphi_{MC}/\varphi_C)$, where φ_{MC} is the potential at the point M where q is located; this potential arises if the conductor C is placed at a potential φ_C, while other conductors in the system, A in this particular case, have zero potential: $\varphi_A = 0$. Likewise, the charge induced on the electrode A is $q_A = -q(\varphi_{MA}/\varphi_A)$, and $\varphi_C = 0$. In the case shown in Fig. 12.2, $\varphi_{MC}/\varphi_C = (d-x)/d$, $\varphi_{MA}/\varphi_A = x/d$. Since $q = -e$, we have $q_C = e(d-x)/d$, $q_A = ex/d$, whence the *elementary* current is

$$i = \dot{q}_A = -\dot{q}_C = e\dot{x}/d = ev/d\,. \tag{12.6}$$

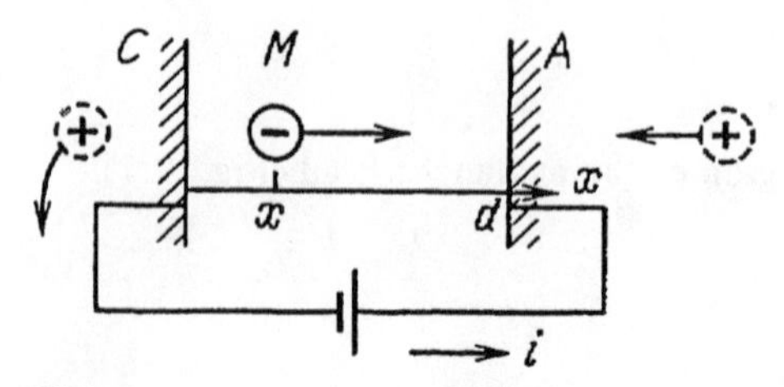

Fig. 12.2. Current flow through the circuit when a charge moves across the gap

The same result can be obtained by phenomenological arguments. An electron passing along a path $dx = v\,dt$ gains from the external field $E = V/d$ an energy $eE\,dx$. This work is done by the power supply source; it equals $iV\,dt$, whence $i = ev/d$. The derivation is elementary but in contrast to the preceding one, it does not specify the mechanism of the effect.

In the case of arbitrary-shape electrodes, that is, of nonuniform field, we obtain

$$i(t) = \dot{q}_A = -q\dot{\varphi}_{MA}/\varphi_A = -q\nabla\varphi_{MA}\dot{\boldsymbol{r}}/\varphi_A = q\boldsymbol{v}\cdot\boldsymbol{E}/V\ , \tag{12.6'}$$

where $\boldsymbol{v}(t) = \dot{\boldsymbol{r}}$ is the velocity of the charge q at a moment t, and $\boldsymbol{E}$ is the field at the point $\boldsymbol{r}$ that it passes through. The ratio $\boldsymbol{E}/V$ is a function of geometry of the field but is independent of field magnitude.

12.2.4 Experimental Studies of Avalanches

A great deal of important information has been obtained in the course of studying single avalanches, series of avalanches, multiplication of avalanches, and avalanche-streamer transformations in the Wilson cloud chamber [12.1]. This chamber makes use of the fact that ions usually serve as centers of condensation of supersaturated vapor. To make avalanches visible, a system of electrodes is placed in a chamber and an admixture of water, alcohol, etc., is added to the host gas. Synchronously with the application of the voltage and with the starting of the avalanche, the gas mixture is expanded adiabatically by 15–20 % so that the vapor becomes supersaturated. The cloud of droplets reproduces the shape of the ionic trail of the avalanche. The visual image is produced by light scattering on droplets, whose density equals (or is proportional to) the density of ions.

The main body of the avalanche has a well-pronounced wedge shape, rounded at the head. The avalanche length (the time of its motion) is controlled by fixing a prescribed length of the rectangular pulse applied to electrodes. The mean electron energy $\bar{\varepsilon}$ can be evaluated using the measured wedge angle (Sect. 12.2.2) and the known ionization coefficient α. The measured length and lifetime of the avalanche yield the drift velocity of electrons, $v_d = x/t$. Photographs of avalanches are also obtained by using the light emission due to the excitation of molecules and atoms. The luminosity being quite faint, photomultipliers and optoelectronic converters are employed for this purpose.

Another useful technique includes measuring with the very small current in external circuit and recording it with oscilloscopes. The result obtained in Sect. 12.2.3 implies that an avalanche produces a current

$$i(t) = N_e(t)ev_d/d = (ev_d/d)\exp\left[(\alpha - a)v_d t\right]\ .$$

Having measured $i(t)$, one can find $\alpha - a$, or α in a gas without attachment. When all the electrons have reached the anode, a much weaker current, that lasts much longer, is recorded; it is caused by the motion of ions. If avalanches multiply, consecutive pulses of electronic current with increasing mean value are observed. The time intervals between pulses indicate whether ions participate in cathode emission or not.

12.2.5 Field Distortion Due to Space Charge

Considerable *space discharge* is generated in an avalanche with high *amplification*, $\exp(\alpha x)$. Space charges produce their own field $\boldsymbol{E}'$ that adds up vectorially with the external field $\boldsymbol{E}_0$ and distort it in the vicinity of the avalanche. This effect becomes strong as charge multiplication continues and influences the subsequent process of ionization. Space charges form a sort of dipole: all the electrons are at the head of the avalanche, while most of the positive ions remain behind. The distance by which the electrons are separated from the main part of ions is determined by the *ionization length* α^{-1} that an electron covers, on the average, before it produces a pair of ions. As long as the external field is only slightly distorted, $\alpha = \alpha(E_0)$. When the amplification becomes high, both α and the spatial distribution of charges become dependent on the resulting field $\boldsymbol{E}$. The field and charge distributions (Fig. 12.3) are described by a set of simultaneous equations that are very difficult to solve.

The fields E' and E_0 in front of the avalanche head add up to give a field stronger than E_0. The fields $\boldsymbol{E}'$ and $\boldsymbol{E}_0$ in the zone between the centers of the space charges of opposite signs point in opposite directions and the resultant field is weaker than E_0. The field also develops a radial component. Assume that the charges of each sign are within a sphere of radius R. The field at its surface is $E' = eN_e/R^2$. The diffusion radius (12.4) can be taken for R, while the amplification and the number of electrons in the avalanche $N_e = \exp(\alpha x)$, are not too great. A preliminary idea of the scales of the quantities can be obtained, if we evaluate the values of r_D and N_e at which E' grows to E_0. For instance, in the field $E_0 = 31.4\,\text{kV/cm}$ that produces breakdown of the air gap of $x = 1\,\text{cm}$ at $p = 1$ atm. and with $\bar{\varepsilon} = 3.6\,\text{eV}$, equation (12.4) gives $r_D = 1.8 \times 10^{-2}\,\text{cm}$. The equality $E' = E_0$ is met when $N_e = \exp(\alpha x) = 0.8 \times 10^8$, $\alpha x \approx 18$. In fact, R at such a high N_e is several times the diffusional value so that $E' < E_0$ (Sect. 12.2.6).

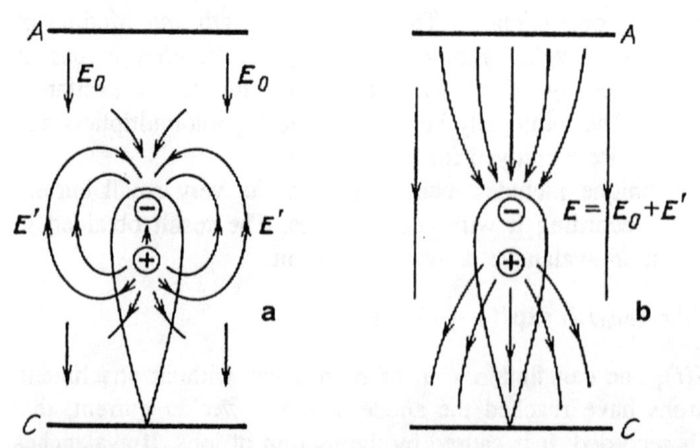

Fig. 12.3. Electric fields in a gap containing an electron avalanche. (a) Lines of force of the external field $\boldsymbol{E}_0$ and of the field of space charge of the avalanche, $\boldsymbol{E}'$, are shown separately. (b) lines of force of the resulting field $\boldsymbol{E} = \boldsymbol{E}_0 + \boldsymbol{E}'$. Circles mark the centers of space charges

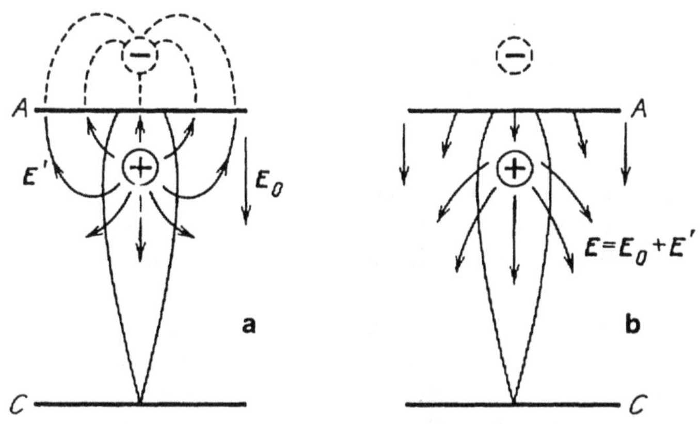

Fig. 12.4. Electric field in a gap after the avalanche has reached the anode and all electrons have sunk into the matal. (a) lines of force E' of the space charge of the trace left by the avalanche and of its mirror image in the anode. (b) lines of force of the resulting field, $E = E_0 + E'$

When the avalanche reaches the anode, the electrons sink into the metal and only the positive space charge of the *ionic trail* remains in the gap (Fig. 12.4). The field is formed by the ionic charge and by its "image" in the anode. The image in the relatively distant cathode plays a rather insignificant role. The field close to the anode is less than E_0, but exceeds it farther off. The field reaches its maximum at the axial distance from the anode of the order of one ionization length α^{-1}. An approximate solution of the corresponding electrostatics problem was given in [12.5].

12.2.6 Repulsion of Electrons

When the number of charges N_e is high, the diffusional spreading of the electron cloud is replaced by their electrostatic repulsion. The rate of spreading due to the latter factor increases with increasing N_e, that is, with t and x, while the rate of diffusion decreases: $dr_D/dt \propto t^{-1/2} \propto x^{-1}$. The rate of expansion of a charged sphere due to repulsion is determined by the drift of electrons in the field of their own space charge:

$$\frac{dR}{dt} = \mu_e E' = e\mu_e R^{-2} \exp(\alpha x) \, , \quad x = \mu_e E_0 t \, .$$

Integration first gives the law of expansion $R(t)$ or $R(x)$, and then the field E' and electron density $n_e = 3N_e/4\pi R^3$:

$$R = \left(\frac{3e}{\alpha E_0}\right)^{1/3} \exp\left(\frac{\alpha x}{3}\right) = \frac{3}{\alpha}\frac{E'}{E_0} \, , \quad n_e = \frac{\alpha E_0}{4\pi e} \, . \tag{12.7}$$

The field E' is proportional to R, and the mean electron density remains unaffected by repulsion accompanied by multiplication. The evaluation shows that diffusional spreading is replaced by repulsion at $N_e = \exp(\alpha x) \sim 10^6$, $\alpha x \approx 14$. The corresponding field E' comes to 2–3 % of E_0. Photographs of avalanches clearly show how the avalanche head broadens abruptly, beginning with a certain length x which corresponds to a certain amplification exponent $\alpha(E_0)x$; presumably, the reason is repulsion. Measurements agree with estimates of the above type [12.1].

Once the electron cloud increases to about one ionization length, $R \approx \alpha^{-1}$, the separation between electrons and ions ceases to be so well pronounced as when $R \ll \alpha^{-1}$. Now the distance between the charges of opposite signs is relatively small and the attractive force due to the positive charge restrictes further repulsive spreading of electrons. The broadening of the avalanche head slows down, or even ceases completely. The maximum transverse size of the avalanche head is thus of the order of $R_{max} \approx \alpha^{-1} \sim 0.1$ cm, because the ionization coefficient in atmospheric-pressure breakdown fields in air is typically about $\alpha(E_0) \sim 10\,\text{cm}^{-1}$. According to (12.7), at the moment when the head ceases to grow we find $E'/E_0 \approx 1/3$. If $E_0 \approx 30$ kV/cm and $\alpha \approx 10\,\text{cm}^{-1}$, then $N_e \approx 7 \times 10^8$, $\alpha x = \ln N_e \approx 20$, $n_e \approx 2 \times 10^{11}\,\text{cm}^{-3}$.

In breakdown fields, $d \ln \alpha / d \ln E \approx 4$. A change in E of 1 % changes α by 4 %. Therefore, the ionization rate at the outer boundary of the electron cloud (on the side of the anode), in the enhanced field $E \approx E_0 + E'$, is several times greater than $\alpha(E_0)$; it is much smaller at the inner boundary, on the side of the ionic trail. In this region we have not just $E \approx E_0 - E'$, but rather $E \approx E_0 - 2E'$, because the field E' of the ionic positive charge is also subtracted from E_0 (Fig. 12.3). Dramatic weakening of the field inside the avalanche head favours the formation there of *quasineutral plasma*, that is, of a streamer (directed to the anode).

12.3 Concept of Streamers

A streamer is a moderately, one can even say, weakly ionized thin channel formed from the primary avalanche in a sufficiently strong electrc field; it grows in one, or both, directions toward the electrodes. In the breakdown of a plane gap, the streamer grows from the anode to the cathode. A streamer, possessing a certain conductivity, can so modify the field upon reaching the electrodes that the degree of ionization and the current may be greatly increased; ultimately, this will lead to a *spark discharge* in the gap. The creation of a streamer (and the resulting gap closure) is not a necessary – but is sometimes a sufficient – condition for breakdown.

For an avalanche to transform into a streamer, it must reach sufficiently high amplification. The space-charge field must increase to a level on the order of the applied field. Otherwise there are no reasons for disturbing the normal evolution of the avalanche. If gaps are not too long and overvoltages are not too high (in comparison with the breakdown voltage), the transformation occurs when the avalanche exhausts its reserves of amplification, that is, when it reaches the anode. The streamer is now initiated at the anode surface, in the region of maximum space charge, and then propagates to the cathode. Such streamers are known as *cathode-directed* or *positive*. The number of charges in the primary avalanche in wide plane gaps and (or) at high overvoltages becomes high even earlier. The avalanche transforms into a streamer before it reaches the anode. In this case the streamer grows toward both electrodes. If the streamer is formed while the avalanche has not yet gone far from the cathode, it mostly grows toward the anode; it is then said to be *anode-directed* or *negative*.

12.3.1 The Mechanism of Formation of Cathode-Directed Streamers

The situation is illustrated in Fig. 12.5. One hypothesis states that the decisive role is played by energetic *photons* that are emitted by atoms excited in the avalanche and produce *photoionization* in the vicinity of the primary avalanche. (Events of production of electrons at the cathode or far from the trail are unimportant in this context because they result in avalanches similar to the primary one.) Electrons produced by photons initiate *secondary avalanches* that are *pulled into the trail* due to the direction of the resulting field (Fig. 12.4). Secondary-avalanche electrons intermix with primary-avalanche ions and form a *quasineutral plasma*. They also excite atoms, so that new photons are emitted. Secondary-avalanche ions enhance the positive charge at the cathode end of the evolved plasma channel. This charge attracts the electrons of the next generation of secondary avalanches, etc. This is how the streamer grows. The process of ionization along the ion trail of the primary avalanche begins at the spot where the positive charge and the field are the highest, that is, at the anode, provided the *degeneration condition* $E' \approx E_0$ has been reached there. This is the situation shown in Fig. 12.5.

In this case the plasma streamer contacts the anode. A streamer is a conductor, so that electrostatically it acts as a metallic "needle" protruding from the surface of the anode (a perfectly conducting needle would be exactly at the anode potential): the field at the end of the streamer is greatly enhanced. The lines of force fan out from this end, which stimulates the attraction of secondary avalanches on all sides of the streamer and hence its growth. The mechanism that somewhat levels off the potential in the needle is the *polarization* of the conductor by the

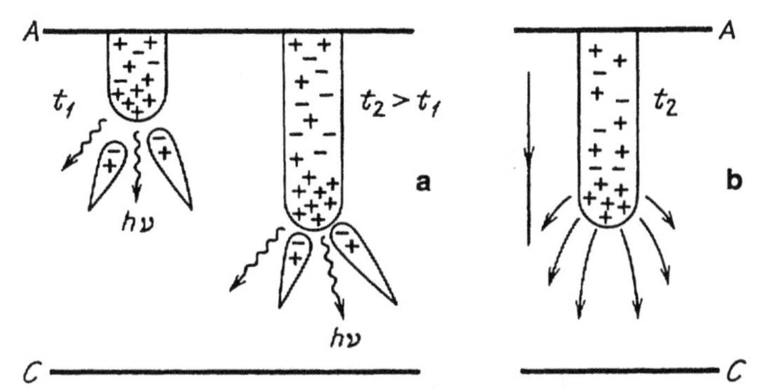

Fig. 12.5. Cathode-directed streamer. (a) Streamer at two consecutive moments of time, with secondary avalanches moving towards the positive head of the streamer; wavy arrows are photons that generate seed electrons for avalanches. (b) Lines of force of the field near the streamer head

external field. Electrons are displaced towards the anode, thus "baring" the positive ions at the cathode end. The field of the resulting dipole, being opposite to the external field, partly cancels the latter and reduces the potential difference along the conductor. In fact, this is the same effect that produces the patterns of Figs. 12.3, 4.

If the source of photons and seed electrons for secondary avalanches is sufficiently great (this seems to be the case), the rate of streamer growth is limited by the rate of neutralization of the positive space charge at the cathode side of the streamer, not by the rate of avalanche production. As for the electrons, they are pulled into this region at the drift velocity corresponding to the field within it. This field is considerably higher than the external one, and is stronger the longer the needle. (This is implied by electrostatics.) Experiments indeed prove that the velocity of propagation (*growth*) of the streamer is greater, the longer it is and the stronger the external field is. To an order of magnitude, the measured velocities are about 10^8 cm/s, while the drift velocities in the external field are about 10^7 cm/s. The streamer channel diameter is comparable with the avalanche head diameter at the stage of maximum expansion, 10^{-2}–10^{-1} cm (Sect. 12.2.6). At any rate, the charge density is not less than the maximum density in the avalanche: presumably, $10^{12}\,\mathrm{cm}^{-3}$ (an estimate of $10^{13}\,\mathrm{cm}^{-3}$ has also been made).

Photons are emitted and absorbed in a random manner; hence, situations are possible in which a new predominant direction appears for the propagation of a great number of secondary avalanches. This is a likely mechanism of generation of experimentally observed zigzag streamers and spark channels.

12.3.2 Formation Criterion

As follows from the preceding presentation, a streamer is born of an avalanche if the field of its space charge reaches a value of the order of the external field. The correspondingly approximate equality,

$$E' = eR^{-2} \exp[\alpha(E_0)x] \approx E_0 \, , \tag{12.8}$$

can be regarded as the *criterion of streamer formation*. It imposes a condition on the parameters of experiment: E_0 and the gap width $d = x$ that is minimal for the given E_0. Numerical results obtained using (12.7) are dependent on the value chosen for the avalanche head radius R. The diffusional radius $r_D(E_0)$ was taken in the earlier work of Loeb and Meek from a formula of the type (12.4). Thus we obtain the well-known Meek breakdown condition, which demands (in the simplified form) that

$$\alpha(E_0)d \approx 18 - 20 \, , \quad N_e = \exp(\alpha d) \sim 10^8 \tag{12.9}$$

(cf. the numerical example of Sect. 12.2.5).

The onset of breakdown was identified in the Loeb-Meek theory with the event of streamer formation. In fact, this is not always so. We know now that if

$d < 5\,\text{cm}$, atmospheric air undergoes breakdown via the Townsend mechanism of avalanche multiplication, not the streamer one. The condition $\alpha d \approx 20$ imposed on the avalanche *enhancement coefficient* by the time the streamer is formed is insensitive to the choice of the value of R because $R \approx \alpha^{-1} \sim 10^{-1}\,\text{cm}$ (Sect. 12.2.6), we find $N_e \sim 10^9$, although $\alpha d \approx 21$ is again quite close to (12.9). Note that the processing of experimental data on the breakdown of various gases at $p \sim 1$ atm. and $d \sim 1$–$10\,\text{cm}$ points to an approximate empirical relation $\alpha(E_0)d \approx 20$, which is equivalent to *Meek's criterion*. The significance of this fact should not be overestimated, because any criterion of type (12.8) implies weak logarithmic dependence of αd on other quantities in the formula.

After an analysis of experiments, Loeb supplemented condition (12.9) with the demand that the electron density in the avalanche at the moment of formation of a streamer be at least $0.7 \times 10^{12}\,\text{cm}^{-3}$, assuming that this level ensures the required rate of emission of photoionization radiation. It cannot be said that the specific mechanism of photoionization of the gas, so important for the streamer process, is clearly understood. Indeed, for a photon to knock out an electron from an unexcited atom, the atom that has emitted this photon, had to be excited to energies above the ionization potential; such events are infrequent. It is assumed that oxygen molecules in the air are ionized by photons emitted by highly excited nitrogen molecules ($I_{N_2} = 15.6\,\text{eV} > I_{O_2} = 12.2\,\text{eV}$), although there is hardly any experimental confirmation of this. Seed electrons may be produced in a complex manner, during the diffusion of the resonance radiation and associative ionization of an excited atom as it combines with an unexcited one.

12.3.3 Anode-Directed Streamer

If the applied field E_0 is such that the condition (12.8, 9) is satisfied at a distance x form the cathode, shorter than the gap width d, the avalanche transforms into a streamer "halfway to the anode". The mechanism of growth towards the cathode remains the same (Sect. 12.3.1). The characteristics of propagation towards the anode are somewhat different from those of the above process because here electrons drift in the same direction as the front of the plasma streamer, not counter to it as in the case of cathode-directed growth. As a result of radition causing photoionization, secondary avalanches are produced in front of the negatively charged streamer head facing the anode (Fig. 12.6). The front electrons of the head, moving rapidly in a strong total field $E_0 + E'$, join the ionic trails of secondary avalanches and together form the plasma. Presumably, in this case a propagation mechanims without photons is also possible. The plasma front propagates at the expense of the electrons of the front, accelerated in the strong field, while the electrons behind the front (in a weak field) do not separate from the ions, that is, the charges form a quasineutral plasma. This resembles the propagation of an ionization wave [12.5, 6]. Some models developed for the description of streamer propagation are outlined in Sect. 12.7.

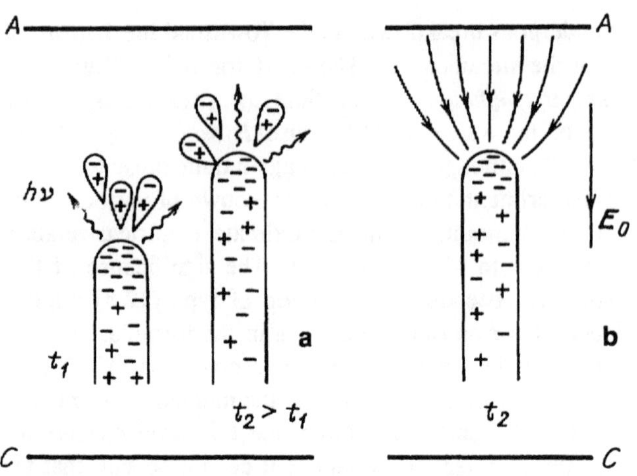

Fig. 12.6. Anode-directed streamer. (a) Photons and secondary avalanches in front of the streamer head at two consecutive moments of time. (b) Field in the vicinity of the head

12.4 Breakdown and Streamers in Electronegative Gases (Air) in Moderately Wide Gaps with a Uniform Field

Attachment of electrons slows down the ionization in avalanches and results in higher breakdown fields and the boundary values of *pd* at which the Townsend mechanism is replaced by the streamer one. In this respect, the analysis of experimentally obtained characteristics for atmospheric air is very illustrative. For obvious reasons, air is of exceptional interest and has been intensively investigated.

12.4.1 Breakdown Fields

Figure 12.7 gives the results of measurements of breakdown voltage in plane gaps in room air. In contrast to Fig. 7.5, where the range is limited to $d = 3$ cm, here it is extended to $d = 30$ cm. The asymptotic tendency of the breakdown field to a constant value of about 26 kV/cm, $E/p \approx 34$ V/cm Torr is evident. This fact is definitely caused by the following: at slightly lower values of E/p the attachment coefficient a is greater than the ionization coefficient α so that the multiplication of electrons is impossible (Sect. 7.2.5). The exact value of $(E/p)_1$ at which the curves of α/p and a/p as functions of E/p intersect is very difficult to determine either experimentally or numerically. One of the more recent sources gives an experimental value $(E/p)_1 = 31$ V/ cm Torr = 23.6 kV/cm atm. It can be regarded as the *theoretical lower limit* for air breakdown in an *ideal* plane gap. In fact, this conclusion holds only for a limited pressure range. At $p = 1$ atm. the breakdown voltage in uniform field does tend to $V_t \approx E_1 d$ as d increases (Fig. 12.7). But beginning at a value of p around 10 atm. (the value depending on d) the breakdown voltage for a given d increases significantly less than linearly

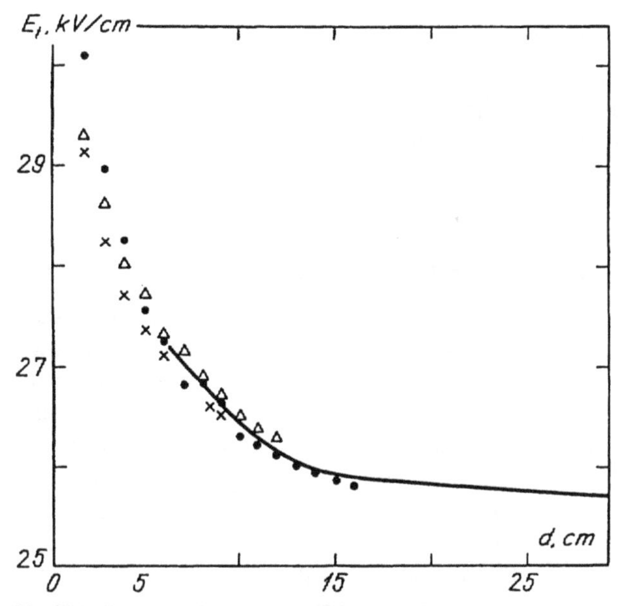

Fig. 12.7. Threshold field for the breakdown of air in a plane gap of length d [12.1]; p = 760 Torr + 10 Torr H_2O, t = 20° C. Data of a number of authors are shown. Spread in thresholds may have been caused by differences in voltage pulse durations [12.1]

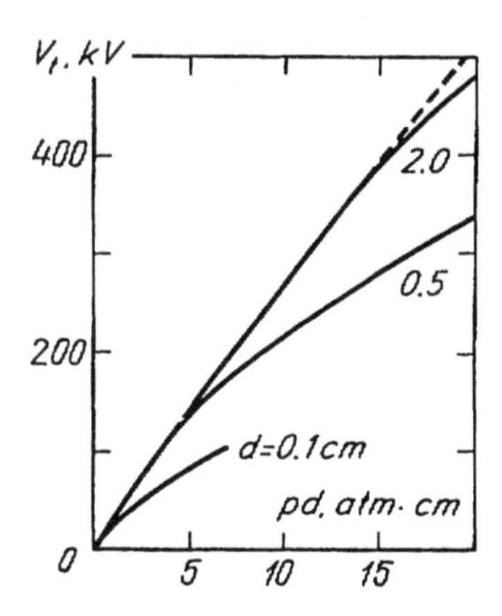

Fig. 12.8. Amplitude of breakdwon voltage (frequency 50 Hz in air in uniform field) as a function of pd, at high gas pressure. Labels to the curves give d in cm [12.7]

with p. Therefore, E_t/p decreases more and more in comparison with $(E/p)_1$ [12.7] (Fig. 12.8). The nature of the effect is not quite clear. There are indications that it is caused by enhanced fields at protrusions on the cathode surface where multiplication is initiated. If this is so, the process occurs as in a gap with a strongly nonuniform field (Sects. 12.8, 9).

12.4.2 Electronegative Gas SF_6

Owing to its high dielectric strength and other suitable properties, SF_6 is used for insulating. Its theoretical lower limit of breakdown threshold, corresponding to the equality $\alpha = a$, is very high: $(E/p)_1 = 117.5$ V/cm Torr = 89 kV/cm atm; see Fig. 12.9. The effect of a decrease in E_t/p with respect to $(E/p)_1$ is also observed (at $p \gtrsim 3$ atm.).

12.4.3 Multiplication of Avalanches or the Streamer?

Table 12.1 lists modern experimental data on the effective multiplication constant $\alpha_{eff} = \alpha - a$ of avalanches in dry air at $p = 1$ atm. in fields that produce breakdown in gaps of various width d. The table gives the numbers of electrons (enhancement factor) in avalanches initiated at the cathode by a single electron and reaching the anode. The table is limited to $d = 3$ cm because the data on α_{eff} at lower E is very unreliable in view of the closeness to the intersection point: $\alpha - a \ll \alpha, a$. We already know that a streamer is formed when the field of the space charge increases to the level of the external field; this event occurs when the critical number of electrons, $N_e \sim 10^8$–10^9, is produced in the avalanche ($\alpha_{eff}d \approx 20$). Table 12.1 shows that in air at $d < 3$ cm this event definitely cannot happen, but if $d \approx 3$ cm, the situation tends to the "critical" one, though the steeply decreasing α_{eff} shifts this point slightly further along the d axis. This conclusion fits the direct data obtained by analyzing oscilloscope traces of the breakdown current: breakdown in air goes via avalanche multiplication if $d < 5$ cm and via the streamer (or leader) mechanism if $d > 6$ cm.

An analysis of the table demonstrates why it is the *Loeb* and *Meek* seem to show conclusively in their early work [12.2, 12.3] that an air gap of $d = 1$ cm undergoes breakdown via the streamer mechanism. These calculations paid no

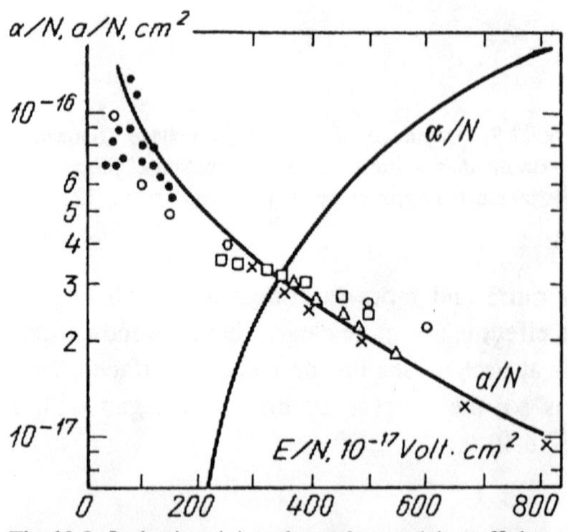

Fig. 12.9. Ionization (α) and attachment (a) coefficients in electronegative gas SF_6. Dots, circles, triangles, and squares represent experimental data. Solid curves plot the calculations using the kinetic equation [12.8]

Table 12.1. Multiplication of electrons in an avalanche at $p = 1$ atm. in fields E_t producing breakdown in gaps of width d [12.9]

d [cm]	pd [10^2 Torr cm]	V_t [kV]	E_t [kV/cm]	E_t/p [V/cm Torr]	$\alpha - a$ [cm^{-1}]	$(\alpha - a)/p$ [10^{-2} cm^{-1} $Torr^{-1}$]	$(\alpha - a)d$	N_e
0.1	0.76	4.54	45.4	59.7	81	10.7	8.1	3.3×10^3
0.3	2.3	11	36.7	48.4	31	4.1	9.3	1.1×10^4
0.5	3.8	17	34	44.7	20.5	2.7	10.2	2.8×10^4
1	7.6	31.4	31.4	41.4	12.4	1.63	12.4	2.4×10^5
2	15	58.5	29.3	38.6	8.0	1.05	16	8.9×10^6
3	23	85.5	28.6	37.6	6.5	0.85	19.5	2.9×10^8

attention to attachment and used for the multiplication coefficient α the results of earlier experiments which also ignored this factor. Thus it was assumed that $\alpha \approx 18\,\mathrm{cm}^{-1}$ at $E = 31.6\,\mathrm{kV/cm}$ ($d = 1\,\mathrm{cm}$). In fact, if attachment is taken into consideration, then $\alpha_{\mathrm{eff}} \approx 12.4$, and this reduces N_e by three orders of magnitude. At the same time, it is difficult to assume that the mechanism of secondary emission from the cathode acts in the interval $d \approx 0.3$–$3\,\mathrm{cm}$. This would require an extremely steep dependence of the secondary emission coefficient γ on E. Indeed, the Townsend breakdown condition (7.1) gives $\gamma \approx \exp(-\alpha_{\mathrm{eff}} d)$, and Table 12.1 indicates that this γ decreases in the corresponding interval of $E \approx 36.7$–$28.6\,\mathrm{kV/cm}$ by 4 orders of magnitude. The mechanism outlined at the end of Sect. 12.1.3 [12.5] may have been at work.

12.4.4 Effect of Overvoltage on Breakdown Mechanism

As a rule, the breakdown field is defined as the minimum field at which breakdown is still observable. Threshold values slightly depend on the initial current (the number of *seed* electrons) and on the *time delay* needed for the discharge to *develop*. Loeb and Meek assume that the breakdown voltage is that at which the breakdown is realized with a time delay of 30 s after the voltage has been applied, provided the initial current density of the cathode is $10^{-13}\,\mathrm{A/cm^2} \approx 1$ electron/μs cm^2. This quantity is known as the *static* breakdown voltage because this process first develops at a slow rate. Since α or α_{eff} depend steeply on E, overvoltage of, say, 10 % is sufficient for a streamer to be produced in a gap in which the breakdown at static voltage occurs by the Townsend mechanism.

Experiments on nitrogen breakdown in uniform field [12.10] are illustrative in this respect. At $p = 400\,\mathrm{Torr}$, $d = 3\,\mathrm{cm}$, and static breakdown field $E/p = 38.5\,\mathrm{V/cm\,Torr}$, when $\alpha d = 9$ (nitrogen has no attachment), the breakdown current increases slowly, with a characteristic rise time of 1–2 μs. The oscilloscope trace shows peaks due to successive multiplying avalanches; photoemission occurs at the cathode. At the same p and d but at overvoltages above 17 %, a streamer is clearly formed. Thus a steep rise in current is recorded at 19 % overvoltage, when $E/p = 45.2$, only 0.2 μs after the peak of the first avalanche. There is competition between the streamer and the avalanche multiplication mechanisms, and its result can be tipped in either direction. An admixture of 2.5 % methane in the above-described experiments with nitrogen reduces to 7–8 % the overvoltage required for the streamer mechanism to dominate. The methane admixture affects α very little, while it reduces the seconary emission coefficient γ by a factor of 10^2, thereby suppressing the process of avalanche multiplication.

12.4.5 Effects of Negative Ions on Streamer Formation

Oscilloscope traces of current in the case of streamer breakdown of air ($d = 9\,\mathrm{cm}$) reveal the following characteristic effect [12.1]. After a short peak due to the first avalanche, a weak "delayed" current flows for a relatively long time, up to 10 μs (this current was proved to be carried by electrons). The trace ends with a stepwise increase in current, that is, breakdown. The lag in the formation of the streamer and in breakdown is caused by attachment. After the first avalanche travels through the gap, the positive charge of the ionic trail is largely compensated for by negative ions. The space charge is not sufficient to produce a streamer. However, negative ions that are mostly concentrated at a short distance $\alpha_{\text{eff}}^{-1} \sim 10^{-1}$ cm from the anode are pulled toward the anode at a drift velocity of order 10^{5} cm/s. The positive ionic charge is bared and the field gradually increases to a level required to form a streamer. The delayed electronic current is explained by detachment of electrons from negative ions in the increasing field of the space charge (we can only guess what the specific detachment mechanism may be; Sect. 12.6.4).

12.5 Spark Channel

The formation of the strongly ionized plasma of a well-developed spark channel is preceded by a poorly understood stage at which the degree of ionization in the zone of the initial streamer channel increases quite rapidly. This is indicated by a dramatic increase of current after the streamer closes the gap in the course of the streamer breakdown. Note that the original streamer channel is incapable of passing a high current. According to Table 2.1, the conductivity of weakly ionized plasma $\sigma \sim 10^{-16} n_e/p\,[\text{atm.}]\,\text{Ohm}^{-1}\text{cm}^{-1}$. At $p = 1$ atm. and $n_e \sim 10^{13}\,\text{cm}^{-3}$ $\sigma \sim 10^{-3}\,\text{Ohm}^{-1}\text{cm}^{-1}$. If the channel diameter $2r_c \sim 10^{-1}$ cm and the field in it $E \sim 10\,\text{kV/cm}$, then the current is negligible: $i = \sigma E \pi r_c^2 \sim 10^{-2}$ A.

12.5.1 Back Wave of Strong Field and Ionization

This process, which seems to initiate the formation of the spark channel, is more pronounced and better understood in the case of leader breakdown in which the gap is closed by a much better conducting channel (Sect. 12.10). Nevertheless, the qualitative picture can be outlined as follows. The potential of the tip, or head, of the streamer growing from the anode to the cathode differs less from the anode potential than that of the nonperturbed field at the same point. An ideally conducting streamer contacting the anode would be entirely at the anode potential. As the tip approaches the cathode, both the fraction of the voltage applied to the electrodes which falls across the nonconducting gap between the streamer tip and cathode, and the field in the gap increase. By the time the tip touches the cathode, the field becomes so strong that electrons liberated by photons from the cathode or from atoms multiply at enormous intensity. The front then propagates from the cathode along the channel of the initial streamer towards the anode; it leaves behind a much stronger ionized plasma. The process looks like a reversed streamer at an ionization considerably higher than in the original one. The highly conducting plasma of the back "streamer" is at a potential close to that of the cathode; hence, an abrupt potential drop and a very strong field appear at the front. Electrons accelerated here are those producing the intensive ionization. The wave front is assumed to propagate to the anode at a velocity on the order of 10^9 cm/s. This is not the velocity of motion of electrons: it is the phase velocity of the propagation of the potential jump (field wave). When this thin highly ionized channel reaches the anode and closes the gap, the foundation is laid for the formation of the "true" spark channel.

12.5.2 Expansion of Spark Channel

The high-density current in a spark channel releases highly localized heat. As a result, the plasma is greatly heated up and thermalized, and its degree of ionization may even increase thermally. The rapid surge in gas temperature, not being compensated for by similarly rapid heat removal, sharply increases the pressure in the current channel. This produces a *cylindrical shock wave,* resembling the explosion of a filament-shaped explosive charge. At the initial stages, the shock wave amplitude is so high that the temperature behind the front is sufficient for the thermal ionization of the gas. The boundary of the current channel is then almost indistinguishable from the shock wave front. Soon, however, the shock wave expanding from the axis weakens, ceases to ionize the gas, and separates from the relatively slowly expanding boundary of the highly ionized region, that is, from the *spark channel.* Now the channel expands only owing to the radial movement of the gas driven by the shock wave, and to heat conduction.

The temperature in the channel reaches 20,000 K; the electron density has been measured and found to reach $n_e \sim 10^{17}\,\mathrm{cm}^{-3}$. The electric conductivity is then determined by Coulomb collisions and is independent of n_e. As follows from (2.9), $\sigma \sim 10^2\,\mathrm{Ohm}^{-1}\mathrm{cm}^{-1}$. The current mostly increases because of the expansion of the channel and the enlarged cross section of the conductor, not because of current density changes. The channel radius grows to $r \sim 1$ cm, the maximum current is $i \sim 10^4$–10^5 A, $j \sim 10^4\,\mathrm{A/cm}^2$, the voltage at the electrodes is substantially lower than the original value, and the field in the channel $E \sim 10^2$ V/cm. A cathode spot seems to form on the cathode. If the power supply source is a capacitor (this is typical of laboratory experiments), the current decreases after the maximum is reached, and several repeated delayed oscillations occur with a half-period of about 10 μs.

This behavior has been studied in a large number of experiments. The first direct evidence of shock waves produced by spark discharges was reported in 1947 [12.11]. Streak photographs of sparks clearly show an advancing shock wave front and a not so rapidly expanding spark channel. At an early stage, 10^{-7}–10^{-6} s after the onset of breakdown, the channel expands at a velocity of about 1 km/s, and then slows down. The theory of gas-dynamics expansion of the spark channel, taking into account the shock wave and the energy release due to time-dependent discharge current, was first developed by *S.I. Drabkina* in 1951 [12.12]. A great deal of experimental data is reviewed in [12.3].

12.6 Corona Discharge

Corona discharges occur only if the field is sharply *nonuniform*. The field near one or both electrodes must be much stronger than in the rest of the gap. This situation typically arises when the characteristic size r of the electrodes is much smaller than the interelectrode distance d. For example, parallel wires of radius r manifest corona discharge in air only if $d/r > 5.85$. Otherwise the increase of voltage between the wires produces a spark between them, and not a corona discharge.

12.6.1 Field Distribution in the Simplest Cases

Exact solutions of electrostatics for simple geometry are indispensable for constructing a theory of corona and for the interpretation of experimental results. The field in the space between coaxial cylinders of radii r (internal) and R at a distance x from the axis is

$$E = V/\left[x \ln(R/r)\right] , \quad E_{\max} = V/\left[r \ln(R/r)\right] , \tag{12.10}$$

where V is the voltage between cylinders. Between concentric spheres of radii r and R we find

$$E = VrR/x^2(R-r) ; \quad E_{\max} \approx V/r \ \text{ if } \ R \gg r . \tag{12.11}$$

In the space between a sphere and a remote plane ($R/r \to \infty$), the field is $E \approx Vr/x^2$. Between a parabolic tip with curvature radius r and a plane perpendicular to it at a distance d, the field at a distance x from the tip along the axis is

$$E = \frac{2V}{(r+2x)\ln(2d/r+1)} , \quad E_{\max} \approx \frac{2V}{r \ln(2d/r)} . \tag{12.12}$$

If a voltage V is applied between parallel wires spaced by a distance d, or between a single wire and a parallel plane at a distance h, the maximum field at a wire of radius r is

$$E_{\max} = V/2r \ln(d/r) \ \text{ and } \ E_{\max} = V/r \ln(2h/r) . \tag{12.13}$$

12.6.2 Ignition Criteria

If the applied voltage V is less than the ignition voltage for corona V_c for the given conditions, a non-self-sustaining current on the order of 10^{-14} A can be detected in the circuit. This current is formed by ions produced by cosmic rays and natural radioactivity. About 10 pairs of ions are produced in air at sea level in $1\,\text{cm}^3$ per 1 s, and the steady-state number of ions in $1\,\text{cm}^3$ is about 10^3. The ignition of corona under laboratory conditions manifests itself not only by a luminous layer around the electrode (which may not be noticed at all) but also by a jump in the discharge current to about 10^{-6} A. The corona discharge belongs to the group of *self-sustained* discharges; the conditions under which it appears reflect the physical mechanism of reproduction of electrons in that region of the enhanced field where ionization occurs. The mechanism of multiplication of electrons is esentially dependent on the *polarity* of the electrode surrounded by the corona.

If this electrode is the cathode (the corona is then said to be *negative*), then *avalanche multiplication* takes place. The secondary process is the emission from the cathode and, possibly, photoionization in the bulk of the gas. In principle, the ignition of a negative corona does not differ from the *Townsend breakdown* and from the ignition of the *dark Townsend discharge* (Sect. 8.3). The *ignition criterion* is an equality of type (7.1), generalized to include nonuniform fields. With attachment effects taken into consideration, we have

$$\int_0^{x_1} [\alpha(x) - a(x)]\, dx = \ln(1 + \gamma^{-1}) \, , \tag{12.14}$$

where γ is the effective coefficient of secondary emission. The integration region in (12.14) stretches from the cathode surface to that point x_1 where $\alpha = a$ and the multiplication of electrons stops. In gases without attachment, the integral is formally extended up to the anode, but in practice it is sufficient to extend it only to the not very remote point x_1 where the field is considerably weakened and $\alpha[E(x_1)] \approx 0$, because $\alpha(E)$ decreases very steeply as we move away from the wire or tip. Molecules are also excited by electrons in the multiplication region. There are practically no electrons beyond this region in an electronegative gas: they form negative ions after having traveled a very short path; in electropositive gases, the field is in any case too weak, electrons are slow, so that the gas is not luminous outside the corona.

If the wire (or tip) is the anode (*positive corona*), the remote large cathode does not participate in multipliation, on account of the weak field in its vicinity. The reproduction of electrons is ensured by *secondary photoprocesses* in the gas around the tip. In contrast to the homogeneous glow of a negative corona, a positive corona displays luminous filaments running away from the tip (Fig. 12.12). These are thought to be streamers. The condition of streamer formation (12.9), also generalized to nonuniform fields, can be chosen as a *criterion of ignition of a positive corona*:

$$\int_0^{x_1} (\alpha - a) dx \approx 18 - 20 \, . \tag{12.15}$$

Despite a possible difference of a factor of 2 to 3 in the enhancement coefficient, that is, in the values of integrals (12.14, 15), this coefficients affects rather insignificantly the value of the critical field at the electrode. The reason is the sharp dependence of α or $\alpha - a$ on E, so that even a small change in $E(x)$ strongly affects the integral. Indeed, experiments show that the ignition voltages of positive and negative coronas differ only a little in a number of gases, including air (V_c of negative corona is lower: presumably in (12.14), $\ln \gamma^{-1} < 20$ because $\gamma \gg 10^{-8}$).

12.6.3 Thresholds in Air

It is clear from the form of criteria (12.14, 15) that they impose constraints mostly on the maximum field at the corona-carrying electrode, which must be above a certain lower limit E_c. The threshold voltage V_c for the ignition of corona is related to E_c by the electrostatic laws of field distribution in the gap, (12.10–13). In 1929 Peek found the following empirical formula for the critical field of corona ignition in air between coaxial cylinders:

$$E_c = 31\delta(1 + 0.308/\sqrt{\delta r})\,\text{kV/cm} . \tag{12.16}$$

Here δ is the ratio of air density to the normal density corresponding to $p = 760\,\text{Torr}$, $t = 25°\,\text{C}$; r is the radius of the internal electrode in cm. The formula describes experiments with a smoothly polished inner electrode, in the ranges $r \sim 10^{-2}$–1 cm, $p \sim 10^{-1}$–10 atm., including experiments in oscillating fields of frequency up to 1 kHz (in this case, E_c stands for field amplitude). Roughness of the electrode surface may lower the threshold value of E_c by 10–20 % because the field is additionally enhanced at the tiniest protrusions. The formula in the generalized form (with different numerical coefficients) is applicable to a number of gases. It can also be used for the wire-plane gap. The ignition voltage V_c is related to $E_c = E_{max}$ by (12.10, 13).

Peek's empirical formula bears the imprint of physical criteria, such as (12.14, 15). This is seen from the calculation of E_c in [12.13] on the basis of the criteria (12.14, 15) with $E(x)$ given by (12.10) and α_{eff} approximated for air at $E/p < 150\,\text{V/cm Torr} = 110\,\text{kV/cm atm.}$ by the formula (employed by electrical engineers):

$$\alpha = 0.14\delta\left\{(E\,[\text{kV/cm}]/31\delta)^2 - 1\right\}\,\text{cm}^{-1} . \tag{12.17}$$

As an exmaple, consider an air gap between a wire of $r = 0.1$ cm and a coaxial cylinder of $R = 10$ cm; $p = 1$ atm., $\delta = 1$. Using Peek's formula, $E_c = 61$ kV/cm. According to (12.10), $V_c = 28$ kV, and the field at the outer electrode is $E(R) = 0.61$ kV/cm. By discharge standards, this field is extremely weak: $E/p = 0.8$ V/cm Torr. Multiplication ends at a radius $x_1 = rE_c/E_1 = 0.25$ cm ($E_1 \approx 24$ kV/cm), that is, the thickness of the layer around the wire where multiplication at the avalanche stage occurs is $x_1 - r \approx 0.15$ cm.

12.6.4 Ignition Lag

The time lag between the moment of application of voltage and the onset of steep rise in current consists of the statistical time of *waiting for a seed electron* at the tip, and the time of *avalanche multiplication* or *streamer formation*. Experiments point to a large, difficult to interpret spread of data for negative coronas; large and not readily controllable effects due to the state of the cathode surface may be responsible for this. In the case of positive coronas, the streamer formation time of 10^{-8} s is evidently too short in comparison with the reported time lags of 10^{-8}–10^{-6} s – which must be assigned to be the waiting time for seed electrons.

They are obviously not produced at the cathode because the lag is typically shorter than the time of drifting from the cathode to the positive tip. Occasional electrons in air convert very rapidly into negative ions. It is assumed that a strong field *liberates electrons* from O_2^- ions colliding with molecules; the efficiency of this process is a maximum at $E/p = 90$ V/cm Torr= 68.5 kV/cm atm. Moist air hydrates ions, forming clusters with water molecules. Ions in clusters are destroyed less efficiently.

12.6.5 Charge Transfer Beyond the Region of Multiplication, and $V - i$ Characteristics

Charge carriers are produced only in the direct vicinity of the corona-carrying electrode surrounded by a strong field. In the remaining part of the gap (*outer region*), the current is carried by charges that are pulled out by the weak field present there. The carriers are positive ions in positive coronas and negative ions in negative coronas (or electrons if the gas is devoid of electronegative components). The current in the outer region of the corona discharge is *non-self-sustaining,* the gas there is not ionized (i.e., suffered no breakdown).

The corona current depends on the applied voltage, or rather, on its excess over the ignition potential V_c. The current is limited by the *space charge* of the charge carriers in the outer region. The region of self-sustained discharge at the corona-carrying electrode is capable of generating a high current and a large number of carriers, but some of them are turned back by the space charge of the same sign and cease moving towards the electrode of the opposite sign. The situation is very similar to the limitation of current by space charge in a vacuum diode (Sect. 6.6); the difference is that here charges drift rather than move freely. These arguments lead to an approximative curve for the $V - i$ characteristic of steady-state corona discharge derived by Townsend, in 1914.

Consider the gap between coaxial cylinders. The current per unit cylinder length across a surface of arbitrary radius x outside the narrow zone of multiplication is $i = 2\pi x e n \mu E = \text{const}$, where n is the charge carrier density and μ is their mobility. Assume that the current, charge density, and distortion of the external field by the space charge are not too large. Then (12.10), corresponding to zero current but also to the actual voltage V, can be retained as a first approximation for the distribution $E(x)$. In this approximation,

$$n = i/2\pi e \mu E x = i \ln(R/r)/2\pi e \mu V = \text{const}\,.$$

Substituting this n into Poisson's equation $x^{-1} d(xE)/dx = 4\pi e n$ and integrating, we arrive at the next approximation to the field distribution. The constant of integration will be so chosen that the product xE in the limit $i \to 0$ be given by (12.10) in terms of the ignition potential V_c:

$$E = \frac{2i \ln(R/r)}{\mu V} \frac{x^2 - r^2}{2x} + \frac{V_c \ln(R/r)}{x}\,, \quad \int_r^R E\, dx = V\,.$$

Integrating E with respect to x, recalling that $x^2 \gg r^2$ in the predominant part of the gap, and solving the resulting equation for i, we find the $V - i$ characteristic of the corona:

$$i = \frac{2\mu V(V - V_c)}{R^2 \ln(R/r)} . \tag{12.18}$$

Formula (12.18) is confirmed by experimental data: the inverse resistance i/V is plotted as a function of V (the curve is known as reduced characteristic), the linearity of the curve giving the confirmation. The extrapolation of the curve to $i/V = 0$ determines the ignition potential V_c in a simpler and more accurate manner than the search for the corona inception point. The mobilities of positive and negative ions being more or less identical, the currents of *positive* and *negative* coronas in an *electronegative* gas at a given V are also *nearly equal*. In a gas *without attachment*, the *negative* corona current, being transported by electrons, is substantially *greater*. An admixture of an electronegative component immediately reduces this current.

Let us make some evaluations for the example considered at the end of Sect. 12.6.3: air at 1 atm., $r = 0.1$ cm, $R = 10$ cm, $V_c = 28$ kV. We set $\mu = 2\,\text{cm}^2/\text{V s}$. The mobilities of O_2^-, O_4^-, O_2^+, O_4^+ in O_2, N_2^+, N_4^+ in N_2 at 1 atm. cluster around this value. Let $V = 40$ kV. We have $i = 4.6\,\mu$A/cm ($1\,\text{A} = 9 \cdot 10^{11}$ V cm/s). The ion density is $n = 2.6 \times 10^8\,\text{cm}^{-3}$. Experiments show that the $i(V)$ dependence of type (12.8) also holds for other geometries. Thus the current in atmospheric air between a needle cathode with tip radius $r \approx 3 \times 10^{-3}$–$3.5 \times 10^{-2}$ mm, and a plane anode perpendicular to the neelde axis at a distance $d \approx 4$–16 mm from the needle tip, is

$$i \approx (52/d^2)V(V - V_c)\,\mu\text{A} , \quad V[\text{kV}] , \quad d[\text{mm}] , \tag{12.19}$$

with $V_c \approx 2.3$ kV being independent of d [12.13].

12.6.6 Corona Losses in High-Voltage Power Lines

This is a problem important in the transmission of electrical energy. The losses due to the corona discharge may be comparable to the Joule heat released in power line conductors. When both electrodes produce a corona discharge, as is the case for two parallel wires of opposite polarity, the current in the outer region separating them is carried by ions of opposite signs moving in opposite directions. The current and energy losses in such *bipolar* corona are much greater than in a *unipolar* corona (with a single corona-producing electrode). As a rule, the total recombination of ions is not achieved, while the mutual *neutralization of space charge* reduces the degree of limitation imposed by the space charge. Furthermore, negative ions arriving at the positive conductor decay in its strong field. This process is an additional source of electrons, somewhat facilitating the ignition of the corona. Using Peek's formula for two parallel wires,

$$E_c = 29.8\delta\left(1 + 0.301/\sqrt{\delta r}\right)\,\text{kV/cm} , \quad r < 1\,\text{cm} , \tag{12.20}$$

which is somewhat less than implied by (12.16) for a single wire. A different empirical formula is used nowadays in electrical engineering: for smooth conductors,

$$E_c = 24.5\delta\left[1 + 0.65(\delta r)^{-0.38}\right]\ \mathrm{kV/cm}\,. \tag{12.21}$$

In the absence of discharges, the field distribution between two parallel wires with a voltage V between them is equivalent to the field distribution between a wire and a plane midway between the two wires, provided the voltage between the plane and a wire is $V/2$. Nevertheless, a corona is easier to start in the former case: one half of V_c in the wire-wire geometry is about 10% lower than V_c in the wire-plane geometry.

$V - i$ characteristics of type (12.18) imply that the power released in a corona discharge is $P = iV \approx \mathrm{const}\, V^2(V - V_c)$. Figure 12.10 shows losses per km of a high-voltage power line with conductors 2.5 cm in diameter. The ignition voltage diminishes considerably in rainy weather and in winter because water droplets and ice crystals deposited on conductors form additional sources of field enhancement. The current and losses sharply increase at the same voltage V.

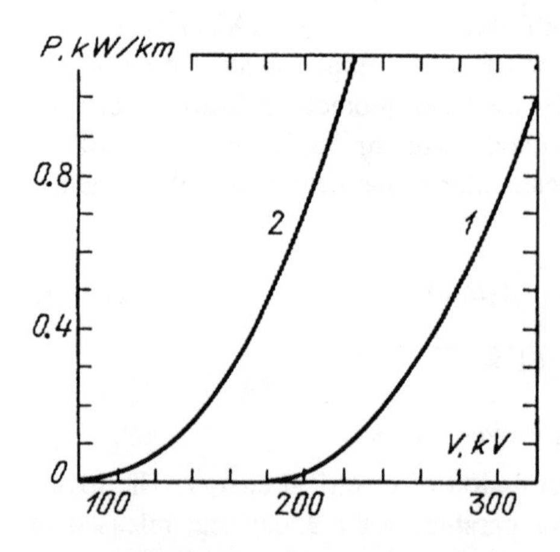

Fig. 12.10. Corona losses from a conductor 2.5 cm in diameter as a function of voltage. *(1)* fine sunny weather, *(2)* slight rain [12.7]

12.6.7 Intermittent Corona Discharge

Under certain conditions, the corona burns in the form of *periodic current pulses*, despite the constancy of voltage; the pulse repetition rate reaches 10^4 Hz if the corona is on the anode, and 10^6 Hz if it is on the cathode. Intermittence was discovered in Loeb's laboratory by Trichel and Kip in 1938. (Loeb and his school greatly contributed to the study of spark and corona discharges.) Periodic phenomena may also prove to be of practical interest. Their frequencies lie in the rf range, so that a corona in a power transmission line may be a source of radio noise. The most convenient source for laboratory study is the corona from a point to a plane electrode in room air.

Positive point. With a point of radius $r = 0.17$ mm at a distance $d = 3.1$ cm from a plane cathode, the corona appears at $V_c \approx 5$ kV and stays intermittent up to $V_1 \approx 9.3$ kV [12.13]. The repetition rate is low at the ends of the interval V_c–V_1, and reaches 6.5 kHz in the middle. The mean current increases to 1 μA at V_1. Such pulses are known as *flashing corona*. Pulses are absent in the interval from V_1 to $V_2 \approx 16$ kV; the current becoming steady. At V_2, the current increases to 10 μA. In the interval from V_2 until the *spark breakdown* of the entire gap at $V_t \approx 29$ kV, the discharge again becomes pulsed, the pulse repetition rate increasing from a low value of 4.5 kHz. The mean current increases with V and reaches 100 μA in the pre-breakdown stage.

If V is only slightly greater than V_c, the flashing corona is associated with so-called *pre-inception streamers*. Electrons produced by ionization flow into the point anode, while a space charge of positive ions accumulates outside the narrow zone where ionization processes take place and streamers appear. The positive layer surrounding the point creates a field that is directed against the external one on the inside of the layer, that is, it reduces the field around the point, thus screening it. The criterion of self-sustainment, (12.5), is thereby violated. New streamers are not formed and the current decreases. As the ions are pulled to the cathode, the strong field at the tip is restored (the point becomes "bare" again). Streamers appear, a new surge of ionization follows, then of current, and the cycle begins anew. Something similar takes place on the second pulsed interval V_2–V_t, but now *pre-breakdown streamers* are involved. In the interval V_1–V_2, the condition of self-sustainment is strictly satisfied and a dc current flows. The current between the pulses of the flashing corona does not drop to zero: a constant component is observed.

Negative point. The current is intermittent in a certain interval of mean currents and voltages, beginning with the ignition point V_c. The pulse repetition rate is greater than in positive coronas, 10^5 Hz at 20 μA, while the pulses are shorter (10^{-7} s) and higher: the peak current reaches 10 mA. The pulses follow in a very regular manner. They are known as *Trichel pulses*. As the voltage is increased, the pulses disappear and a steady-state corona is sustained until the spark breakdown of the discharge gap.

In principle, the factors causing the nonsteady behavior here are the same as in the case of positive coronas, although some differences are observed. When an avalanche grows from the cathode point, the positive space charge lies close to the point, while the negative charge is at a somewhat greater distance. If this charge is formed by electrons, they are rapidly pulled away to the anode, so that the negative charge density is very low, i.e., it produces no screening of the field of the point. The presence of positive ions in the immediate vicinity of the tip only enhances the field there. As a result, Trichel pulses are *not observed* in the *electropositive* gases N_2 and Ar. Air is different: once an electron is far from the point, it finds itself in a weaker field and becomes attached to a molecule. The space charge of negative ions weakens the field of the point, the multiplication of avalanches is suppressed, and the current decays. As the negative ions are pulled towards the anode and positive ions move to the point, the external field is restored and favorable conditions for a new pulse build up.

The corona discharge is not necessarily harmful, as it is in power transmission lines. It is usefully employed in electric filters and separators, and is fundamental for Geiger-Müller counters, developed in 1929 to detect nuclear particles. A recent good review is available on the physics of corona discharges [12.13].

12.7 Models of Streamer Propagation

Is the Head of a Cathode-Directed Streamer Insulated from the Anode? The insulation degree of and the extent to which the anode potential is transferred by the streamer channel to its positive tip constitute one of the *main questions of the theory,* because the answer determines whether the streamer channel will be transformed into a spark channel after the gap is closed (Sect. 12.5), that is, whether the streamer breakdown may be realized. Until recently, two models of the streamer process, based on *extreme* assumptions, were considered: *absolute insulation* of the positive streamer head from the anode[2] and *ideal conduction* of the streamer channel.

12.7.1 Model of Self-Sustaining Streamers

This model was developed by *Dawson* and *Winn* in 1965 [12.14]. The head of a cathode-directed streamer is a sphere of radius r_0, containing N_+ positive ions. As it moves, it leaves behind a quasineutral ionized channel; however, it is assumed that its conductivity is negligibly low and the head is not connected to the anode. The problem then is to find the conditions under which the head is absolutely *autonomous,* that is, moves regardless of the external field, producing and absorbing avalanches by its own field.

Let a photoelectron that initiates the avalanche be liberated at a distance x_1 from the center of the positive sphere, in the direction of motion (Fig. 12.11). An electron avalanche develops at this point in the field $E = eN_+/x^2$ and moves towards the head. The number of electrons produced in it at a point x_2 is $N_e = \exp \int_{x_1}^{x_2} \alpha \, dx$. Let the avalanche radius r_D increase by the diffusion law

$$dr_D^2/dt \approx 4D_e \, , \quad r_D(x_2) = \left\{ \int_{x_1}^{x_2} 4 \frac{D_e/\mu_e}{E(x)} dx \right\}^{1/2} .$$

When the "spherical" electron head of the avalanche enters the ionic head of the streamer, quasineutral plasma is formed at this point, and the ionic trail of the avalanche transforms into the newly-advanced streamer head. For this process to be continuous and steady, it is necessary that at the moment when the head of the

[2] A slight refinement must be added here: what is meant is that insulation sets in after the very first electrons sink into the anode at the moment of formation of the streamer.

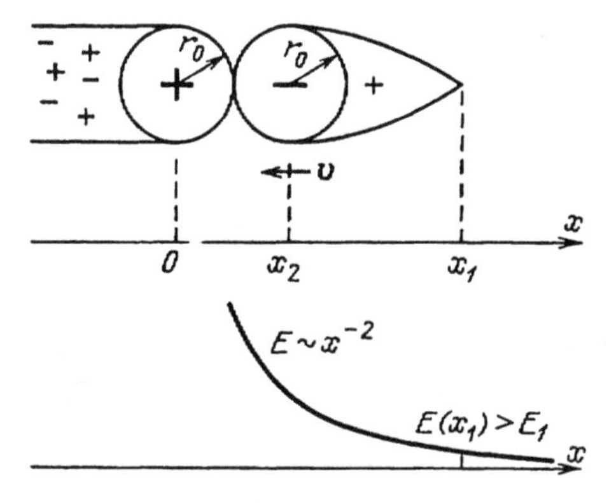

Fig. 12.11. Model of self-sustaining streamer

avalanche touches the head of the streamer, that is, at $x_2 = r_0 + r_D$, the numbers of charges in the two heads and their radii be equal: $N_e = N_+ \equiv N_0$, $r_D = r_0$. The point of avalanche inception, x_1, has been fixed artificially, to correspond to the free path length of photons, but so as to meet the condition that the field $E(x_1)$ in air be sufficient for ionization: $E(x_1) \gtrsim 30\,\mathrm{kV/cm}$.

The self-sustained process air is possible with the following values of parameters: photon "path length" $x_1 \approx 2 \times 10^{-2}\,\mathrm{cm}$, number of ions in the streamer head $N_+ \equiv N_0 \approx 10^8$, head radius $r_0 \approx 2.7 \times 10^{-3}\,\mathrm{cm}$. The streamer propagation rate is $v = x_2/t$, where t is the time in which the avalanche progresses from x_1 to x_2; $v \approx 10^7\,\mathrm{cm/s}$. These figures do not go beyond reasonable limits, but they do not in fact seem to be realistic. The streamer radius r_0 is too small, and the charge density in it is correspondingly too high: $n_e = n_+ \approx 3N_0/4\pi r_0^3 \approx 2.6 \times 10^{15}$ cm^{-3}. The neglected repulsion of electrons in the avalanche increases its radiuss, the field $E \approx eN_0/r_0^2$ decreases, ionization slows down significantly, that is, the process may become "disbalanced".

12.7.2 Is Strictly Steady Propagation of Streamers Possible in the Absence of External Fields?

The answer to this question is "it is not", because energy is needed for the ionization of the new volume of gas absorbed by the streamer channel, and this energy can be drawn only from outside. Actually, the process in question is not steady but merely quasi-steady, in the sense that the parameters change by only a little during one steep of propagation of the head, over a distance about equal to the size of the head. Estimates show [12.14] that the initial amount of energy in the head (electrostatic energy $e^2N_0^2/2r_0 \approx 2.7 \times 10^{11}\,\mathrm{eV}$), taking into account the expenditure on ionization and excitation, is sufficient for moving the streamer by about 3 cm. This result fits remarkably well the experiment of the inventors of the model, who measured the distance covered by the streamer after the external voltage was switched off (finite-length pulses with abrupt rise and fall edges were applied to the electrodes).

12.7.3 Positive Streamer Insulated from the Anode in a Nonuniform Field (Streamer Corona)

The Dawson-Winn model was considerably improved by *Gallimberti* in 1972 [12.15], who was taken into account consistently the energy balance of processes involving the external field. Gallimberti also introduced an approximate description of photoionization, so as to eliminate the arbitrariness in prescribing the distance x_1 at which the avalanche is initiated. As in [12.14], it was assumed that only one "equivalent" avalanche is started. The positive sphere was again assumed insulated from the anode. The parameters of the streamer (N_0, r_0) change as it propagates in the nonuniform field. The energy balance equation expresses the fact that the work done by the external field on electrons compensates for the energy spent by electrons on ionization, excitation, attachment, on transfer to molecules, and also on the change in the potential electrostatic energy of the positive charge of the streamer head (both intrinsic and in the external field).

The fairly complicated equations were solved numerically for the specific conditions of corona in air. Figure 12.12 demonstrates excellent agreement with experimental results. The length of the streamer is about 11 cm, which the head covers in 10^{-7} s. The streamer moves over the first 5 cm (the region of especially strong field of 15 to 5 kV/cm) at a velocity of about 2×10^8 cm/s. The streamer stops where the field drops to 2 kV/cm. The number of positive charges in the head reaches the maximum $N_0 \approx 1.6 \times 10^9$ ($eN_0 \approx 2.5 \times 10^{-10}$ C).

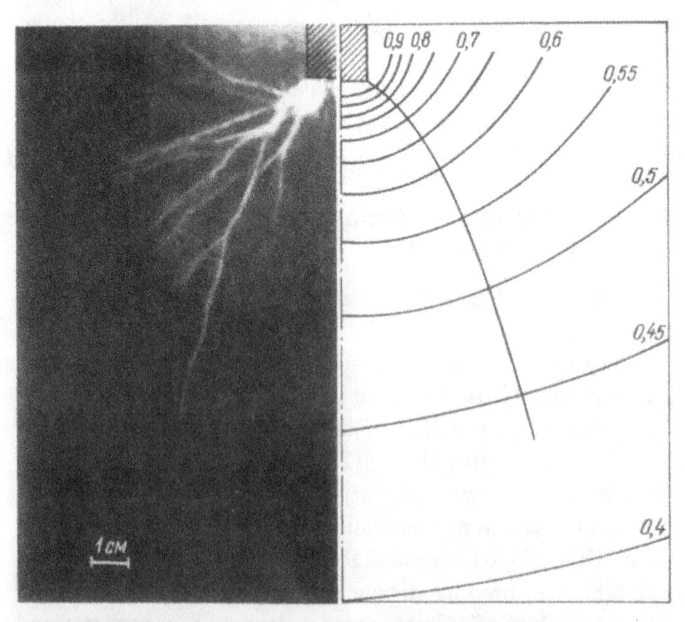

Fig. 12.12. Streamer moving from a positive rod 2 cm in diameter to a plane at a distance of 150 cm [12.15]. Constant voltage, 125 kV. *Right:* results of calculations; equipotential surfaces are shown; numbers on the curves give the fraction of applied voltage measured from the plane electrode. *Left:* photograph of streamers under the conditions of calculations

A numerical Monte Carlo procedure was developed for the approximation of two equivalent avalanches that are randomly initiated at different points of the angle sector near the streamer head [12.16]. This computer simulation produced a branching zigzag pattern of streamers under the same conditions; it resembled the photograph even more.

12.7.4 External Field Necessary for the Propagation of a Streamer

It is clear from the arguments above that the growing streamer must be supplied with energy; this function is fulfilled by the applied field, which does work on the electrons. This can be interpreted macroscopically as the energy released by the *streamer current*. The streamer current in the approximation of a concentrated charged head moving at a velocity $\boldsymbol{v}$ is given by formulas (12.6′, 6). If $N_0 \sim 10^9$, $v \sim 10^8$ cm/s in a long discharge gap of $d \sim 1$ m, then the average current is $i \sim 10^{-4}$ A, and if the gap is narrow, $d \sim 10$ cm, then $i \sim 10^{-3}$ A. The power $iV = eN_0vE$ is spent on ionization, on electron and vibrational excitation of molecules, and an attachment compensation. If the average energy spent on producing one electron is w, then $eN_0vE = n\pi r_0^2 vw$, where the charge density is $n \approx 3N_0/4\pi r_0^3$. The external field required to supply the energy to sustain the streamer is $E_s \approx w/er_0$, which satisfies the obvious condition $eE_s r_0 \approx w$.

The radius r_0 of the charged head and of the streamer channel is found from the condition of self-sustainment of the head in the strong field it creates and depends very little on the external field. If in air $w \approx 50$ eV and $r_0 \approx 10^{-2}$ cm, we obtain an estimate $E_s \approx 5$ kV/cm. A detailed calculation [12.15] for the case of a homogeneous field yielded $E_s = 7$ kV/cm.

These figures are in reasonable agreement with experimental data. According to [12.17], in dry atmospheric-pressure air we find $E_s \approx 4$ kV/cm. As reported in [12.18], the average external field that was necessary to sustain the steady growth of a streamer in air in the experiment was $E_s \approx 4.7$ kV/cm, almost independently of gap width and field nonuniformity. In technical grade nitrogen (up to 2 % O_2), $E_s \approx 1.5$ kV/cm; in Ar, $E_s \approx 0.4$ kV/cm (all at 1 atm.). The value of E_s is very sensitive to attachment, which removes electrons from the process. To multiply electrons up to the required level N_0, high energy expenditure is necessary.[3] Thus an admixture of O_2 to Ar increases E_s from 0.4 to 2.3 kV/cm at an O_2 content of 10 %. On the other hand, if air is heated to 1000 K, electrons are liberated from negative ions and E_s falls from 4.7 to 0.7 kV/cm [12.18]. The necessary field also increases as the humidity is raised. At a water vapour content of 2×10^{-5} g/cm^3, E_s is greater by a factor of 1.5 than in dry air [12.17].

---

[3] This may reflect the reduction of conductivity in the streamer channel (see Sect. 12.7.7), although this certainly has no place in the theory of the self-sustaining streamer heads.

12.7.5 Model of an Ideally Conducting Streamer Channel

This assumption was first suggested by Loeb in early work on the streamer theory, and was incorporated into the constructed one-dimensional theory of the ionization wave (modeling the streamer) in the field [12.19]. The two-dimensional model was developed in [12.5] for the process of streamer growth from the midpoint of a plane gap towards both electrodes. The surface of the ideally conducting body of the streamer is equipotential. Surface charges induced by the external field are distributed over this surface (the plus sign on the side facing the cathode, the minus sign towards the anode). It was assumed that the streamer body forms an ellipsoid of revolution elongated parallel to the field, the velocity of motion of the surface at each point being directed along the outward normal and equal in magnitude to the electron drift velocity in the appropriate field (Sect. 12.3). The calculated velocities of the anode and cathode ends of the streamer coincided with the results of measurements in neon at $p = 1$ atm. in external fields of $E = 10$–$15\,\mathrm{kV/cm}$. The velocities increase as E and the streamer length l increase; for $l \approx 0.5\,\mathrm{cm}$, $v \sim 10^8\,\mathrm{cm/s}$.

Let us note a useful formula for estimating the enhanced field E_m at a rounded tip of radius r at the end of a conducting rod of length l placed along the external field E_0:

$$E_m/E_0 = 3 + 0.56(l/r)^{0.92}\,, \quad 10 < l/r < 2000\,.$$

This formula was obtained by approximating the calculated fields [12.18].

If the ideally conducting channel begins at the anode, it is as a whole at that potential, $\varphi = V$ (at the cathode, $\varphi = 0$). There is no longitudinal field in the channel. We denote the potenital at a point x of the channel that is produced by the applied voltage in the absence of the streamer by $\varphi_0(x)$. The coordinate x is measured along the channel from the anode. The additional potential $\varphi_1(x) = V - \varphi_0(x)$ is mostly produced by the positive charge concentrated on the channel surface. The charge per cm of its length is $q(x) = C\varphi_1(x)$, where C is the conductor capacitance per unit length. It can be evaluated by the formula $C = [2\ln(R/r_0)]^{-1}$, valid for a conductor of radius r_0 in a grounded coaxial cylinder. For its radius R, can take the distance X to grounded cathode (if X is less than the streamer length l or l if $X > l$. If $r_0 = 10^{-2}\,\mathrm{cm}$, $R = 30\,\mathrm{cm}$, $C = 0.062 = 0.07\,\mathrm{pF/cm}$.[4] Far from the anode, $\varphi_1 \sim V$. If, for example, $\varphi_1 = 50\,\mathrm{kV}$, then $q = 3.5 \times 10^{-9}\,\mathrm{C/cm}$. The current of an ideally conducting streamer of length $x = l$ is created by the continuous sinking of electrons into the anode and by the formation of a new region of positive charge at the end of the conductor: $i = q(l)v$. If $q = 3.5 \times 10^{-9}\,\mathrm{C/cm}$ and $v = 10^8\,\mathrm{cm/s}$, then $i = 0.35\,\mathrm{A}$. This value is significantly greater than in an insulated streamer head.

[4] The measurements [12.20] best agree with calculations for $C = 0.05\,\mathrm{pF/cm}$.

12.7.6 Compensation Zone

Actually, the concept of a charged sphere with a moderate-conductivity electroneutral channel stretching out behind it (as in the Dawson-Winn model) contains a contradiction.

The contradiction was pointed out by *Griffiths* and *Phelphs*.[5] Let the entire positive charge eN_0 be concentrated in a sphere of radius r_0. The potential φ_1 created by this charge falls off away from the sphere, including the direction along the channel toward the anode; the maximum of φ_1 equals $\varphi_s = eN_0/r_0$ (for $N_0 = 10^9$, $r_0 = 10^{-2}$, $\varphi_s = 14\,\text{kV}$). A very strong additional field, $E_1 \sim eN_0/r_0^2 \sim 10^3\,\text{kV/cm}$, will draw electrons from the immediate neighborhood of the channel to the head.

The positive charge thus gets distributed not within the sphere but over a certain length L at the end of a "nonconducting" channel so as to level off the potential $\varphi(x)$ at about $\varphi_0(l-L)$. Beyond the right end of the region $l-L < x < l$ (the "compensation zone"), the potential drops sharply to the level of the external potential, $\varphi_0(l)$. If the potential jump at the end of the streamer is $\varphi_1 \approx \varphi_0(l-L) - \varphi_0(l) \approx E_0(l)L \sim 10\,\text{kV}$, and the external field is $E_0 \sim 10\,\text{kV/cm}$, then $L \sim 1\,\text{cm}$. For $C \sim 10^{-13}\,\text{F/cm}$, $q(l) \sim 10^{-9}\,\text{C/cm}$ and the total charge of the extended tip of the streamer $Q \approx q(l)L/2 \sim 10^{-9}\,\text{C}$ is somewhat greater than in the case of the spherical head.

12.7.7 Which Model Is Closer to the Truth?

In fact, this question opened Sect. 12.7. The answer can be obtained experimentally or found in the theory which takes into account the finite conductivity of the streamer channel. Even in very simplified physical formulations, the streamer process is described by equations so complicated that hopes of a solution may be associated only with numerical methods, and such publications are now appearing. Some experimental results have also been obtained. In experiments [12.18], the charge transfer to the cathode in the process of the propagation of a single streamer from the anode was recorded using an oscilloscope; the streamer current was then calculated. The current in technical-grade nitrogen at $p = 1$ atm. in a gap of $d = 1$ m, between the positive rod and a plane during the streamer propagation time $t \sim 10^{-5}\,\text{s}$, is $i \sim 10^{-3}\,\text{A}$. The current transfers the charge $Q \sim 10^{-8}\,\text{C}$. This is much greater than that found in the spherical head theory. If we assume that what is measured is the conduction current through the streamer channel, the conductance for unit length is found from Ohm's law $i = \gamma E$, for the external field, E to be $\gamma = \pi r_0^2 \sigma \sim 10^{-7}$–$10^{-6}\,\text{cm/Ohm}$.

Unfortunately, the channel radius is unknown. If $r_0 = 10^{-2}\,\text{cm}$, then $\gamma = 10^{-6}$ leads to $\sigma = 3 \times 10^{-3}\,(\text{Ohm}\cdot\text{cm})^{-1}$, which corresponds to the plasma density in the channel $n_e \approx 4 \times 10^{13}\,\text{cm}^{-3}$. If $r_0 \approx 3 \times 10^{-3}\,\text{cm}$, then $n_e \approx 4 \times 10^{14}$; this is apparently an upper bound. The measured values of i and Q are too high for the insulated sphere model but too low for the model of an ideally conducting

[5] Their arguments are presented in a detailed review [12.21] on the spark breakdown of long gaps.

channel. Under the conditions of the experiment, $V \sim 10^2$ kV, the charge on the tip of an ideally conducting streamer that has traveled half the gap length is $q \approx CV \sim 10^{-8}$ C/cm, so that at $v \sim 10^7$–10^8 cm/s the current is $i = qv \sim 0.1$–1 A. The real situation is intermediate between the two models; and presumably, it corresponds to a considerable insulation of the streamer tip from the anode and to a compensation zone in that region.

These conclusions are supported by numerical calculations. Numerical results do not point to the nonmonotonic potential $\varphi(x)$ implied by the sphere model, or to the vanishing of the field in the channel, as in the case of ideal conduction. A two-dimensional problem was treated in [12.22], on the propagation of the plasma channel in nitrogen at 1 atm. in a plane gap of $d = 0.5$ cm at $V = 30$ kV, $E_0 = 60$ kV/cm. Self-sustainment via initiation of secondary avalanches due to photoionization was not discussed. Instead, an initial electron background was postulated, $n_e = 10^8\,\mathrm{cm}^{-3}$. An initial plasma source was defined at the anode with $n_e \sim 10^{14}\,\mathrm{cm}^{-3}$. The plasma channel with $n_e \sim 10^{14}\,\mathrm{cm}^{-3}$ grows from it toward the cathode at a veloctiy $v \approx 2 \times 10^8$ cm/s. The field at its end reaches 150 kV/cm, but the field inside the channel is reduced by a mere 20 to 30 % in comparison with the external field. The initial channel radius, arbitrarily fixed in the calculation (about 10^{-2} cm), was preserved in the course of channel growth; the streamer radius thus could not be obtained from the equations (neither was it possible to measure it experimentally).

This is a weak spot in the calculations [12.23] for air, as well, although in other respects this work is more comprehensive: the process in the channel and around the streamer head has been analyzed with photoionization taken into account. The calculation was quasi-one-dimensional, the radius was imposed artificially: $r_0 = 10^{-2}$ cm. Such parameters of the streamer as velocity and density of plasma at the tip and the field at the tip on the side of the channel, are determined by the increase of potential $\varphi_1(l)$ at the head over the initial level. The quantity $\varphi_1(l)$ is a function of external conditions and conductivity in the channel, which is determined by electronic kinetics. For example, if $\varphi_1(l) = 9$ kV, we obtain $v = 1.4 \times 10^8$ cm/s, $E = 4$ kV/cm, and $n_e \approx 10^{15}\,\mathrm{cm}^{-3}$.

The unit-length conduction of the streamer channel in air far from the tip does not exceed 10^{-10} cm/Ohm owing to attachment. This is seen from the negligibly low current, less than 10^{-6} A, that flows after the streamer touches the cathode and "closes" the gap [12.18]. Attachment is unimportant only in short streamers, and at early stages of the propagation of long ones. In air at $p = 1$ atm., the attachment frequency is $\nu_a \approx 10^7\,\mathrm{s}^{-1}$ so that at $v \sim 10^8$ cm/s the concentration n_e in the channel decreases considerably at a distance $x \sim v/\nu_a \sim 10$ cm. According to the calculation [12.23], the concentrations in the streamer channel of $l \approx 8$ cm are $n_e \approx 2 \times 10^{14}\,\mathrm{cm}^{-3}$ and $n_- \sim 1.5 \times 10^{13}\,\mathrm{cm}^{-3}$.

A relatively high conductivity of the streamer channel has been recorded in argon cleaned of electronegative impurities [12.18]: $\gamma > 10^{-5}$ cm/Ohm, $i \sim 10^{-2}$ A. The streamer channel was observed to transform into a spark channel after the cathode was reached. Generally, streamer (not leader; see Sects. 12.8, 9) breakdown has been observed only in inert gases. It seems that if the model of the ideally conducting streamer remains meaningful at all, it may hold only for inert gases.

12.7.8 Plasma Decay and the Radius of a Streamer Channel

The conductivity of the channel in air decreases rapidly as we move away from the streamer head. Actually, the plasma in the channel decays even in nitrogen (where attachment does not occur) if the external field is not much greater than the limiting field for the streamer propagation. $E_s \approx 1.5\,\text{kV/cm}$ at $p = 1$ atm. The field behind the head does not exceed the external field and the corresponding $E/p = 2\,\text{V/cm Torr}$ is too low for the ionization of the cold nitrogen (Sect. 8.7.7). Over the period $t \sim 10^{-5}\,\text{s}$ during which a streamer passes across a meter-wide gap, electrons recombine to a density $n_e \sim (\beta t)^{-1} \sim 10^{12}\,\text{cm}^{-3}$ from an arbitrarily higher initial value. A decrease in the conductivity of the channel (the more important, the longer the channel and the more uniform the external field) prevents the streamer breakdown of long uniform gaps.

Despite a certain similarity, a streamer channel with weakly ionized nonequilibrium plasma cannot be likened to the positive column of a glow discharge. Ionization processes in the discharge column balance out the loss of charge and the plasma in the column is self-sustained. The streamer channel is rather a "passive" plasma trace left behind the advancing self-sustained streamer head. The trace radius (channel radius) r is presumably determined by ambipolar diffusional expansion of the plasma from its initial size, that is, from the streamer head radius $r_0 \approx 3 \times 10^{-3}$–$10^{-2}\,\text{cm}$ (Sect. 12.7.1). The seemingly unchanging "thinness" of a long channel creates the false impression of an inherent channel radius. In reality, the trace simply has not had enough time to expand significantly: at $D_a \approx 2\,\text{cm}^2/\text{s}$, the increase of the radius over a time $t \sim 10^{-5}\,\text{s}$ is $r - r_0 \approx \sqrt{D_a t} \approx 4 \times 10^{-3}\,\text{cm} \approx r_0$.

The preservation of the radius of the initial perturbation of the growing plasma channel in the calculation [12.22] (mentioned in Sect. 12.7.7) is caused by the field build-up in the neighborhood of the channel; the spreading due to diffusion is negligible. The authors choose such a high value of the external field, $E \approx 40E_s$, that the ionization in the channel greatly exceeds the recombination.

12.8 Breakdown in Long Air Gaps with Strongly Nonuniform Fields (Experimental Data)

12.8.1 Effect of Field Nonuniformity on Breakdown Voltage

Field nonuniformity reduces the breakdown voltage for a given distance between the electrodes. This is illustrated in Fig. 12.13, where the measurements were conducted at the industrial frequency of 50 Hz. The nonsteady nature of the field makes practically no difference, because the half-period of $10^{-2}\,\text{s}$ is long in com-

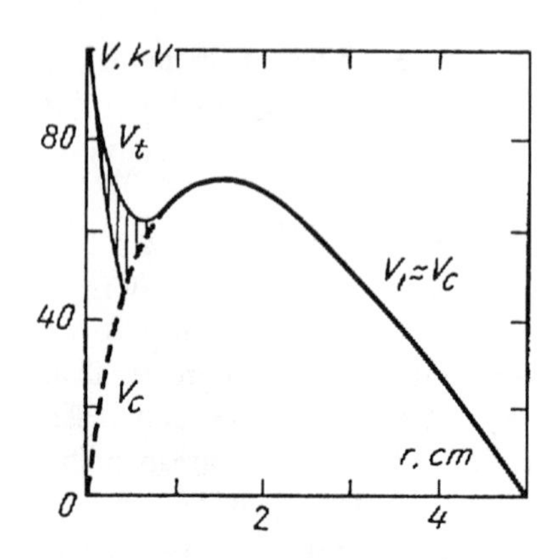

Fig. 12.13. Voltages of corona ignition V_c and of breakdown V_t in an air gap between concentric cylinders as functions of radius of the inner electrode; outer electrode radius $R = 5\,\mathrm{cm}$. (Amplitude of 50 Hz ac voltage is plotted.) The region of enhanced sprad in breakdown voltages is shaded [12.9]

parison with the time scale of breakdown.[6] Since the polarity of electrodes is alternating, the breakdown occurs at the polarity that facilitates it (see below). The tangent to the curve $V_t(r)$ at $r = R$, with the slope of 32 kV/cm, roughly corresponds to the breakdown of plane gaps of the same size $d = R - r$ in uniform field. As r decreases, that is, as the degree of nonuniformity is enhanced, the threshold curve $V_t(r)$ deviates more and more downwards from the tangent. The mean breakdown field $E_{av} = V_t/(R - r)$ diminishes monotonically in comparison with the level 32 kV/cm. The reason for this effect of nonuniformity is that any breakdown criterion includes the coefficient of enhancement of primary avalanches, $\int \alpha\, dx$, which is very sensitive to the distribution $E(x)$ owing to the steeply climbing curve $\alpha(E)$. The distribution of the field in comparison with the uniform picture increases the enhancement at preserved $\int E\, dx$, or decreases the potential difference at constant enhancement. We have already encountered this effect several times (Chap. 8).

Figure 12.13 is illustrative in another aspect, as well. The range of voltages $V_c < V < V_t$ over which a corona burns contracts as r increases, and the degree of field nonuniformity decreases. If the field is not too nonuniform, $r/R \gtrsim 0.1$, no corona develops: increasing the voltage on the electrodes leads straight to the breakdown of the gap. If, however, the radius of the electrode carrying the corona is very small (very high nonuniformity), the difference between the corona initiation and breakdown potentials, $V_t - V_c$, becomes large.

The effect of the degree of field nonuniformity on breakdown voltage is also revealed by the fact that it is easier to produce breakdown between a rod and a plane than between two rods, at the same separation d. The corresponding threshold voltages (also at 50 Hz) are plotted in Fig. 12.14 for gaps of d up to 10–12 m. The rod was a rectangular metal bar of square cross section, of $1/2''$ sides. For the same d, the capacitance of the rod-plane gap is greater than that of the rod-rod gap, because the volume occupied by the field is greater. Therefore,

[6] In order to eliminate the effect of the rate of voltage build-up during the "switch-on" period in measurements of dielectric strength of a gap for dc or for time-dependent (not pulsed) fields, the electrode voltage (or the amplitude of the 50-Hz voltage) is raised gradually, over a time of up to several minutes (Sect. 12.8.4).

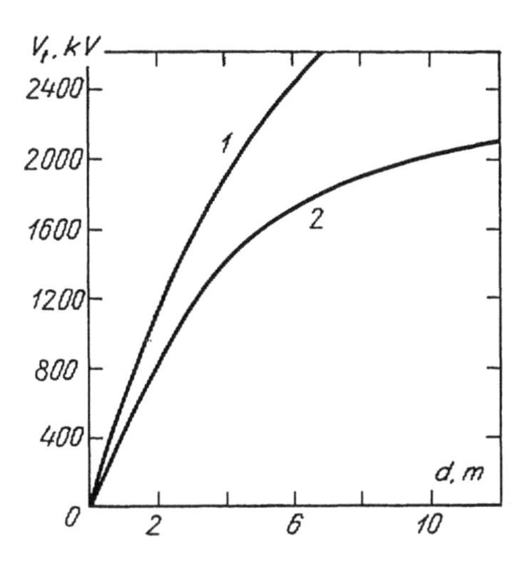

Fig. 12.14. Amplitude of breakdown voltage at $f = 50\,\mathrm{Hz}$ in an air gap of length d. *(1)* rod–rod, *(2)* rod–plane gap [12.7]

the electric charge on the rod is greater at the same voltage in the former case; furthermore, the field at the tip and in most of the gap is higher.

From the standpoint of field distribution, the rod-rod gap of d at V is equivalent to a rod-plane gap of $d/2$ at $V/2$. This factor affects the values of the respective breakdown parameters in the conditions of complete symmetry produced by the oscillating field. Thus the rod-rod gap of $d = 6\,\mathrm{m}$ has $V_t = 2400\,\mathrm{kV}$, while the rod-plane gap of $d = 3\,\mathrm{m}$ has $V_t = 1200\,\mathrm{kV}$. At smaller separations this equivalence rule does not hold as strictly: $V_t = 1850\,\mathrm{kV}$ in the rod-rod gap of $d = 4\,\mathrm{m}$, and $V_t = 850\,\mathrm{kV}$ in the rod-plane gap of $d = 2\,\mathrm{m}$. Nevertheless, the deviation is not large.

12.8.2 Effect of Polarity

The breakdown threshold in a constant field depends very strongly on the polarity of the "active" electrode (Fig. 12.15; the same rod geometry). In the case of a negative rod, the breakdown voltage is roughly twice that of the positive one. This is a result of the difference between the conditions for the development of avalanches and streamers at the active electrode. The avalanches at the *rod anode* travel to it from the outside; as they come nearer, they enter the region of

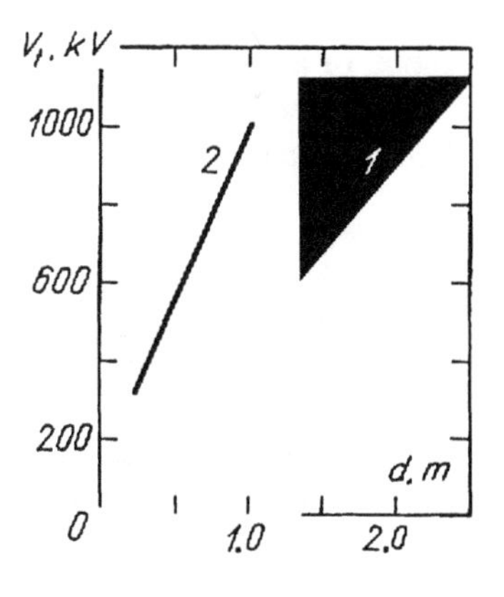

Fig. 12.15. Threshold voltages in air gaps of length d between a rod and a plane. *(1)* positive rod (anode), *(2)* negative rod [12.9]

progressively stronger field. This factor *facilitates* the multiplication of electrons and *stimulates the avalanche-streamer transition*. In the case of the *rod cathode*, multiplying avalanches move farther from the electrode into the region of progressively weaker field. The multiplication process is therefore *slowed down* and the *avalanche-streamer transition is inhibited*. Moreover, in the case of the positive rod electrons sink into the metal, leaving behind a noncompensated positive space charge, which enhances the field at the electrode. In the case of the negative rod, however, the field of the corresponding negative space charge is somewhat compensated for by the field of positive ions, all of which stay in the gas.

The data presented above are typical of that reported laboratories doing research into high-voltage equipment. This information is used in designing power transmission lines and other high-voltage structures.

12.8.3 Mean Breakdown Field Is Low: Important Implications

Although the voltages required for the breakdown of long gaps in air are enormous (millions of volts if $d \gtrsim 10\,\text{m}$), the *mean* electric field in the gap, $E_{av} = V_t/d$, is puzzlingly *low*. If $d \approx 1$–$2\,\text{m}$, then $E_{av} \approx 10\,\text{kV/cm}$ for negative and $4.5\,\text{kV/cm}$ for positive points (Fig. 12.15); at $d \approx 10\,\text{m}$ and industrial frequency, the mean amplitude is $E_{av} \approx 2\,\text{kV/cm}$ (Fig. 12.14); if $d \approx 30\,\text{m}$, it is $1\,\text{kV/cm}$. This is the level of field in the predominant part of the gap; the minimum values (at the plane) are even lower. We know, however, that the ionization of air by electron impact requires at least $E_1 \approx 24\,\text{kV/cm}$. The conclusion is unambiguous: once the ionized channel or leader in a long air gap has been created at the electrode at which the field is concentrated, it breaks through to the other electrode with no ionizing assistance from the external field.

The role played in this case by the applied voltage is not directly to sustain an intensive multiplication of electrons in any region within the gap. This would be required for the breakdown in an ideal plane gap at vanishingly small overvoltage. The important factor here is the ionization at each point in the path of the avalanche. Its every segment must contribute to enhancement so as to make it sufficient for the reproduction of secondary avalanches (involving cathode emission), or for the avalanche–streamer transition. If, however, the potential difference is nonuniformly distributed in space, it only has to create a field sufficient for intensive multiplication in the neighborhood of a single electrode. Once a *plasma channel* is initiated, it *grows*, energizing the required ionization in the gas mostly by its *own field*, or rather, by the *field of its charged tip*.

12.8.4 Effect of the Rate of Voltage Build-up

The description above *does not imply* that the voltage is determined by the field required only to *generate* the leader. As the channel grows longer, it is more and more difficult to sustain this growth, so that it is desirable to *constantly increase* the voltage in order to sustain the advancement of the leader over long distances and to make it reach the opposite electrode. Obviously, breakdown can be achieved at a constant voltage $V_{t,const}$ but this voltage needs to be higher than the final voltage at the closing of the gap by the channel in the case of a gradual increase of V. For each length d, there is an *optimal rate of voltage growth, dV/dt*. At this rate, the final (i.e. the breakdown) voltage has a minimum value. $V_{t,min}$. Breakdown is obstructed both by too slow and by too rapid a voltage increase. The optimal time of voltage build-up rate on a rod electrode above the grounded plane is found from an empirical formula $\tau_{opt} \approx 50d\,[\mathrm{m}]\,\mu\mathrm{s}$; it has been experimentally verified up to $d \approx 30\,\mathrm{m}$.

Since the regime in which the leader channel grows under gradually increasing voltage is the most favorable for breakdown, it is natural to interpret the experimental value τ_{opt} as the time scale for the leader bridging the gap. This gives us the velocity of motion of leader tip, $v_l \approx d/\tau_{opt} \approx 2 \times 10^6\,\mathrm{cm/s}$. Direct measurements give similar values (Sect. 12.9.4). For values of $d \approx 10\,\mathrm{m}$, the quantities $V_{t,min}$ are found to be 30 % lower than $V_{t,const}$. Breakdown also occurs after pulses shorter than τ_{opt}. As an example, consider the so-called "standard storm pulse" of $50\,\mu\mathrm{s}$ duration and front rise time of $1.2\,\mu\mathrm{s}$. This calls for higher voltages than $V_{t,min}$. For example, for $d = 4\,\mathrm{m}$, the voltage must be twice $V_{t,min}$ ($V_{t,min} \approx 1.1\,\mathrm{MV}$); the latter corresponds to $\tau_{opt} \approx 200\,\mu\mathrm{s}$. Roughly speaking, in this case it is necessary to have the leader traverse the gap at four times the speed.

12.9 Leader Mechanism of Breakdown of Long Gaps

12.9.1 Insufficiency of the Streamer Process for the Realization of Breakdown

In Sect. 12.5 we discussed the simplest scheme of *streamer breakdown*, such as appeared realistic in the early days of spark theory. A lone streamer grows from the anode; having reached the cathode, it triggers the propagation of a return wave of intense ionization; the result is the formation of the spark channel. Gradually it became clear that something of this sort *is* realized; but only if the gap is *not too long*, the *degree of nonuniformity of the field is not too high*, and *attachment is absent* (Sect. 12.7.7). The point is that the conductivity in the streamer channel is not high enough effectively to transfer the anode potential towards the cathode, in order that a strong field and the subsequent return wave of intense ionization can be generated. Moreover, as the streamer advances, its head enters the region of progressively weaker fields, thereby setting the limit to further advance. These processes become much more severe in electronegative gases (Sect. 12.7).

In a strongly nonuniform air gap (in fact, it is virtually unrealistic to hope to create uniform field in very long gaps), where the conditions for streamer growth are unfavourable, the streamers *stop without reaching the opposite electrode*. This is clearly seen in the photographs of streamer corona at a point (Fig. 12.12) but the same occurs in breakdown. The streamer length depends on the applied voltage and on a number of other conditions. It may be 10–100 cm; streamers in electropositive gases are longer, up to 1 m.

We know that streamers become longer as the voltage increases; hence, we would hypothesize that the breakdown of the entire gap occurs when the field $E_{av} = V_t/d$ grows to reach E_s, so that the streamer is given the possibility to reach the electrode it was traveling towards. In fact, long gaps undergo breakdown at much lower voltages and mean fields: if d = 10–30 m, E_{av} = 2–1 kV/cm (Sect. 12.8.3), while E_s = 4.7 kV/cm (Sect. 12.7.4). Nature prefers a *different way out;* it is seen in high-speed photographs of long laboratory sparks and lightning (historically, the work began with lightning).

12.9.2 Growth of Leader Channel

A thin, highly ionized, highly conductive channel grows from the active electrode (from the strong field region) along the path prepared by the preceding streamers; this channel transfers the electrode potential to its tip to a much greater extend than streamers do. This channel is known as a *leader*. The leader channel "extends" the tip of the electrode and carries it, at a high potential relative to the opposite electrode, toward this electrode. The leader head, like a metallic tip, is a source of exceptionally strong field and thus sends out streamers that fan out and prepare the initial (far from low) electron density. Electrons ionize the gas intensively in the strong field of the leader head, thereby creating a "new" head and thus supporting the advance of the strongly ionized channel (Fig. 12.16). When the channel closes the gap, the return wave starts from the second electrode (the plane), which triggers the transformation of the leader channel into the spark.

Here we more or less repeat the description given in Sects. 12.3, 12.5 of the streamer breakdown mechanism. Indeed, the differences between the streamer and leader is not so much qualitative as quantitative, i.e., in the degree of ionization and in the strength of the field produced. A streamer absorbs avalanches, a leader absorbs streamers, parallelling Peter Brueghel's famous drawing "Large fishes gobbling up smaller ones". One more link is added to the already described chain of processes that precede the spark breakdown: *avalanches-streamer-return wave* is replaced by *avalanches-streamer-leader-return wave.*

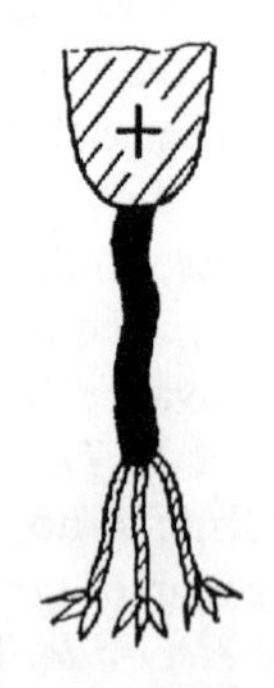

Fig. 12.16. A leader grows from a positive tip along the path prepared by streamers which, in their turn, absorb avalanches

12.9.3 On the Mechanism of Leader Formation

The foremost condition of leader formation in air is an increase in gas temperature, at least to the extent necessary to suppress a decrease in conductivity owing to electron attachment. When streamers start at a positively charged tip (*streamer corona*), they normally branch off a single *stem*. It has been suggested that the total electron current of all the streamers converging at this channel deposits enough Joule heat to raise the temperature of the gas to a level sufficient for thermal ionization [12.24]. The gas is thereby strongly ionized, electrons sink into the anode, and a large positive charge is left behind in the channel; this charge producing a strong field. The cycle is then repeated, with the streamer corona starting now not at the anode, but at the head of the new segment of the ionized channel. Thus starts the growth of the leader. However, evaluations made to support this hypothesis [12.9] indicate that the streamer current heats the air in the channel only to 3000 K. This is certainly insufficient for the onset of thermal ionization of air: about 8000 K would be needed. Spectroscopic data also indicate that the temperature in the stem of the streamer head is not high enough.

A mechanism of low-temperature transition from streamers to leader in air was suggested in [12.25]. This model also postulated the Joule heating of air in the stem of the streamer corona, but the mechanism of conductivity growth is the intense liberation of electrons from earlier-formed negative ions. To destroy O_2^- ions in dry air, a temperature $T \approx 1500\,K$ is sufficient; detachment takes about 10^{-7} s. A slightly higher temperature, up to 2000 K, is required for appreciable detachment in humid air.

In moist air, hydrated ions $O_2^-\ [H_2O]_n$ ($n = 1, 2, 3$) are formed. In these ions, the electron binding energy I_n^- increases with n while the bonding energy of the H_2O molecule, $E_{n\,H_2O}$ decreases. For example, $I_3^- = 2.65\,eV$, while for O_2^-, $I_0^- = 0.44\,eV$; $E_{3\,H_2O} = 0.23\,eV$. Hydrated ions are progressively decomposed by successive separation of H_2O by successive molecular impacts, after which the electron is lost. The time required for detachment is 10^{-6}–10^{-5} s at $T \approx 1500$–$2000\,K$, and is much greater at lower temperatures. We may recall in this connection the observations described in Sect. 12.7.4 concerning the experimentally observed sharp rise in conductivity of the streamer channel in response to "artificial" heating of the air. The rate of temperature increase with current also increases. At low temperatures, Joule heat is stored in molecular vibration; as T increases, the energy passes more and more rapidly to translational (and rotational) degrees of freedom. Thus the vibrational relaxation time at nearly normal air humidity of $0.8 \cdot 10^{-5}\,g/cm^3$ and $T = 300\,K$ is $\tau_{VT} \approx 60 \times 10^{-5}$ s; at 1000 K, it is 8×10^{-5} s, and at 2300 K, it is 10^{-5} s.

An increase in the conductivity due to an increase in temperature will result in the narrowing of current flow to a thin channel, that is, in something like the contraction of Sect. 9.11.

12.9.4 Experimental Data on Leader Parameters

Information on the leader can be extracted by deciphering the data obtained from high-speed photography using optoelectronic equipment, concurrent electrical measurements in the external circuit (current and voltage), by measuring the field in the channel using auxiliary electrodes or by other methods, performing spectroscopic measurements, and so forth [12.21, 12.26].

The following figures give an idea of leader parameters [12.21]. The mean velocity of leader propagation in an air gap of $d = 10\,\text{m}$ between a conical anode and a plane cathode, with voltage rising to $V = 1.6\,\text{MV}$ during $t \approx 2 \cdot 10^{-4}\,\text{s}$, is $v_l \approx 2 \times 10^6\,\text{cm/s}$. During the first $10^{-4}\,\text{s}$, the leader channel expands at a radial velocity of $v_r \approx 10^4\,\text{cm/s}$; by $t \approx 10^{-4}\,\text{s}$, the expansion slows down considerably to $v_r \approx 2 \times 10^2\,\text{cm/s}$ and the channel radius is $r \approx 0.2\,\text{cm}$. The current during the passage of the leader across the gap is $i_l \approx 1\,\text{A}$ (for $t \approx 5 \times 10^{-4}\,\text{s}$), and the power released is $P \approx 1\,\text{MW}$. The mean field in the channel falls as the leader propagates; when half of the gap has been crossed, it levels off (by estimate) at about $E \approx 1\,\text{kV/cm}$. These figures imply that the channel conductivity is $\sigma \approx 10^{-2}\,(\text{Ohm}\,\text{cm})^{-1}$. Calculations show the vibrational relaxation to take $10^{-4}\,\text{s}$ and a temperature $T \approx 5000\,\text{K}$ to be established in the gas; at the same time, $T_e \approx 2 \times 10^4\,\text{K}$. The density of electrons, estimated from σ and the gas density at this temperature and $p = 1\,\text{atm.}$, is $n_e \approx 10^{13}\,\text{cm}^{-3}$.

This estimate of n_e is comparable with the results of calculations and of measurement data analysis for attachment in the streamer channel close to its head, where not all the electrons have yet been removed. Using the above data, we can estimate the linear charge of the leader channel at the tip, $q = i/v_l \approx 5 \times 10^{-7}\,\text{C/cm}$. In a channel of length $l \approx d/2$, the voltage drop is $El \approx 500\,\text{kV}$; at $V = 1.6\,\text{MV}$, the potential of the leader head is $\varphi \approx 1\,\text{MV}$. Hence, the linear capacitance is $C \approx q/\varphi \approx 0.5\,\text{pF/cm}$. Other measurements give $C \approx 0.25\,\text{pF/cm}$ [12.27]. The linear capacitance of a leader is about five times that of a single streamer. This finds an explanation in the relatively thick "sheath" surrounding the high-conductivity thin leader channel; the sheath is formed of weakly conducting plasma and the space charge injected into it by streamers.

Numerous streamers fan out forwards from the leader tip. The streamer corona in front of the leader head (Fig. 12.16) in a gap of $d = 10\,\text{m}$ extends to a distance about 1 m. When these streamers reach the cathode plane, the advance of the leader is accelerated. This stage is called the *final jump*. At this stage the leader current increases jumpwise to $i \sim 100\,\text{A}$, the ionization in the channel increases considerably, and the field drops to $E \sim 100\,\text{V/cm}$. The transition to the final jump, which may start very early in not too long gaps, reliably leads to breakdown. Before the final jump the breakdown may be prevented by lowering the external voltage. Sometimes, as the leader head approaches the electrode, the greatly enhanced field between them starts a counterleader from the electrode.

12.9.5 What Contributes to the Leader Current

A leader fuses the currents of numerous streamers that start from the tip, into a single channel. Only the *joint efforts* of streamers can heat the gas to the required temperature. Short, newborn streamers which still retain ahigh conductivity contribute via their conduction current, and long streamers which have lost electrical contact with the leader owing to attachment contribute via displacement current (as described in Sect. 12.2.3). The long streamers are much more numerous because of their longer lifetimes. Before the final jump, their heads travel at the streamer velocity $v_s \gg v_l$ over a distance $l \sim 1$ m to a spot where the field of the leader tip falls below the level $E_s \approx 4.7$ kV/cm (Sect. 7.4). There the streamers slow down and for some time travel at the leader velocity. This is because the leader carries forward the point where $E = E_s$ at the velocity v_l.

If a long streamer carries a charge Q_s (independent measurements give $Q_s \approx 5 \times 10^{-10}$ C in air), the entire leader current $i_l \approx 1$ A can be generated solely at the expense of the displacement current only if streamers appear at a frequency $\nu_b \approx i_l/Q_s \approx 2 \times 10^9\,\mathrm{s}^{-1}$ [12.27]. If the deceleration time over a length $l \sim 1$ m is $\tau_{dec} \sim l/v_s \sim 10^{-5}$ s, the number of long streamers in the corona in front of the leader tip is $N_s \sim \nu_b/\tau_{dec} \sim 2 \times 10^4$. The loss of conductivity in a streamer is, say, 20-fold over a time $\tau_{cond} \sim 3\nu_a^{-1} \sim 3 \times 10^{-7}$ s, that is, the number of short (still conducting) streamers is $N_{s,cond} \sim \nu_b \tau_{cond} \sim 10^3$. If the current due to conducting streamers produces a predominant contribution to i_l, then the frequency ν_b needs redefinition; nevertheless, independent experiments [12.18] support the estimate given above.

The behavior during the final jump is different, a fact used in [12.28] to explain the observed rise in current. Roughly speaking, now each longer streamer hits the cathode without slowing down, at a velocity v_s, and brings a charge Q_s. While the leader current i_l before the final jump is proportional to v_l, during the final jump $i_l' \sim v_s$. Hence, $i_l'/i_l \approx v_s/v_l \approx 10^2$. For details, see [12.28].

12.9.6 Why Voltage Has to Be Raised During Leader Progress in Order to Facilitate Its Advance

As in the case of a streamer, the leader velocity and current are determined by the potential φ_l of its tip. A rough empirical formula is known: $v_l = 1.2 \times 10^6 \times (\varphi_l\ [\mathrm{MV}])^{1/2}$ cm/s [12.27]; if $\varphi_l < 200$–300 kV, no leader can form in a long gap: $i_l = C\varphi_l v_l$, where the linear capacitance of a channel C is a quantity less dependent on parameters (Sects. 12.9.4, 12.7.5). If V is the instantaneous voltage between the electrodes and E_l is the average field in the leader channel of length l, then $V = \varphi_l + E_l l$. If the potential φ_l is insufficient, the current is very low and the field E_l is high as a result of poor conductivity; hence an exceedingly high voltage V is needed to sustain the progress of a leader. If φ_l is too high, an excessively high voltage is again required despite the smallness of the drop $E_l l$. For each length l, there is an optimal potential φ_l at which V is minimal. This is a fairly stable quantity, approximately equal to 1 MV. It is to sustain the optimal regime at a moderate advance velocity $v_l \sim 1.5 \times 10^6$ cm/s and current $i_l \approx 1$ A (before a final jump) that the voltage needs to be continuously increased as the leader increases in length.

12.9.7 What Is the Main Difference Between a Leader and a Streamer? Why Does an Intermediate Streamer Stage Appear in the Leader Process?

These are questions of principal importance and we will complete the discussion of the streamer and leader mechanisms by trying to give as clear answers to these questions as possible. The leader and the streamer are formations *of the same type*. Both are plasma channels which grow in a self-sustaining manner in an external field which is too low for ionization. The electron density and radius of the channel heads are comparable, differing by one to two orders of magnitude at the most. The principal difference lies in the *tendencies* of the plasma. The streamer plasma tends to lose conductivity because of attachment, especially in air, while the leader plasma does not. A streamer is thus insufficient for the breakdown in a long gap. The difference in tendencies is connected with the gas temperature at the channel head, and this temperature depends on the current i and is determined by the power released per unit length, $W = iE$ (compare with formulas (10.15), (10.33)). The elevated temperature in the leader head suppresses the attachment in air. This opens the way to a further increase in temperature and ionization (as in the contraction of the glow discharge).

When the current is sufficiently high to heat the gas, avalanches cannot directly be transformed into a leader channel, since this process is preceded by the streamer-avalanche transition. Once the number of electrons reaches a certain, not extremely high, level and the avalanche radius reaches a very low value $r_0 \sim 10^{-2}$ cm, the avalanche immediately transforms into a weak-current streamer of this radius, with low power per unit length. Additional growth of the avalanche is needed for the "avalanche-leader" transition; it is prevented since the avalanche-streamer transitions *stops the growth.* As a result, a sufficiently high-current leader channel can be formed only by the merging of streamers.

We also mention an unusual theory of leader propagation [12.29] in which the ionization of the gas around the leader head is assumed to originate in the heating of the gas by a shock wave. The latter is produced by the release of the Joule heat in the leader head. This mechanism is similar to that of light detonation, supported by high intensity laser beam (see [12.6]). In spite of the attractiveness of the theory [12.29] its quantitative arguments do not seem quite convincing. On *negative* leaders, see Sect. 12.12.

12.10 Return Wave (Return Stroke)

When the leader head reaches the electrode, its charge is immediately neutralized and the channel head acquires the potential of the electrode. If this is the cathode, electrons are knocked out of it by photons or the strong field and a cathode spot is formed; if it is the anode, the electrons of the leader head sink into the metal. A *return wave* is then sent through the leader channel towards the point, this is a wave of decay of potential and of neutralization of the linear charge of the leader channel and of its surroundings (the sheath); the neutralization is not necessarily complete. The physical mechanism of the return stroke is best understood in terms of discharging a charged long line connected to ground. This interpretation has been used in numerical calculations [12.26].

Let an electric charge, q per cm, be distributed over a long wire. The capacitance, inductance, and ohmic resistance per cm are denoted below by C, L, and R. The potential U (with respect to "ground") and current i, treated as functions of time t and coordinate x along the wire, are described by the equations (in absolute system of units)

$$-\frac{\partial U}{\partial x} = \frac{L}{c^2}\frac{\partial i}{\partial t} + Ri\,, \quad -\frac{\partial i}{\partial x} = C\frac{\partial U}{\partial t}\,, \tag{12.22}$$

which have a fairly lucid physical meaning (see any textbook on electricity). If the conductor is ideal ($R = 0$), (12.22) reduces to the wave equation

$$\partial^2 U/\partial t^2 - v^2 \partial^2 U/\partial x^2 = 0\,, \tag{12.23}$$

where the wave velocity is $v = c/\sqrt{LC}$.

If a uniformly charged line with $q_0 = CU_0 = \text{const}$ and $i_0 = 0$ is connected at the initial instant $t = 0$ to ground at $U = 0$, a wave of potential decay and line discharge travels from the ground contact at a velocity v (Fig. 12.17). The current behind the wave is $i = -U_0/\varrho = -q_0 v$, where $\varrho = c^{-1}\sqrt{L/C}$ is the so-called wave resistance (characteristic impedance). In this idealized scheme, the potential at the wave front changes abruptly from U_0 to 0, and the field is "infinite". For a wire of radius r in vacuum, surrounded at a large distance H by a grounded cylinder, $L = C^{-1} = 2\ln(H/r)$ in absolute values and v equals the speed of light c.

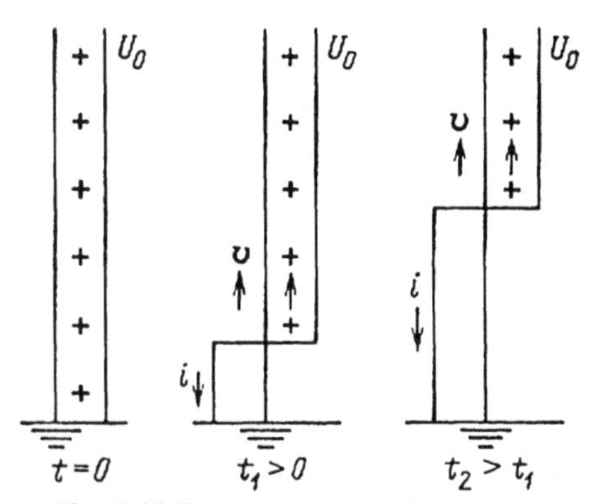

Fig. 12.17. Discharge of an ideally conducting charged line to ground; the diagram models the return wave after a positive leader arrives at the cathode. $t = 0$, the moment the line connects to ground; t_1 and t_2, two stages of wave propagation; diagrams of current i and potential V are shown

In reality, when a conductor possesses active resistance and a charge in a sheath around the leader channel, the wave front has a finite width, the field in it is finite, and the velocity $v \sim 0.1c$ (this follows from experiments and evaluations [12.26]). It is this field at the potential "jump" that imparts high energy to the electrons. In macroscopic terms, a large amount of Joule heat $W = iE$ W/cm is released at the wave front, where both the current and the field are high. It is this *wave of high energy release* rushing through the channel that results in the formation of the *spark channel*, as described in Sect. 12.5. Some quantitative characteristics will be given later with regard to the lightning discharge (Sect. 12.11.5).

12.11 Lightning

12.11.1 Thunderclouds

The cause of electric discharges in the atmosphere is the formation and spatial separation of positive and negative charges, resulting in an electric field. If the field reaches the value sufficient for the breakdown of air, the discharge occurs. In almost 90% of cases, the negative charge is found at the bottom part of the cloud and the main positive charge is at the top. For the cloud shown in Fig. 12.18, the potential difference V between the centers of the positive and negative charges q, and between the lower part of the cloud and the ground, is on the order of $V \sim q/L \sim 10^8$ V (q = 40 C = 3.6×10^{13} V/cm, $L \approx 2.5$ km), the mean electric fields being on the order of $E_{av} \sim V/L \sim 300$ V/cm. Fields of this order of magnitude are also measured on the ground under heavy clouds, and inside clouds (on aircraft).

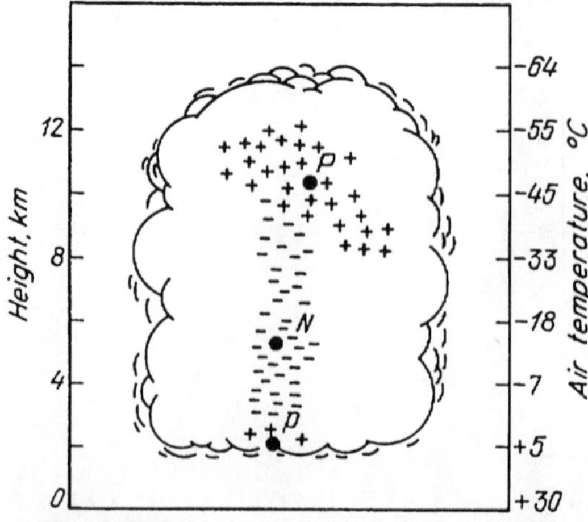

Fig. 12.18. Probable charge distribution of a thundercloud. Black dots mark centroids of charge clouds. According to measurements of electric fields around clouds, P = +40 C above, p = +10 C below, and N = −40 C. [12.30]

The current concepts of the origin of atmospheric electricity were largely developed in 1940s by *YaI. Frenkel* [12.31]. Charges appear in the atmosphere as a result of ionization of molecules and the knocking out of electrons from macroscopic particles by cosmic rays. Electrification of water droplets when they break into fragments during their fall is also possible, the separating fine droplets carrying away mostly negative charge. Negative ions generated by the ionization of air tend to stick to water droplets. Polar molecules of water in the surface layer of a droplet become aligned with their positive ends oriented inwards and negative ends pointing outwards, thus forming an electrical double layer. The field formed inside the layer points outwards, towards the air. There is no field beyond the layer. Hence, the potential jump $\Delta\varphi$ takes place within the layer, the value of φ inside the droplet being higher than in the surrounding medium; measurements give $\Delta\varphi = 0.26$ V. A droplet absorbs negative ions until the potential difference between liquid and air vanishes. The number of absorbed charges, N, is given by the equality $eN/r = \Delta\varphi$, where r is the droplet radius. If $r = 10^{-3}$ cm, $N \approx 2 \times 10^3$. The separation of charges on the scale of a cloud is caused by the downward drift of negatively charged droplets due to gravity; most positively charged particles retain molecular form and do not move downwards.

In general, lightning discharges inside clouds are more frequent than discharges between the lower part of clouds and the ground. For obvious reasons, however, experimental study has concentrated on the latter species, the one to be discussed in the sections below.

12.11.2 Experimental Techniques

A camera for high-speed filming of lightning was constructed by C.V. Boys in 1900. The film was placed under a disk with two objective lenses installed on the disk diameter on two sides of the center. The disk revolved, producing two photoscans of the process. In 1929 Boys designed a camera in which fixed objective lenses projected an image onto a moving film via prisms. The film was clamped to the inner surface of a revolving drum.[7] The first photoscans of lightning (sometimes called *Boys' patterns*) were successfully obtained only in 1928. Current notions of the evolution of the lightning discharge are based on the research of B.F.J. Schonland (1935–1940), who worked with redesigned Boys' cameras.

Photography is supplemented with measurements of the discharge current. This is best accomplished using a "lightning receiver," that is, by recroding the oscillogram of current on the spot receiving the lightning stroke. Simultaneously obtained, synchronized lightning photoscans and current oscillograms are especially valuable. Electric and magnetic perturbations due to the discharge are picked up by antennae on other devices, although it is far from easy to extract information on lightning current from these data. At the early stages of light-

[7] In a typical modern high-speed photorecorder, the objective lens and the film are fixed, and the beam moves along the film, being reflected from a rotating mirror.

ning research, *klydonographs* were widely used, that is, instruments which form Lichtenberg figures (Sect. 12.1.1). In the USSR, I.S. Stekolnikov and co-workers carried out much work on lightning discharges (from the 1930s). Their success mostly stemmed from the careful combination of lightning studies and the analysis of long sparks produced in the laboratory [12.26]. On methods and equipment, see also [12.32].

12.11.3 Sequence of Events

On the average, a lightning flash lasts for 200 ms. It consists of several *pulses* each of some 10 ms separated by about 40 ms. Each pulse begins with the growth of a *leader channel* from the cloud to the ground (Fig. 12.19). With the exception of its head, the channel is rather dark. The leader transports negative charge (from the negatively charged cloud); the current at this stage is about 100 A. As the leader approaches ground, its channel begins to branch out, the branches developing in a zig-zag manner. When the main leader reaches the ground, or collides with the counter-leader (Sect. 12.9.4), a highly luminous channel (return wave) starts to travel on its path towards the cloud at an enormous velocity, 0.1–0.3 of the speed of light. This phenomenon is called the *return stroke* or the *main stage* of the lightning discharge. The lightning current then reaches its maximum of about 100 kA. The most dangerous effects of the lightning stroke are connected with this peak current (e.g., overvoltage surges in transmission lines). Then the negative charge of the cloud is drained to the ground through the spark channel within 40 ms, by a low current (about 200 A). The outflow of electric charge from the large volume of a cloud about 1 km in size is possible only because electrons are released from negatively charged macroscopic particles and ions by strong electric fields.

In response to heat release in the channel, the pressure rises, sending out a *shock wave*. The acoustic wave to which the shock wave decays at a large distance is perceived as *thunder*. The energy released in one lightning discharge is $qV/2 \sim 10^9$–10^{10} J, corresponding to the explosion of about a tonne of explosive.

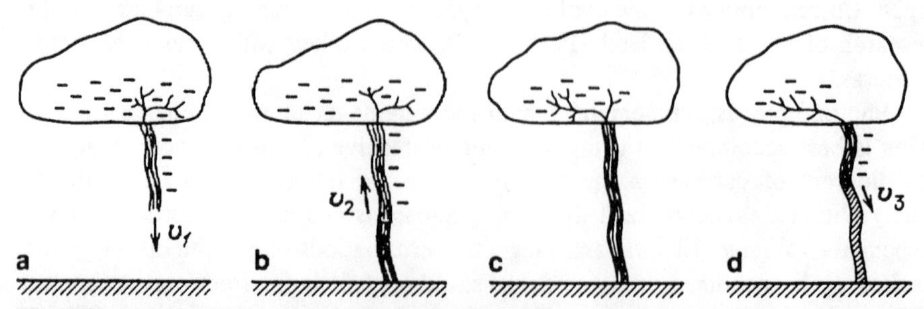

Fig. 12.19. Lightning discharge: **(a)** primary (step) leader travels to the ground at a velocity v_1; **(b)** return stroke wave travels upward at a velocity v_2; **(c)** intracloud breakdown occurs from the return stroke channel to the left-hand part of the cloud. The charge of the right-hand part has been drained through the spark channel; **(d)** secondary leader (dart leader) travels at a velocity v_3 through the partly decayed plasma of the spark channel, and so forth

According to spectroscopic data, the temperature in the spark channel is about 25,000 K and the electron density is $(1-5) \times 10^{17}\,\mathrm{cm}^{-3}$ [12.32], which on the whole corresponds to total single ionization of atoms.

One lightning pulse transfers to the ground only a fraction of the cloud charge contained in the neighborhood of the location reached by the tip of the return stroke channel. The supply of charges from distant locations to this region takes time, so that discharge to the ground quietens down for a while. Charges from the farther parts of the cloud arrive to the channel tip by way of *intracloud* breakdowns; once the paths to the distant regions have been traced, the next lightning pulse is triggered. The conductivity of the first-stroke spark channel has decreased by this time, so that a new leader propagates through the residual channel from the cloud to the ground: it increases ionization in the old channel. The second (and subsequent) leaders do not branch. When the second leader reaches the ground, the second return stroke occurs, and the process is repeated several more until the entire negative charge of more and more distant cloud regions is transferred to the ground. The positive charge located very far away (very high in the cloud) seems to remain there, since the available voltage is too low for the breakdown across such a large distance.

12.11.4 Dissimilarity of the First and Subsequent Leaders

Boy's patterns show (Fig. 12.20) that the motion of the first negative leader towards the ground is *intermittent*. A bright streak (step) lights up on the photograph, corresponding to a length of 50 m; this is repeated regularly. The impression is that the leader moves forward by violent dashes separated by pauses. This leader is known as a *stepped leader*. The average velocity of advancement of the leader's front towards the ground, calculated from the slope of the envelope of bright streaks on photoscans is $(1-2) \times 10^7$ cm/s. Judging by photographs, the leader radius is definitely much greater than that of the truly highly conduc-

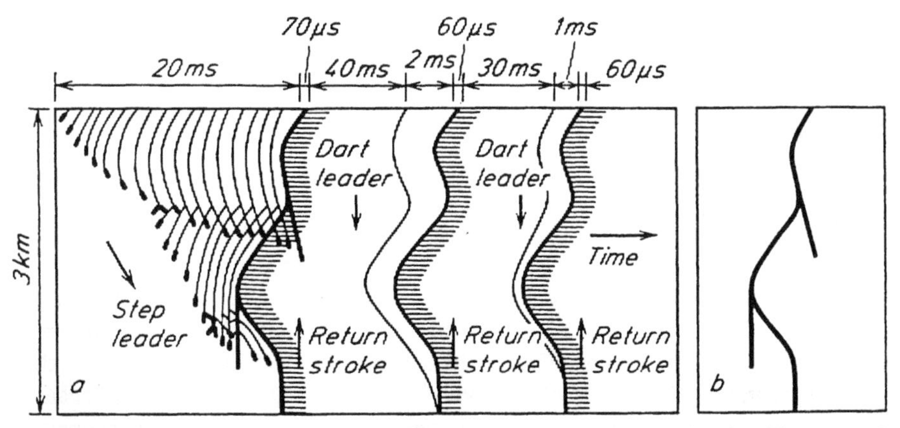

Fig. 12.20. Boys' pattern obtained by photographing a lightning flash on moving film. Time proceeds from left to right; for the sake of convenience, the time scale is nonlinear. (a) time scan, (b) the same flash, photographed without scannning [12.32, 33]

Table 12.2. Main parameters of lightning from a negatively charged cloud to the ground[a]

	Lowest	Typical	Highest
1. *Flash*			
Duration [s]	0.01	0.2	2
Number of pulses	1	3–4	26
Interval between pulses [ms]	3	40	100
total charge transferred [C]	3	25	90
2. *Step leader*			
Step length [m]	3	50	200
Pause between steps [μs]	30	50	125
Average velocity [cm/s]	1×10^7	1.5×10^7	2.6×10^8
Charge transferred [C]	3	5	20
3. *Dart leader*			
Velocity [cm/s]	1×10^8	2×10^8	2.1×10^9
Charge transferred [C]	0.2	1	6
4. *Return stroke*			
Channel length [km]	2	5	14
Velocity [cm/s]	2×10^9	5×10^9	1.4×10^{10}
Peak current [kA]		10–20	110
Peak duration at half height [μs]	10	40	250
Charge transferred, except of continuous current [C]	0.2	2.5	20

[a] Presented systematically in [12.32] using the data of numerous measurements

tive zone. The second – and subsequent – stroke leaders propagate continuously, without steps and at a much higher velocity, 10^8–10^9 cm/s. Certain similarities of the image on Boy's patterns with darts led to the term *dart leader*.

Time intervals between subsequent stages of discharge can be estimated from Fig. 12.20, and their characteristic parameters can be found in Table 12.2. The leader current is $i \sim q_l v/d$ [see (12.6)], where q_l is the charge transferred, v is its propagation velocity and d is the length on the order of cloud height. If $q_l = 5$ C, $d = 3$ km, $v = 1.5 \times 10^7$ cm/s, then $i \approx 250$ A. The leader charge per unit length is $q_1 = q_l/d \sim 10^{-5}$ C/cm; $i \sim q_1 v$. It appears that the field in the leader channel is of the order of 100 V/cm.

12.11.5 Return Stroke

This process was discussed in Sect. 12.10. The wave propagating upwards from the ground is followed by a high current that carries to the ground the charge from the leader channel and its surroundings. The potential behind the wave front approaches that of the ground. The potential in front of the wave stays close to the leader potential, and hence to cloud potential, at practically zero current because the leader current vanishes, together with the termination of leader advancement. Judging by photoscans, the return wave of the lightning discharge rushes upward at a velocity $v \approx (0.1$–$0.3)c \approx 3 \times 10^4$–$10^5$ km/s, that is, it lasts for about 10^2 μs.

The transitional wave front region in which the potential drop occurs and the field is strong, and which is correspondingly highly luminous, is shown by the scan to be $l \approx 25$–110 m long. Hence, the field there is $E \sim V/l \sim 10$ kV/cm. This is a very high field for a strongly ionized plasma: the field in arc channels in air is $E \sim 10$ V/cm. If the return stroke current is $i = 20$ kA, the energy released in the front region is $W = iE = 2 \times 10^8$ W/cm (that in the arc is $W \sim 10^3$ W/cm). Clearly, the plasma channel containing this tremendous heat release rapidly expands and forms a well-developed spark channel. Energy is spent mostly on heating, dissociation, and thermal ionization of the surrounding layers of air, not on any significant rise in temperature or ionization of the very zone of heat release where the gas has already been totally single ionized.

12.12 Negative Stepped Leader

A careful look at a Boys' pattern shows that bright lines (steps) do not just follow one another on the way to the ground, but overlap: each begins with a "recoil," starting from the midpoint of the preceding one. This behavior is at odds with a simple interpretation of the intermittent picture: a rapidly moving luminous object which draws a streak on the film but then stops for while. The question about the meaning carried by the term "stepped" is tacitly avoided in the monographs [12.2, 12.3, 12.5, 12.32]. Beginning with Schonland, various hypotheses were proposed concerning the step mechanism, but all arguments and estimates were extremely "fuzzy" and vague; it is highly unlikely, therefore, that it is by accident that their description of these contributions in a very solid and detailed monograph by *M.A. Uman* in [12.32] appears to be devoid of any physical meaning.[8]

Another difficulty seems to be the fact that the experimental analysis of such a remote, random, and unwieldy object as lightning does not provide adequate information for the profound understanding of the details of the extremely complex spark discharge phenomena. Indeed, although laboratory studies of several-meter-long sparks emitted from negative points (in which leaders also reveal 3–5 steps of $d = 6$ m [12.26]) have not yet resulted in clear physical interpretations, their better resolution yielded at least a much more detailed sequence of processes to be modelled.

The following picture emerges from the analysis of electro-optical scans and simultaneous oscillograms of the leader current (both obtained with very high resolution) [12.26], see Fig. 12.21. In the case of a positive point, or of its "extension," that is, a positive leader (both attract avalanches), streamers are formed most easily at the tips; in contrast, the avalanche–streamer transition in the case of a negative point (repulsing the avalanches) can take place only at some distance from the head of a point or leader. A brightly luminous plasma "blob" of about

[8] Uman, speaking about lightning theory, emphasized the meaninglessness of using the terms "pilot leader" and "streamer," which only replace the physical interpretation of phenomena.

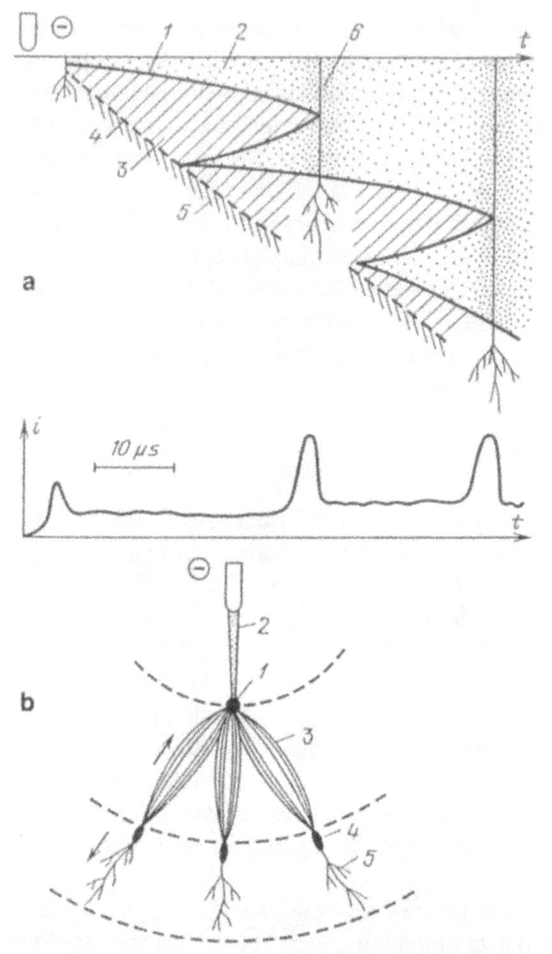

Fig. 12.21. Evolution of a negative step leader constructed using by interpretation of the photoscan of the process. The scan (a) is compared to the leader current trace (b); shown below is a spatial diagram giving a kind of "instantaneous photograph" of the process at some moment of time. Numbers on the diagram mark the same elements that they mark on photoscan (a). *(1)* head of the leader channel, *(2)* leader channel, *(3)* cathode-directed streamers starting from a plasma blob (hypothetical leader propagating in both directions), *(4)* plasma "blob", *(5)* anode-directed streamers, *(6)* a step of the leader [12.26]

$1\,\mathrm{cm}^3$ in volume appears at some distance from the tip of the negative leader. It sends out streamers (cathode-directed to the leader tip and anode-directed to the plane electrode). It can be hypothesized that a secondary positive leader propagates from the plasma towards the main negative leader, as a result of the polarization of the plasma blob by the external field and owing to the enhancement of the field between the positive charge of the dipole and the leader tip. When the leaders collide, the entire path from the point to the opposite end of the secondary leader becomes highly conducting; this leads to the tip potential building up in the channel and in a current surge which heats up the channel and produces the flash (Fig. 12.21). The head of the main leader jumps to the former plasma blob, thus producing a step. The model is certainly not quite conclusive, especially because the origin of the plasma blob remains unclear. There can be no doubt, though, that understanding will come only through such carefully designed laboratory experiments.

The understanding of the physics of long sparks, which is one of the most complicated fields in gas discharges, does more than just satisfy our inquisitiveness. It helps to solve practical problems in high-voltage equipment and lightning protection. It is no accident that an international European association of scientists, Les Renardieres Group, was created to conduct research in this field (a report on its recent work appears in [12.34]).

13. Capacitively Coupled Radio-Frequency Discharge

13.1 Drift Oscillations of Electron Gas

13.1.1 Introductory Remarks

In a capacitively coupled rf (ccrf) discharge, an rf voltage is applied to (typically) plane-parallel electrodes (Fig. 7.20). The electrodes may be in contact with the discharge, or they may be insulated from it by solid dielectrics. In this sense, we refer to discharges as *electrode* or *electrodeless*, respectively. Our presentation will mostly presume the "industrial" frequency $f \approx 13.6\,\mathrm{MHz}$[1] and medium pressures 10–100 Torr that are currently of considerable interest, partly in view of the application of ccrf discharges to high-power lasers (Chap. 14). As a rule, the plasma of such discharges is weakly ionized, nonequilibrium, and resembles the glow discharge plasma. In order to understand better the results of observations and measurements (Sect. 13.4), it is desirable to have a clear picture of some general features of ccrf discharge processes. They are given in a qualitative scheme (Sect. 13.1.2) and in a simplified *electrical-engineering-type* theory (Sect. 13.2).

The electron collision frequency ν_m at a pressure $p \sim 10\,\mathrm{Torr}$ exceeds the value $\omega = 2\pi f = 0.85 \times 10^8\,\mathrm{s}^{-1}$ by a factor of about 10^3, so that electrons in an oscillating field $E = E_a \sin \omega t$ perform *drift oscillations* (Sect. 3.1.3) with velocity and displacement amplitudes

$$v_{da} = \mu_e E_a\,, \quad A = \frac{v_{da}}{\omega} = \frac{\mu_e E_a}{\omega} = \frac{\mu_e p}{\omega}\frac{E_a}{p}\,. \tag{13.1}$$

For $E_a/p \sim 10\,\mathrm{V/cm\,Torr}$, which is typical of a nonequilibrium weakly ionized plasma of molecular gases, and for the value of ω given above, the oscillation amplitude is $A \approx 0.1\,\mathrm{cm}$. It is small in comparison with discharge gap lengths L typical in experiments, $L \sim 1$–$10\,\mathrm{cm}$. Drift velocities and oscillation amplitudes of ions are less by a factor of $\mu_e/\mu_+ \sim 10^2$, so that in many cases the oscillatory motion of the ions can be ignored. Even for a very low density of electrons, $n_e = 10^8\,\mathrm{cm}^{-3}$, and for the electron temperature $T = 1\,\mathrm{eV}$ characteristic of a collisional plasma, the Debye radius $d_D = 0.05\,\mathrm{cm} \ll L$. Therefore, the plasma

[1] The corresponding wavelength $\lambda = 22\,\mathrm{m}$ lies in the short-wave range. High-power rf units generate noise, interfering with radio and communications. To avoid this, several narrow frequency intervals were assigned to this work by international agreements (including the above frequency).

is *electrically neutral* over most of the discharge gap. However, the electron gas at the boundaries of the plane gap "swings" with respect to the relatively "fixed" ions and thus periodically bares the positive charges, just like receding surf leaves behind bare sand. This behavior is the reason for the appearance of *space-charge layers at the electrodes*. The same term will be applied below to the layers formed at *dielectrics* that insulate electrodes from the plasma.

13.1.2 Distribution of Space Charge, Field, and Potential in Plane Gaps

Assume the electrodes to be bare. Those electrons whose equilibrium positions were spaced from the electrode by a distance less than the oscillation amplitude sink into the metal "for good" [13.1]. Layers of noncompensated ionic charge remain on both sides in the equilibrium state. On the whole, the gas is *positively charged.*[2] During subsequent swings, the electron gas merely "touches" the electrodes (we neglect the slow diffusion process). Figure 13.1 shows schematically the pattern of *swings* of the electron gas under the assumption that ions are absolutely unmoving and are uniformly distributed over the gap length, while electrons perform no diffusional motion. Actually diffusion does blur the boundary between plasma and ionic layers. Figure 13.2 plots field and potential distributions corresponding to Fig. 13.1, for the same times, each quarter of the period. The field E in the uniform electroneutral part of the gap is constant over its length. To be more specific, we measure the potential from the left-hand electrode. We can assume it to be grounded, with the oscillating voltage being applied to the right-hand electrode. It can be assumed that the magnitude and direction of electric current are described by the vector $\boldsymbol{E}$, because the *conduction current* in the plasma itself is usually greater than the *displacement current.*

Various experiments have demonstrated that electrons incident on the surface of dielectrics (glass, etc.) get *attached* to the surface. The dielectric then becomes charged more and more negatively until the field produced by this charge

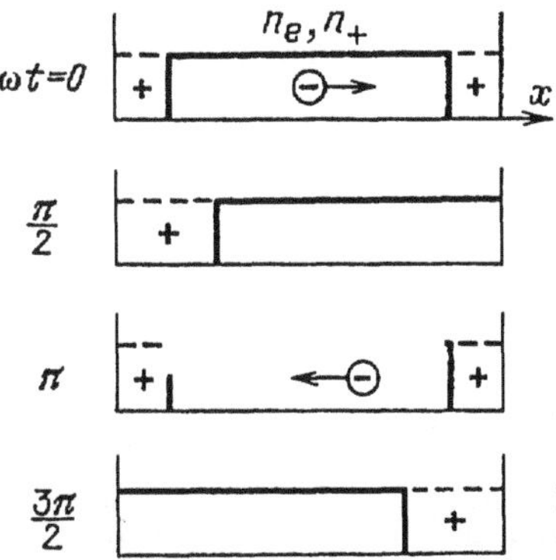

Fig. 13.1. Swings of the electron gas. The gas of ions is assumed to be fixed and uniform. Time is measured from the moment when electrons pass through the equilibrium position when moving to the right. Distributions $n_e(x)$ are shown for every quarter of the period

[2] Note that the gas in ordinary glow discharges is also charged positively as a whole, owing to the existence of the cathode layer (the negative charge of the anode layer is much smaller).

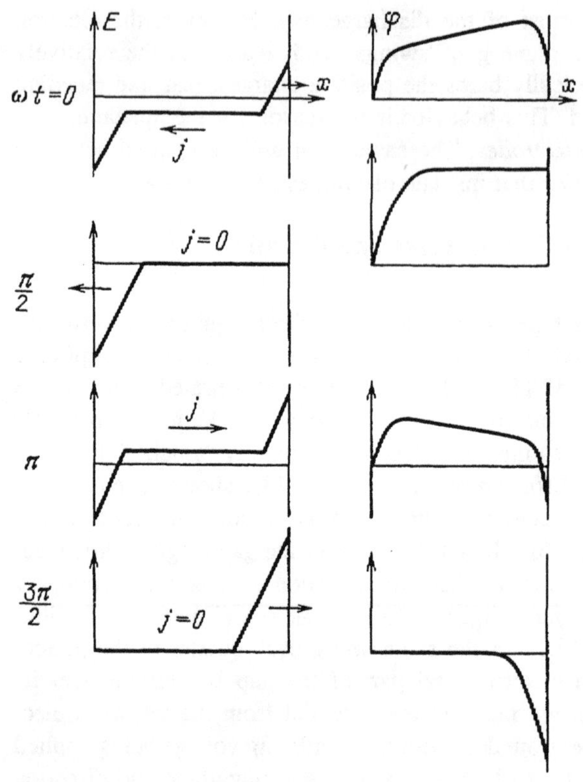

Fig. 13.2. Field and potential distributions in the gap corresponding to Fig. 13.1

repels all new electrons (Sect. 8.1.7). This also takes place in ccrf discharges with insulated electrodes, the surface of the insulator being capable of accepting all the electrons that would be contained in the electrode layer. In other words, Fig. 13.1, 2 holds equally well for ccrf discharges with insulated electrodes. Although this effect has not been directly observed, some indirect experimental confirmation is known (Sect. 13.6.4). In fact, experiments do not show appreciable differences between electrode and electrodeless ccrf discharges.[3]

[3] If a constant (d.c.) breakdown voltage is applied to such a system with insulated electrodes, breakdown occurs in the gas. After this, electrons from the gap attach to the dielectric that covers the positive electrode. The field of this charge is directed in the gap toward the applied field and thus partly cancels the latter. The field in the gap decreases to a level insufficient for sustaining the discharge, or even drops to zero, so that the process stops. Experiments show that after this the *dielectric cell* can withstand the above-breakdown voltage without repeated breakdowns and without sustaining an appreciable current.

13.2 Idealized Model of the Passage of High-Frequency Current Through a Long Plane Gap at Elevated Pressures

13.2.1 Equations of the Electrical Process in the Electrodeless Case

Consider a plane discharge gap of sufficiently large length L between solid surfaces (dielectric or metallic). Assume that the process is one-dimensional, that is, is independent of the transverse coordinates. It will soon be clear why we make the qualifications of long gap and elevated pressure. Let the electrodes be insulated from the plasma by dielectric plates, with dielectric constant ε and thickness l (Fig. 13.3a). For the plasma we assume $\varepsilon = 1$. The continuity equation for the space charge density ϱ, the equation of electrostatics, and Ohm's law for the conduction current in a plasma, j_1,

$$\frac{\partial \varrho}{\partial t} + \frac{\partial j_1}{\partial x} = 0 \,, \quad \frac{\partial E}{\partial x} = 4\pi\varrho \,, \quad j_1 = \sigma E \,. \tag{13.2}$$

imply that the sum of the conduction and displacement currents (Sect. 3.3.2),

$$j_1 + \frac{1}{4\pi}\frac{\partial E}{\partial t} = j_1 + j_{\text{dis}} = j(t) \,, \tag{13.3}$$

is independent of the coordinate x at each moment of time.

We assume the plasma to be electrically neutral throughout the entire gap, except in the thin layers at the boundaries where charges due to polarization are concentrated. The charges are regarded as *surface charges*. This is equivalent to assuming the homogeneity of plasma in space. Indeed, (13.2, 3) imply that if $\varrho = 0$, then E, j_1, σ are independent of x. Conversely, once we assume the homogeneity of the gap, we conclude that the plasma in it is electrically neutral. We denote by q_1 and q the surface charge density in the plasma and on the electrodes, respectively. The latter charges are "collected" by the applied e.m.f. from the external circuit. The dielectric layers do not let them through

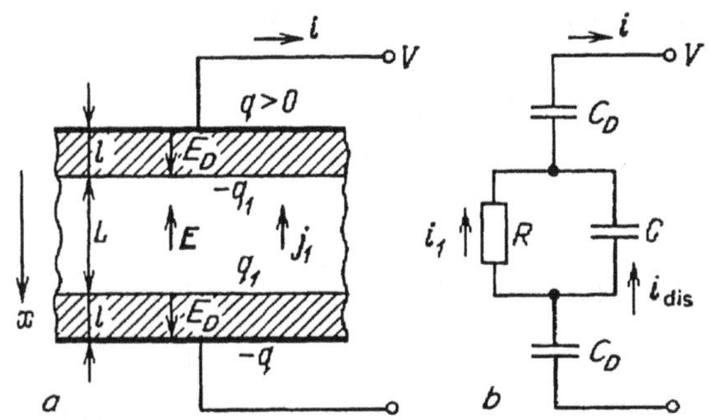

Fig. 13.3. Electrodeless ccrf discharge. (a) Section of discharge chamber. Signs of surface charges and the directions of field and current correspond to the phase in which the charge of the upper electrode is positive and decreasing in magnitude. (b) Equivalent electric circuit of the discharge device

into the plasma (see also Sect. 13.2.2). We assign directions to the field E_D in the dielectrics, and signs to the charges q and q_1 at a certain time, in accord with the instantaneous situation shown in Fig. 13.3a. In subsequent calculations, the equations automatically control the signs of the terms denoting the fields, charges, and currents.

Applying general electrostatics equations (3.16, 17) to the electrode-insulator and insulator-plasma interfaces and taking into account that the field in the metal is negligible owing to its very high conductivity, we find

$$\varepsilon E_D = 4\pi q \, , \quad E - \varepsilon E_D = -4\pi q_1 \, . \tag{13.4}$$

Applying the continuity equation to the same boundaries, we obtain

$$\dot{q}_1 \equiv dq_1/dt = j_1 \, , \quad \dot{q} = i/S \, , \tag{13.5}$$

where i is the current in the external circuit and S is the area of the electrodes. A combination of (13.4, 5) implies that the current $i(t)$ in the external circuit coincides with the total current in each dielectric and in the gap, jS [by virtue of (13.3)]; this gives the *law of conservation* of the total current of conduction and displacement along the entire electric circuit.

Potential drops across layers do not exist within the framework of the assumption of the surface nature of charges at the plasma boundaries, that is, of "zero" thickness of these space-charge layers. For the time being we put off the discussion of real layers and potential drops; note only that neglecting any phenomenon at the electrodes (and the corresponding potential differences) can be justifiable to a certain extent if these differences are small in comparison with the voltage $V_1 = EL = (E/p)(pL)$ across the homogeneous plasma, and the length of this plasma is the greater part of the total gap length. Actual potential drops averaged over one period do not exceed 200 V at medium pressures, and the thickness of nonuniform electrode layers is of the order of 1 cm (Sect. 13.4). Therefore, the analysis is not meaningless if pL is sufficiently high (recall the stipulation made at the beginning of this chapter). It fits the glow discharge theory for high pL quite well. In this case, attention can be focused on the positive column, which has its analogue in the homogeneous plasma of this section. Hereafter we refer to this plasma as "positive column."

Under the assumptions made above, the instantaneous voltage across the electrodes is

$$V = 2E_D l + EL \, . \tag{13.6}$$

Eliminating the fields from (13.6) and (13.4) and from the equation $j_1 = \sigma E$, we arrive at the system of equations

$$V = \frac{8\pi l}{\varepsilon} q + \frac{j_1 L}{\sigma} \, , \quad j_1 = 4\pi\sigma(q - q_1) \, , \quad \dot{q}_1 = j_1 \, , \quad \dot{q} = j \, , \tag{13.7}$$

which determines the current. It must be supplemented by the equation of the external circuit, that is, an analogue of (8.1) that relates V to i.[4] System (13.7) can be given an "electrical engineering" form by introducing currents $i = jS$ and $i_1 = j_1 S$ instead of densities, charges $Q = qS$, $Q_1 = q_1 S$, ohmic resistance of discharge $R = L/\sigma S$, capacitance of dielectrics $C_D = \varepsilon S/4\pi l$, and that of the plasma space $C = S/4\pi L$. We obtain

$$V = \frac{2Q}{C_D} + i_1 R\,, \quad i_1 R = \frac{Q - Q_1}{C}\,, \quad \dot{Q}_1 = i_1\,, \quad \dot{Q} = i\,. \tag{13.8}$$

The equivalent electric circuit to these equations is shown in Fig. 13.3b. The conduction current passes through the active resistance and the displacement current passes through the plasma capacitance, with $i_1 + i_{dis} = i$.

13.2.2 Equations in the Case of Bare Electrodes

If the insulating plates are absent, the capacitances C_D in Fig. 13.3b must be removed (Fig. 13.4). This gives

$$V = i_1 R\,, \quad i_1 R = \frac{Q - Q_1}{C}\,, \quad \dot{Q}_1 = i_1\,, \quad \dot{Q} = i\,. \tag{13.9}$$

The problem of current closing at the interface of the plasma and metal conductors is not considered in this approximation. The conduction currents on the different sides of the interface are unequal, $i \neq i_1$, and there is surface charge on the interface, $Q - Q_1 \neq 0$, owing to the difference between the conductivities of the plasma, σ, and the metal ('$\sigma = \infty$').

Equations (13.8, 9) describe not only rf but also all non-steady-state discharges, such as pulsed ones, provided that the formulated assumptions are not violated over their characteristic durations. For instance, the displacement of the electrons must be small in comparison with L.

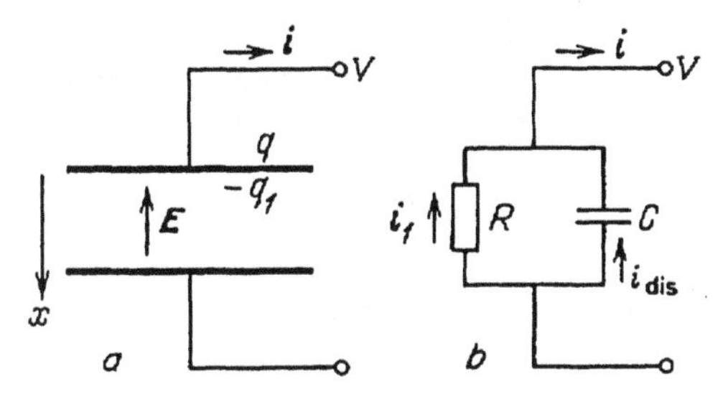

Fig. 13.4. (a) Electric circuit of the device, and (b) equivalent circuit of electrode ccrf discharge, similar to that of Fig. 13.3

[4] It depends on the method of connecting the rf oscillator to the circuit; here these details are ignored.

13.2.3 Solution for the Case of Insulated Electrodes

The plasma in the discharge gap is sustained by the field; since the field oscillates, the periods of predominant production of electrons alternate with the periods of their predominant removal. Assume that a sine voltage $V = V_a \sin \omega t$ is applied to the electrodes. Let the plasma decay to only a small extent in one half-period of oscillation. This condition holds at the postulated frequency $f \sim 10^7\,\mathrm{s}^{-1}$: the characteristic frequencies of the removal of electrons in glow discharge plasmas do not exceed $10^5\,\mathrm{s}^{-1}$. Then the *depth of modulation* of n_e, σ, and R (at a frequency $2f$) is negligible and $R(t) \approx$ const. Equations (13.8) can be integrated, regarding the plasma resistance as a constant parameter (though this is not yet known; Sect. 13.3). The results of this simple calculation will now be given. The voltage across the plasma V_1, the field in the plasma E and the conduction current i_1 are proportional to one another and are given by the formulae

$$V_1 = EL = i_1 R = \frac{V_a}{1 + 2C/C_D} \frac{\omega\tau}{\sqrt{1+\omega^2\tau^2}} \sin(\omega t + \varphi) , \qquad (13.10)$$
$$\tau = R(C_D/2 + C) , \quad \varphi = \arctan(\omega\tau) .$$

The current in the external circuit (the discharge current) is

$$i = \frac{V_a}{R(1 + 2C/C_D)} \frac{\omega\tau\sqrt{1+\omega^2\tau_\sigma^2}}{\sqrt{1+\omega^2\tau^2}} \sin(\omega t + \varphi + \Delta\varphi) , \qquad (13.11)$$
$$\tau_\sigma = RC = 1/4\pi\sigma , \quad \Delta\varphi = \arctan(\omega\tau_\sigma) .$$

The amplitude of the voltage across the plasma, V_1, is the lower in comparison with the applied voltage V, the smaller the capacitance of the insulators (the lower $\omega\tau$). The insulators then carry the correspondingly greater part of the voltage. The voltage V_1 is shifted in phase with respect to V because only the displacement current passes through the dielectric layers. For this same reason, and owing to the displacement current in the plasma, the discharge current i is phase-shifted relative to V. The time τ characterizes the *inertial behavior* of the entire capacitor device in Fig. 13.3. If it is charged and then its electrodes are shorted to discharge the system, the charge relaxes, obeying the law $Q = Q(0)\exp(-t/\tau)$, provided the plasma has not yet decayed ($R =$ const).

The time τ_σ characterizes the rate of *dispersal of space charges* in the plasma itself (Sect. 9.2.2). The quantity $\omega\tau_\sigma = \omega/4\pi\sigma$ that we already know from Sect. 3.5.6 determines the ratio of amplitudes of the displacement and conduction currents in the plasma:

$$i_{dis,a} = \frac{i_a \omega\tau_\sigma}{\sqrt{1+\omega^2\tau_\sigma^2}} , \quad i_{1a} = \frac{i_a}{\sqrt{1+\omega^2\tau_\sigma^2}} , \quad \frac{i_{dis,a}}{i_{1a}} = \omega\tau_\sigma . \qquad (13.12)$$

These currents are shifted in phase by $\pi/2$, so that not the amplitudes but their squares are added: $i_a^2 = i_{1a} + i_{dis,a}^2$.

13.2.4 The Version with Bare Electrodes

By analogy with the above analysis, we find from (13.9) that

$$V_1 = EL = i_1 R = V = V_a \sin \omega t \ , \tag{13.13}$$

$$i = (V_a/R)\sqrt{1+\omega^2\tau_\sigma^2} \sin(\omega t + \Delta\varphi) \, . \tag{13.14}$$

The phase shift $\Delta\varphi$ between the discharge current i and the field E in the plasma (or the conduction current i_1 in it) is the same as with insulated electrodes; it is given by (13.11). Formulas (13.12) remain valid. However, now there is no difference between the voltage on plasma V_1 and that across the electrodes V. Indeed, we have neglected the possible insulation of electrodes by the electrode layers which were assumed to have zero thickness.

13.3 $V - i$ Characteristic of Homogeneous Positive Columns

The solution of Sect. 13.2 demonstrates the field and current arising in a homogeneous extended plasma with a specified conductivity if rf voltage is applied to the electrodes. Another side of the question is what value of conductivity is produced in the discharge gap in a given field or what field is required to sustain a given electron density. The answers depend on the electronic processes in the plasma, and the resulting relation between n_e and E, or j and E, is the $V - i$ characteristic of the positive discharge column. If other conditions are known, this characteristic can be transformed into the $V - i$ curve of the discharge device as a whole. The conditions for the sustainment of nonequilibrium homogeneous plasma in an rf discharge do not differ, in principle, from those for the positive column of the glow discharge (Sect. 8.6): ionization must compensate for the loss of electrons.

13.3.1 Frequency of Ionization by an RF Field

The ionization frequency is determined by the electron energy spectrum. The spectrum depends on the field frequency. The rate of its buildup is characterized by the electron energy loss frequency $\nu_u = \nu_m\delta$ (Sect. 2.3.7). If $\omega \gg \nu_m\delta$, the spectrum responds too slowly to oscillations of the rf field. As we have explained in Sect. 5.5.2, the frequency of ionization in the rf field, $\nu_{i,rf}$ does not differ from the ionization frequency $\nu_i(E)$ in a constant field E whose magnitude equals the effective value E_{eff} of the oscillating field. Formula (5.33) gives E_{eff} in terms of E_a, ω, and ν_m. In the frequency range where $\nu_m \gg \omega \gg \nu_m\delta$, E_{eff} is equal to the root-mean-square field $E_a/\sqrt{2}$.

In the opposite limit of $\omega \ll \nu_m\delta$, the spectrum "follows" the field oscillations, constantly adjusting itself to the instantaneous value of $E(t)$, while the instantaneous frequency of lionization is equal to the ionization frequency $\nu_i(E)$ in a constant field of the same magnitude. hence, we simply have to average $\nu_i[E(t)]$ in time:

$$\nu_{i,rf} = \frac{\omega}{2\pi}\int_0^{2\pi/\omega} \nu_i\left(E_a|\sin\omega t|\right) dt = \frac{2}{\pi}\int_0^{\pi/2} \nu_i(E_a \sin\varphi) d\varphi\,. \qquad (13.15)$$

Since $\nu_i(E)$ is a very steep function, ionization occurs by short surges when the field reaches its maximum value. For the rest of the period, electron losses predominate. The situation in molecular gases (N_2, air, $CO_2 + N_2 + He$ laser mixture) at $p \gtrsim 30$ Torr is close to the second limiting case. The electron losses are characterized by the effective recombination coefficient β_{eff} (Sect. 8.8). If the ionization coefficient $\alpha(E)$ is given in the Townsend form (4.5), the expansion of $1/\sin\varphi$ in the neighborhood of $\varphi = \pi/2$ yields a formula useful for calculations:

$$\nu_{i,rf} = (2E_a/\pi Bp)^{1/2}\nu_i(E_a)\,,\quad \nu_i(E) \propto \exp(-Bp/E)\,; \qquad (13.15')$$

it expresses $\nu_{i,rf}$ in terms of the ionization frequency in a constant field equal to the amplitude of the rf field. The coefficient multiplying $\nu_i(E)$ characterizes the effective fraction of time during which ionization occurs. Formula (13.15′) is more accurate, the smaller this fraction.

13.3.2 Example of Calculation of the $V - i$ Characteristic

Consider the positive column of a discharge at the frequency $f = 13.6$ MHz in the laser mixture $CO_2 : N_2 : He = 1 : 6 : 12$ at $p = 30$ Torr, which corresponds to conditions in high-power CO_2-lasers [13.2, 13.3] (Sects. 14.4, 5). In this plasma, $\beta_{eff} \approx 4 \times 10^{-6}\,cm^3/s$. To maintain the electron density n_e, the ionization frequency required is $\nu_{i,rf} = \beta_{eff} n_e$; for instance, $\nu_{i,rf} = 4 \times 10^4\,s^{-1}$ for $n_e = 10^{10}\,cm^3$. This quantity, characterizing also the rate of plasma decay, is so small in comparison with f that the oscillations of n_e around the mean value are negligible. In the mixture specified here, $\delta \approx 0.8 \times 10^{-2}$ and the rate of build-up of the electron spectrum, $\nu_m\delta \approx 6.4 \times 10^8\,s^{-1}$, is almost an order of magnitude greater than $\omega = 0.85 \times 10^8\,s^{-1}$. We can, therefore, consider the spectrum to oscillate with the field, and $\nu_{i,rf}$ to be given by (13.15). Figure 13.5 shows the dependence of E_a/p on n_e, that is, a sort of "$V - i$ characteristic" of the positive column. For comparison, the dashed curve plots the same dependence found under the

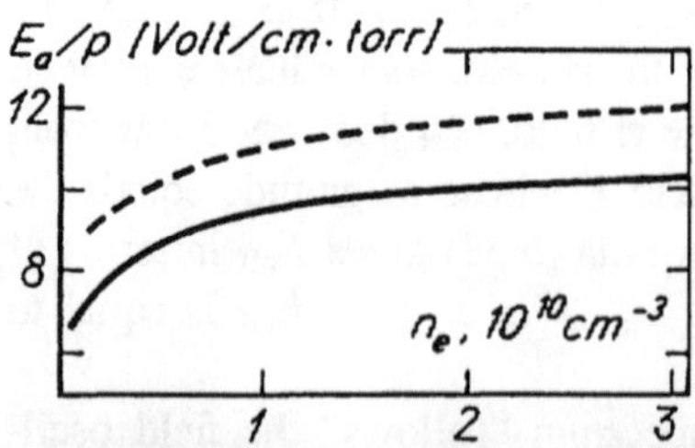

Fig. 13.5. Calculated dependence of field amplitude on electron density in the positive-column plasma of ccrf discharge (equivalent of the $V - i$ curve of the column) with laser gas mixture $CO_2 : N_2 : He = 1 : 6 : 12$, $p = 30$ Torr, $\beta_{eff} = 4 \times 10^{-6}\,cm^3/s$. *Solid curve*, electron spectrum tracks field oscillations (the low-frequency limit). *Dashed curve*, the spectrum is too slow to oscillate; it corresponds to a constant field equal to the rms value. The true dependence is close to the solid curve at the frequency 13.6 MHz [13.4]

assumption that the electron spectrum undergoes no oscillations and corresponds to the constant field $E = E_a/\sqrt{2}$, as we saw in the case of $\nu_m \gg \omega \gg \nu_m \delta$. If the inequality $\nu_m^2 \gg \omega^2$ is definitely satisfied, the true curve lies between those in Fig. 13.5, but closer to the solid curve, in accord with the estimate $\omega \ll \nu_m \delta$. As we see, the interval between the limiting versions is not great. This is natural because the ionizing field is $E_a/\sqrt{2} = 0.7 E_a$ in one limit, and slightly less than E_a in the other.

13.3.3 Example of Calculations of the Electric Parameters of an Electrodeless Discharge

To illustrate the "electrical engineering" model of Sect. 13.2, we perform the calculation for the same gas at the same p and f as in Sect. 13.3.2, for the geometric parameters of the setup of [13.3], (see Sect. 14.4.5): $L = 5.5$ cm, $l = 1$ cm, $\varepsilon = 5$. Let us concentrate on the case of $n_e = 10^{10}\,\mathrm{cm}^{-3}$. If $\nu_m = 0.8 \times 10^{11}\,\mathrm{s}^{-1}$, then the conductivity is $\sigma = 3.5 \times 10^{-5}\,\mathrm{Ohm}^{-1}\mathrm{cm}^{-1}$. The curves of Fig. 13.5 show that plasma with this conductivity is sustained by $E_a/p = 9.4$ V/cm Torr. The plasma parameter $\omega\tau_\sigma = 0.23$, $j_{1a}/j_{\mathrm{dis},a} = 4.3$ and $j_{1a} = 0.95 j_a$. The plasma is therefore dominated by the conduction current; its current density amplitude is $j_a \approx \sigma E_a \approx 10\,\mathrm{mA/cm}^2$. The energy input into the plasma is $\sigma E_a^2/2 = 1.4\,\mathrm{W/cm}^3$.

The voltage across the column is $V_1 = 1.55$ kV. The ratio of the capacitances is $C_D/2C = 14$ and the parameter $\omega\tau = 3.6$. The amplitude V_{1a} of the voltage across the plasma is 0.9 of the applied voltage amplitude V_a, so that $V_a \approx 1.75$ kV. The phase shift φ between E, V_1, and V is $\varphi = 74.5°$, and that between the current in the external circuit and the voltage (i and V) is $\varphi + \Delta\varphi = 87.5°$.

The above calculation is based on the assumption that the positive column (homogeneous plasma free of the influence of electrode effects) is the main element of plasma as far as the voltage drop and the extension along the field are concerned. This is true for large values of pL. Experimenters frequently work with moderately long discharges and moderate pressures; the potential drops at the electrodes are then substantial, and the positive column does not occupy the predominant part of the entire gap length. On some calculations concerning cases of this sort, see Sect. 13.8.

13.4 Two Forms of CCRF Discharge Realization and Constant Positive Potential of Space: Experiment

It has been experimentally established that ccrf discharges are possible in *two greatly different forms*. The visual differences lie in the glow distribution along the gap, the essential difference consists in the processes in the electrodes. S.M. Levitsky was the first to point this out in 1957. His papers [13.1, 13.5] were important in the evolution of the current concepts and understanding of the nature of ccrf discharge at moderate pressures.

13.4.1 Jumps on Ignition Potential Curves

Experiments [13.5] were devoted to discharges in hydrogen and argon in a glass vessel with parallel bare disk electrodes 4.2 cm in diameter placed in the vessel. The interelectrode spacing L was varied up to 10 cm; the frequency of the 100 W rf oscillator was $f \approx$ 1–70 MHz, and the pressure was $p \sim 10^{-2}$–10^{1} Torr. Discharge ignition potentials V_t were measured at different f, p, L, most attention being paid to the near-jump region of the ignition curves $V_t(p)$ at given ω and L.

The jumps that had already been discovered in the experiments of the 1930s (Sect. 7.7.2) appear at combinations of parameters at which the range of the electrons' oscillations at the moment of breakdown $[2A = (\mu_e p)V_{ta}/\omega pL]$ becomes equal to the interelectrode distance. If ω and L are fixed, then at higher pressures $2A < L$, so that electrons oscillating from the mid-plane of the gap do not reach the electrodes. Their losses are therefore lower (only diffusional) and V_t decreases with respect to the case of lower p at which $2A > L$ and drifting electrons strike the electrodes. This effect is well pronounced in H_2 but smeared in Ar, presumably owing to stronger diffusion. Figure 13.6 plots the ignition characteristic (lower solid curve) which clearly shows the jump.

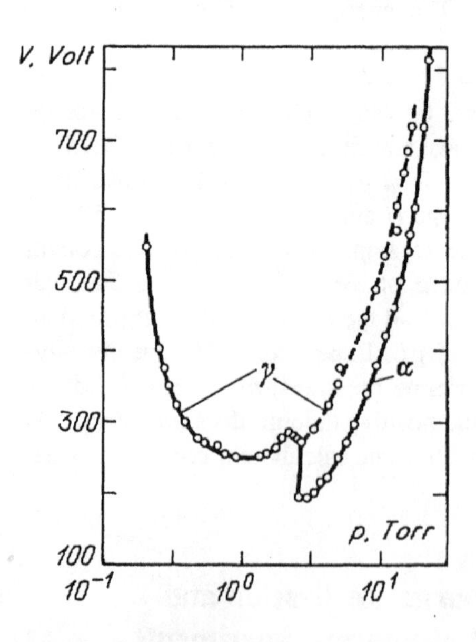

Fig. 13.6. Ignition potential of an electrode ccrf discharge as a function of pressure with hydrogen, f = 3 MHz, L = 2.64 cm. Left-hand branch up to the jump is the γ discharge; the right-hand branch (*solid curve*) is the α discharge. *Dashed curve* plots the secondary ignition potential (transition of the α to γ mode) [13.5]

13.4.2 α and γ Discharges

Observations have shown that discharges ignited at pressures to the "left" and "right" of the jump are not identical. If the discharge is on the right, a diffuse glow appears in the middle of the gap, while the gas at the electrodes is dark. The voltage on the electrodes changes very little upon ignition, indicating weak conductivity of the ionized gas and a low discharge current. In discharges to the left of the jump, an intensive glow is localized at the electrodes and consists of alternating layers whose color and sequence resembles those of the layers in the cathode region of a dc glow discharges. The electrode voltage appreciably decreases after ignition, indicating a considerable current is the discharge.

Levitsky gave the following interpretation to these features: the current in the neighborhood of the electrodes in the weak-conductivity discharge is mostly capacitively coupled and remains as *displacement current*, as before the ignition. Hence, the ignition of the discharge does not affect the behavior of electrodes that continue neuther to emit nor to absorb charges. In high-conductivity discharges, to the left of the jump, an ionic current flows into the electrode that is negative at a given moment, *secondary emission* takes place, and a *cathode layer* is formed at the "cathode" for some time, as in glow discharge, until the polarity is reversed. Now the circuit from the middle of the gap to the electrodes (which are alternating as cathodes) is closed by *conduction currents*. Levitsky referred to discharges with nonconducting layers as "α" and to discharges with conducting layers as "γ" discharges, γ symbolizing the role of secondary emision (γ processes). When the voltage applied to a burning α discharge is increased, it suddenly transforms to the γ form, like a *secondary ignition*. The $V - i$ curve of secondary ignition extends that of the γ discharge to the left of the jump. It is shown by the dotted curve in Fig. 13.6.

13.4.3 Constant Spatial Potential in CCRF Low-Pressure Discharge

It was noticed as early as the 1930s that a low-pressure ccrf discharge leads to the *sputtering* of electrodes, so that a layer of metal is deposited on the walls of a glass tube. The only explanation was the existence of a *constant field*, directed from the midpoint to the two electrodes and accelerating positive ions to high energies. These ions strike the electrodes and cause the effect known in dc glow discharge as *cathode sputtering*. Alternating fields, even of high amplitudes, are inefficient in this respect, because the energies of the oscillatory motion of ions are too low. Probe methods detected a *constant potential positive* with respect to the electrodes in the middle of the discharge gap. Figure 13.7 shows the results of the probe measurements in the setup already described in Sect. 13.4.1. The

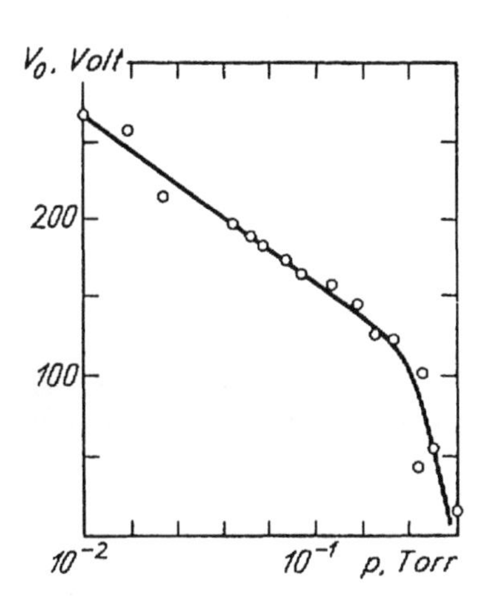

Fig. 13.7. Probe measurement results of constant plasma potential as a function of pressure. Hydrogen, f = 68 MHz, L = 10 cm [13.1]

constant potential V_0 for $f = 68$ MHz, $L = 10$ cm, $p = 3.8 \times 10^{-2}$ Torr increases almost linearly with increasing electrode voltage and is less than its amplitude V_a by a factor of about 2. Measured energy spectra revealed that the mean energy of ions is comparable with V_0, and the maximum energy, with V_a.

Levitsky's idea was that constant positive potential in a ccrf discharge is produced by the buildup of an excess of ions in the gap as a result of the absorption of electrons by the electrodes (Sect. 13.1). Using the model of oscillations of the electron gas illustrated in Fig. 13.1, he was able to evaluate the constant potential V_0 (Sect. 13.6.2). On the whole, these arguments correctly identified some key points. Each of the factors taken into consideration (the removal of a part of electrons from the gap, and the oscillations of the electron gas) may in itself generate a constant positive potential of the plasma. The first mechanism is obvious. If a body as a whole is charged positively, it possesses a positive potential. But *even if the gas is electrically neutral* (as it would be if the electrons did not attach to insulators in the electrodeless discharge), it is nevertheless at a positive potenital with respect to the boundaries in the sense of a time average (Sect. 13.6.4). Indeed, the positive charge layer is essentially thicker than the layer of the equal negative charge, because electrons are forced by the field to "huddle up" to the insulator, and the voltage drop in the negative layer is smaller.

The presented model is clearly valid only for the α discharge. If a cathode layer, just like that in ordinary glow discharge, is formed in the neighborhood of the temporary cathode in the γ discharge, then the positive potential with respect to the cathode at the layer-plasma interface must be of the order of the normal cathode fall V_n. Since no anode fall accompanies it, or this fall is small, V_0 averaged over one period must be of the order of V_n; this is indeed observed at higher pressures (Sect. 13.4.4).

Figures 13.8,9 give the results of recent measurements, carried out by the probe method in glass tube 5 to 8 cm in diameter, with disk electrodes. The potential V_0 decreases with increasing pressure, other conditions being equal. The paper does not specify the form of the discharge (α or γ), but it is worthy of note that a certain dependence of V_0 on electrode material and gas species was observed. Namely, V_0 was greater for those combinations of metal and gas

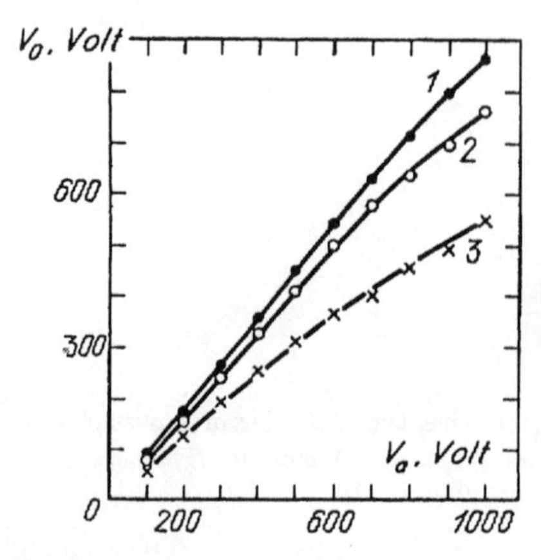

Fig. 13.8. Constant plasma potential as a function of amplitude of electrodes voltage. Discharge in neon, with titanium electrodes: $f = 6$ MHz, *(1)* $p = 0.3$ Torr, *(2)* $p = 0.5$ Torr, *(3)* $p = 1.1$ Torr [13.6]

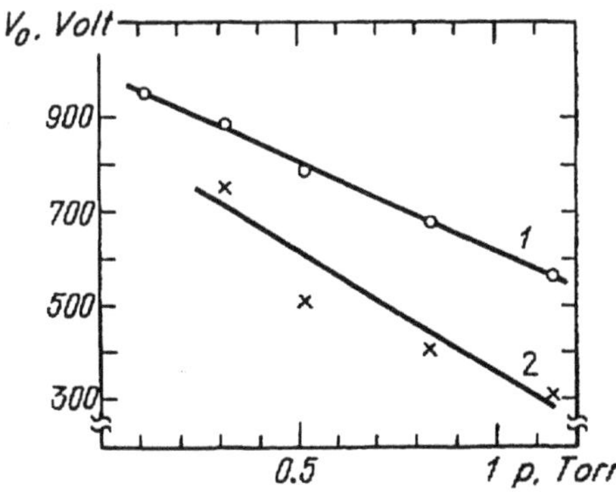

Fig. 13.9. Constant plasma potential as a function of pressure: Neon-titanium, $V_a = 1000\,V$; *(1)* $f = 6\,MHz$, *(2)* 1.5 MHz [13.6]

for which the normal cathode fall was higher, although V_0 was still far from V_n. The paper reported a certain, mostly increasing, dependence of V_0 on frequency (in the range from 1 to 12 MHz).

The positive potential does accelerate ions towards the electrodes but it also accelerates electrons towards the plasma. Indeed, beams of fast electrons with energies of 100 eV were recorded in experiments with helium at $p = 0.66$ Torr, $f \approx 1$–$15\,MHz$, $V_a \approx 390\,V$ [13.7]. This was discovered in the course of measuring the electron distribution function by probe techniques. The energy of the beam electrons was close to V_0.

13.4.4 Weak-Current and High-Current (α and γ) Medium-Pressure CCRF Discharges

The fact that two forms of ccrf discharges exist, the properties of these forms, and the characteristics of transition of one form into the other have been thoroughly studied since 1978 by N.A. Yatsenko, who succeeded in greatly clarifying the nature of these phenomena. His chief interest was focused on the medium pressure range of $p \sim 10$–100 Torr (it was largely ignored before). One of the important results obtained by Yatsenko was: a direct experimental proof of the fact that α and γ discharges differ in current density by more than an order of magnitude; and that the electrode layers of the γ discharge possess high conductivity. For these reasons, Yatsenko referred to the γ discharge as the high-current, and to the α discharge as the weak-current modes. The measurements [13.8] were made in a vessel of large volume (60 liters) with a pair of water-cooled disk-shaped brass electrodes, 10 cm in diameter. The gap between them could be increased up to 10 cm. Sometimes electrodes were insulated from the plasma by 1 to 3 mm thick layers of glass, Teflon, etc. The discharge was ignited by a 3 kW generator, mostly at $f = 13.6\,MHz$.

Figure 13.10 gives static $V-i$ characteristics of discharge: i.e., the rf voltage V on the electrodes as a function of the rf current i in the steady-state regime (rms values). In the range of the lowest voltages and currents, V remains almost constant as the current increases. Under these conditions, the discharge does not cover the entire surface area of the electrodes. Its diameter in the interelectrode

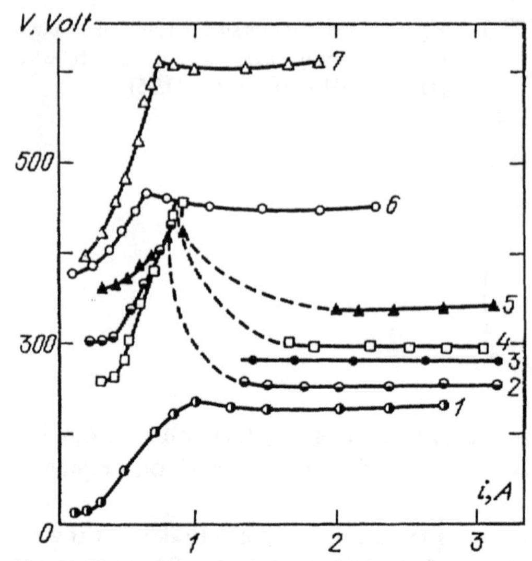

Fig. 13.10. $V-i$ characteristic of ccrf discharge at the frequency 13.6 MHz (rms values) : *(1)* helium, p = 30 Torr, L = 0.9 cm; *(2)* air, 30 Torr, 0.9 cm; *(3)* air, 30 Torr, 3 cm; *(4)* CO_2, 30 Torr, 0.9 cm; *(5)* CO_2, 15 Torr, 3 cm; *(6)* air, 7.5 Torr, 1 cm, electrodes are insulated with glass; *(7)* air, 7.5 Torr, electrodes are insulated with teflon [13.8]

gap is close to the diameter of the spot on the electrodes. The glow is concentrated at the middle of the gap, fading away at the electrodes (in space-charge layers) (Fig. 13.11, curve a). This is a typical weak-current (α) discharge with nonconducting electrode layers. Coating the electrodes with a dielectric does not change the picture.[5] When the current increases at this stage, the discharge broadens transversely and fills the electrode surface. The current density at the electrode remains constant, that is, the phenomenon of a *normal current density* exists in a weak-current discharge when it covers the electrode only partially. The current density depends on the gas species, on pressure, and – surprisingly – on the interelectrode spacing L (Fig. 13.12, Sect. 13.9.1).

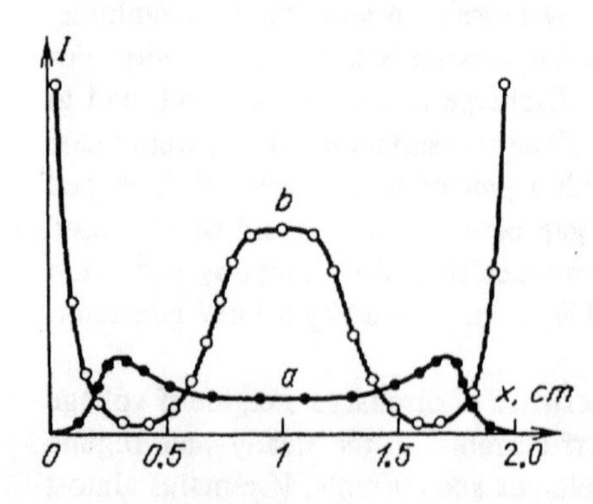

Fig. 13.11. Glow intensity distribution across the gap (in the apparatus of [13.8]): 10 Torr air, brass electrodes, L = 2 cm, (a) weak-current mode, (b) high-current mode. Rms voltage on the electrodes is close to 300 V in both cases

[5] The electrodes can be completely removed from the discharge tube: such experiments have been carried out on a number of occasions.

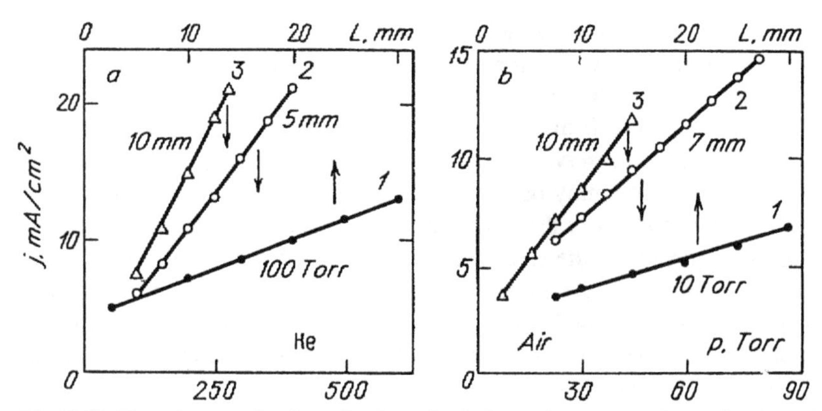

Fig. 13.12. Normal current density at the electrodes in the weak-current mode as a function of pressure and interelectrode spacing. (a) Helium, (b) air; $f = 13.6\,\text{MHz}$: *(1)* $j(L)$ at $p = 100\,\text{Torr}$ in He and at 10 Torr in air; *(2)* $j(p)$ at $L = 0.5\,\text{cm}$ in He and at $L = 0.7\,\text{cm}$ in air; *(3)* $j(p)$ at $L = 1\,\text{cm}$ [13.9]

When the electrode surface is totally covered by the current and the discharge diameter increases to the electrode diameter, further increase in current requires higher voltage (as it does in an abnormal glow discharge), although the layers here remain dark and nonconducting. Their thickness in the normal regime is $d \approx 0.2$–$0.6\,\text{cm}$. To the accuracy of a low-ionic saturation current (as in the case of a negative probe), the current is coupled to the electrode by the displacement current. The low conductivity of the discharge was checked by passing through it a dc current or (which is the same) a 50 Hz ac current.

When a sufficiently high electrode voltage is reached, the α discharge is sharply transformed into the high-current γ type. This is the secondary ignition observed in [13.5] (Sect. 13.4.2) and is indicated by jump or bend in the $V - i$ characteristic. The glow in the gap undergoes redistribution; layers resembling the *layers of a glow discharge* are formed at each electrode. The *negative glow* and the *Faraday dark space,* and also the *positive column* in the middle, are clearly pronounced. The distribution of luminance is plotted in Fig. 13.11 by the curve *b*. The total thickness of "near-cathode" regions up to the positive column in air at $p = 15\,\text{Torr}$ is 1.2 cm. The current density at the electrode increases sharply: at $p = 30\,\text{Torr}$ in air, by a factor of 20 from 1.2×10^{-2} to $0.24\,\text{A/cm}^2$ (the transition results in a transversal contraction of the discharge). The conductivity sharply increases. The resistance to dc or low-frequency current drops by 2 orders of magnitude. The constant potential V_0 across the discharge space in the high-current regime is 100 to 200 V; it is practically independent of pressure and roughly corresponds to the normal cathode fall of the glow discharge.[6] The thickness of the space-charge layer at the electrodes, d, becomes much smaller: in air at $p = 15\,\text{Torr}$, $d \approx 0.4\,\text{cm}$ in the weak-current mode and $d \approx 0.03\,\text{cm}$ in the high-current mode. These figures are obtained by probe measurements of the

[6] It is not clear, therefore, why the potential V_0 dropped abruptly in [13.5] to a negligible value of several volts at $p \sim 1\,\text{Torr}$.

distribution of constant potential $V_0(x)$ as a function of the distance x from the electrode. The layer thickness in the high-current mode is comparable with that of the cathode layer of ordinary glow discharges.

The high-current mode is also characterized by a *normal current density*; here it is similar to the effect observed in the ordinary glow discharge, but the current density in the rf discharge may be substantially larger, owing to the contribution of the *displacement current* in the "cathode" layer, where the field is very strong. As the current is increased, the electrodes fill up with the glow, and the diameter of the current spot on the electrodes and that of the plasma column (these are not very different) both grow. This stage corresponds to the last constant-voltage segment on the $V-i$ characteristic of Fig. 13.10, which corresponds to the *normal* mode of the glow discharge.

Recent measurements [13.10] make it possible to evaluate the frequency dependence of the collapse parameters of the α-discharge and its transformation into the γ-form (V_{tr}, i_{tr}). The measurements were carried out in nitrogen between two bare aluminum electrodes in the ranges of $f = 15$–55 MHz, $p \approx 10$–50 Torr, $L \approx 0.6$–1.2 cm. The $V-i$ characteristics are similar to those shown in Fig. 13.10. As the frequency increases at constant pressure, V_{tr} decreases while i_{tr} increases. Thus if $L = 0.6$ cm and $p = 35$ Torr, we have

$$\begin{aligned}
&\text{for } f = 15.4 \text{ MHz} : \quad V_{tr} = 440\,\text{V (rms)} \quad i_{tr} = 45\,\text{mA (rms)} \\
&\text{for } f = 20.4 \text{ MHz} : \quad V_{tr} = 405\,\text{V} \quad i_{tr} = 60\,\text{mA} \\
&\text{for } f = 29.25\,\text{MHz} : \quad V_{tr} = 305\,\text{V} \quad i_{tr} = 100\,\text{mA}\,.
\end{aligned}$$

As the pressure increases at constant frequency, V_{tr} increases and i_{tr} decreases. If $L = 0.6$ cm and $f = 29.25$ MHz, we have

$$\begin{aligned}
&\text{for } p = 25\,\text{Torr} : \quad V_{tr} = 270\,\text{V} \quad i_{tr} = 134\,\text{mA} \\
&\text{for } p = 55\,\text{Torr} : \quad V_{tr} = 320\,\text{V} \quad i_{tr} = 95\,\text{mA}\,.
\end{aligned}$$

In the latter case, the α-discharge occupied a whole area 1.5 cm^2 of the electrode, which corresponds to $j_{tr} \approx 60$ mA/cm^2; in the γ-mode of discharge, $j \approx 1000$ mA/cm^2.

13.4.5 The Domain of Weak-Current Discharge

Experiments demonstrate that the weak-current discharge can burn only at values of pL smaller than a certain critical value $(pL)_{cr}$ (Fig. 13.13). This value depends on the electrode material and the gas species. If $pL \approx (pL)_{cr}$, the weak-current discharge becomes unstable and either transforms into the high-current form or dies out. If $pL > (pL)_{cr}$, the former mode cannot be ignited at all, so that only the high-current discharge can be realized. Both regimes are possible if $pL < (pL)_{cr}$. Critical values in molecular gases are not large; in air, $(pL)_{cr} \approx 40$ Torr·cm. Typically, the high-current form is realized in molecular lasing mixtures CO_2 + N_2 + He at medium pressures and gap length of several centimetre. The weak-

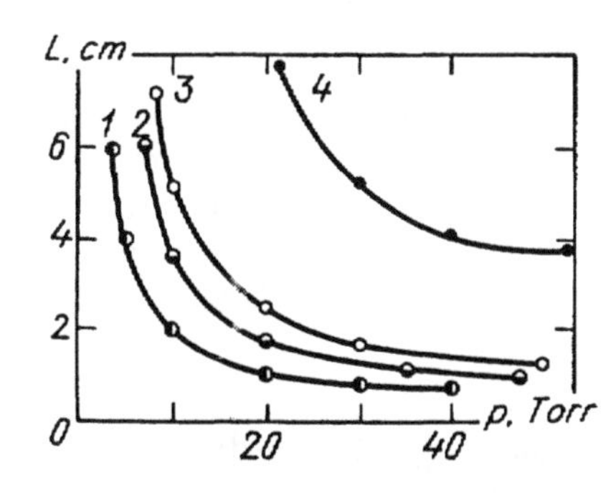

Fig. 13.13. Boundaries of domains of weak-current discharge. Only high-current mode is possible *above* and to the *right* of the curve (brass electrodes, f = 13.6 MHz): *(1)* nitrogen; *(2)* air; *(3)* CO_2; *(4)* helium [13.8]

current form transforms into the high-current form by way of the *breakdown* of the space-charge layer [13.8]. It appears that the dependence of the weak-current domain size and $(pL)_{cr}$ on frequency has never been studied experimentally.

13.4.6 Electrodeless Electrode Discharge and Electrode Electrodeless Discharges

Although this sounds like wordplay, it reflects the essentials of the process: a discharge with bare electrodes in the weak-current mode does not differ at all from a discharge with inusulated electrodes, so that it is electrodeless in the sense that the electrodes do not make their presence felt. As in the case of dielectrics, the current through the space-charge layers closes the circuit only via the displacement current. However, the "electrodeless" discharge with insulated electrodes is found to be sustainable in the high-current mode, with all the attributes of *secondary emission* and *multiplication of electrons* in the *cathode layer* (negative glow, dark Faraday space) found in the ordinary glow and high-current rf discharge with bare electrodes. This discharge does not differ from the latter one either in visual appearance or in current density. The same is indicated by Fig. 13.10 which also shows the $V-i$ characteristics of insulated-electrode cases. The voltage amplitude on the electrodes is then naturally higher than the ordinary value, because part of the voltage is developed across the dielectric layers.

When the ionic current of the dc glow discharge flows from the cathode layer into the metal, some electrons are ejected from the letter; most of them neutralize the ions at the surface, while the others create the emission electron current. The loss of electrons from the surface layer of the cathode is balanced out by their inflow from the inside, at the expense of those entering the anode. The cathode layer is charged positively; hence, the cathode of a constant-current discharge carries a constant excess electron charge. Bare electrodes of ccrf discharges also carry an excess negative charge (Sect. 13.1.2), but it changes as the cathode and anode phases alternate. When the electrodeless discharge is ignited, the electrons of the gas attach to the surface of the dielectrics, no inflow of charges occurring from the inside of the body. In the cathode phase of the high-current mode, the attached electrons neutralize the incident ions and are partly liberated by them, while in the anode phase their store is replenished.

If a capacitance equivalent to that of the dielectric insulating the plasma is introduced into the external circuit of the discharge with bare electrodes, the discharges with bare and insulated electrodes must behave equivalently provided the surfaces of the metal and of the dielectric are identical vis-a-vis *electron emission* and *neutralization of ions*. This is conclusively demonstrated by the results of numerical modeling of the structure of ccrf discharges (Sect. 13.8).

13.5 Electrical Processes in a Nonconducting Electrode Layer and the Mechanism of Closing the Circuit Current

In the approximate solution of Sect. 13.2, *space-charge layers* at the boundaries of the discharge gap were regarded as *discontinuity surfaces* on which surface charges were concentrated and the field changed abruptly. Let us look now at the nonconducting layer in the weak-current discharge as a *finite-thickness* element with a space charge and continuous field, and go into details of how the current closes the circuit between the plasma and the electrodes. From the arguments given above it is fairly clear that the closing in the α-discharge is made by the displacement current, so that it is useful to pause and consider on the physical content of this concept.

13.5.1 What Is the Displacement Current?

In trying to formulate a system of equations to describe electromagnetic phenomena (Sect. 3.3), Maxwell met with the following difficulty (this happened in the 1860s): if the fields are not constant, the equations of the magnetic field valid for dc currents and the equation of continuity for charges are incompatible. To remove the contradiction, Maxwell postulated, without any experimental justification, that the magnetic field is generated not only by the motion of charges, but also by changes in the electric field, just like the electric field is generated when the magnetic field is varied. The quantity $(1/4\pi)\partial \boldsymbol{D}/\partial t$, which Maxwell added to the conduction current, he called the *displacement current*. The phenomenon of electromagnetic induction was supplemented with a magnetoelectric analogue, while equations (3.13, 14) acquired a spectacular symmetry. Thus one of the most fundamental laws of nature, implying the existence of electromagnetic waves, was discovered in a purely speculative manner.

In fact, a connection with magnetic phenomena is not necessary for the quantity known as displacement current to appear in the theory of electricity. The purely electrical equations

$$\partial \varrho/\partial t + \operatorname{div} \boldsymbol{j} = 0\,, \quad \operatorname{div} \boldsymbol{D} = 4\pi\varrho \tag{13.16}$$

imply that

$$\operatorname{div}\left(\boldsymbol{j} + \frac{1}{4\pi}\frac{\partial \boldsymbol{D}}{\partial t}\right) = 0\,. \tag{13.17}$$

The vector under the divergence symbol in (13.17) has all the properties characterizing the dc current density which obeys the equation $\operatorname{div} \boldsymbol{j} = 0$. The vector has no sources, its lines of current are closed even if the electric circuit contains nonconducting segments (a line of current may also go to infinity, closing the circuit as if via infinity); the product of the magnitude of the vector and the cross-sectional area of the tube of current at any moment in time is identical along the entire circuit. The vector can therefore be interpreted as the *total* current density (conduction plus displacement). But this interpretation is to a great extent a matter of convention, because of the *two* components of displacement current (Sect. 3.3.2), the quantity $(1/4\pi)\partial \boldsymbol{E}/\partial t$ is not a current in the literal sense of the word, in contrast to the polarization current $\partial \boldsymbol{P}/\partial t$.

If we forget about the truly physical meaning of the quantity $(1/4\pi)\partial \boldsymbol{E}/\partial t$ as the source of *vortices of magnetic field,* it does not reflect any other physical reality, has no other physical content. However, when ac current is considered in a *quasipotential* electric field generated predominantly by charges,[7] we completely ignore the existence of a magnetic field. When dealing with gaseous conductors, that is, discharges, we can often neglect the tiny polarization current because $\varepsilon \approx 1$. Does this mean that by dealing with pure displacement current $(1/4\pi)\partial \boldsymbol{E}/\partial t$, we work with a concept devoid of physical content?

It evidently does. The mechanism of ac current in a quasipotential field is *exhaustively* interpreted in terms of the *motion of charges* and *electrostatics* using equations of type (13.16) (with Ohm's law) without resorting to a symbolic displacement current. Its extremely useful role arises because the interpretation of the integral of (13.16) or (13.17) as the *law of conservation of the total current* greatly simplifies the calculations and allows the representation of all results in concise, convenient, and elegant form. If, however, we wish to comprehend the real process, it is necessary to discern actual physical processes using phrases like "the displacement current flows through the capacitor." One has to build up a picture of how the e.m.f. of the power supply displaces the electrons of the external circuit towards one plate of this capacitor, where they stop, not being allowed any further by the poor-conductivity medium; how the departed electrons bare a layer of positive charge on the opposite side of the capacitor; how an electric field arises between charges of the opposite signs; how this field changes with time while charges are accumulated (or dispersed); how a strong current passes through "wires" at almost zero field, while there is no current in the poor-conductivity medium and at the same time the field is strong and

[7] A time-dependent electric field is never strictly potential: it inevitably contains an element of vortex field. If, however, the characteristic size L of the system is small compared with the wavelength λ, the field is approximately potential. Indeed, (3.14) implies $E_{\mathrm{vort}} \sim \omega H L/c$; (3.13) implies $H \sim 4\pi j L/c$, whence $E_{\mathrm{vort}} \sim 4\pi\omega L^2 j/c^2$. On the other hand, (3.16) implies $E_{\mathrm{pot}} \sim 4\pi \varrho L$ and $\omega\varrho \sim j/L$, whence $E_{\mathrm{pot}} \sim 4\pi j/\omega$. Therefore

$$E_{\mathrm{vort}}/E_{\mathrm{pot}} \sim (\omega L/c)^2 = (2\pi L/\lambda)^2 = (L/\bar{\lambda})^2 .$$

For example, we find for ccrf discharge at $L \sim 10\,\mathrm{cm}$ and $\bar{\lambda} = 22\,\mathrm{m}$ that $E_{\mathrm{vort}}/E_{\mathrm{pot}} \sim 10^{-3}$.

changing; both the chain of arguments and the electric circuit are closed. The example is clearly trivial. But such a detailed picture must be worked out for every situation that we wish to analyze thoroughly.

13.5.2 Field Within Layers

Let us examine quantitatively the situation shown in Figs. 13.1, 2. As a first approximation, we make use of solution (13.13, 14) of Sect. 13.2. We only have to remember that the phase is shifted in the figures relative to (13.13, 14) so that the field E in the plasma and the discharge current i are equal now to

$$E = -E_a \cos \omega t \ , \quad i = -i_a \cos(\omega t + \Delta\varphi) \tag{13.18}$$

for the same amplitude values and $\Delta\varphi$ implied by (13.11). If we neglect the diffusional spreading of the plasma-ionic layer interface,[8] the field in the layer obeys the equation $\partial E_l/\partial x = 4\pi e n_+$. For simplification, we assume, as in Fig. 13.1, that $n_+(x) = \text{const}$. In the left-hand layer, on which we fix our attention,

$$E_l(x,t) = E(t) - 4\pi e n_+ [d(t) - x] \ , \tag{13.19}$$

where $d(t)$ is the coordinate of the interface or the layer thickness. At the electrode surface,

$$E_0(t) \equiv E_l(0,t) = E(t) - 4\pi e n_+ d(t) \ . \tag{13.20}$$

The plasma interface is formed by "edge" electrons that oscillate in the field E at the drift velocity $\dot{d} = -\mu_e E$. Taking into account that $d(0) = A$, we have

$$d(t) = A(1 + \sin \omega t) \ , \quad A = \mu_e E_a/\omega \ . \tag{13.21}$$

Substituting (13.18, 21) into (13.20) and recalling that n_+ is equal to n_e in the plasma and that the conductivity of the plasma is $\sigma = e n_e \mu_e$, we find the field at the electrode:

$$E_0 = -\frac{E_a}{\omega\tau_\sigma}\sqrt{1 + \omega^2\tau_\sigma^2}\,\sin(\omega t + \Delta\varphi) - \frac{E_a}{\omega\tau_\sigma} \ . \tag{13.22}$$

If the plasma conductivity is high, $\omega\tau_\sigma \ll 1$, the field at the electrode is almost always large in comparison with the field at the plasma boundary of the layer. Both its constant component and the amplitude of the variable component equal $E_{0a} = E_a/\omega\tau_\sigma$, the phase shift of the ac components of E_0 with respect to E being equal to $\pi/2$.

[8] Diffusion effects were evaluated in [13.4].

13.5.3 Charges and Currents

The field E_0 is created by the surface charge q on the electrode, $E_0 = 4\pi q$. Its variable component is caused by the discharge current $i = S\dot{q}$ in the external circuit, and the constant component, by the charge of those electrons that went into the solid from the gas. Indeed, the constant component of charge density on the electrode is equal in magnitude to the layer charge at the moment when the electrons pass through the midpoint or, which is the same, to the time-averaged charge

$$q_0 = -E_a/4\pi\omega\tau_\sigma = -en_+A\,, \quad E_{0,\text{const}} = -4\pi en_+A\,.$$

The discharge current calculated using (13.22),

$$i = -S\sigma E_a\sqrt{1+\omega^2\tau_\sigma^2}\cos(\omega t + \Delta\varphi)\,, \quad S\sigma E_a = V_a/R\,, \tag{13.23}$$

coincides with solution (13.18, 14) found in the first approximation. If $\omega\tau_\sigma \ll 1$, the plasma conduction current $i_1 = S\sigma E$ is not very different from the discharge current i. As we see from (13.22), the positive charge on the electrode ($E > 0$) appears for a shorter part of the period, the lower the value of $\omega\tau_\sigma$.

13.5.4 Reality and Symbols

The picture described in Sect. 13.2 by the discontinuity approximation thus has not been destroyed by replacing the discontinuity with an insulating space charge layer between the plasma and electrode. At the same time, the mechanism of closing the circuit at the plasma-electrode interface, qualitatively clarified in Fig. 13.1, now becomes quantitatively understandable. For example, at the stage when the negative charge accumulates at the left-hand electrode (the current flows leftward), the charge repels the electrons of the plasma progressively more strongly. The plasma boundary defined by the edge electrons shifts away from the electrode, while locally noncompensated ionic charge remains in the layer. The total positive charge of the layer also piles up, not because of enhanced concentration but owing to new positively charged regions joining the layer. The positive charge is supplied to the layer by the conduction current from the plasma (electrons leave it): $\dot{q}_1 = en_+\dot{d} = -en_e\mu_e E = i_1/S$. Of course, no current flows within the layer, but the symbolic quantity – the displacement current – coincides with the discharge current. As follows from (13.19–23),

$$\frac{1}{4\pi}\frac{\partial E_1}{\partial t} = \frac{\dot{E}_0}{4\pi} = -\sigma E_a\sqrt{1+\omega^2\tau_\sigma^2}\cos(\omega t + \Delta\varphi) = \frac{i}{S}\,.$$

The plasma conductivity being constant, the varying negative charge q of an electrode is not given enough time to be completely neutralized by the positive charge of the layer. The noncompensated charge in the region of contact of the plasma and metal conductors, $q - q_1 = E/4\pi$, is the source of field E inside the plasma, while the difference between the accumulation rates of q and q_1,

$\dot{q} - \dot{q}_1 = \dot{E}/4\pi$, that generates the variable field component in the plasma is symbolically interpreted as the displacement current i_{dis} in the plasma itself:

$$\dot{i} - \dot{i}_1 = S\dot{E}/4\pi = i_{dis}\,, \quad i_1 + i_{dis} = i\,.$$

13.5.5 Insulated Electrodes

Figures 13.1, 2 and all relations obtained above remain valid in this case also. The constant charge q_0 now attaches to the dielectric surface facing the gas. The field in the gas at the insulator surface is $E_0 = 4\pi(q + q_0)$. Its constant (time-averaged) component $4\pi q_0$ is determined by the deposited charge, and the variable $4\pi q$ is related to the variable charge q on the electrode. According to (13.22), $|q_0| = \sigma E_a/\omega \approx j_a/\omega$. In the weak-current mode at $f \sim 10$ MHz, we have $j_a \sim 10\,\mathrm{mA/cm^2}$, whence $|q_0| \sim 10^{-10}\,\mathrm{C/cm^2}$. About 10^9 electrons attach to $1\,\mathrm{cm^2}$ of the dielectric surface. This is not much in comparison with about 10^{16} atoms/cm^2 on the surface, so that the surface is not very densely covered with the deposited electrons.

13.6 Constant Positive Potential of the Weak-Current Discharge Plasma

13.6.1 Definition

If a sinusoidal voltage is applied to the electrodes, the time-averaged potential difference between them is zero. The time-averaged potential difference between any point in the plasma and the electrode would also vanish if the field were everywhere harmonic. The field is indeed nearly sinusoidal in the electroneutral plasma and in the insulators but it is not harmonic in the electrode layer, so that the plasma acquires a constant potential V_0, with respect to the electrode, equal to the potential difference, averaged over a period, between an arbitrary point in the piasma and the electrode. The averages of potential differences over regions of harmonic field (plasma, insulators) vanish. As a result, the constant potential V_0 of the plasma with respect to the electrode, which is measured in experiments, is equal to the time-averaged voltage V_C across the near electrode layer. In the case of a symmetric (plane) gap, both electrodes are equivalent, on the average. We will now calculate the constant potential $V_0 = \langle V_C(t)\rangle$ in several situations.

13.6.2 Electrons Perform Drift Oscillations; the Gas Is Charged

This second condition signifies that electrons from boundary layers were absorbed into the metal or attached to the dielectric. The potential of the plasma boundary of the layer with respect to the electrode is

$$V_C = -\int_0^d E_l\,dx = -\frac{E_0 + E}{2}d \approx -\frac{E_0 d}{2} = \frac{E_{0a}A}{2}(1 + \sin\omega t)^2\,.$$

We have used here (13.18–22). The neglect of E in comparison with E_1 in this expression corresponds to the case of a highly-conducting plasma, $\omega\tau_\sigma \ll 1$. In this case the potential is positive nearly all the time. The constant potential calculated by averaging V_C at an arbitrary $\omega\tau_\sigma$ is

$$V_0 = \frac{3}{4}\frac{E_a}{\omega\tau_\sigma}A = 3\pi e n_e A^2 = \frac{3\pi e(\mu_e p)^2 n_e}{\omega^2}\left(\frac{E_a}{p}\right)^2 . \qquad (13.24)$$

A formula of type (13.24) was first derived by *Levitsky* [13.1], where, presumably by mistake, the value of V_0 was twice as great. On the whole, (13.24) gives the correct order of magnitude for V_0. For instance, if $n_e = 10^{10}\,\text{cm}^{-3}$ and $A = 0.1\,\text{cm}$, then $V_0 = 140\,\text{V}$. A more detailed comparison with experiments is not easy owing to the scarcity of experimental data for the conditions relevant for (13.24). Nevertheless, the calculation of V_0 using (13.24) is definitely useful for better understanding the physics of the processes and for further elaboration of the theory.

13.6.3 Effect of Inhomogeneity in Ionic Distribution

The approximation $n_+(x) = \text{const}$ is quite crude. True, it does permit the successful completion of all the calculations in a simple and clear analytic form. This in fact means a lot, because a better way of uncovering the essentials of a process does not exist. Certainly, we cannot expect the obtained formulas to be highly accurate. Actually, n_+ in weak-current layers decreases in a monotonic fashion from the plasma to the electrode or dielectric boundary, undergoing a severalfold reduction. The layers are thicker than in the $n_+ = \text{const}$ approximation, the field at the boundaries is weaker, but the voltage across the layers and the constant plasma potential are almost twice as large (Sect. 13.8.2).

13.6.4 Drift Oscillations; the Gas Is Neutral

Let us turn to the approximation $n_+(x) = \text{const}$ and assume that electrons do not attach to insulators. At the moments of equilibrium (if $E = -E_a \cos(\omega t)$, these are $\omega t = 0, \pi$) the layers are absent, and at the moments $\omega t = \pi/2,\ 3\pi/2$ electrons are forced by the field towards the surface (exactly to the surface if diffusion is absent). Formulas (13.18) for the field in the plasma, and also (13.19, 20), describing the left-hand layer in the cathode half-period when the layer is positively charged, still hold. But since now $d(0) = 0$, we have $d = A \sin \omega t$ instead of (13.21). The field at the insulator surface is determined by (13.22) without the second term because now the field E_0 has no constant component. The thickness of the negative charge layer is small during the anode half-period even though the field E_0 has the same magnitude. Consequently, the integral over the anode half-period can be neglected when the potential difference in the layer is time-averaged. As a result,

$$V_0 = \left\langle -\frac{E_0 + E}{2} d \right\rangle = \frac{V_{C,\max}}{4} = \frac{\pi}{2} e n_e A^2 . \qquad (13.25)$$

The potential V_0 in an electrically neutral gap is less by a factor of 6 than in a charged gap. Experiments do not point to large differences in V_0 for discharges with insulated and bare electrodes. This can be regarded as an indirect proof of the attachment of electrons to insulators.

13.6.5 Low Pressures

In this case the amplitude of electron oscillations (not necessarily drift oscillations) may be comparable with the interelectrode spacing, so that any extended positive column is out of the question. The layer of positive space charge may fill nearly the entire gap in the course of electron swings. A very simple estimate of V_0 is then possible. When electrons shift to the right, a poorly conducting region occupied by ions is formed between the left-hand electrode and the electrons: almost the entire applied voltage $V(t)$ falls across this region. The approximate instantaneous potential of the plasma driven to the right-hand electrode (with respect to the left-hand one) is equal precisely to this fall. The potential difference between the plasma squeezed at the left-hand electrode and this electrode becomes small in the next half-period. Hence,

$$V_0 \approx \frac{1}{2\pi}\int_0^{\pi} V_a \sin \omega t \, d(\omega t) = \frac{V_a}{\pi} \,. \qquad (13.26)$$

The constant positive potential of the plasma is approximately one-third of the amplitude of the applied voltage. This result follows as a limiting case from a more detailed theory of the collisionless rf discharge constructed in [13.11].[9] it is compared with experimental data in Fig. 13.14.

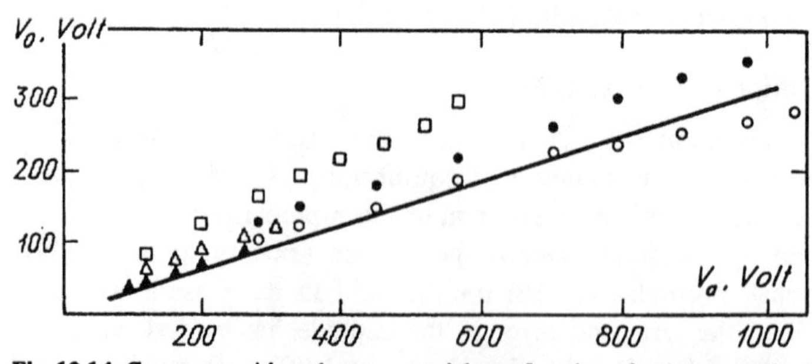

Fig. 13.14. Constant positive plasma potential as a function of voltage amplitude on electrodes at low pressure. *Solid curve:* $V_0 = V_a/\pi$. *Dots* represent experimental data for He and H_2 at low pressures [13.11]

[9] The notion of constant positive potential resulting from the rectification of rf voltage by nonlinear conduction of the layers is developed in [13.11] and in some other papers. In fact, this radio-electronics analogy must not upstage the real cause of the effect, namely, the oscillation of the electron gas with respect to the ionic residue and a positive charge of the gas volume.

13.7 High-Current Mode

13.7.1 Common Features with Weak-Current Modes

Electrode layers of positive space charge are formed in ccrf discharges of both forms. The reason is the oscillation of the electron gas over a range much greater than that of the ionic oscillations, and the resulting absorption of electrons by the electrodes. In any case, if plasma conductivity is sufficiently high, the field moves the ions towards the electrodes for the larger part of the period. An ion reaching the electrode may cause secondary emission; an emitted electron then produces numerous electron-ion pairs in the strong field of the layer. Some electrons manage to cross a considerable voltage difference without collisions and thus reach high energy. We conclude that despite the reversal of sign of the applied voltage every half-period, the electrode behaves in both regimes mostly as the cathode.[10]

Whatever the form of discharge, the field in the layers has a variable component and displacement currents are present. As for the region far removed from the electrodes, where the electrodes phenomena are not felt (the positive column if there is enough space for it), the processes of production and transfer of charges in it are of similarly local nature. The localization is caused by the smallness of the electron displacements: $A \ll L$. Possible quantitative differences in columns of different modes arise because of the differences in electron and current densities.

13.7.2 Essential Aspects of the Differences

All the best-pronounced features of the two forms of burning of ccrf discharges are determined by whether the process of multiplication of electrons due to secondary emission on the electrode is self-sustained, whether the Townsend-type condition (7.1) of reproduction of charged particles is satisfied in the layer, in fact, whether the gas in the layer suffers *breakdown*. Breakdown does occur in the high-current mode, but not in the weak-current mode. The self-sustained conduction current grows in the former case to a comparatively high value, nearly the same as in the normal glow discharge. In the latter case, the ionic current to the electrode is low and is mostly related to the thermal motion (diffusion) of ions. These small losses of ions are balanced out, most likely, at the expense of the ionization of atoms by electrons during that part of the period when electrons of the plasma approach the electrode, and partly at the expense of the flux of ions from the positive column.

We repeat this as it is an essential point. It may be said that in the high-current mode each electron emitted from the electrode (or from the dielectric)

[10] Obviously, only if $A = \mu_e E_a/\omega \ll L$. If $A \gg L$ the discharge within each half-period resembles the constant-current discharge, and the electrode serves alternatingly as cathode and then as anode. Such low-frequency discharges (known as ac-current discharges) are now employed in laser technology (Sect. 14.4.6).

and each ion absorbed into it are *reproduced* by avalanche multiplication *directly in the electrode layer*. In the weak-current mode, the *entire discharge* participates in the balance of production and removal of charges, and the multiplication in the layer due to secondary emission *is not balanced*. We will additionally clarify at the end of Sect. 13.7.4 why the current is "high" in the high-current mode and"weak" in the weak-current mode.

The available experimental data indicate that layers very similar to the *cathode layer of the normal glow discharge* are formed in the *high*-current mode at both electrodes. The experimentally measured layer thickness d and constant potential V_0, equal to the average voltage across the layer, are found to be close to the normal values d_n and V_n. As we find for d_n and V_n, $d \propto p^{-1}$ and is independent of frequency, and V_0 is independent of p and ω. The normal current density is also satisfactorily explained (Sect. 13.7.4). Typical regions of *negative glow* and *Faraday dark space* also appear. In the *weak*-current discharge, there can be no self-sustained conduction current in the layer because the layer thickness is greater than d_n by an order of magnitude and is so high that the measured voltage is too low to satisfy the Townsend condition at the actual values of pd. The layer thickness d is proportional to the electron drift velocity in the plasma, while d_n is independent of it. Furthermore, $d \propto \omega^{-1}$ and is almost independent of pressure. All this indicates that the scale that determines d is the amplitude $A = \mu_e E_a/\omega$, as we assumed in the above model that required no ionic current.

13.7.3 Secondary Ignition

The transition from the weak- to high-current mode is a result of the Townsend breakdown of the electrode layers in which the electrodes act as cathodes and the plasma column serves as an anode (i.e., it receive electrons). The transition is realized when the voltage across the layer (obviously, it is intermediate between the maximum voltage $V_{C,max}$ and the average V_0) reaches the value approximately equal to the Paschen breakdown threshold $V_t(pd)$. The value of pd in a weak-current discharge in molecular gases at medium pressures is much higher than the value corresponding to the minimum of the Paschen curve, which is close to pd_n. The breakdown voltage is appreciably higher than V_n.[11] After the breakdown, the initially thick ionic layer forms a normal cathode layer as in the initiation of a glow discharge, that is, the zone of positive space charge and the voltage across it contract to values that are optimal for the self-sustainment of the values d_n, V_n, while the remaining part of the former layer is converted into electrically neutral plasma.

This approximate condition of secondary ignition, $V_0 \approx V_t(pd)$, implies a restriction on the maximum possible current density in the weak-current mode [13.12]:

[11] If $pL < 10^{-1}$ cm Torr, the discharge gap belongs to that part of the left-hand branch of the Paschen curve where the breakdown voltage is high, regardless of the gas species. Possibly, this is why no sign of the high-current mode (conduction current) has been observed in experiments with very low pressures ($p \sim 10^{-2}$ Torr).

$$j \sim \omega V_0/4\pi d < \omega V_t(pd)/4\pi d\ , \qquad (13.27)$$

which is generally supported by experimental data. The higher the frequency, the higher the limiting current density and the plasma density $n \equiv n_e$ at which the weak-current discharge is still sustainable. A more detailed calculation of breakdown in a layer [13.13] gives the following approximate formula for evaluating the critical plasma density corresponding to the transition from the weak-current mode in nitrogen at $p = 15$ Torr to the high-current mode:

$$n_{cr} \approx 1.7 \times 10^8 (f[\mathrm{MHz}])^{3/2}\,\mathrm{cm}^{-3}\ ,\quad f \sim 1-30\,\mathrm{MHz}\ .$$

Only the high-current mode is realized at a given frequency if $n > n_{cr}$, or a given n and $f < f_{cr}$. Thus the weak-current mode is practically unfeasible in ac discharges at $f = 10$ kHz.

13.7.4 Current Density

The normal cathode current density in a glow discharge is $j_n = C_1 p^2$, where C_1 is a constant, specific for each gas-metal pair (Sect. 8.4). Measurements have demonstrated [13.14] that the high-current ccrf discharge also involves the *phenomenon of normal current density* in which a change in discharge current produces a proportional change in the area occupied by the discharge on the electrodes (including insulated ones). It was found, however, that the current density may considerably exceed j_n, because of the inevitable displacement current added to the conduction current at the electrodes. The conduction current must be nearly equal to j_n because the thickness d and voltage V of the space-charge layer are close to the normal values d_n, V_n. The displacement current in the layer is $j_{dis} \approx \omega V/4\pi d$. If we take into consideration that $d \approx d_n = C_2 p^{-1}$, where C_2 is also a constant for a given pair of materials and $V \approx V_n$ is independent of pressure, then $j_{dis} \approx (\omega V/4\pi C_2)p \propto p$. Since the conduction current j_n and displacement current j_{dis} are different functions of p, their relative contributios to the total current j change with p. As j_n and j_{dis} are shifted in phase by $\pi/2$, the rms value of j is

$$j = \sqrt{j_n^2 + j_{dis}^2} = C_1 p^2 \sqrt{1 + (\omega V/4\pi C_1 C_2 p)^2}\ . \qquad (13.28)$$

Calculations using this formula are compared with experimental data in Fig. 13.15 taken from [13.14]. For the brass-air combination used for the measurements (obtained with the setup described in Sect. 13.4.4), $C_1 = 2 \times 10^{-4}$ A/cm^{2}Torr, $C_2 = 0.23$ cm Torr. The frequency was $f = 13$ MHz. For the voltage V on the layer, the value 320 V was used, which is the rms rf-voltage at the electrodes when they have been brought together so closely that the positive column is practically absent ($L \approx 1$–2 cm, filled with layers and parts of Faraday spaces). A large fraction of the applied voltage is developed across the layers. The experimental rms values of j fit the calculated curve 2 quite well. The curve as a whole corresponds only to the high-current mode. Note that j_{dis} at relatively

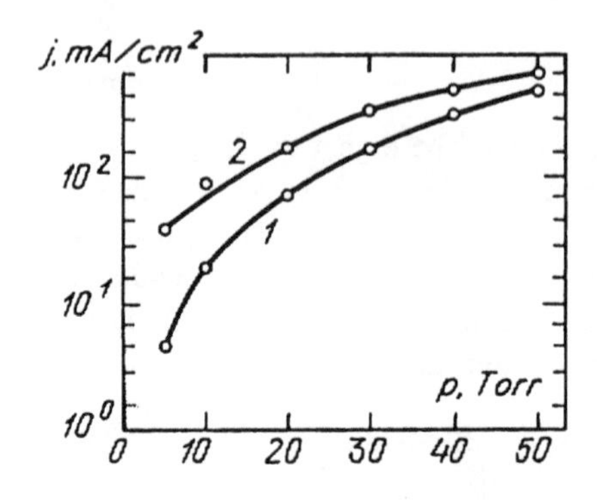

Fig. 13.15. Electrode normal current density in high-current mode as a function of pressure (air, brass electrodes, $f = 13\,\text{MHz}$). *(1)* j_n for normal dc glow discharge in air, with brass cathode. *(2)* Calculations using (13.28), taking into account displacement current. *Dots* represent measured values [13.14]

low pressures ($p \approx 5\,\text{Torr}$) is greater than j_n by an order of magnitude, while these quantities are almost equal to $p \approx 50\,\text{Torr}$. The high current of the *high-current mode* at *low pressure* thus also closes the circuit via the *displacement current*. But the very possibility of letting through a high current – indicating a fairly high degree of ionization in the gas – rests on the multiplication of electrons in layers after breakdown, where optimal conditions, corresponding to a normal cathode layer, are created.

Let us emphasize again that the *difference* between the high-current and weak-current modes is *not in the ratio* of conduction and displacement currents on the electrodes. We did see that the latter is again much greater than the former in the low-pressure high-current mode. The main result of the breakdown of electrode layers is not the build-up of a high ionic current, but an abrupt *drop in layer thickness*. If the voltage remains within the same order of magnitude, the field at the electrode sharply increases, so that high-density displacement current can flow. A thick weak-field layer of the weak-current discharge cannot let through a high (displacement) current, while a thin high-field layer of high-current discharge can. This is the reason for the "high-current" behavior.

13.7.5 The Simplest Model for the Functioning of "Cathode" Layers

Assume that the ion density n_+ in the layer is constant, even though it differs from that in the positive column. Let the plasma boundary oscillate, touching the boundary for only a short time. The plasma conductivity is high, and the field amplitude in the plasma is much less than at the electrode. Finally, we assume that the ionic current to the electrode exists but is much lower than the displacement current. Then the field E_0 at the left-hand electrode, the layer thickness d, and the voltage V_C on the layer, $V_C = E_0 d/2$, are described by the same formulas as in Sect. 13.5.2. In terms of the amplitude of the discharge current density, $j = \dot{E}_0/4\pi = -j_a \cos \omega t$, we can write E_0, d, and V_C in the form

$$E_0 = -(4\pi j_a/\omega)(1 + \sin \omega t)\,,$$
$$d = (j_a/e\omega n_+)(1 + \sin \omega t)\,, \quad V_C = (2\pi j_a^2/e\omega^2 n_+)(1 + \sin \omega t)^2\,. \tag{13.29}$$

If the voltage drop across the plasma is low, in contrast to Sect. 13.2, the potential difference at the electrodes is $V = (8\pi j_a^2/e\omega^2 n_+) \sin \omega t$.

The flux of ions to the electrode, $n_+\mu_+E_0$, is balanced out by their production in the layer. In avalanche multiplication by secondary emission, charges are mostly produced far from the cathode, and if the characteristic displacement amplitude of ions $A_+ = \mu_+E_{0a}/\omega \ll d_a$, an ion needs a considerable number of periods to cover the distance from the point of production to the electrode (E_{0a}, d_a, V_{Ca} denote the maximum (amplitude) values). The correlation of the phases of production of an ion and of its arrival at the electrode thereby vanishes. Let us ignore the effect of delay in the arrival of ions and write a condition, equivalent to (8.9), *of self-stustaiment of the current of charges* in the layer [13.15]:

$$\langle n_+\mu_+E_0(t)\rangle = \left\langle \gamma n_+\mu_+E_0(t)\left\{\exp\left(\int_0^{d(t)} \alpha[E_1(x,t)]dx\right) - 1\right\}\right\rangle . \qquad (13.30)$$

The averaging here is carried out over one period. The field $E_1 = E_0(1 - x/d)$ is substituted into the ionization coefficient; γ is the secondary emission coefficient. The main contribution to the multplication integral is made by the short time interval around the time moment $\omega t = \pi/2$ when $|E_0|$ and d are maximal.

Equation (13.30) relates V_{Ca} and d_a or V_{Ca} and j_a because $E_{0a} = 8\pi j_a/\omega$. The "$V - i$ characteristic" of the layer, $V_{Ca}(j_a)$, has a minimum. By analogy to the theory of von Engel and Steenbeck (Sect. 8.4), it is natural to assume that precisely this "normal state" is realized. One would expect that calculations using (13.30, 29) should give *normal* values of V_{Ca} and d_a that are quite close to those obtained from (8.9) at the same A and B in (4.5) for α, and the same γ. In the example calculated in [13.15] (air, $A = 8.6$, $B = 254$, $\gamma = 0.1$, $\mu_+p = 1.4 \times 10^3$), we find that at $f = 13.6$ MHz and $p = 30$ Torr, the normal amplitude of the cathode fall is $(V_{Ca})_n \approx 280$ V, the mean layer thickness is $d_n \approx 0.02$ cm, and the normal current density is $j_{an} \approx 50$ mA/cm^2. The ion density is defined as $n_+ = E_{0a}/4\pi e d_a$. The ionic current amplitude $j_{ia} = en_+\mu_+E_{0a}$, calculated in the next approximation, comes to within 15 % of j_{an}.

Similarity laws hold for normal parameters: $(V_{Ca})_n$ is independent of pressurie, $(d_a)_n \propto p^{-1}$, $(E_{0a})_n \propto p$, the current is approximately equal to the displacement current, $j_{an} \propto p$, and the ionic current is $j_{in} \propto p^2$. Hence, the ionic current at sufficiently low pressures is indeed a small fraction of the total. But if we extrapolate $j_{dis} \propto p$ and $j_i \propto p^2$ to high pressures, j_i can be expected to exceed j_{dis}. According to the concepts of [13.15], the condition $j_i \geq j_{dis}$ corresponds to the weak-to-high current $(\alpha - \gamma)$ transition, and the entire theory relevant to the case of $j_i \ll j_{dis}$ describes the weak-current discharge.

This conclusion is wrong and at variance with experiment. The theory is correct but holds fo the *high-current*, not the *weak-current* mode. This is indicated by using the Townsend condition of self-sustainment of conduction current in the layer, (13.30), and by choosing the normal parameters at the minimum of the $V - i$ curve of the cathode layer. The experimental curve in Fig. 13.15, showing $j_i \ll j_{dis}$ at low pressurs, belongs entirely to the high-current mode by layer thickness, glow distribution, and current-density magnitude. The transition is always accompanied by jumps of an order of magnitude jumps in d and j. The small layer thickness calculated above, $d \approx 0.02\,\text{cm}$, is typical of the high-current discharge. In the weak-current discharge at the same ambient conditions (air, $p = 30\,\text{Torr}$), $d \approx 0.3\,\text{cm}$, $j_a \approx 7\,\text{mA/cm}^2$ [13.8, 13.12].

The ionic current at lower pressures is small compared with the displacement current because an ion needs considerable time (many periods) to cross the d_{an}-thick layer from the point of most copious production to the electrode. A kind of pause develops between the relatively fast removal of ions into the electrode from the contiguous region, $A_+ \approx \mu_+ E_{0a}/\omega$, and the arrival of new ones to replace them, after which a deficiency of ions appears at the electrode. The ionic current becomes weak, and is compensated for by the displacement current. Indeed, the ratio of currents to the electrode is characterized by the ratio of the lengths indicated:

$$\frac{j_{ia}}{j_{disp,a}} \approx \frac{en_+\mu_+E_{0a}}{\omega E_{0a}/4\pi} = \frac{4\pi en_+}{E_{0a}} \cdot \frac{\mu_+E_{0a}}{\omega} = \frac{A_+}{d_a} .$$

At sufficiently high pressurs, when $d_{an} \propto p^{-1}$ decreases to the displacement amplitude A_+ which is independent of p, the removal of ions is immediately balanced out by the arrival of ions produced in the same period, and the conduction current becomes predominant. Experiments do *not* reveal *any jumps and transitions* in this situation.

13.8 The Structure of a Medium-Pressure Discharge: Results of Numerical Modeling

13.8.1 Equations

The charge density and field distributions in the discharge gap and their evolution in time are described in the one-dimensional approximation, with diffusion neglected and attachment absent, by the following system of equations:

$$\begin{aligned}
&\frac{\partial n_e}{\partial t} + \frac{\partial \Gamma_e}{\partial x} = \alpha|\Gamma_e| - \beta n_e n_+ , \quad \Gamma_e = -n_e\mu_e E , \\
&\frac{\partial n_+}{\partial t} + \frac{\partial \Gamma_+}{\partial x} = \alpha|\Gamma_e| - \beta n_e n_+ , \quad \Gamma_+ = n_+\mu_+ E , \qquad (13.31)\\
&\frac{\partial E}{\partial x} = 4\pi e(n_+ - n_e) .
\end{aligned}$$

The boundary conditions on electrodes or insulting surfaces indicate that when the field points towards the surface, secondary emission is produced, $\Gamma_e = -\gamma\Gamma_+$; when the field points away from the surface, the ionic current vanishes: $\Gamma_+ = 0$.

An rf voltage V is applied to the electrodes.[12] If the electrodes are insulated, V is divided between the gas-filled gap and the insulators. At the beginning of the discharge, the initial electron density must be specified.

The solution of this problem shows that at $L \gg d$ a homogeneous positive column is formed after some time in the greater part of the gap of length L, where the *ionization-recombination equilibrium* is maintained *on average over a period*. To study the structure of electrode layers, it is sufficient to specify the plasma density far from the solid surface, $n = n_e = n_+$, and the field E in it, $E = E_a \cos \omega t$. Its amplitude E_a is related to n by the condition $\langle \alpha \mu_e E \rangle = \beta n$ (Sect. 13.3.2 and Fig. 13.5). This is equivalent to fixing not electrode voltages, but the *discharge current* whose amplitude j_a is expressed in terms of n and $E_a(n)$. Thus $j_a \approx en\mu_e E_a$ when the displacement current in the plasma is low.

The calculations in [13.13] were carried out with the following set of data:

$$\alpha = 12p \exp(-342p/E)\,\mathrm{cm}^{-1} \quad \text{for} \quad E/p > 100\,\mathrm{V/cm\,Torr}\,,$$

$$\alpha = 2.4p \exp(-155p/E)\,\mathrm{cm}^{-1} \quad \text{for} \quad E/p < 100\,\mathrm{V/cm\,Torr}\,,$$

$$\mu_e = 4.4 \times 10^5/p\,, \quad \mu_i = 10^3/p\,\mathrm{cm}^2/\mathrm{Vs} \quad (p \text{ in Torr})\,,$$

$$\beta = 2 \times 10^{-7}\,\mathrm{cm}^3/\mathrm{s}\,, \quad \gamma = 0.01\,,$$

which is often used in modeling discharges in nitrogen; $f = 13.6\,\mathrm{MHz}$, $p = 15\,\mathrm{Torr}$. Calculations were continued until a strictly periodic solution was obtained; this needed about 10^2–10^3 periods.

13.8.2 Results

Depending on the specified V for a gap of given length L, or on the specified n (or j_a) in the positive column, the distributions typical of the weak- or high-current modes set in *automatically*. A *change in structure* occurs at $n_{cr} \approx 8.72 \times 10^9\,\mathrm{cm}^{-3}$, $j_{a,cr} \approx 9.2\,\mathrm{mA/cm}^2$.[13] Experimentally the transition was observed in air at 30 Torr at $j_{a,cr} \approx 17\,\mathrm{mA/cm}^2$ [13.8], but [13.12] predicts j at $p = 15\,\mathrm{Torr}$ (the pressure used in the calculation) to be less by a factor of 1.5; the agreement with measurements is thus quite good.

Figure 13.16 shows calculated density distributions of ions and electrons at the left-hand electrode (cf. Fig. 13.1) for a typical weak-current mode: $j_a = 4\,\mathrm{mA/cm}^2$, $n = 3.7 \times 10^9\,\mathrm{cm}^{-3}$ (in the positive column), $E_a/p = 14\,\mathrm{V/cm\,Torr}$ (in fact, the value of E_a/p was definitely overestimated considerably since gas heating was neglected). The discharge current is closed at the electrode almost

[12] Equations of this type were used in [13.16] to model non-self-sustaining rf discharges with external ionization (in the weak-current mode), and in [13.17], to model ac discharges (Sect. 14.4.6).

[13] The corresponding quantity $(E_a/p)_{cr} = 15\,\mathrm{V/cm\,Torr}$ is less indicative since the actual $V-i$ curve of the column may greatly differ from the postulated one owing to gas heating, stepwise ionization, etc. The typically measured critical voltage V_{cr} across the electrodes (Sect. 13.4.4) is even less indicative. It depends on discharge gap width L, on the lengths of the dielectrics (if the electrodes are insulated), and, of course, on E_a/p and on the method of cooling the gas.

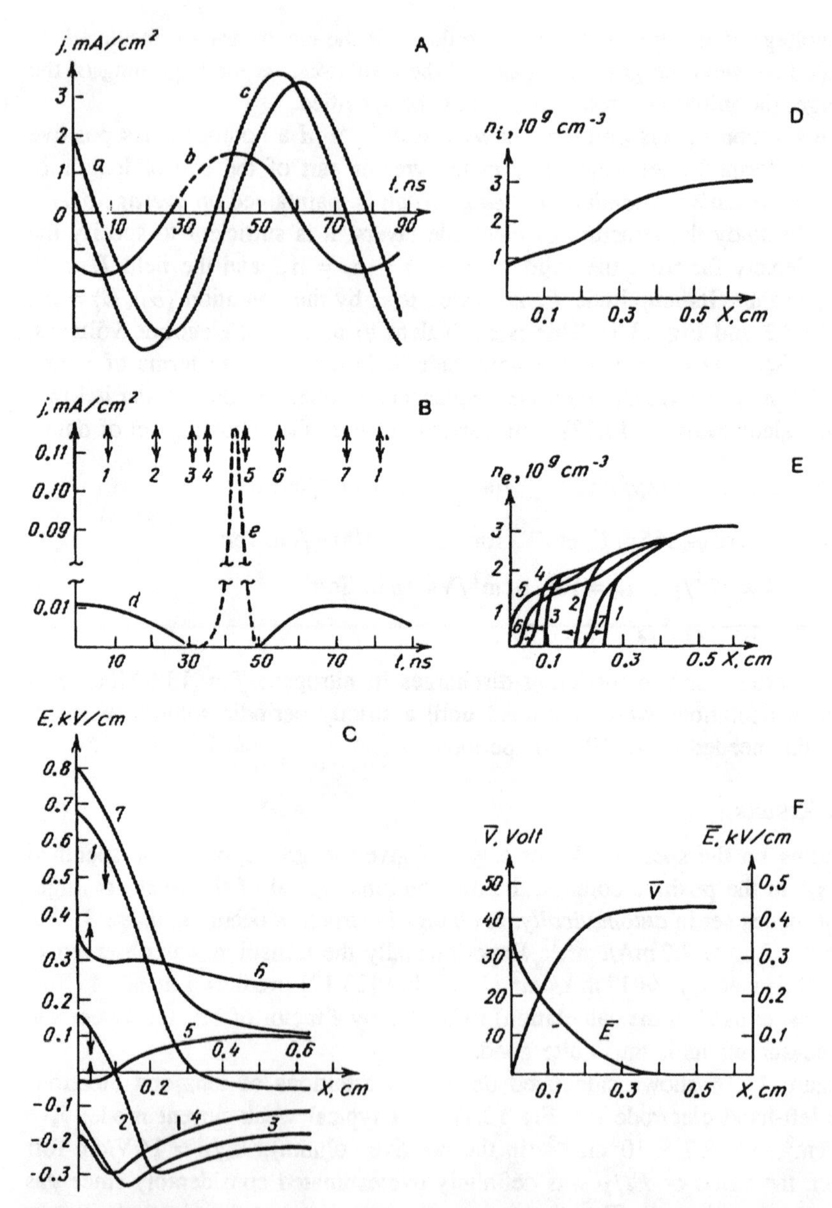

Fig. 13.16. Calculated results for the weak-current mode (N_2, p = 15 Torr, f = 13.6 MHz, discharge current amplitude j_a = 4 mA/cm²). (**A**), (**B**) evolution of components of currents during one period. The position of the time origin is arbitrary. *Curve a* conduction current in plasma, *curve b* displacement current in plasma, *curve c* displacement current of the left-hand electrode, *curve d* ionic current and *curve e* electronic current at the left-hand electrode. In (**B**) *arrows* with numbers indicate times at which instantaneous distributions of n_e and E at the left-hand electrode are given in parts (**E**) and (**C**). *Arrows* at n_e curves indicate the direction of motion of the plasma boundary. Fields $E > 0$ directed towards the electrode, $E < 0$ away. *Arrows* at $E(x)$ curves indicate that the field increases (↑) or decreases (↓). (**D**) density of ions, n_i. (**F**) constant field $\bar{E}(x)$ and constant potential $\bar{V}(x)$ with respect to the electrode [13.13]

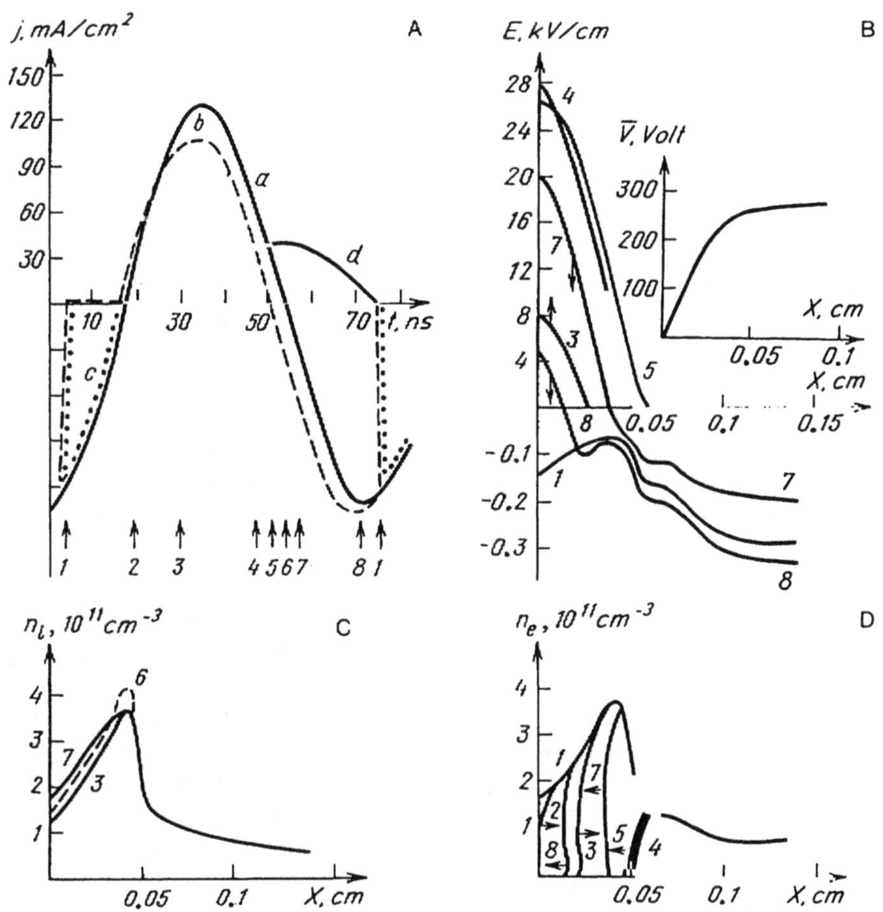

Fig. 13.17. Calculated results for high-current mode (N_2, p = 15 Torr, f = 13.6 MHz, j_a = 0.12 A/cm²). (A) evolution of various components of currents during one period. The position of the origin time is arbitrary. *Curve a* conduction current in plasma, *curve b* displacement current of the left-hand electrode, *curve c* electronic current and *curve d* ionic current at the left-hand electrode; the displacement current in the plasma is to low to be shown on this figure. *Arrows* with numbers indicate times at which instantaneous distributions of n_i, n_e, and E at the left-hand electrode are given in parts (B), (C), (D). (B) fields $E > 0$ (directed towards the electrode) and $E < 0$ (away from the electrode). *Scales* above and below the abscissa are different. *Arrows* at $E(x)$ curves indicate that the field increases (↑) or decreases (↓). $\bar{V}(x)$ is constant potential with respect to the electrode. (C), (D) density of ions, n_i, and electrons, n_e. *Arrows* at n_e curves indicate the direction of motion of plasma boundary [13.13]

exclusively by the displacement current. The maximum value of ion current at the electrode is less by a factor of 400: 10^{-2} mA/cm². The "anode" phase, at which electrons are squeezed by the field to the electrode, lasts for only one tenth of a period. The maximum value of the electron current to the electrode is correspondingly 10 times greater than the ionic current, 10^{-1} mA/cm², because equal amounts of positive and negative charges are deposited on an electrode during a period. The discharge structure is qualitatively the same as in Fig. 13.1, 13.2. The differences are connected only with the fact that in the electrode layer $n_+(x) \neq$ const but decreases from the positive column towards the electrode (as a result of removal of a certain number of ions to the electrode). The density of ions fluctuates so little in time that a figure drawn to this scale cannot show these varitions. The constant plasma potential $V_0 = 45$ V is greater by a factor of 1.7 than that given by (13.24). The maximum layer thickness of 0.3 cm fits quite well the value measured in air, 0.35 cm [13.12], and is twice that given by the model of Fig. 13.1: $2A = 2\mu_e E_a/\omega \approx 0.14$ cm.

Plots of $E(x, t)$ and other quantities for the version under consideration here can be found in [13.13] .

Figure 13.17 shows similar results for the high-current mode. In contrast to Fig. 13.16, we give here the complete picture of the process because it is not as trivial as in the case of weak-current mode. The calculations refer to $j_a = 0.12$ A/cm^2 ($n = 9 \times 10^{10}$ cm^{-3}, $E_a/p = 18.6$ V/cm Torr; as in the preceding case, the value of E/p has been greatly overestimated but this is unimportant for the structure of the electrode layer).

Similar current densities were measured in air at 15 Torr: 0.16 A/cm^2 in [13.12], 0.11 A/cm^2 in [13.14]. The ionic current to the electrode is considerable even though it is lower than the displacement component. The entire current becomes electronic at the "anode" stage. According to [13.12], the measured conduction current in the layer is one-third of the total, which agrees with the calculations. The layer is considerably thinner than in the weak-current mode; its maximum value is 0.05 cm, and n_+ has a maximum inside the layer. As the "cathode" layer is being formed, the first oscillations are followed by the formation and propagation of the ionization wave from the plasma to the electrode; the wave leaves the plasma behind and squeezes the space-charge layer to the electrode. This also happens when a d.c. glow discharge is formed after breakdown.

13.8.3 Nonlocal Effects

These effects are ignored in (13.31) (see Sect. 8.4.10), so that the calculation gives only a model behavior of the transition zone between the space-charge layer and the positive column (Sect. 8.5). In fact, experiments in the high-current mode definitely reveal a Faraday dark space, which is contiguous to the zone of negative glow (see Fig. 13.11). Figure 13.11 implies that the maximum of light emission in the weak-current mode lies at the boundary separating the electrode layer and the positive column. Note that this fact is supported by calculations based on equations (13.31). The maximum of ionization rate $\langle\alpha|\Gamma_e|\rangle$, and hence, of excitation rate of atoms, falls within the transition region between the layer and the column.

13.9 Normal Current Density in Weak-Current Mode and Limits on the Existence of this Mode

13.9.1 Physical Explanation

Yatsenko, who discovered these effects [13.8] (Sect. 13.4.4), gave them the following qualitative interpretation. The measured $V-i$ characteristic of the positive column is falling. It can be given the symbolic form $V_{col} = CpLj^{-m}$, where C and m are positive constants. The $V-i$ curve of the electrode layer, where the displacement current flows, is rising, $V_C = 4\pi dj/\omega$, because the layer thickness d is independent of current. In view of the phase shift between V_c and V_{col}, the voltage on the electrodes is $V = (V_{col}^2 + V_C^2)^{1/2}$. The current-voltage characteristic $V(j)$ of the discharge has a minimum at $j = j_n = (\omega CpL\sqrt{m}/4\pi d)^{1/(m+1)}$. One is inclined to assume that the minimum voltage V_{min} and normal current density j_n are realized, as in a dc discharge, when the electrodes are not totally covered by the discharge. The normal current density is the higher, the greater pL (as in Fig. 13.12). But as pL increases, j_n and V_C grow to critical values at which layer breakdown occurs and the high-current mode sets in. This gives the upper limit (in pL) of the existence of the weak-current mode.

13.9.2 On the Extension of the Domain of the Weak-Current Mode

This extension would be very desirable, because such a regime would be favorable for use in high-power lasers (Sect. 14.4.7). No power loss occurs in nonconducting layers, nor do they constitute sources of perturbation for the development of instability.

For the layers to remain nonconducting, the discharge current density j must not exceed the critical value j_{cr} of the transition. As follows from the arguments of Sect. 13.9.1, this is possible only if $pL < (pL)_{cr}$. However, the reason for the rising curve $j(pL)$ is the falling $V-i$ characteristic of the positive volumn, which in all likelihood results from the heating of the gas (Sect. 8.7). If the heating is partially suppressed by intensely pumping the gas through the discharge region, the constraint on pL becomes less severe. The weak-current mode will then survive at higher values of pL than those reported in [13.8]. Indeed, calculation of the $V-i$ characteristic taking into consideration the convective heat transfer [13.18] does support this expectation though, of course, the decisive proof will be given by experiment. Calculations predict the weak-current discharge domain to extend slightly at higher frequencies.

Note that the calculation of the parameters of the $\alpha-\gamma$-transition in the framework of the layer breakdown model [13.13], for the current-voltage characteristic of the positive column and of the entire discharge with gas heating taken into account as in [13.18], gave satisfactory agreement with the experimental results [13.10] concluding Sect. 13.4.4. The phenomenon of normal current density in the α-mode may vanish depending on the frequency and the role of heating in a specific experiment; on other words, the α-discharge may be anomalous. All these remarks point to the importance of detailed calculation taking into account the heating of the gas in high-frequency electrical discharges.

13.9.3 Remark on Low-Pressure Discharges

Interest in this field was recently revived in connection with etching and deposition applications in the manufacturing of semiconductor materials for the microelectronics industry. Industrial reactors often employ plane-parallel electrodes separated by distances $L \sim 10\,\text{cm}$. The operation is conducted in the frequency range from 50 kHz to 50 MHz at pressures from 10^{-2} to 1 Torr; $V \lesssim 1000\,\text{V}$, $j \sim 1\,\text{mA/cm}^2$. Low pressures are characterized by a very low voltage drop across the positive column plasma, much lower than the ionization potential of gases [13.19]. Consequently, the production of electrons is ascribed entirely to the electrode layers, on which practically the entire voltage falls. However, the analysis of this picture in [13.19] does not reveal the mechanism of the process. A special issue of IEEE Transactions [13.20] was devoted to ccrf discharges and their applications to plasma-chemistry technologies (see also [13.21] and the recent review [13.22]).

A detailed and exhaustive consideration of the capacitive rf discharges is presented in a recently published book (see Further Reading [1]). The experimental data, physics, theory, computer simulation, experimental methods and measurement technique, applications for CO_2 lasers and plasma processing are given. There are chapters on rf discharges and technology in the new book (see Further Reading [2]).

14. Discharges in High-Power CW CO_2 Lasers

14.1 Principles of Operation of Electric-Discharge CO_2 Lasers

Electric discharge CO_2 lasers have the highest power among all existing types of cw lasers and are technologically the most advanced and best adjusted to cutting, metal welding, thermal hardening of parts, and to a number of other technological operations. Their cw power has increased in the two decades since the development of the first one by Patel in 1964 from a milliwatt to many kilowatts, and industrial lasers of up to several kW of power are now commercially available [14.1]. Systems with powers above 10 kW also exist. It can be said that no other applications-oriented problem has stimulated so much progress in gas discharge science as the problem of developing high-power cw CO_2 lasers.

14.1.1 Laser Transition in the CO_2 Molecule

CO_2 lasers employ the transition between two vibrational (rather, vibrational-rotational) levels of the electronic ground state of the CO_2 molecule. The lasing wavelength $\lambda = 10.6\,\mu m$ ($\hbar\omega = 0.117\,eV$) lies in the far infrared. The linear CO_2 molecule undegoes vibrations of three types (called vibration *modes*; see Fig. 14.1). The frequency ν_1 corresponds to symmetric vibrations, ν_2 to deformation vibrations, and ν_3 to asymmetric vibrations. A photon of lasing energy is emitted in the transition from the state 001 to 100; the numbers indicate the vibrational quantum numbers in the modes ν_1, ν_2, ν_3, respectively. The transition $001 \rightarrow 020$, $\lambda = 9.4\,\mu m$ is also possible, but it is normally much weaker.

The energy level of a molecule is determined by the state of not only the vibrational but the rotational motion as well. Thermodynamic equilibrium among rotational states sets in rapidly. Hence, while the rotational state from which the laser transition occurs is depleted, new molecules from other rotational states transfer into it, so that all rotational states form a sort of common energy pool. If necessary, a specific vibrational-rotational line can be singled out. The photon energy is 41 % of the energy of the upper level 001. This is the *principal limit* for the efficiency of a CO_2 laser. If the entire energy consumed to create the lasing medium went only to excite the 001 level of CO_2 molecules and each excitation event was followed by the emission of a photon, the energy of the laser emission would equal 41 % of the absorbed energy. The actual efficiency is typically several times lower.

14.1.2 Mechanism of Creating Population Inversion

In electric-discharge CO_2 lasers the upper laser level is pumped by *electron-impact excitation of vibrations* in the positive-column plasma of a glow discharge. The emission power is greatly increased by an admixture of *nitrogen* to CO_2. The energy of the first vibrational level of the N_2 molecule, which is efficiently excited by electron impact in the glow discharge and then slowly relaxes, is quite close to the energy of the upper 001 level of CO_2 (Fig. 14.1). Hence, fast resonance transfer of vibrational energy from N_2 molecules to CO_2, directly populating the necessary level, becomes feasible. For the N_2, frequency of the transfer of vibrational to translational energy (the rate of VT relaxation) is low even in collisions with water molecules (which are exceptionally active in this respect), namely $\nu_{VT} = 175 p_{H_2O}\,s^{-1}$,[1] while the frequency of energy transfer with CO_2 molecules is $\nu_{tr} = 1.8 \times 10^4 p_{CO_2}\,s^{-1}$. Molecules in the upper laser level 001 either emit photons (on the average, if the lower laser level has a lower population) or undergo the VT relaxation at a frequency $\nu_{VT} = 365 p_{CO_2} + 110 p_{N_2} + 85 p_{He}\,s^{-1}$ (in a mixture with helium).

High population inversion is a result not only of intense *pumping* to the upper level, but also a sufficiently fast *depletion* of the lower 100 level, which is constantly being populated by molecules that have emitted a photon of laser energy. Molecules must rapidly "leave" the 100 level for other states in order to "clear" the level and maintain a considerable difference between populations, $N_{001} - N_{100}$ (the inversion, which determines the amplification and radiation power). The lower 100 state of CO_2 is depleted by resonance transfer of vibrational energy of the ν_1 mode into that of the ν_2 mode (the 100 → 020 transition) followed by VT deactivation of vibrations (020 → 010 → 000), and by the two-step VT deactivation (100 → 010 → 000).

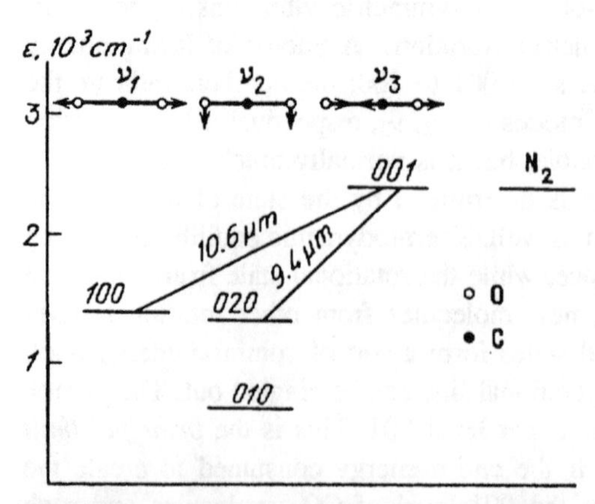

Fig. 14.1. Lower vibrational levels of the CO_2 and N_2 molecules. *Above:* directions of atomic motions of the CO_2 molecule in different types of vibrations

[1] At room temperature. In collisions of N_2^* with N_2 and O_2, $\nu_{VT} = 0.035 p_{N_2} + 0.28 p_{O_2}\,s^{-1}$; p [Torr].

14.1.3 Unacceptability of Considerable Heating

Simultaneously with deactivating collisions, reverse processes are present which excite the lower laser level 100. As a result of direct and reverse processes, its population is not very different from the equilibrium value corresponding to the translational gas temperature T. Therefore, T must be sufficiently low to maintain high population inversion and heating the gas to more than 150–200° C is undesirable. If $T > 450$–$500\,\mathrm{K}$, the inversion and output power are greatly reduced, and at $T > 600$–$650\,\mathrm{K}$ laser action becomes impossible. Sufficiently fast heat removal is the necessary condition for the operation of the laser. Indeed, if the efficiency is 10–20 %, then 80–90 % of the electric power input goes into heating the gas.

A number of factors reduce the efficiency in comparison with the ideal of 41 %. Not all of the energy acquired by electrons from the field is spent on exciting the vibrations, although about 80 % of the electrocal power is converted into vibrations of N_2 and 001 CO_2 at the optimal gas composition and optimal E/p [about several V/cm Torr]. Some 001 CO_2 molecules undergo relaxation instead of emitting a photon. Energy is also partly lost in reflections at mirrors, and so on. It appears that an efficiency higher than 24 % has never been achieved; a realistic figure is 10–15 %.

14.1.4 Laser Mixture

Experiments demonstrated that an admixture of *helium* is very useful in CO_2 lasers. Being light and possessing high thermal conductivity, helium enhances the heat removal from the discharge and improves the quality of the discharge itself. Presumably, helium atoms increase the rate of deactivation of the lower laser level of CO_2. The laser mixture is usually composed of three ingredients: CO_2, N_2, and He. Various compositions are used to optimize the gas for a given design, for instance, $CO_2 : N_2 : He = 1 : 1 : 8$ or $1 : 6 : 12$ (the ratios are in numbers of molecules), etc. However, helium is an expensive gas, while its consumption is high, because the mixture has to be constantly renewed. The cost factor being important for industrial equipment, the tendency is to reduce the helium content. Some systems are free of helium, as a result of multiparameter optimization.

14.2 Two Methods of Heat Removal from Lasers

14.2.1 Diffusion-Cooled Lasers

These are systems in which the heat is carried out of the discharge zone by heat conduction. All early designs, and modern ones for powers $P \lesssim 1\,\mathrm{kW}$, nearly always employ the classical scheme of a glow discharge in a tube (Fig. 14.2). Annular electrodes (so as not to interfere with the light beam) are placed in a long glass tube of radius $R \approx 1$–$3\,\mathrm{cm}$. For example, a mixture of $CO_2 : N_2 :$

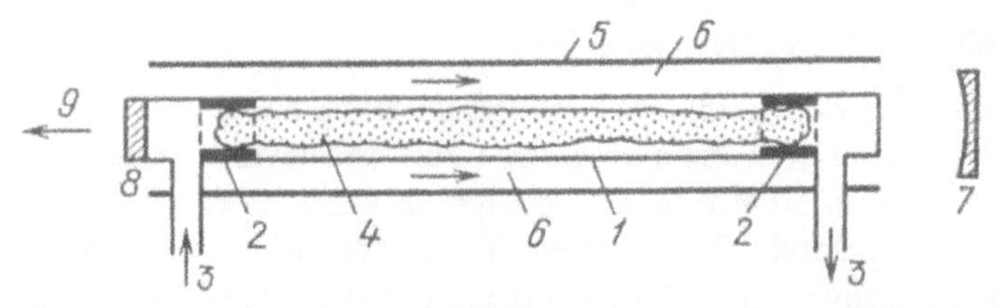

Fig. 14.2. Low-power diffusion-cooled CO_2 laser: *(1)* discharge tube, *(2)* ring electrodes, *(3)* slow pumping of laser mixture, *(4)* discharge plasma, *(5)* outer tube, *(6)* flow of water collant, *(7)* opaque mirror, *(8)* semitransparent output mirror, *(9)* output radiation

He = 1 : 1 : 8 can be employed at $p \approx 20$ Torr. In order to continuously refresh the mixture, the gas is slowly pumped through the tube: the discharge causes some molecules to dissociate, and undesirable by-products are formed. The heat flux from the discharge transfers heat to the tubes cooled by water flow. One of the mirrors (metallic) is placed outside the tube, the radiation reaching it through a salt window. The second mirror (semi-transparent) forms the window at the opposite end; it allows the exit of the beam and is made of NaCl, ZnSe, or GaAs [14.2].

To reduce the laser length at high power levels, when the tube length reaches tens of meters, the tube is broken into elbow-jointed sections with reflectors at the elbow points (Fig. 14.3a). Electric power is fed into the individual, not too long (meter-long) sections, otherwise voltage would have to be too high. Discharge parameters are typical of a noncontracted, *diffuse* positive columns: $n_e \approx 10^{10}\,\text{cm}^{-3}$, $E/p \approx 2$–4 V/cm Torr, $j \approx 10\,\text{mA/cm}^2$, $w \approx 0.5\,\text{W/cm}^3$. Lasers of this type are reliable, robust, easy to operate, and are produced by industry in various versions of tens and hundreds of watts, up to a kilowatt (sometimes even higher).

Unfortunately the power output of this design can be increased only by increasing the tube length, to avoid the *unacceptable heating of the gas* by more than, say, $\Delta T \approx 200°$ C. The equality of energy release and heat removal rates, (8.24), gives

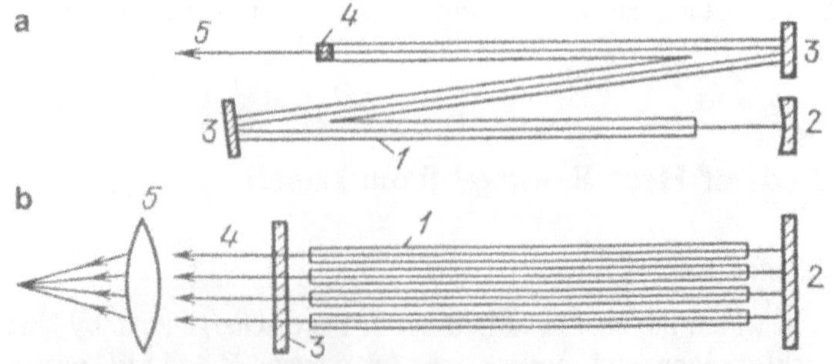

Fig. 14.3. Multiple-section laser tubes for powers of 1 kW and higher. (a) Series connection of tubes: *(1)* discharge tubes, *(2)* opaque mirror, *(3)* elbow mirrors, *(4)* semitransparent output mirror, *(5)* output beam. (b) Parallel connection of tubes in the optical resonator: *(1)* discharge tubes, *(2)* opaque mirror, *(3)* semitransparent output mirror, *(4)* output beams, *(5)* focusing lens

$$jE = Nc_{p1}\Delta T/\tau\ ,\quad \tau \approx \Lambda^2/\chi = Nc_{p1}\Lambda^2/\lambda\ , \tag{14.1}$$

where τ is, in this case, the heat-condution time for heat removal from the discharge region and $\Lambda_T \approx R/2.8$ (Sect. 8.7.1). Hence,

$$j\pi R^2 E = iE = (2.8)^2\pi\lambda\Delta T\ , \tag{14.2}$$

that is, a given ΔT is produced by a given power input per unit length of the tube, which is independent of both pressure [thermal conductivity $\lambda(p) \approx$ const] and tube radius R. It is thus impossible to increase the current arbitrarily by, for example, increasing R or to increase the voltage by increasing p : the result is overheating and the termination of lasing. One cannot input more than approximately 1 kW of electric energy per meter of tube, corresponding to an output of 70–100 W of radiation from the same length.

The principle of *diffusional cooling* found the another realization also [14.3]. A large number of small-diameter tubes are arranged into a compact bundle and "organized" into a resonator, not in series, but in parallel (Fig. 14.3b). The system consists of a large number of independent lasers, each with its own power supply but with common resonator and common cooling. The bundle of parallel light beams (in 37 [14.3])each carrying about 100 W of power can be focused by one large-diameter lens into a single focal spot, where the total power is concentrated. An industrial laser was designed on the basis of this model.

14.2.2 High-Flowrate Lasers with Convective Cooling

The time τ of heat removal from the discharge volume is effectively reduced by *intense pumping* of the gas through the discharge; this increases the density of energy release [see (14.1)] and hence increases the density of the radiative output of energy. Considerable success was achieved in this way in laser design, and *multi-kilowatt* lasers were developed with moderate size active regions. If the length of the discharge zone along the gas flow is L and the flow velocity is u, then $\tau \approx L/u$. For example, if $L = 30$ cm and $u = 100$ m/s, $\tau = 3 \times 10^{-3}$ s, while the diffusional heat removal at $p = 20$ Torr and $R = 3$ cm gives $\tau \approx 5 \times 10^{-2}$ s ($\chi \approx 30\,\mathrm{cm^2/s}$).

The energy released per second in 1 g of a gas of density ϱ pumped through the discharge zone is jE/ϱ. In accordance with the first formula of (14.1), the temperature of the output gas increases in a time τ by ΔT. The temperature constraint imposes a limitation on the admissible *specific energy input* into the discharge, that is, on the amount of energy that can be deposited in 1 g of the gas, $q = P_E/Q$ J/g, where P_E (in W) is the power input into the discharge, $Q = \varrho u S$ g/s is the mass flowrate of the gas, and S is the cross-sectional area of the flow. The inequality $\Delta T < \Delta T_{max}$ imposes limitations on the parameters of the system:

$$q = \frac{P_E}{Q} = \frac{jESL}{\varrho u S} = \frac{jE\tau}{\varrho} < \frac{c_p\Delta T_{max}}{1-\eta} = q_{max}\ . \tag{14.3}$$

Here we have introduced the efficiency $\eta = P/P_E$, because the radiated power P does not participate in the heating of the gas.

If $\eta \approx 15\,\%$, then $q_{max} \approx 700$–$800\,\mathrm{J/g}$. For instance, if we aim at a radiative output $P = 10\,\mathrm{kW}$, this requires the input $P_E = P/\eta \approx 70\,\mathrm{kW}$, demanding the flowrate $Q > P_E/q_{max} \approx 100\,\mathrm{g/s}$; a more realistic estimate is twice as high. One hour of work would need nearly a tonne of the laser mixture! Such high flowrates in high-power lasers designed for long-term operation make *closed cycling* unavoidable. The laser mixture circulates through the system, passing through the discharge chamber many times (Fig. 14.4). The gas is pumped through by compressors, pumps, and the like [14.1]. The system always includes a *heat exchanger*, in which the heat acquired in the discharge is removed from the gas. As with laser tubes, the mixture has to be gradually renewed.

The discharge chamber is often a *wide plane channel*. The field and current are either perpendicular to the flow (transverse discharge) or point along it (longitudinal discharge), as shown schematically in Fig. 8.1. A beam in a resonator is sent along the larger width of the channel at right angles to the gas flow and current. In order to collect energy from the maximum possible volume of the active medium and to make the resonator longer, the resonator is often made multi-pass (Fig. 14.5). The two end mirors form the resonator, while the intermediate (metal) mirrors send the beam back into the discharge zone. The

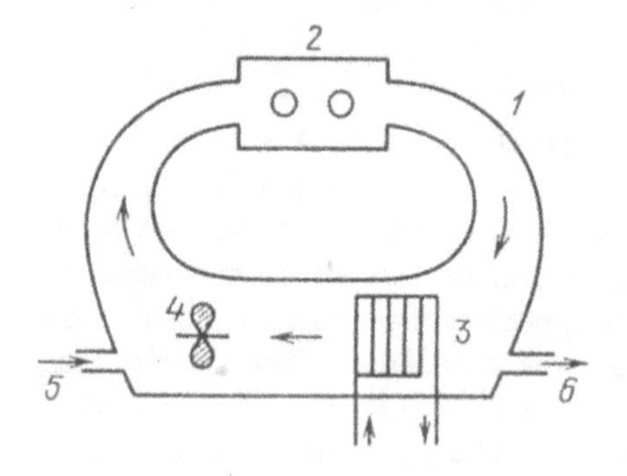

Fig. 14.4. High-flowrate, closed-cycle laser. *(1)* gas contour, *(2)* discharge chamber with resonator, *(3)* heat exchangeer, *(4)* ventilator, *(5)* fresh mixture inlet, *(6)* outlet for removing the used mixture

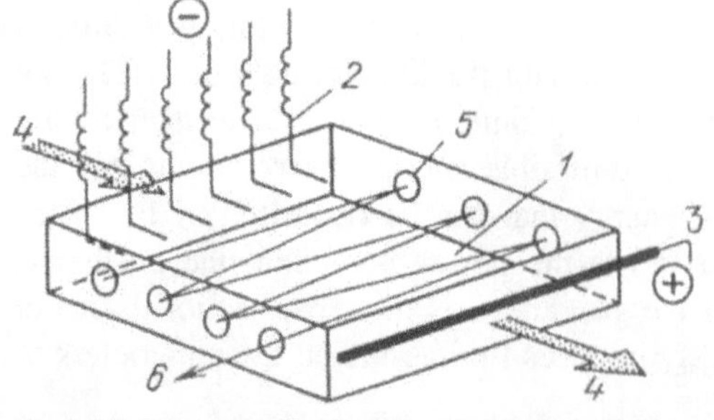

Fig. 14.5. Discharge chamber of a high-flowrate laser with longitudinal discharge pumping in which the multi-pass resonator is employed. *(1)* discharge channel, *(2)* cathode rods, connected to voltage source through individual ballast resistors (Sect. 14.3.2), *(3)* tube anode, *(4)* gas flow, *(5)* mirrors; one end mirror is semitransparent for output of laser beam *(6)* [14.4]

beam emerges either through a semitransparent mirror or through an annular window surrounding a metal mirror. Tube lasers with high-flowrate longitudinal gas pumping have become quite popular in recent years.

14.3 Methods of Suppressing Instabilities

14.3.1 The Problem of Reaching the Theoretical Limit of Energy Input

The problem of the energy input limit reaching is an acute one for high-flowrate lasers. The constraint (14.3) may be said to set a *theoretical limit* to the maximum current and electric power at a given pressure and mass flowrate of the gas. This limitation cannot be circumvented since the temperature of the gas cannot be appreciably increased. For a given time of flight $\tau = L/u$, the limiting energy input corresponds to pressure-independent maximum admissible electron and current densities. Indeed,

$$q \approx jE\tau/\varrho \propto n_e(E/p)^2\tau\ , \quad j \propto n_e(E/p)\ , \tag{14.4}$$

and E/p is only weakly dependent on p. Numerically, for typical laser mixtures (say, 1 : 6 : 12) and for $\tau \approx 3 \times 10^{-3}$ s and $E/p \approx 8$ V cm Torr, the estimates are $n_{e,max} \approx 3 \times 10^{10}\,\mathrm{cm}^{-3}$, $j_{max} \approx 20\,\mathrm{mA/cm^2}$, and $w_{max} \approx 0.16p\,\mathrm{W/cm^3}$.

These values do not at first appear to be excessive: they are achieved in laboratory experiments time and again in discharges in tubes (generally in small volumes and at low pressures). However, the problem of reaching these levels in large volumes (channels about 100 cm wide 20–40 cm long in the flow direction, 5–10 cm high) at pressures of tens or even a hundred Torr, was found to be very difficult, and complex. As the current and power (j and w) are increased, the discharge becomes unstable long before the temperature limit (14.3) is reached, and develops inhomogeneities (Chap. 9). As a rule, *contraction* occurs (filamentation; see Fig. 14.6), which results in the complete termination of lasing. The *suppression of discharge instabilities* is the principal problem in developing high-power lasers. The buildup of instabilities is greatly stimulated by restricting the gas to a closed circuit, where instability-triggering products accumulate, and by the natural tendency to increase pressure.

Indeed, increasing the gas density is the obvious way to generate high power output and at the same time retain moderate overall dimensions of the unit. This is clear from (14.3) if we ask the question how can the flowrate $Q = \varrho uS$ be increased? Increasing the velocity above 200-250 m/s leads to considerable technical obstacles, and the cross section S is scaled up, all the dimensions increase. We are left with only one attractive approach: increase the density (pressure) and thereby make better use of the volume. The higher the pressure, however, the more easily the homogeneous discharge breaks down. Therefore, we will now discuss some methods of stabilizing the discharge and thus increasing the limiting current and power above which the discharge becomes unstable.

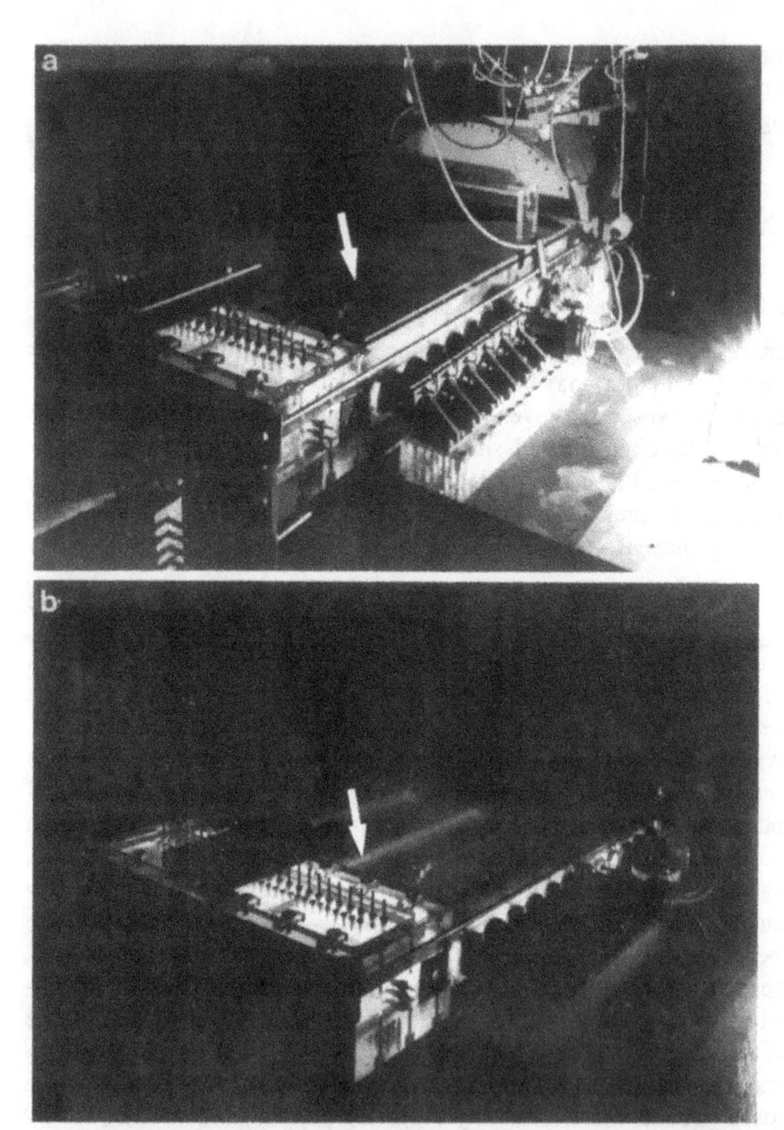

Fig. 14.6. Discharge in a large, high-flowrate closed-cycle laser (see Sect. 14.4.5). The upper plate covering the discharge channel is made of glass, allowing a good view of the discharge (shown by arrow). *Upper* photograph shows a discharge burning uniformly on the whole volume. *Lower* was taken at the moment of formation of current filaments (shown by arrow). Both the gas and the current flow in the direction of the bright columns [14.5]

14.3.2 Segmentation of the Cathode

When a current-conducting filament is formed between two points on large electrodes, the current jumps up, the voltage between the electrodes drops sharply [see (8.1)], and the field at other points becomes too low to sustain a steady discharge. This process can be partially prevented (and the discharge can therefore be somewhat stabilized) by segmenting the electrodes (in the particular cathode). The voltage is then applied to each segment independently through an individual ballast resistor. The resistors themseles have a stabilizing effect on the discharge, by suppressing the fluctuations of current. As a result, when a filament with enhanced current is formed on one of the cathodes, this does not lead to a catastrophic drop of voltage on other discharge paths leading to different cathode segments.

Segmented cathodes are employed in a number of systems. Another important point is that the normal current density at the cathode, j_n, is 10 to 100 times greater at elevated pressures than the admissible current density in the positive column. If the density in the column, according to the estimate of Sect. 14.3.1, is on the order of $10\,\text{mA/cm}^2$, then at, say, $p = 40\,\text{Torr}$, we have $j_n \approx 2 \times 10^{-4} p^2\,\text{A/cm}^2 \approx 300\,\text{mA/cm}^2$. It is best to exploit cathodes in the regime on the border between the normal and abnormal, i.e., with complete coverage of the electrode area. This regime helps to stabilize the discharge owing to the rising $V - i$ characteristic of the abnormal cathode layer. Consequently, the cathode area has to be made much smaller than the current cross-sectional area in the main part of the discharge chamber. Hence to distribute the cathode surface more uniformly over the larger current cross section of the chamber, the cathode segments are often widely spread out.

In transverse discharges, the cathode segments are spread uniformly over the dielectric plate, "flush" to the surface, in order not to disturb the flow (Fig. 8.1a). In longitudinal discharges, a number of cathode rods are arranged at the entrance to the discharge channel (Fig. 8.1b). Their shapes and numbers are adjusted empirically. As a rule, segmented cathodes are placed at the inlet, upstream of the flow, since the most troublesome instabilities are those generated at the inlet, not those at the outlet, the latter being more rapidly carried away from the discharge zone.

14.3.3 Flow Control

Stabilization of the discharge is facilitated by making the velocity field, uniform especially in systems with a longitudinal discharge. The important point is the minimization of possible nonuniformities from which instability may later develop. An increase in velocity, that is, the reduction of the time of residence of a gas particle in the discharge zone, is generally a measure suppressing the evolution of instabilities, although an excessive flow velocity may result in the reverse [14.1,2]. The creation of intensive *small-scale turbulence* has a favorable effect because it produces, as do diffusion and heat conduction, a damping of the incipient perturbations. The required turbulence scale and the design of turbulence-producers (rows of rods, grids, etc.) are empirically adjusted.

It must be said that when one attempts to stabilize the discharge and improve its characteristics, much is done on the basis of purely intuitive experimental study, not via a theoretically justified program. In this area experiment is clearly ahead of the theory, and the implementation of many ideas depends on the experience, physical intuition, and skill of the experimenter, which in the most outstanding cases border on art. This is in fact typical of much work with gas discharges, which reveal a host of elusive and subtle properties.

14.3.4 Applications of Non-Self-Sustaining Discharges

Both the methods outlined above and some other techniques (typically used in combination) have given very positive results. An example is a 5 kW laser with a self-sustained transverse discharge (Sect. 14.4.1). A power output of 20 kW was achieved in a laser with a self-sustained longitudinal discharge by segmenting the electrode and using complex aerodynamic devices [14.6]. But the most revolutionary step was the application of a *non-self-sustaining discharge* for laser pumping, with a *beam of fast electrons* as an external ionizing agent (UV light can also be used for this purpose). This was first realized (1970) in pulsed CO_2 lasers and is now widespread in pulsed laser systems. In addition, considerable experience has already been gained in using electron beams in cw lasers [14.7, 8].

A non-self-sustained discharge has an advantage over a self-sustained discharge in that it allows us to break the chain of evolution of instability, (9.8), in the only link which permits such a termination. If the ionization is sustained by an external source which is immune (or nearly immune) to plasma processes, a change in ionization rate by "intrinsic" electrons due to a change in E/N becomes relatively unimportant. Indeed, the ionization by electrons accelerated in the field E is of minor importance. The most "dangerous" (namely, *thermal*) *instability* is thereby paralyzed.

When ionization is produced by an external source, n_e and the conductivity are determined by its intensity and are independent of the applied voltage. At a given intensity of the source and a given pressure, the current and energy deposition rate can be increased only by increasing the voltage applied to electrodes, that is, E/N. Once E/N gets close to the value corresponding to a self-sustaining discharge, the rate of ionization by discharge-produced electrons becomes comparable to that due to the external source and all factors leading to instability undergo a revival. The method thus has its limitations.

A non-self-sustaining discharge is also attractive because it separates the functions of sustaining ionization and of imparting to electrons the energy necessary for the efficient excitation of N_2 and 001-CO_2 vibrations. In self-sustaining discharges, both functions are served by the same applied field. In a non-self-sustaining discharge, the field serves only the latter function. Hence, the value of E/N can be optimized for laser pumping, while the power density $jE \propto n_e N(E/N)^2$ can be raised not by increasing voltage and E/N, but by increasing $p \propto N$. The limitation (14.3, 4) on heating and n_e still remains in force.

14.4 Organization of Large-Volume Discharges Involving Gas Pumping

14.4.1 Transverse Self-Sustaining Discharge

The most usual circuit for a transverse discharge is shown in Fig. 8.1a. A number of specially designed cathode segments (it may run into hundreds of elements) are connected to individual ballast resistors and installed above a common anode plate. The resultant ballast resistance must come to 20–30 % of the discharge resistance. Unfortunately, this fraction of electric power is then lost irretrievably. An expanding current stream starts at each cathode; all streams overlap at a distance of 1–2 cm, and the common plasma stretches to the anode.

Analysis demonstrates, however [14.1, 14.2, 14.9], that this common plasma of positive column is still not uniform, and that this is typical of electronegative gases under the conditions of short residence time of a particle in the discharge zone. Not enough active molecules accumulate in the plasma to detach electrons from negative ions. As a result, the balance of attachment and detachment is not set up. Electrons are lost on the way to the anode by attachment, and these losses are not compensated for either by ionization or by detachment. The conductivity decreases towards the anode, and the field increases correspondingly (since the current is constant). The situation in the anode region is reversed: the ionization rate in the enhanced field exceeds the rate of charge removal in the bulk, and an excess of positive ions is introduced into the plasma. In contrast to the classical model of the positive column, therefore, we find here no local balance of production and removal of electrons, so that the plasma column is not self-sustaining. The same is found for short longitudinal discharges, though the effect is absent in electropositive molecular gases, where the local balance is realized. Inhomogeneity transverse to the current (i.e. inhomogeneity along the flow) also exists: the temperature increases along the flow, the density decreases, and the field is reduced by a factor up to 1.5–2. The $V - i$ characteristic of this discharge is almost horizontal or slightly rising (in contrast to the falling one of the discharge in a tube without gas flow).

The discharge described above was thoroughly investigated, including the aspects of stability, formation of domains, and contraction, in [14.9]. This led to the development of the first series of closed-cycle high-flowrate commercial lasers manufactured in the USSR, which are now operating in a number of plants [14.10]. Some specifications of this laser model (using a three-pass resonator) are: power $P \approx 5$ kW, discharge length along the flow $L = 20$ cm, width $a = 90$ cm, channel height (the interelectrode spacing) $h = 4$ cm, $p \approx 50$ Torr, $u \approx 80$ m/s, the gas mixture ratio $CO_2 : N_2 : He = 1 : 20 : 20$, $w \approx 5.5$ W/cm^3, $q \approx 320$ J/q. The same principle was implemented in a 10 kW laser [14.1, 14.11] and newer, improved models are being developed.

14.4.2 Longitudinal Self-Sustained Discharge

The construction of such a laser is shown in Fig. 8.1b and also in Fig. 14.5, referred to earlier, which illustrates a multipass resonator [14.4]. High-flowrate pumping with a closed cycle is also used in tube designs similar to that of Fig. 14.3a. In a certain sense, this discharge can also be regarded as a longitudinal one. The same concept is utilized in the Soviet commercial 1.5 kW *Karat* laser [14.1, 14.12]. Industrial lasers manufactured by Spectra Physics employ a discharge arranged in an unusual way; these lasers have power of 1.2, 2.5, or 5 kW. The cathode is a tube at the entrance to a plane channel (like the anode in Fig. 14.5) and the anode lies on the bottom plane of the channel (as in Fig. 8.1a), being segmented into strips along the flow. The current is thus "curved," "transverse" at the anode and "longitudinal" at the cathode.

14.4.3 Non-Self-Sustained Discharge with Ionization by an Electron Beam

In this type of laser, a beam of electrons emitted by a hot metal surface and accelerated in a vacuum by a high voltage, $V_1 \sim 100\,\mathrm{kV}$, is sent into the discharge chamber. The chamber is separated from the accelerator by a thin membrane (Fig. 14.7). Having traversed a grid electrode, the beam penetrates the gas. The voltage V_1 is so adjusted that the free path length of electrons at the pressure of the laser mixture is comparable with the distance to the other electrode. Then the electron energy eV_1 is used almost completely and the volume is ionized more uniformly. A dc current is passed through the electrodes and the ionized gas; the energy released by it providing the laser pumping. Fast electrons ionize the gas not so much in a direct way, but rather via secondary, slower electrons. If the excitation of electron states is taken into account, the average energy spent on forming one pair of ions is $\varepsilon_i \approx 50\,\mathrm{eV}$.

Let us estimate the source of fast electrons needed to sustain the electron density n_e in a volume Ω of the active medium. Let S_1 electrons be produced per second in $1\,\mathrm{cm}^3$; they are lost as a result of attachment and bulk recombination, with effective coefficient β_{eff} (Sect. 8.8). Under steady-state conditions, $S_1 = \beta_{\mathrm{eff}} n_e^2$; if we ignore losses, the required source power is $p_e = \varepsilon_i S_1 \Omega = i_1 V_1$, where i_1 is the beam current. For example, if $n_e = 10^{11}\,\mathrm{cm}^{-3}$, $\Omega = 10^3\,\mathrm{cm}^3 = 1$ liter, and $\beta_{\mathrm{eff}} = 10^{-6}\,\mathrm{cm}^3/\mathrm{s}$, we find $S_1 = 10^{16}\,\mathrm{cm}^{-3}\mathrm{s}^{-1}$, $P_e \approx 80\,\mathrm{W}$; if $V_1 = 100\,\mathrm{kV}$, $i_1 = 0.8\,\mathrm{mA}$.

The most important advantage of electron-beam ionization lasers lies in that the stability of the discharge permits their operation at about atmospheric pressures (pulsed lasers work at even higher pressures). The dimensions of the active zone are thus dramatically reduced. The power expenditure on maintaining ionization is negligible in comparison with the dc power P_E that can be injected into the volume for pumping. Thus for $p = 300\,\mathrm{Torr}$ and the voltage producing the value of E/p favorable for pumping [$\approx 5\,\mathrm{V/cm\,Torr}$], we have $P_E = \sigma E^2 \Omega \approx 75\,\mathrm{kW}$ ($\sigma \approx 10^{-13} n_e/p\,\mathrm{Ohm}^{-1}\mathrm{cm}^{-1}$). Ten-kilowatt cw lasers have been developed and operated successfully with a closed cycle [14.7, 14.8]. Nevertheless, such systems have not become widespread, in contrast to pulsed

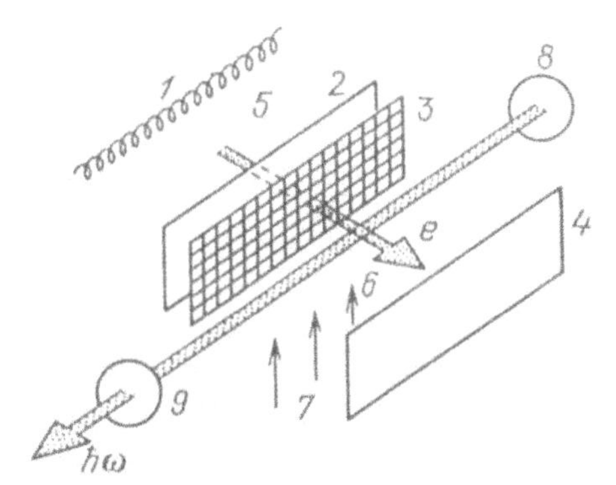

Fig. 14.7. Laser using ionizing electron beam: *(1)* emitter of electrons, *(2)* membrane separating the high-vacuum part *(5)* of the accelerator from the discharge chamber *(6)*, *(3)* grid electrode for non-self-sustained discharge, *(4)* second electrode, *(7)* direction of gas flow, *(8)* opaque mirror, *(9)* output mirror

lasers. The limitation stems from technical factors: constant use destroys the membranes, personnel have to be shielded from the laser because of the dangerous x-ray radiation, and so forth. There is, therefore, a stimulus for the continued search for alternative, more suitable methods of arranging a non-self-sustaining (or nearly so) type of discharge.

14.4.4 Combined Discharge with DC and RF Fields

A cw radiative power of 27 kW due to the input of 160 kW of electric power has been achieved in a large closed-cycle laser with a dc longitudinal discharge using a system employing an auxiliary capacitively coupled rf (ccrf) discharge [14.13]. The research preceding the development of this laser [14.14] and its successful outcome greatly stimulated the studies of ccrf discharge physics (Chap. 13). We shall now describe some parameters of this unity, whose basic form is shown in Fig. 14.8: L = 53 cm, a = 244 cm, h = 6.3 cm, u = 140 m/s; the gas ratio (CO_2 : N_2 : He) was 1 : 6.4 : 12.6 at p = 30 Torr. Without the rf field only 60 kW (0.74 W/cm^3) could be supplied to the dc discharge. With the injection of 60 kW of rf power, the dc power could be raised to 100 kW, giving a total power of 160 kW (2 W/cm^3). Regarding this discharge as non-self-sustaining would be wrong: the rf and dc powers are comparable so that both fields (rather, their vector sum) serve both functions simultaneously (they ionize the gas and impart energy to electrons). However, no additional information on further developments in this field (or on the 20 kW laser [14.6]) has appeared in the literature. The system may have proved to be too complicated for general use.

14.4.5 Non-Self-Sustaining Discharge with Gas Ionization by Repeated Capacitively Coupled Pulses

This truly non-self-sustaining discharge was realized in a large closed-cycle laser [14.5]. The arrangement of the discharge device resembles that of Fig. 14.8. The essenitai element is that the electrodes of the capacitor are installed within the discharge chamber; they are insulated from the plasma by thin heat-resistant glass (Fig. 14.9). This greatly increases the capacitance of the insulators and reduces the voltage across them, so that the ionizing field in the plasma is enhanced

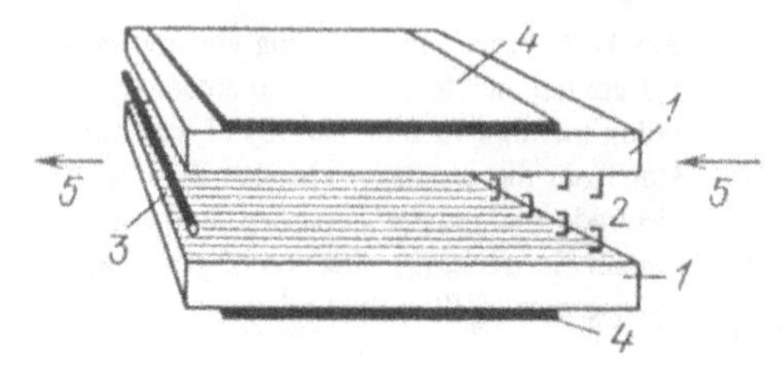

Fig. 14.8. High-flowrate laser with longitudinal dc discharge and auxiliary transverse ccrf discharge; *(1)* dielectric plates forming the discharge channel, *(2)* rows of cathode segments (360 of them), *(3)* anode (4 parallel tubes), *(4)* electrode plates for rf voltage, *(5)* direction of gas flow. Mirrors for multipass resonator, placed as indicated in Fig. 14.5, are not shown [14.13]

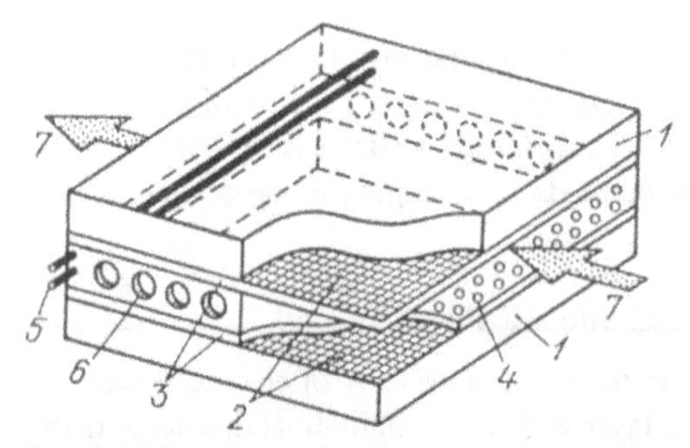

Fig. 14.9. High-flowrate laser of Fig. 14.6: *(1)* thick transparent glasses forming the channel, *(2)* metal grids to which high-voltage ionizing pulses are fed, *(3)* thin heat-resistant insulating glasses, *(4)* cathode segments (50 of them), *(5)* two anode tubes, *(6)* cavities for installing the mirrors of the multipass resonator. *Arrows* show the direction of gas flow [14.5]

(Sect. 13.2; Fig. 13.3). The discharge chamber volume is 27 liters: $L = 63\,\mathrm{cm}$, $a = 76\,\mathrm{cm}$, $h = 5.5\,\mathrm{cm}$, the flow velocity $u \leq 230\,\mathrm{m/s}$. Pulses of about $10^{-7}\,\mathrm{s}$ duration at a voltage of about 15 kV are fed to the capacitor plates at frequencies up to 100 kHz. They sustain an extremely uniform plasma in the volume at $n_e \approx 10^{10}\,\mathrm{cm}^{-3}$. The pulse duty factor is sufficiently low for the average power not to be high, about 1 kW. At the same time, the plasma hardly decays during the $10^{-5}\,\mathrm{s}$ between pulses. This cc-type electrodeless discharge is in itself completely *self-sustained*, and its electrical processes have common features with ccrf discharge [14.5, 14.15].

The high power consumed for laser pumping is introduced by the longitudinal dc field, and the $V-i$ curve of dc current is almost linear (Fig. 14.10) because n_e and the plasma resistance are determined by an external ionizer and are almost independent of the dc voltage.[2] This picture holds as long as the current is much lower than the self-sustaining value. On approach to self-sustainment, the discharge undergoes filamentation. The photographs of Fig. 14.6 were taken in

[2] Cf Fig. 8.22 for the characteristics of a self-sustaining discharge in the same setup. The $V-i$ characteristic is horizontal and independent of velocity.

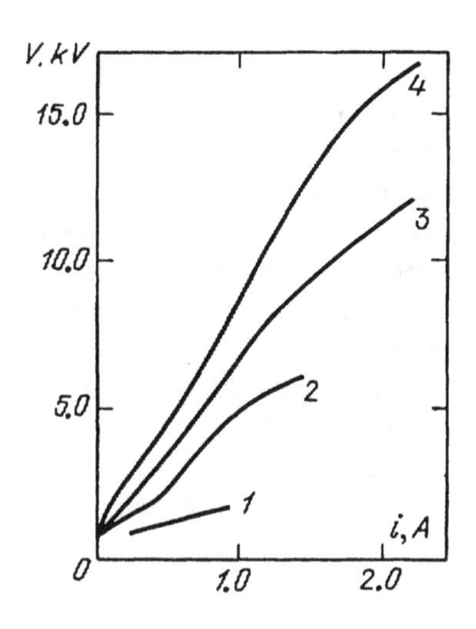

Fig. 14.10. $V - i$ characteristic of non-self-sustaining dc discharge in the apparatus of Fig. 14.9, with ionizing pulse frequency 100 kHz; laser mixture $CO_2:N_2:He = 1:7:12$; $p = 30$ Torr. *(1)* zero flow velocity, *(2)* $u = 60$ m/s, *(3)* 130 m/s, *(4)* 220 m/s [14.5]

the apparatus described above. The realization of an external ionizer producing repeated pulses made it possible to scale up the dc power by more than an order of magnitude, raising it to $P_E \approx 70$ kW (2.6 W/cm^3) at $p = 50$ Torr in the $1:6:12$ mixture and achieving the output $P = 6$ kW of beam power (reaching as high as 10 kW in the improved model of this laser). A compact combined-ionization industrial laser has been developed utilizing the arrangement outlined above [14.16]. It can operate both in the cw and in the pulsed regimes, generating light pulses at a frequency up to 250 Hz. This last regime is useful for welding, and makes it possible to punch numerous small holes, etc.[3]

Note that high-voltage pulses separated by relatively long gaps sustain the same inization under the same conditions at much lower average power than the rf field ($n_e \approx 10^{10}$ cm^{-3} at 0.05 W/cm^3 and 1.4 W/cm^3, respectively). Owing to the steep dependence of ionization frequency on field, it is sufficient to raise the field by a factor of 2–3 in order to reduce the time of its application and the power required tens of times without reducing the number of electrons produced. As a result, pulses at a well-chosen frequency and duty factor ionize the gas much more efficiently than the nearly non-stop sinusoidal field in which the ionization occurs only within short intervals around the maximum value of the field. Obviously, the high rf power is not lost in the laser, being spent on laser pumping, just as dc energy is. But it is more difficult to inject nearly a hundred kilowatts of rf power into the discharge than the same wattage of dc power. This is the advantage of combining a high-power, non-self-sustaining dc discharge current with low-power ionizing pulses, instead of using a comparable high-power, self-sustaining ccrf discharge, if the latter is arranged. On the mechanism of discharge stabilization by pulses and rf fields, see Sect. 9.4.4 and [14.7].

[3] One need to say that in the latest laser structures with longitudinal discharge the stability is reached in the other way avoiding cathode segmentation which introduces some technical complications.

14.4.6 AC Discharge

By ac discharge we mean the electrode or electrodeless capacitively coupled discharge at a frequency $f \leq 10\,\mathrm{kHz}$. Low frequencies up to 10 kHz have an important advantage in comparison with rf: they are produced by generator systems which are simpler, cheaper, and more reliable than high-power rf oscillators. Compared with dc power, low frequencies have the same important advantage that rf has: the discharge can be stabilized using reactive (capacitive) ballast elements in which Joule heat is not released and power is not lost. An ac discharge possesses the stabilization qualities of a non-self-sustaining discharge but at the same time requires no auxiliary ionization source. The same power source serves both ionization and laser pumping, but these two functions are separated in time. Electrons are produced in short bursts at the moments when the ionizing field nears its amplitude values, while the rest of the time energy is pumped into the plasma at the recombination stage when it is not subject any more to thermal instability.

The advisability of utilizing the ac discharge in CO_2 lasers was first pointed out in [14.18], which started a detailed analysis of this process. It occupies an intermediate position between dc and rf discharges. Regardless of whether the electrodes are bare or insulated, breakdown must take place at each of them in the appropriate half-period, and the cathode layer must be formed. A low-frequency, and hence weak, displacement current is too small to close the circuit at the electrode. In this respect the discharge resembles the ***high-current*** (i.e. γ) ccrf discharge mode. The ac discharge can also be employed in high-flowrate lasers and in diffusion-cooled tube lasers. The latter version being better investigated has received more attention, a multi-channel ac laser has been developed [14.1, 14.2] whose design resembles that of Fig. 14.3b, but the electrodes are placed outside the tubes: the resulting simplification is considerable, allowing the use of very narrow tubes (only 5 mm in diameter). A meter-long laser of overall diameter 7.5 cm encloses 61 tubes and has an output of 1.2 kW of radiative power. The ballast elements are the capacitances of the glass areas under the electrodes. An industrial laser has been developed, based on this model.

14.4.7 Self-Sustaining CCRF Discharge

The ccrf discharge has a number of advantages over both dc and ac discharges: It is more stable than the dc discharge, and allows a greater energy input. Ballast resistances, which always produce a favorable effect on discharge stability, can be replaced by reactive (capacitive) elements; this eliminates energy losses which come to about 30 % of electrical power in ordinary ballast resistors. Obviously, rf discharges are not better than low-frequency discharges in this respect, since capacitive ballast is utilized in both cases. One of the essential advantages of the ccrf discharge is the possibility of *eliminating cathode layers*, which are inherent in both dc and ac discharge currents: Firstly, some electrical energy is lost in the cathode layers, and secondly, this is the place where perturbations – the seeds of instability – are born. In this respect the *weak-current* (i.e. α) *mode* of the ccrf discharge, in which the plasma current is coupled to the electrodes only by displacement currents (Chap. 13), corresponding to the absence of cathode layers, is more favorable for lasers. This is a drawback because it implies a limitation on current density and, in fact, on gap length and pressure (Sects. 13.4, 9).

Another advantage of ccrf discharges (actually, it also characterizes capacitively coupled low-frequency discharges) is the practical convenience of working with long tubes. Discharges with high-flowrate and axial pumping of gas have high stability and homogeneity. The latter factor is very important for the quality of radiation. As in tube lasers, a constant longitudinal field calls for voltages of tens of kV. If, however, a dielectric tube is placed between two strip electrodes (so-called strip-line design), very moderate rf voltages of about 1 kV (or even less) become sufficient. The intensity of radiation is easily modulated by modulating the applied rf voltage.

A ccrf discharge has been successfully utilized in low-power lasers.[4] it is especially convenient in small waveguide systems. Electrodes are placed above and below a tube of square cross-section (about $2 \times 2\,\mathrm{mm}^2$ [14.21]). A considerable number of publications have appeared in recent years on CO_2 lasers utilizing ccrf discharges: Some ccrf CO_2 lasers of 1 kW (and higher) power have been reported [14.22, 14.23], and several systems of this type were mentioned in [14.24–26]. It has been claimed that a new generation of industrial CO_2 lasers (ccrf systems) that are more compact, light, simple in design, and more convenient than the preceding generation of dc lasers, has been developed recently [14.27]. At any rate, several firms already manufacture ccrf CO_2 lasers of power 1 kW and higher. A compact waveguide slit CO_2 laser of medium power of 100 to 200 W [14.28] is also very efficient; the rf discharge is excited in it not in a tube but in a zigzag extended slit. Considerable expectations are associated with the rf approach in laser design.

The detailed consideration of the CO_2 lasers with rf pumping (physical principles, slit and other devices, principles of the parameter choice and optimization) are presented in the new book (see Further Reading [1]).

[4] For information on one of the first ccrf CO_2 lasers, see [14.20].

Appendix

Selected formulas, constants, and relations between units that are frequently encountered and often used in gas discharge physics.

Fundamental Constants

Velocity of light	$c = 2.998 \cdot 10^{10}$ cm/s
Planck's constant	$h = 6.625 \cdot 10^{-27}$ erg·s
	$\hbar = h/2\pi = 1.054 \cdot 10^{-27}$ erg·s
Electron charge	$e = 4.802 \cdot 10^{-10}\, \mathrm{g}^{1/2} \cdot \mathrm{cm}^{3/2}/\mathrm{s}$
Electron mass	$m = 9.109 \cdot 10^{-28}$ g
Proton mass	$M_{\mathrm{p}} = 1.672 \cdot 10^{-24}$ g
Atomic mass unit	$M_0 = 1.660 \cdot 10^{-24}$ g
Boltzmann's constant	$k = 1.380 \cdot 10^{-16}$ erg/K
Universal gas constant	$R = 8.314 \cdot 10^{7}$ erg/K·mole $= 1.986$ cal/K·mole
Avogadro's number	$N = 6.023 \cdot 10^{23}$ 1/mole

Atomic Constants

Bohr radius $a_0 = h^2/4\pi^2 me^2 = \hbar^2/me^2 = 0.529 \cdot 10^{-8}$ cm

Ionization potential of the hydrogen atom

$$I_{\mathrm{H}} = e^2/2a_0 = 2\pi^2 e^4 m/h^2 = e^4 m/2\hbar^2 = 13.60\,\mathrm{eV}$$

Rydberg constant

$$\mathrm{Ry} = I_{\mathrm{H}}/h = 2\pi^2 e^4 m/h^3 = 3.290 \cdot 10^{15}\,\mathrm{s}^{-1}$$

Electron velocity in the first Bohr orbit

$$v_0 = 2\pi e^2/h = e^2/\hbar = 2.187 \cdot 10^{8}\,\mathrm{cm/s}$$

Classical electron radius $r_0 = e^2/mc^2 = 2.818 \cdot 10^{-13}$ cm

Compton wavelength $\lambda_0 = h/mc = 2.426 \cdot 10^{-10}$ cm

$$\lambda\!\!\!^{-}_0 = \lambda/2\pi = \hbar/mc = 3.862 \cdot 10^{-11}\,\mathrm{cm}$$

Electron rest energy	$mc^2 = 511\,\mathrm{keV} = 8.185 \cdot 10^{-7}\,\mathrm{erg}$
Fine structure constant	$\hbar c/e^2 = 137.0$
Thomson cross section	$\Phi = 8\pi r_0^2/3 = 6.65 \cdot 10^{-25}\,\mathrm{cm}^2$
Ratio of proton to electron masses	$M_\mathrm{p}/m = 1836$
Electric field due to the nucleus at the first Bohr orbit	$e/a_0^2 = 5.14 \cdot 10^9\,\mathrm{V/cm}$
Area of spectral line with unit oscillator strength	$\pi e^2/mc = 2.65 \cdot 10^{-2}\,\mathrm{cm^2 s^{-1}}$
Atomic unit of cross section (area of first Bohr orbit)	$\pi a_0^2 = 0.880 \cdot 10^{-16}\,\mathrm{cm}^2$

Formulas

Radiation flux from a blackbody surface at temperature T

$$S = \sigma T^4 = 5.67 \cdot 10^{-5}\{T[\mathrm{K}]\}^4 = 1.03 \cdot 10^{12}\{T[\mathrm{eV}]\}^4\,\mathrm{erg/cm^2 s}\ .$$

Spectral intensity of equilibrium radiation

$$I_{\nu,\mathrm{eq}} d\nu = cU_{\nu,\mathrm{eq}} d\nu/4\pi = 2h\nu^3 c^{-2}\left[\exp(h\nu/kT) - 1\right]^{-1} d\nu$$
$$\mathrm{erg/(cm^2 s \cdot sr)}\ \ (\text{maximum at } h\nu = 2.822kT)\ .$$

Saha's formula relating equilibrium densities of electrons, n_e, ions, n_+, and neutral atoms, N

$$\frac{n_\mathrm{e} n_+}{N} = A\frac{g_+}{g_\mathrm{a}} T^{3/2} \exp(-I/kT)\ ,$$

$$A = 2\left(\frac{2\pi mk}{h}\right)^{3/2} = 4.85 \cdot 10^{15}\,\mathrm{cm^{-3}K^{-3/2}} = 6.06 \cdot 10^{21}\,\mathrm{cm^{-3}eV^{-3/2}}\ ,$$

n is in cm^{-3}, g_+ and g_a are statistical weights .

Maxwell distribution function normalized to unity

$$f(v_x, v_y, v_z) dv_x dv_y dv_z = \left(\frac{m}{2\pi kT}\right)^{3/2} \exp\left[-\frac{m(v_x^2 + v_y^2 + v_z^2)}{2kT}\right] dv_x dv_y dv_z$$

$$\varphi(v)\, dv = 4\pi\left(\frac{m}{2\pi kT}\right)^{3/2} \exp(-mv^2/2kT) v^2 dv$$

$$n(\varepsilon)\, d\varepsilon = \frac{2}{\sqrt{\pi}} \frac{\sqrt{\varepsilon}}{(kT)^{3/2}} \exp(-\varepsilon/kT)\, d\varepsilon$$

Velocity of an electron with energy ε : $v_e = 5.93 \cdot 10^7 \sqrt{\varepsilon\,[\mathrm{eV}]}\,\mathrm{cm/s}$.

Velocity of a particle with relative atomic mass A

$$v = 1.38 \cdot 10^6 \sqrt{\varepsilon[\mathrm{eV}]/A}\,\mathrm{cm/s}$$

Mean thermal velocity of an electron

$$\bar{v}_e = (8kT/\pi m)^{1/2} = 6.21 \cdot 10^5 \sqrt{T\,[\mathrm{K}]} = 6.71 \cdot 10^7 \sqrt{T\,[\mathrm{eV}]}\,\mathrm{cm/s}$$

Mean thermal velocity of a particle with relative atomic mass A

$$\bar{v} = 1.45 \cdot 10^4 \sqrt{T\,[\mathrm{K}]/A} = 1.56 \cdot 10^6 \sqrt{T\,[\mathrm{eV}]/A}\,\mathrm{cm/s}$$

Effective cross section σ in terms of the mean number of collisions P per 1 cm path at 1 mm Hg and 0° C

$$\sigma = 2.83 \cdot 10^{-17}\,P\,\mathrm{cm}^2\,, \quad P = 3.53 \cdot 10^{16} \sigma\,\mathrm{cm}^{-1}\mathrm{Torr}^{-1}$$

Debye radius

$$d = \left(\frac{kT_e}{4\pi e^2 n_e}\right)^{1/2} = 6.88 \left\{\frac{T_e[\mathrm{K}]}{n_e}\right\}^{1/2} = 742 \left\{\frac{T_e[\mathrm{eV}]}{n_e}\right\}^{1/2} \mathrm{cm}$$

Plasma frequency

$$\omega_p = \left(\frac{4\pi e^2 n_e}{m}\right)^{1/2} = 5.65 \cdot 10^4 \sqrt{n_e}\,\mathrm{s}^{-1}$$

Critical electron density for electromagnetic wave of frequency $\omega = 2\pi f$ (λ is wavelength)

$$n_{cr} = m\omega^2/4\pi e^2 = 1.24 \cdot 10^4\,\{f[\mathrm{MHz}]\}^2\,\mathrm{cm}^{-3} = 1.11 \cdot 10^{13}\{\lambda[\mathrm{cm}]\}^{-2}\,\mathrm{cm}^{-3}$$

The factor in the electrostatics equation $\mathrm{div}\,\boldsymbol{E} = 4\pi e(n_+ - n_e)$:

$$4\pi e = 1.81 \cdot 10^{-6}\,\mathrm{V \cdot cm}\,, \quad e = 1.44 \cdot 10^{-7}\,\mathrm{V \cdot cm}$$

Electronic conductivity of an ionized gas for a current of frequency ω (ν_m is the collision frequency of electrons for momentum transfer)

$$\sigma = \frac{e^2 n_e \nu_m}{m(\omega^2 + \nu_m^2)} = 2.53 \cdot 10^8 n_e \frac{\nu_m}{\omega^2 + \nu_m^2}\,\mathrm{s}^{-1} = 2.82 \cdot 10^{-4} n_e \frac{\nu_m}{\omega^2 + \nu_m^2}$$
$$\mathrm{Ohm}^{-1}\mathrm{cm}^{-1}\,, \quad \nu_m[\mathrm{s}^{-1}]\,, \quad \omega[\mathrm{s}^{-1}]$$

Dielectric permittivity of an ionized gas

$$\varepsilon = 1 - \frac{4\pi e^2 n_e}{m(\omega^2 + \nu_m^2)} = 1 - \frac{\omega_p}{\omega^2 + \nu_m^2}$$

Electromagnetic wave absorption coefficient in an ionized gas

$$\mu_\omega = \frac{4\pi e^2 n_e \nu_m}{mc(\omega^2 + \nu_m^2)} = \frac{4\pi\sigma}{c} = 0.106 n_e \frac{\nu_m}{\omega^2 + \nu_m^2}\ \mathrm{cm}^{-1}$$

The rate of electron energy increase in an oscillating field considering only elastic collisions with atoms of mass M ($E = E_a/\sqrt{2}$ where E_a is amplitude)

$$\frac{d\varepsilon}{dt} = (\Delta\varepsilon_E - \Delta\varepsilon_{el})\,\nu_m = \left[\frac{e^2 E^2}{m(\omega^2 + \nu_m^2)} - \frac{2m}{M}\varepsilon\right]\nu_m$$

Electron energy acquired from the field in one collision

$$\Delta\varepsilon_E = \frac{e^2 E^2}{m(\omega^2 + \nu_m^2)} = \frac{1.75 \cdot 10^{15}\{E[\mathrm{V/cm}]\}^2}{\omega^2 + \nu_m^2} = \frac{6.34 \cdot 10^{17} S[\mathrm{W/cm^2}]}{\omega^2 + \nu_m^2}\ \mathrm{eV}$$

Electron mobility

$$\mu_e = \frac{e}{m\nu_m} = \frac{1.76 \cdot 10^{15}}{\nu_m}\ \mathrm{cm^2/(s \cdot V)}$$

Skin layer thickness

$$\delta = \frac{c}{\sqrt{2\pi\sigma\omega}} = \frac{5.03}{\{\sigma[\mathrm{Ohm^{-1}cm^{-1}}]f[\mathrm{MHz}]\}^{1/2}}\ \mathrm{cm}$$

Flux of rf energy into a conductor placed inside a solenoid coil (H_a and i_a are amplitudes, σ is conductivity, n is turns number per cm)

$$S = \frac{cH_a^2}{16\pi}\left(\frac{\omega}{2\pi\sigma}\right)^{1/2} = 9.94 \cdot 10^{-2}\{i_a n[\mathrm{A \cdot t/cm}]\}^2 \left\{\frac{f[\mathrm{MHz}]}{\sigma[\mathrm{Ohm^{-1}cm^{-1}}]}\right\}^{1/2} \frac{\mathrm{W}}{\mathrm{cm^2}}$$

Magnetic field inside a solenoid coil

$$H = (4\pi/c)in = 1.26 in[\mathrm{A \cdot t/cm}]\ \mathrm{Oe}$$

Characteristic diffusion length Λ determining the diffusion loss "frequency" $\nu_d = D/\Lambda^2$

$$\frac{1}{\Lambda^2} = \left(\frac{2.4}{R}\right)^2 + \left(\frac{\pi}{L}\right)^2 \quad \text{– cylinder of radius } R \text{ and length } L$$

$$\frac{1}{\Lambda^2} = \left(\frac{\pi}{L_1}\right)^2 + \left(\frac{\pi}{L_2}\right)^2 + \left(\frac{\pi}{L_3}\right)^2 \quad \text{– parallelepiped with sides } L_1,\ L_2, \text{ and } L_3$$

$$\frac{1}{\Lambda^2} = \left(\frac{\pi}{R}\right)^2,\quad \Lambda = \frac{R}{\pi} \quad \text{– sphere of radius } R$$

Ambipolar diffusion coefficient at $T_e \gg T$

$$D_a = \mu_+ \left[\text{cm}^2/(\text{s}\cdot\text{V})\right] T_e[\text{eV}]\,\text{cm}^2/\text{s}$$

Rms field of a wave in vacuum, $E = 19\sqrt{S[\text{W/cm}^2]}$ V/cm.

Frequency of ionization by electron impact, for the Maxwell spectrum and for the cross section being a linear function of ε at the ionization threshold I, $\sigma = C(\varepsilon - I)$:

$$\nu_i = C(I + 2kT)\bar{v}_e N \exp(-I/kT) \approx 2.2\cdot 10^7 \left\{C\,[\text{cm}^2/\text{eV}]\cdot 10^{17}\right\} \times \left\{T\,[\text{eV}]\right\}^{1/2} I[\text{eV}]\,p[\text{Torr}]\exp(-I/kT)\,\text{s}^{-1}$$

Relations Between Units

Energy in Electron Volts

The energy $\varepsilon = 1\,\text{eV} = 1.602\cdot 10^{-12}$ erg corresponds to

temperature $\varepsilon/k = 11{,}610\,\text{K}$,
frequency $\varepsilon/h = 2.418\cdot 10^{14}\,\text{s}^{-1}$,
wavelength $hc/\varepsilon = 1.240\cdot 10^{-4}\,\text{cm} = 12{,}400\,\text{Å}$;
wave number $\varepsilon/hc = 8067\,\text{cm}^{-1}$;

1 eV per molecule corresponds to 23.05 kcal/mol = $9.65\cdot 10^{11}/\mu$ [erg/g] = $96.5/\mu$ [kJ/g] (μ is the relative molecular mass).

Electric Units

Charge: 1 C = $3\cdot 10^9$ CGSE units = $6.25\cdot 10^{18}$ electron charges = $9\cdot 10^{11}$ V·cm

Electron charge: $e = 1.6\cdot 10^{-19}$ C = $1.44\cdot 10^{-7}$ V·cm (in the formula $E = e/r^2$)

Current: 1 A = 1 C/s = $3\cdot 10^9$ CGSE = $6.25\cdot 10^{18}$ electron charges per second = $9\cdot 10^{11}$ V·cm/s

Voltage: 1 V = 1/300 CGSE

Electric field strength: E [V/cm] = $300E$ [CGSE].

Resistance: 1 Ohm = $1/(9\cdot 10^{11})$ CGSE = $1/30c$, $1/c = 30$ Ohm

Conductivity: $\sigma\,[\text{Ohm}^{-1}\text{cm}^{-1}] = 1/(9\cdot 10^{11})\sigma\,[\text{s}^{-1}]$, $\sigma\,[\text{s}^{-1}] = 9\cdot 10^{11}\sigma\,[\text{Ohm}^{-1}\text{cm}^{-1}]$

Capacitance: 1 Farad = $9\cdot 10^{11}$ CGSE (cm) , 1 picofarad (pF) = 0.9 cm

Inductance: 1 Henry = 10^9 CGSE (cm); 1 μH = 1000 cm

Magnetic field strength: 1 Oersted = 1 CGSE

Energy: 1 Joule = 10^7 CGSE (erg)

Power: 1 Watt = 10^7 CGSE (erg/s)

Barometric Units and Corresponding Relations

Normal atmospheric pressure: 1 atm = $1.013 \cdot 10^6$ dyne/cm^2 or erg/cm^3
At 0° C, 1 atm = 760 mm Hg = 760 Torr = $1.013 \cdot 10^5$ Pascal $\approx$ 0.1 MPa. 1 Torr = 133.3 Pa, 1 kPa = 7.5 Torr

The number of molecules in 1 cm^3 at 0° C at 1 atm (Loschmidt's number): $2.687 \cdot 10^{19}\,\text{cm}^{-3}$

The number of molecules in 1 cm^3 at 20° C (room temperature) and 1 Torr: $3.295 \cdot 10^{16}\,\text{cm}^{-3}$

$$E/p\,[\text{V/cm} \cdot \text{Torr}] = 3.30 \cdot 10^{16} E/N\,[\text{V} \cdot \text{cm}^2] = 0.33 E/N\,[\text{Td}]$$

$$E/N\,[\text{V} \cdot \text{cm}^2] = 3.03 \cdot 10^{-17} E/p\,[\text{V/cm} \cdot \text{Torr}] = 10^{-17} E/N\,[\text{Td}]$$

The unit for E/N : 1 Townsend (Td) = 10^{-17} V·cm^2

References

Chapter 1

1.1 A. von Engel, M. Steenbeck: *Elektrische Gasentladungen. Ihre Physik und Technik*, Vol. II (Springer, Berlin 1934)
1.2 L.B. Loeb: *Basic Processes of Gaseous Electronis* (University of California Press, Berkeley CA 1960)
1.3 N.A. Kaptsov: *Electrical Phenomena in Gases and in Vacuum* (Gostekhizdat, Moscow 1950)
1.4 A. von Engel: *Ionized Gases* (Clarendon, Oxford 1965)
1.5 S.C. Brown: *Basic Data of Plasma Physics* (MIT Press, Cambridge MA 1959)
1.6 G. Francis: *Ionization Phenomena in Gases* (Butterworths, London 1960)
1.7 V.L. Granovsky: *Electric Current in Gas (Steady Current)* (Nauka, Moscow 1971)
1.8 Yu.P. Raizer: *Fundamentals of Modern Physics of Gas Discharge Processes* (Nauka, Moscow 1980)
1.9 A.M. Howatson: *An Introduction to Gas Discharges* (Pergamon, Oxford 1976)
1.10 B.E. Cherrington: *Gaseous Electronics and Gas Lasers* (Pergamon, Oxford 1982)
1.11 Yu.P. Raizer: *Physics of Gas Discharges* (Nauka, Moscow 1987) (in Russian)
1.12 Ya.B. Zel'dovich, Yu.P. Raizer: *Physics of Shock Waves and High-Temperature Hydrodynamic Phenomena* (Academic, New York 1966)
1.13 L.M. Biberman, V.S. Vorobyev, I.T. Yakubov: *Kinetics of Nonequilibrium Low-Temperature Plasma* (Plenum, New York 1987)
1.14 M.N. Hirsh, H.J. Oskam (eds.): *Gaseous Electronics*, Vol. 1, *Electrical Discharges* (Academic, New York 1978)
1.15 E. Kunhardt, L. Luessen (eds.): *Electrical Breakdown and Discharges in Gases* (Plenum, New York 1983)
1.16 J. Dutton: *A Survey of Electron Swarm Data*, J. Phys. Chem. Ref. Data **4** (1975) No. 3, pp. 577–856 (Handbook on electron processes in plasma)
1.17 M. Mitchner, C.H. Kruger: *Partially Ionized Gases* (Wiley, New York 1973)
1.18 L.A. Artsimovich, R.Z. Sagdeev: *Plasma Physics for Physicists* (Atomizdat, Moscow 1979)
1.19 V.E. Golant, A.P. Zhilinsky, S.A. Sakharov: *Fundamentals of Plasma Physics* (Wiley, New York 1979)

Chapter 2

2.1 S.C. Brown: *Basic Data of Plasma Physics* (MIT Press, Cambridge MA 1959)
2.2 A.D. MacDonald: *Microwave Breakdown in Gases* (Wiley, New York 1966)
2.3 R.B. Brode: Rev. Mod. Phys. **5**, 257 (1933)
2.4 E.W. McDaniel: *Collision Phenomena in Ionized Gases* (Wiley, New York 1964)
2.5 R.A. Nielsen: Phys. Rev. **50**, 950 (1936)
2.6 R.A. Nielsen, L. Colldi: Rev. Sci. Instr. **23**, 39 (1952)
2.7 R.A. Nielsen, N.E. Bradbury: Phys. Rev. **51**, 69 (1937)
2.8 L. Spitzer: *Physics of Fully Ionized Gases* (Wiley, New York 1962)

2.9 J.E. Allen, P.C. Stangeby: J. Phys. D: Appl. Phys. **6**, 224 (1973)
2.10 J. Dutton: *A Survey of Electron Swarm Data*, J. Phys. Chem. Ref. Data **4** (1975) No. 3, pp. 577–856
2.11 R.W. Crompton, L.C. Huxley, D.J. Sutton: Proc. Roy. Soc. **A 218**, 507 (1953); R.W. Crompton, D.J. Sutton: Proc. Roy. Soc. **A 215**, 467 (1952)
2.12 J.H. Parker, J.J. Lowke: Phys. Rev. **181**, 290, 302 (1969)
2.13 E.D. Lozansky, O.B. Firsov: *Theory of Sparks* (Atomizdat, Moscow 1975)
2.14 W.H. Cramer, J.H. Simons: J. Chem. Phys. **26**, 1272 (1957)
2.15 W.H. Cramer: J. Chem. Phys. **28**, 688 (1958); **30**, 641 (1959)
2.16 M.A. Biondi, L.M. Chanin: Phys. Rev. **94**, 910 (1954)
2.17 J.A. Hornbeck: Phys. Rev. **84**, 615 (1951)
2.18 A.D. Barkalov, A.A. Samokhin: Preprint IAE No. 4147/6 (1985)
2.19 A.M. Soroka, G.I. Shapiro: Pisma Zh. Tekh. Fiz. **5**, 129 (1979)
2.20 Ya.B. Zel'dovich, Yu.P. Raizer: *Physics of Shock Waves and High-Temperature Hydrodynamic Phenomena* (Academic, New York 1966)

Chapter 3

3.1 Yu.P. Raizer: *Laser-Induced Discharge Phenomena* (Consultants Bureau, New York 1977)

Chapter 4

4.1 J.B. Hasted: *Physics of Atomic Collisions* (Butterworths, London 1964)
4.2 E.D. Lozansky, O.B. Firsov: *Theory of Sparks* (Atomizdat, Moscow 1975)
4.3 A. von Engel: *Handbuch der Physik* Bd. 21 (Springer, Berlin 1956)
4.4 A.L. Ward: J. Appl. Phys. **33**, 2789 (1962)
4.5 A. von Engel: *Ionized Gases* (Clarendon, Oxford 1965)
4.6 D.R. Long, R. Geballe: Phys. Rev. **A 1**, 260 (1970)
4.7 L. Vriens: Phys. Lett. **8**, 260 (1964)
4.8 B.M. Smirnov: *Ions and Excited Atoms in Plasma* (Atomizdat, Moscow 1974)
4.9 Ya.B. Zel'dovich, Yu.P. Raizer: *Physics of Shock Waves and High-Temperature Hydrodynamic Phenomena* (Academic, New York 1966)
4.10 G.J. Schulz: Phys. Rev. **128**, 178 (1962)
4.11 C.E. Melton: J. Chem. Phys. **57**, 4218 (1972)
4.12 L.M. Chanin, A.V. Phelps, M.A. Biondi: Phys. Rev. Lett. **2**, 344 (1959)
4.13 A.N. Prasad, J.D. Graggs: Proc. Phys. Soc. **76**, 223 (1960)
4.14 D.R. Bates (ed.): *Atomic and Molecular Processes* (Academic, New York 1962)
4.15 V.S. Fomenko, I.A. Podchernyaeva: *Emission and Adsorption Properties of Materials. Handbook* (Atomizdat, Moscow 1975)
4.16 L.N. Dobretsov, M.V. Gomoyunova: *Emission Electronics* (Nauka, Moscow 1966)
4.17 E. Guth, C.J. Mullin: Phys. Rev. **61**, 339 (1942)
4.18 E.L. Murphy, R.H. Good: Phys. Rev. **102**, 1464 (1956)
4.19 T.H. Lee: J. Appl. Phys. **30**, 66 (1959)
4.20 S.C. Brown: *Basic Data of Plasma Physics* (MIT Press, Cambridge MA 1959)
4.21 G. Francis: *Ionization Phenomena in Gases* (Butterworths, London 1960)
4.22 F. Llevellyn-Jones, A.B. Parker: Proc. Roy. Soc. **A 213**, 185 (1952)
4.23 J.M. Meek, J.D. Graggs: *Electrical Breakdown of Gases* (Clarendon, Oxford 1953)

Chapter 5

5.1 N.L. Aleksandrov, A.M. Konchakov, E.E. Son: Fiz. Plazmy **4**, 169 (1978)
5.2 N.L. Aleksandrov, F.I. Vysikailo, R.Sh. Islamov, I.V. Kochetov, A.P. Napartovich, V.G. Pevgov: Teplofiz. Vys. Temp. **19**, 22 (1981)

5.3 N.L. Aleksandrov, E.E. Son: in *Plasma Chemistry*, ed. by B.M. Smirnov (Atomizdat, Moscow 1988) No.7,p.35
5.4 J.P. Boeuf, E. Marode: J. Phys. D., Appl. Phys. **15**, 2169 (1982)
5.5 V.A. Shveigert, I.V. Shveigert: Fizika Plazmy **14**, 347 (1988)
5.6 Yu.P. Raizer, M.N. Shneider: *Contributed Papers of the 19th ICPIG*, Belgrade, 1989, Vol. 1, p. 188; Fizika Plasmy **15**, 318 (1989); Teplofiz. Vys. Temp. **29**, 22 (1991)
5.7 M.J. Kushner, H. Pan, J.V. DiCarlo: *Invited Papers of the 19th ICPIG*, Belgrade, 1989, p. 52

Chapter 6

6.1 Yu.M. Kagan, V.I. Perel: "Probe Techniques of Plasma Analysis", Usp. Fiz. Nauk **81**, 409 (1963)
6.2 Yu.M. Kagan: in *Spectroscopy of Gas Discharge Plasma* (Nauka, Leningrad 1970)
6.3 R.H. Huddlestone, S.L. Leonard (eds.): *Plasma Diagnostic Techniques* (Academic, New York 1965)
6.4 W. Lochte-Holtgreven (ed.): *Plasma Diagnostics* (North-Holland, Amsterdam 1968)
6.5 V.L. Granovsky: *Electric Current in Gases* (Gostekhizdat, Moscow 1952)
6.6 E.O. Johnson, L. Malter: Phys. Rev. **80**, 59 (1950)
6.7 L.M. Biberman, B. Panin: Zh. Tekh. Fiz. **21**, 12 (1951)
6.8 O.V. Kozlov: *Electric Probes in Plasma* (Atomizdat, Moscow 1969)

Chapter 7

7.1 J.M. Meek, J.D. Graggs: *Electrical Breakdown of Gases* (Clarendon, Oxford 1953)
7.2 S.C. Brown: *Basic Data of Plasma Physics* (MIT Press, Cambridge MA 1959)
7.3 M. Knoll, F. Ollendorff, R. Rompe: Gasentladungstabellen (Springer, Berlin 1935)
7.4 J.G. Trump, R.J. Van de Graaff: J. Appl. Phys. **8**, 327 (1947)
7.5 A.D. MacDonald: *Microwave Breakdown in Gases* (Wiley, New York 1966)
7.6 A.D. MacDonald, D.U. Gaskell, H.N. Gitterman: Phys. Rev. **130**, 1841 (1963)
7.7 Yu.P. Raizer: *Laser-Induced Discharge Phenomena* (Consultants Bureau, New York 1977)
7.8 P.D. Maker, R.W. Terhune, C.M. Savage: Quantum Electronics III, ed. by P. Grivet, N. Bloembergen (Columbia University Press, New York 1964)
7.9 R.G. Meyerand, A.F. Haught: Phys. Rev. Lett. **11**, 401 (1963)
7.10 D.H. Gill, A.A. Dougal: Phys. Rev. Lett. **15**, 845 (1965)
7.11 Yu.P. Raizer: "Optical Discharges" (review paper), Usp. Fiz. Nauk **132**, 549 (1980)
7.12 P. Woskoboinikow, W.J. Milligan, H.C. Praddaude, D.R. Cohn: Appl. Phys. Lett. **32**, 527 (1978)
7.13 M.J. Soileau: Appl. Phys. Lett. **35**, 309 (1979)
7.14 V.N. Parfenov, L.N. Pakhomov, V.Yu. Petrunkin, V.A. Podlevsky: Pisma Zh. Tekh. Fiz. **2**, 731 (1976)
7.15 G. Francis: *Ionization Phenomena in Gases* (Butterworths, London 1960)
7.16 E.W.B. Gill, A. Engel, von: Proc. Roy. Soc. **A 192**, 446 (1948)

Chapter 8

8.1 V.L. Granovsky: *Electric Current in Gas (Steday Current* (Nauka, Moscow 1971)
8.2 A. Engel, M. Steenbeck: *Elektrische Gasentladungen. Ihre Physik und Technik*, Vol. II (Springer, Berlin 1934)
8.3 A. Engel, von: *Ionized Gases* (Clarendon, Oxford 1965)
8.4 S.C. Brown: *Basic Data of Plasma Physics* (MIT Press, Cambridge MA 1959)
8.5 V.Yu. Baranov, A.A. Vedenov, V.G. Nizyev: Teplofiz. Vysok. Temperatur **10**, 1156 (1972); A.A. Vedenov: *Physic of Electric Discharge CO_2-Lasers* (Energoizdat, Moscow 1982)

8.6 G.G. Gladush, A.A. Samokhin: Prikl. Mekh. i Tekh. Fiz. No.5, 15 (1981)
8.7 V.N. Melekhin, N.Yu. Naumov: Pis'ma v Zh. Tekh. Fiz. **12**, 99 (1986)
8.8 Yu.P. Raizer, S.T. Surzhikov: Pis'ma v Zh. Tekh. Fiz. **13**, 452 (1987); Teplofiz. Vysokikh Temperatur **26**, 428 (1988)
8.9 Yu.P. Raizer: *Invited papers of the 17th Internat. Conf. on Phenomena in Ionized Gases*, Budapest, July 1985, p. 111; Teplofiz. Vys. Temperatur **24**, 984 (1986)
8.10 A.L. Ward: Phys. Rev. **112**, 1852 (1958)
8.11 S.Ya. Bronin, V.M. Kolobov: Fizika Plasmy **9**, 1088 (1983)
8.12 J.P. Boeuf, E. Marode: J. Phys. D, Appl. Phys. **15**, 2169 (1982)
8.13 P. Gill, C.E. Webb: J. Phys. D, Appl. Phys. **10**, 229 (1977)
8.14 Yu.P. Raizer, M.N. Schneider: *Contributed Papers of the 19th IC PIG*, Belgrade, 1989 Vol. 1, p. 188; Fizika Plasmy **15**, 318 (1989); Teplofiz. Vys. Temp. **29**, 22 (1991)
8.15 V.A.Shveigert, I.V. Shveigert: Fizika Plazmy **14**, 347 (1988)
8.16 G.S. Solntsev, A.I. Orlov, V.A. Dovzhenko: Radiotekhnika i Elektronika **9**, 1980 (1970)
8.17 Yu.M. Kagan, C. Cohen, P. Avivi: J. Appl. Phys. **63** (1), 60 (1988)
8.18 A.V. Nedospasov: Zh. Tekh. Fiz. **26**, 1202 (1956)
8.19 G. Ecker, K.G. Emeleus: Ann. Phys. **15**, 53 (1965)
8.20 A.J. Davies, J.G. Evans, E. Marode, P. Segur: in *ICDGA*, Edinburgh, 1980, Vol. 2, p. 59
8.21 L.S. Polak, I.A. Sergeev, D.I. Slovetsci: Teplofiz. Vysok Temperatur **15**, 15 (1977)
8.22 Yu.B. Golubovsci, V.M. Teleshko: Teplofiz. Vysok Temperatur **22**, 428 (1984)
8.23 Yu.S. Akishev, K.V. Baidze at all: Fizika Plasmy **11**, 999 (1985)
8.24 H. Brunet, Rossa-Serra: J. Appl. Phys. **57**, 1574 (1985)
8.25 A.H. Mnazakanjan, G.V. Naidis: in *Plasma Chemistry*, ed. by B.M. Smirnov (Energoatomizdat, Moscow 1987), No. 14, p. 227
8.26 A.V. Berdishev, I.V. Kochetov, A.P. Napartovich: Fizika Plasmy **14**, 741 (1988)
8.27 L.J. Denes, J.J. Lowke: Appl. Phys. Lett. **23**, 130 (1973)
8.28 J.J. Lowke, A.V. Phelps, B.W. Irwin: J. Appl. Phys. **44**, 4664 (1973)
8.29 W.L. Nighan, W.J. Wiegand: Phys. Rev. **A10**, 922 (1974)
8.30 A.P. Napartovich, V.G. Naumov, V.M. Shashkov: Fizika Plasmy **1**, 821 (1975)
8.31 Yu.P. Raizer, G.I. Shapiro: Fizika Plasmy **4**, 810 (1978)
8.32 A.C. Eckbreth, P.R. Balszuk: AIAA Paper No. 72–723 (1972)
8.33 S.V. Pashkin, P.I. Peretyatko: Kvantovaya Elektronika **5**, 1159 (1978)
8.34 N.A. Generalov, V.D. Kosynkin, V.P. Zimakov, Yu.P. Raizer, D.I. Roitenburg: Fizika Plasmy **6**, 1152 (1980)
8.35 E.P. Velikhov, V.S. Golubev, S.V. Pashkin: "Glow Discharge in Gas Flow" (review paper), Usp. Fiz. Nauk **137**, 117 (1982)
8.36 G. Francis: "The Glow Discharge at Low Pressure", in *Encyclopedia of Physics*, ed. by S. Flugge, Handbuch der Physik, Vd. XXII (Springer, Berlin 1956) pp. 53–208
8.37 Yu.S. Akishev, A.P. Napartovich, P.I. Peretyatko, N.I. Trushkin: Teploffiz. Vysok Temperatur **18**, 873 (1980)

Chapter 9

9.1 R.A. Haas: Phys. Rev. **A8**, 1017 (1973)
9.2 W.L. Nighan: in *Principles of Laser Plasma*, ed. by G. Bekefi (New York 1976) Chap. 7
9.3 E.P. Velikhov, V.D. Pismennyi, A.T. Rakhimov: Usp. Fiz. Nauk **122**, 419 (1977)
9.4 A.V. Eletsky, A.T. Rakhimov: in *Plasma chemistry*, ed. by B.M. Smirnov (Atomizdat, Moscow 1977) No. 4, p. 123
9.5 A.P. Napartovich, A.N. Starostin: in *Plasma chemistry*, ed. by B.M. Smirnov (Atomizdat, Moscow 1978) No. 5, p. 153
9.6 G. Ecker, W. Kroll, O. Zoller: Phys. Fluids **7**, 2001 (1964)
9.7 N.G. Basov, E.M. Belenov et al.: Pis'ma v Zh. Eksp. Teor. Fiz. **14**, 421 (1971)

9.8 Yu.P. Raizer, G.I. Shapiro: Fizika Plazmy **4**, 810 (1978)
9.9 L.J. Denes, J.J. Lowke: Appl. Phys. Lett. **23**, 130 (1973)
9.10 G.D. Melnikov, A.P. Napartovich: Fizika Plasmy **1**, 891 (1975)
9.11 W.L. Nighan, W.J. Wiegand: Phys. Rev. **A10**, 922 (1974)
9.12 L. Pekarek: Usp. Fiz. Nauk **94**, 463 (1968)
9.13 A.V. Nedospasov: Usp. Fiz. Nauk **94**, 439 (1968)
9.14 A. Garscadden: "Ionization Waves", in *Gaseous Electronics I, Electrical Discharges*, ed. by M.N. Hirsh, H.J. Oskam (Academic, New York 1978)
9.15 G. Francis: "The Glow Discharge at Low Pressure", in *Encyclopedia of Physics*, ed. by S. Flugge, Handbuch der Physik, Bd XXII (Springer, Berlin 1956) pp.53–208
9.16 A.V. Nedospasov, Yu.B. Ponomarenko: Teplofiz. Vysok Temperatur **3**, 17 (1965)
9.17 L.D. Tsendin: Zh. Tekh. Fiz. **40**, 1600 (1970)
9.18 W.P. Allis: Physica **C82**, 43 (1976)
9.19 A.B. Stewart: J. Appl. Phys. **27**, 911 (1956)
9.20 L.D. Tsendin: Zh. Tekh. Fiz. **52**, 635, 643 (1982)
9.21 Yu.B. Golubovsky, A.K. Zinchenko, Yu.M. Kagan: Zh. Tekh. Fiz. **47**, 1478 (1977)
9.22 C. Kenty: Phys. Rev. **126**, 1235 (1962)
9.23 N.A. Generalov, V.D. Kosynkin, V.P. Zimakov, Yu.P. Raizer, D.I. Roitenburg: Fizika Plasmy **6**, 1152 (1980)

Chapter 10

10.1 W. Finkelnburg, H. Maecker: "Elektrische Bogen und thermisches Plasma", *Handbuch der Physik*, Vol. XXII (1956) S.254–444
10.2 M.F. Zhukov et al: *Electrode Processes in Arc Discharges* (Nauka, Novosibirsk 1982)
10.3 W.D. Davis, H.C. Miller: J. Appl. Phys. **40**, 2212 (1969)
10.4 J.M. Lafferty (ed.): *Vacuum Arcs. Theory and Application* (Wiley, New York 1980)
10.5 G.A. Lyubimov, V.I. Rakhovski: "Cathode Spot of a Vacuum Arc" Usp. Fiz. Nauk **125**, 665 (1978)
10.6 V.L. Granovsky: *Electric Current in Gas (Steady Current)* (Nauka, Moscow 1971)
10.7 I.I. Beilis: Dokl. Acad. Nauk **298**, 1108 (1988); Teplofiz. Vys. Temp. **25**, 1224 (1988); Pis'ma Zh. Tekh. Fiz. **14**, 1124 (1988)
10.8 T.H. Lee: J. Appl. Phys. **30**, 166 (1959)
10.9 I.I. Beilis: Pis'ma Zh. Tekh. Fiz. **16**, N10, 28 (1990)
10.10 Yu.D. Korolev, G.A. Mesyats: *Field-Emission and Explosive Processes in Gas Discharges* (Nauka, Novosibirsk 1982)
10.11 C.W. Kimblin: IEEE Trans. **PS-2**, 310 (1974)
10.12 A. von Engel: *Ionized Gases* (Clarendon, Oxford 1965)
10.13 S.V. Dresvin: *Physics and Technology of Low Temperature Plasmas* (Iowa State Univ. Press, Ames IA 1977)
10.14 J.M. Yos: *Revised Transport Properties of High Temperature Air and its Components* (AVCO, T.R. Nov. 1967)
10.15 Yu.P. Raizer: *Laser-Induced Discharge Phenomena* (Consultants Bureau, New York 1977)

Chapter 11

11.1 Yu.P. Raizer: *Laser-Induced Discharge Phenomena* (Consultants Bureau, New York 1977)
11.2 L.D. Landau, E.M. Lifshitz: *Electrodynamics of Continuous Media* (Pergamon Press, Oxford 1960)
11.3 R.E. Rovinsky, V.A. Gruzdev, A.P. Sobolev: Prikl. Mekh. i Tekh. Fiz. No. 1, 143 (1967)
11.4 R.E. Rovisnky, V.A. Gruzdev, T.M. Gutenmakher, A.P. Sobolev: Teplofiz. Vys. Temperatur **5**, 557 (1967)

11.5 L.M. Blinov, V.V. Volodko, G.G. Gontarev, G.V. Lysov, L.S. Polak: in *Generators of Low-Temperature Plasma* (Energiya, Moscow 1969)
11.6 P.L. Kapitsa: Zh. Eksp. Teor. Fiz. 57, 1801 (1969)
11.7 Yu.P. Raizer: Pis'ma v Zh. Eksp. Teor. Fiz. **11**, 195 (1970)
11.8 N.A. Generalov, V.P. Zimakov, G.I. Kozlov, V.A. Masyukov, Yu.P. Raizer: Pis'ma v Zh. Eksp. Teor. Fiz. **11**, 447 (1970)
11.9 Yu.P. Raizer: Pis'ma v Zh. Tekh. Fiz. **7**, 938 (1981)
11.10 N.A. Generalov, V.P. Zimakov, G.I. Kozlov, V.A. Masyukov, Yu.P. Raizer: Zh. Eksp. Teor. Fiz. **61**, 1434 (1971)
11.11 D.R. Keefer, B.B. Hendriksen, W.F. Braerman: J. Appl. Phys. **46**, 1080 (1975)
11.12 G.I. Kozlov, V.A. Kuznetsov, V.A. Masyukov: Zh. Tekh. Fiz. **49**, 2304 (1979)
11.13 J. Uhlenbusch: Invited Papers of the 16th Internat. Conf. on Phenomena in Ionized Gases, Düsseldorf, 1983, p. 119
11.14 Ya.B. Zel'dovich, Yu.P. Raizer: *Physics of Shock Waves and High-Temperature Hydrodynamic Phenomena* (Academic Press, New York 1966)
11.15 Yu.P. Raizer, A.Yu. Silantyev: Kvantovaya Elektronika **13**, 593 (1986)
11.16 Yu.P. Raizer, S.T. Surzhikov: Kvantovaya Elektronika **11**, 2301 (1984); Teplofiz. Vys. Temperatur **23**, 29 (1985)
11.17 Yu.P. Raizer: "Optical Discharges" (review paper), Usp. Fiz. Nauk **132**, 549 (1980)
11.18 Yu.P. Raizer: *Fundamentals of Modern Physics of Gas Discharge Processes* (Nauka, Moscow 1980)
11.19 S.V. Dresvin: *Physics and Technology of Low Temperature Plasmas* (Iowa State Univ. Press, Ames 1977)
11.20 S.V. Kononov, M.I. Yakushin: Prikl. Mekh. i Tekh. Fiz. No. 6, 67 (1966)
11.21 M.V. Gerasimenko, G.I. Kozlov, V.A. Kuznetsov, V.A. Masyukov: Pis'ma v Zh. Tekh. Fiz. **5**, 954 (1979)
11.22 Yu.P. Raizer, A.Yu. Silantyev, S.T. Surzhikov: Pis'ma v Zh. Tekh. Fiz. **12**, 134 (1986); Teplofiz. Vys. Temperatur **25**, 454 (1987)
11.23 K.G. Guskov, Yu.P. Raizer, S.T. Surzhikov: Kvant. Elektron. **17**, 937 (1990)

Chapter 12

12.1 H. Raether: *Electron Avalanches and Breakdown in Gases* (Butterworths, London 1964)
12.2 L.B. Loeb: *Basic Processes of Gaseous Electronics* (University of California Press, Berkeley CA 1960)
12.3 J.M. Meek, J.D. Craggs: *Electrical Breakdown of Gases* (Wiley, New York 1978)
12.4 I.S. Marshak: "Electrical Discharge of Gases at Nearly Atmospheric Pressure" (review paper), Usp. Fiz. Nauk **71**, 631 (1960)
12.5 E.D. Lozansky, O.B. Firsov: *Theory of Sparks* (Atomizdat, Moscow 1975)
12.6 Yu.P. Raizer: *Laser-Induced Discharge Phenomena* (Consultants Bureau, New York 1977)
12.7 D.V. Razevig (ed.): *High-Voltage Engineering* (Energiya, Moscow 1976)
12.8 I.M. Bortnik, A.N. Kushko, A.N. Lobanov: Abstracts of Reports to the 2nd USSR Conf. on Physics of Electrical Discharge of Gases, Tartu 1984, Part II, p. 270
12.9 M.V. Kostenko (ed.): *High-Voltage Engineering* (Vysshaya Shkola, Moscow 1973)
12.10 H. Tholl: Zs. Naturforsch. **19a**, 346, 704 (1964)
12.11 I.S. Abramson, N.M. Gegechkori, S.I. Drabkina, S.L. Mandelshtam: Zh. Eksp. Teor. Fiz. **17**, 862 (1947)
12.12 S.I. Drabkina: Zh. Eksp. Teor. Fiz. **21**, 473 (1951)
12.13 M. Goldman, N. Goldman: "Corona Discharges", in *Gaseous Electronics I, Electrical Discharges*, ed. by M.N. Hirsh and H.J. Oskam (Academic Press, New York 1978)
12.14 G.A. Dawson, W.P. Winn: Zs. Phys. **183**, 159 (1965)
12.15 I. Gallimberti: J. Phys. D, Appl. Phys. **5**, 2179 (1972)

12.16 S. Badaloni, I. Gallimberti: in Proc. 11th Int. Conf. on Phenomena in Ionized Gases, Prague 1973, p. 196
12.17 R.F. Griffiths, C.T. Phelps: Quart. J. R. Math. Soc. **102**, 419 (1976)
12.18 E.M. Bazelyan, A.Yu. Goryunov: Abstracts of reports to the 2nd USSR Conf. on Physics of Electrical Breakdown of Gases, Tartu 1984, Part I, p. 42; Elektrichestvo No. 11, 27 (1986)
12.19 R.D. Klingbeil, A. Tidman, R.F. Fernsler: Phys. Fluids **15**, 1969 (1972)
12.20 E. Marode: J. Appl. Phys. **46**, 2005 (1975)
12.21 I. Gallimberti: "The Mechanism of the Long Spark Formation" (review paper), J. de Physique **40**, Coll. C7, suppl. au No. 7 (1979) p. 193
12.22 S.K. Dhali, P.F. Williams: Phys. Rev. **A31**, 1219 (1985)
12.23 A.S. Gaivoronsky, I.M. Razhansky: Zh. Tekh. Fiz. **56**, 1110 (1986)
12.24 G.N. Aleksandrov: Zh. Tekh. Fiz. **35**, 1225 (1965)
12.25 I. Gallimberti: Electra **76**, 5799 (1977)
12.26 E.M. Bazelyan, B.N. Gorin, V.I. Levitov: *Physical and Engineering Foundations of Lightning Protection* (Gidrometeoizdat, Leningrad 1978)
12.27 E.M. Bazelyan, I.M. Razhansky: *Spark Discharge in Air* (Nauka, Novosibirsk 1988)
12.28 E.M. Bazelyan: Elektrichestvo No. 5, 20 (1987)
12.29 M.M. Kekez, P. Savic: in *Electrical Breakdown and Discharges in Gases*, ed. by E. Kunhardt, L. Luessen (Plenum, New York 1983) p. 419
12.30 D.J. Malan: Ann. Geophys. **8**, 385 (1952)
12.31 Ya.I. Frenkel: *Theory of Atmospheric Electricity Phenomena* (Gostekhizdat, Moscow 1949)
12.32 M. Uman: *Lightning* (McGraw-Hill, New York 1969)
12.33 B.F. Schonland: in *Handbuch der Physik* Bd. 22, S. 576, (Springer, Berlin 1956)
12.34 Les Renardiers Group: IEEE Proc. 133, PtA No. 7, 395–483 (1986)

Chapter 13

13.1 S.M. Levitsky: Zh. Tekh. Fiz. **27**, 1001 (1957)
13.2 C.O. Brown, J.W. Davis: Appl. Phys. Lett. **21**, 480 (1972)
13.3 N.A. Generalov, V.P. Zimakov, V.D. Kosynkin, Yu.P. Raizer, D.I. Roitenburg: Pis'ma v Zh. Tekh. Fiz. **1**, 431 (1975); Fizika Plasmy **3**, 626, 634 (1977); *ibid.* **6**, 1152
13.4 Yu.P. Raizer: Fizika Plasmy **5**, 408 (1979)
13.5 S.M. Levitsky: Zh. Tekh. Fiz. **27**, 970 (1957)
13.6 A.A. Kuzovnikov, V.L.Kovalenkov, V.P. Savinov: Vestnik MGU, Ser. 3 **24**, No. 4, 28 (1983)
13.7 A.A. Kuzovnikov, V.P. Savinov: Radiotekh. i Elektronika **18**, 816 (1973)
13.8 N.A. Yatsenko: Zh. Tekh. Fiz. **51**, 1195 (1981)
13.9 N.A. Yatsenko: Zh. Tekh. Fiz. **52**, 1220 (1982)
13.10 P. Vidaud, S.M.A. Durrani, D.R. Hall: J. Phys. D: Appl. Phys. **21**, 57 (1988)
13.11 V.A. Godyak, A.A. Kuzovnikov: Fizika Plazmy **1**, 496 (1975)
13.12 N.A. Yatsenko: Teplofiz. Vys. Temperatur **20**, 1044 (1982)
13.13 Yu.P. Raizer, M.N. Shneider: Fizika Plazmy **13**, 471 (1987); *ibid* **14**, 226 (1988)
13.14 N.A. Yatsenko: Zh. Tekh. Fiz. **50**, 2480 (1980)
13.15 A.S. Smirnov: Zh. Tekh. Fiz. **54**, 61 (1984)
13.16 A.S. Kovalev, A.T. Rakhimov, V.A. Feoktistov: Fizika Plazmy **7**, 1411 (1981)
13.17 A.D. Barkalov, V.D. Gavrilyuk, G.G. Gladush *et al.*: Teplofiz. Vys. Temperatur **16**, 265 (1978)
13.18 Yu.P. Raizer, M.N. Shneider: Teplofiz. Vys. Temperatur **25**, 1008 (1987); Fizika Plazmy **16**, 878 (1990)
13.19 M.G. Gill: Vacuum **34**, 357 (1984)
13.20 *IEEE Transactions on Plasma Science*, Special Issue on the Physics of RF Discharges for Plasma Processing **PS-14**, No. 2 (1986)
13.21 B.N. Chapman: *Glow Discharge Processes – Sputtering and Plasma Etching* (Wiley, New York 1980)
13.22 D.B. Graves: AIChE Journal **35**, 1 (1989)

Chapter 14

14.1 G.A. Abilsiitov, E.P. Velikhov, V.S. Golubev, A.G. Grigoryants, F.V. Lebedev, G.A. Nikolaev: High-Power CO_2-Lasers and Their Applications in Technology (Nauka, Moscow 1984)
14.2 A.A. Vedenov: Physics of Electrical-Discharge CO_2-Lasers (Energoizdat, Moscow 1982)
14.3 G.I. Kozlov, V.A.Kuznetsov, V.A. Masyukov: Pis'ma v Zh. Tekh. Fiz. **4**, 129 (1978)
14.4 A.C Eckbreth, J.W. Davis: Appl. Phys. Lett. **19**, 101 (1971)
14.5 N.A. Generalov, V.P. Zimakov, V.D. Kosynkin, Yu.P. Raizer, D.I. Roitenburg: Pis'ma v Zh. Tekh. Fiz. **1**, 431 (1975); Fizika Plasmy **3**, 626, 634 (1977): *ibid.* **6**, 1152 (1980)
14.6 A.E. Hill: Appl. Phys. Lett. **18**, 194 (1971)
14.7 E. Hoag, H. Pease, J. Staal, J. Zar: J. Quantum Electronics QE-**9**, 652, (1973)
14.8 N.G. Basov, I.K. Babaev, V.A. Danilychev *et al.*: Kvantovaya Elektronika **6**, 772 (1979)
14.9 E.P. Velikhov, V.S. Golubev, S.V. Pashkin: "Glow Discharge in Gas Flow" (review paper), Usp. Fiz. Nauk **137**, 117 (1982)
14.10 F.K. Kosyrev, N.P. Kosyreva, E.I. Lunev: Avtomat. Svarka **9**, 72 (1976); V.M. Andriyakhin: Nauka i Zhisn, No. 11, 86 (1977)
14.11 G.A. Abilsiitov, L.I. Antonova, A.V. Artamonov *et al.*: Kvantovaya Elektronika **6**, 204 (1979)
14.12 V.S. Aleinikov, V.V. Bibikova, O.S. Lysogorov: Elektronaya Promyshlennost No. 5–6, 71 (1981)
14.13 C.O. Brown. J.W. Davis: Appl. Phys. Lett. **21**, 480 (1972)
14.14 A.C. Eckbreth, P.R. Blaszuk: AIAA Paper No. 72–723 (1972)
14.15 Yu.P. Raizer: Fizika Plasmy **5**, 408 (1979)
14.16 N.A. Generalov, V.P. Zimakov, V.D. Kosynkin, Yu.P. Raizer, N.G. Solovyev: Kvantovaya Elektronika **9**, 1549 (1982)
14.17 Yu.P. Raizer, G.I. Shapiro: Fizika Plazmy **4**, 810 (1978)
14.18 V.D. Gavrilyuk, A.F. Glova, V.S. Golubev, F.V. Lebedev: Kvantovaya Elektronika **4**, 2034 (1977)
14.19 V.V. Antyukhov, A.I. Bondarenko, A.F. Glova, V.S. Golubev *et al.*: Kvantovaya Elektronika **8**, 2234 (1981)
14.20 V.Kh. Goikhman, V.M. Goldfarb: Zh. Prikl. Spektrosk. **21**, 456 (1974)
14.21 D. He, D.R. Hall: Appl. Phys. Lett. **43**, 726 (1983)
14.22 O.S. Akirtava, V.D. Dzhikiya *et al.*: Pis'ma v Zh. Tekh. Fiz. **43**, 1231 (1981); O.S. Akirtava, V.S. Golubev, *et al.*: in Abstracts of reports to the USSR Conf. on *Laser Applications in National Economy*, Zvenigorod, May 1985 (Nauka, Moscow 1985)
14.23 W. Schock, H. Hugel, P. Hoffman: Electro Opt. Laser **13**, 76 (1981)
14.24 W. Waidelich (ed.): Proceedings of the 7th Internat. Congress Laser 85 Optoelectronics (Springer, Berlin 1986)
14.25 H. Hügel: in *Gas Flow and Chemical Lasers* Proc. 6th. Intern. Sympos. Jerusalem, 1986, (Springer, Berlin, Heidelber 1987) p. 258
14.26 S. Yatsiv: ibid., p. 252
14.27 P. Hoffmann: in Proceedings of the 2nd Internat. Confer. Lasers Manuf., Birmingham, March 1985 (Kempston, Amsterdam 1985)
14.28 P.P. Vitruk, N.A. Yatsenko: Pis'ma Zh. Tekh. Fiz. **15**, 1 (1989)

Further Reading

1. Yu.P. Raizer, M.N. Shneider, N.A. Yatsenko: *Radio-Frequency Capacitive Discharges* (CRC Press, Boca Raton, Florida 1995)
2. M.A. Lieberman, A.J. Lichtenberg: *Principles of Plasma Discharges and Materials Processing* (Wiley, New York 1994)

Subject Index

CPSIA information can be obtained
at www.ICGtesting.com
Printed in the USA
LVHW04s1926180818
587394LV00007B/62/P